# Wheat and Barley: Molecular Approaches to Crop Improvement

# Wheat and Barley: Molecular Approaches to Crop Improvement

Editor: Penny Bell

Callisto Reference
www.callistoreference.com

**Callisto Reference,**
118-35 Queens Blvd., Suite 400,
Forest Hills, NY 11375, USA

Visit us on the World Wide Web at:
www.callistoreference.com



ISBN: 978-1-64116-823-6 (Hardback)



**Cataloging-in-Publication Data**

Wheat and barley : molecular approaches to crop improvement / edited by Penny Bell.
p. cm.
Includes bibliographical references and index.
ISBN 978-1-64116-823-6
1. Wheat. 2. Barley. 3. Wheat--Genetics. 4. Barley--Genetics. 5. Crops--Genetic engineering.
6. Crop improvement. I. Bell, Penny.
SB191.W5 W44 2023
633.11--dc23

# Table of Contents

# Preface

The world is advancing at a fast pace like never before. Therefore, the need is to keep up with the latest developments. This book was an idea that came to fruition when the specialists in the area realized the need to coordinate together and document essential themes in the subject. That's when I was requested to be the editor. Editing this book has been an honour as it brings together diverse authors researching on different streams of the field. The book collates essential materials contributed by veterans in the area which can be utilized by students and researchers alike.

Wheat and barley both are important domesticated crops which belong to the grass family. Barley is a significant source of dietary fiber whereas wheat is a significant source of carbohydrates. Wheat is used to make biscuits, pasta, cookies, bread, noodles, crackers, bulgur and various other products. It is also used to manufacture animal feed and beverages. Barley is primarily utilized to make beer and other alcoholic beverages. It is also widely used as livestock feed. Wheat and barley genetics have undergone a revolutionary change with advancements in genomics. Marker-assisted selection (MAS), marker-assisted backcrossing (MABC) and pyramiding of genetic factors influencing important traits have accelerated the discovery and use of rare allelic variants in traditional breeding schemes. This book provides significant information on the molecular approaches to improvements in wheat and barley. It will also provide interesting topics for research, which interested readers can take up. This book, with its detailed analyses and data, will prove immensely beneficial to professionals and students involved in this area at various levels.

Each chapter is a sole-standing publication that reflects each author's interpretation. Thus, the book displays a multi-facetted picture of our current understanding of application, resources and aspects of the field. I would like to thank the contributors of this book and my family for their endless support.

**Editor**

# Preface

1

# Identification and Characterization of PLATZ Transcription Factors in Wheat

**Yuxin Fu [1,2], Mengping Cheng [1,3], Maolian Li [1,3], Xiaojiang Guo [1], Yongrui Wu [2] and Jirui Wang [1,3,4,*]**

1 Triticeae Research Institute, Sichuan Agricultural University, Chengdu 611130, China; yuxinfu_anna@163.com (Y.F.); 71250@sicau.edu.cn (M.C.); 71296@sicau.edu.cn (M.L.); xiaojiang9945@gmail.com (X.G.)

2 National Key Laboratory of Plant Molecular Genetics, CAS Center for Excellence in Molecular Plant Sciences, Institute of Plant Physiology and Ecology, Shanghai Institutes for Biological Sciences, Chinese Academy of Sciences, Shanghai 200032, China; yrwu@sibs.ac.cn

3 Key Laboratory for Crop Genetic Resources and Improvement in Southwest China, Sichuan Agricultural University, Ministry of Education, Chengdu 611130, China

4 State Key Laboratory of Crop Gene Exploration and Use in Southwest China, Sichuan Agricultural University, Chengdu 611130, China

* Correspondence: wangjirui@gmail.com

**Abstract:** The PLATZ (plant AT-rich protein and zinc-binding protein) transcription factor family is a class of plant-specific zinc-dependent DNA-binding proteins. PLATZ has essential roles in seed endosperm development, as well as promoting cell proliferation duration in the earlier stages of the crops. In the present study, 62 *TaPLATZ* genes were identified from the wheat genome, and they were unequally distributed on 15 chromosomes. According to the phylogenetic analysis, 62 *TaPLATZ* genes were classified into six groups, including two groups that were unique in wheat. Members in the same groups shared similar exon-intron structures. The polyploidization, together with genome duplication of wheat, plays a crucial role in the expansion of the *TaPLATZs* family. Transcriptome data indicated a distinct divergence expression pattern of *TaPLATZ* genes that could be clustered into four modules. The *TaPLATZs* in Module b possessed a seed-specific expression pattern and displayed obvious high expression in the earlier development stage of seeds. Subcellular localization data of *TaPLATZs* suggesting that they likely perform a function as a conventional transcription factor. This study provides insight into understanding the structure divergence, evolutionary features, expression profiles, and potential function of PLATZ in wheat.

**Keywords:** wheat; transcription factor; PLATZ; seed-specific expression; duplication

---

## 1. Introduction

Transcription regulation of genes in response to developmental and environmental changes, mediated by the DNA-binding transcription factors (TFs), is an important regulatory mechanism in plants [1]. Transcription factors bind to the cis-acting elements upstream of gene promoters to activate or repress gene expression.

During the evolution of transcription factor families, the TFs between plants and animals or yeast do not always correspond. In Arabidopsis, over 5% of its genome is devoted to encoding more than 1500 transcription factors, approximately 45% of which are specific to plants [2]. Some classes of transcription factors appear to have specifically evolved in plants, such as the WRKY, NAC, and AP2/EREBP families, which strictly integrate gene regulatory networks in plant growth processes, including metabolism, hormone signaling, pathogen defense and senescence [3–5].

The PLATZ (plant AT-rich sequence and zinc-binding) TF family is a class of plant-specific zinc-dependent DNA-binding proteins whose first member PLATZ1 was isolated in pea. The pea PLATZ1 acts as a transcriptional repressor by nonspecifically binding to A/T-rich sequences [6]. Multiple sequence alignments show that PLATZ proteins possess two highly conserved zinc finger motifs comprising several cysteine and histidine residues, with C-$x_2$-H-$x_{10}$-C-$x_2$-C-x$_{(4–5)}$-C-$x_2$-C-x$_{(3–7)}$-H-$x_2$-H at N-terminal region and C-$x_2$-C-x$_{(10–11)}$-C-$x_3$-C [6], both of which have been described as being necessary for zinc-dependent DNA binding.

PLATZ proteins participate in environmental stress responses. The transcription levels of *GmPLATZ1*(*Glycine max PLATZ1*) were upregulated by ABA and drought stress in soybean. Germination in *GmPLATZ1*-overexpression transgenic Arabidopsis was retarded in response to ABA and osmotic stress [7].

The biological functions of PLATZ proteins were not well identified until the maize *Fl3* gene was cloned from a classic endosperm semi-dominant mutant [8]. *Fl3* (*ZmPLATZ12*) is specifically expressed in maize endosperm starchy cells, encoding a PLATZ protein, Fl3, that is required for transcription of transfer RNAs (tRNAs) and 5S ribosomal RNA (5S rRNA), through interaction with the RNA polymerase III (RNAPIII) subunits RPC53 and TFC1, which participate in the development of endosperm and filling of storage materials in maize seeds [3].

PLATZ proteins are involved in leaf growth and senescence. *ORESARA15* (*ORE 15*) encodes a PLATZ TF from Arabidopsis, promotes leaf growth by enhancing the rate and duration of cell proliferation in the leaf earlier development stage, and functions as a negative regulator of leaf senescence in the later stage by modulating the GRF/GIF regulatory pathway [9].

The *PLATZ* gene LOC_Os06g45540 from rice was mapped as a major QTL locus (*GL6*), identified from flanking sequence analysis of Japonica T-DNA insertion mutant *sg6* almost simultaneously, which function in regulating grain sizes. *GL6* determines grain length and spikelet number by affecting cell proliferation through gene expression regulation via the RNA polymerase III transcription machinery [10]. *SG6* determines grain size by interacting with the core cell cycle machinery DP protein and regulating spikelet hull cell division [11].

PLATZ family members have been identified in several plant species, including 13 genes in the *Arabidopsis* genome, eight genes in *Glycine max*, nine genes in *Gossypium hirsutum*, 15 genes in *Oryza sativa*, and 17 genes in *Zea mays*. The PLATZ proteins from rice, maize, and *Arabidopsis* were clustered into five subfamilies based on their amino acid structures [12]. However, the characterization and function of PLATZ in wheat have not yet been elucidated.

Wheat (*Triticum aestivum* L., BBAADD) is a staple food crop for more than one-third of the global human population and provides approximately 20% of the calories consumed by humans globally. Wheat is an allohexaploid species originated from two hybridizations: first between *Triticum urartu* (AA) and *Aegilops speltoides* (SS), and second between allotetraploid wild emmer wheat [*Triticum turgidum* ssp., BBAA] and *Aegilops tauschii* (DD) [13–15]. Recently, high-quality genome sequencing and assembly have been applied to wheat [16,17]. The genome sequences serve as references for underpinning genome scans for phenotypic associations and gene cloning from wheat and their wild progenitors for the accelerated development of improved wheat varieties [18,19]. The availability of these reference sequences allows the identification of genes within genetic intervals [20].

Recent advances in wheat genomics have presented an opportunity to identify wheat PLATZ family genes in hexaploid wheat. Identification and functional characterization of *PLATZs* in wheat will provide an avenue for exploring and verifying the conserved and specific mechanism of PLATZ members in plant development. Here, we characterized wheat *PLATZs* with respect to the following aspects: (1) genome-wide identification of PLATZ genes in wheat; (2) analysis of conserved motifs and cis-elements of PLATZ genes; (3) identification of the chromosomal distribution and the gene duplications; (4) phylogenetic and evolutionary relationship of *PLATZ* in wheat associate with rice and maize; (5) elucidation of expression profiles of *PLATZ* genes; and (6) subcellular localization of *PLATZ* genes. The results provide conditions for further study of *PLATZ* genes concerning flower and fruit growth in wheat.

## 2. Results

### 2.1. Identification of the PLATZ Genes in Wheat

According to the Hidden Markov Model (HMM) profile (PF04640), a total of 50 candidate sequences were found in the IWGSC v1.1 high-confidence (HC) and low-confidence (LC) peptide database using the HMM search program. This dataset was simplified by keeping the first splice variant from each transcript for further analyses. An additional 13 sequences were identified using a BLAST search in the unannotated IWGSC v2.0 genomic database. Subsequently, the PLATZ domain in the identified sequences was confirmed in the Pfam database, SMART database, and NCBI CDD program. By getting rid of sequences with incomplete PLATZ domains, 62 sequences were identified as putative *PLATZ* family members for further analysis (Table S1). The number of *PLATZs* in wheat is approximately three times greater than that of Arabidopsis (13 *AtPLATZs*), maize (17 *ZmPLATZs*), and rice (15 *OsPLATZs*) [12,21]. These *TaPLATZs* have 145 to 275 amino acids, with an average of 221 amino acids. The molecular weights of TaPLATZ were between 16.67 kDa and 29.76 kDa. The predicted pI values of TaPLATZ ranged from 5.21 to 9.68. The 62 identified *PLATZ* genes contain the conserved cysteine and histidine residues enrichment regions: C-x2-H-x(11–12)-C-$x_2$-C-x$_{(4–8)}$-C-$x_2$-C-x$_{(3–4)}$-H-$x_2$-H and C-$x_2$-C-x$_{(10–11)}$-C-$x_3$-C. The N-domain and C-domain conserved structure of PLATZ proteins are crucial for their zinc-binding ability [6].

### 2.2. Phylogenetic Tree and Conserved Motif Characterization of TaPLATZ Genes

Multiple sequence alignment data of 62 *TaPLATZ* genes were used to construct a neighbor-joining phylogenetic tree and further explore the similarity and diversity of motif compositions. *TaPLATZs* were classified into six groups (Groups I-VI) based on the topological structure of the phylogenetic tree (Figure 1A,B). Group VI was the largest one, containing 16 *TaPLATZs*. Both Groups I and III had 12 *TaPLATZs*, followed by Group II and V with 9 *TaPLATZs*. In contrast, Group IV contained the least, with four *TaPLATZs* (Table S5).

A neighbor-joining phylogenetic tree was carried out based on *PLATZs* from wheat associated with maize, and rice (Figure 1A). In most groups, gene phylogeny followed species phylogeny [22]. Only wheat *PLATZs* could be found in groups V and VI. More members of wheat *PLATZs* could be found than that from maize and rice in groups I, III, and IV. Some *TaPLATZ* genes were orthologs to *ZmPLATZ* and *OsPLATZ* with 100% bootstrap value, such as *TaPLATZ33*, *TaPLATZ40*, *TaPLATZ46* orthologs to *ZmPLATZ9* and LOC_Os02g09070 (Group III), revealing the sequences conservation during species evolution.

A total of 12 motifs, named motif-1 to motif-12, were detected using the MEME online program (Table S4). The highly conserved motif distribution in each group ensures the classification of genes accurately and regulation of downstream genes precisely. However, these motifs showed distinct divergence among six groups (Figure 1C). Motif-2 and motif-8 matched to the conservation cysteine and histidine residues in the N-terminal of PLATZ protein, which was found in all groups. Motif-5 and motif-4 had the conservation cysteine residues in the C-terminal of PLATZs that were detected in all groups. Motif-7 represents the termination region of PLATZs in all groups, indicating the integrity of most PLATZs. Motif-10 was detected in all groups except Group V. Motif-6 was only detected in Group IV, Group V, and Group VI; and motif-11 was specific to Group II and V. Motif-12, located in the starting position of translation, was only observed in Group I. All *TaPLATZs* contained PLATZ-conserved domains in the central region (Figure 1D), providing the ability for zinc-dependent DNA binding.

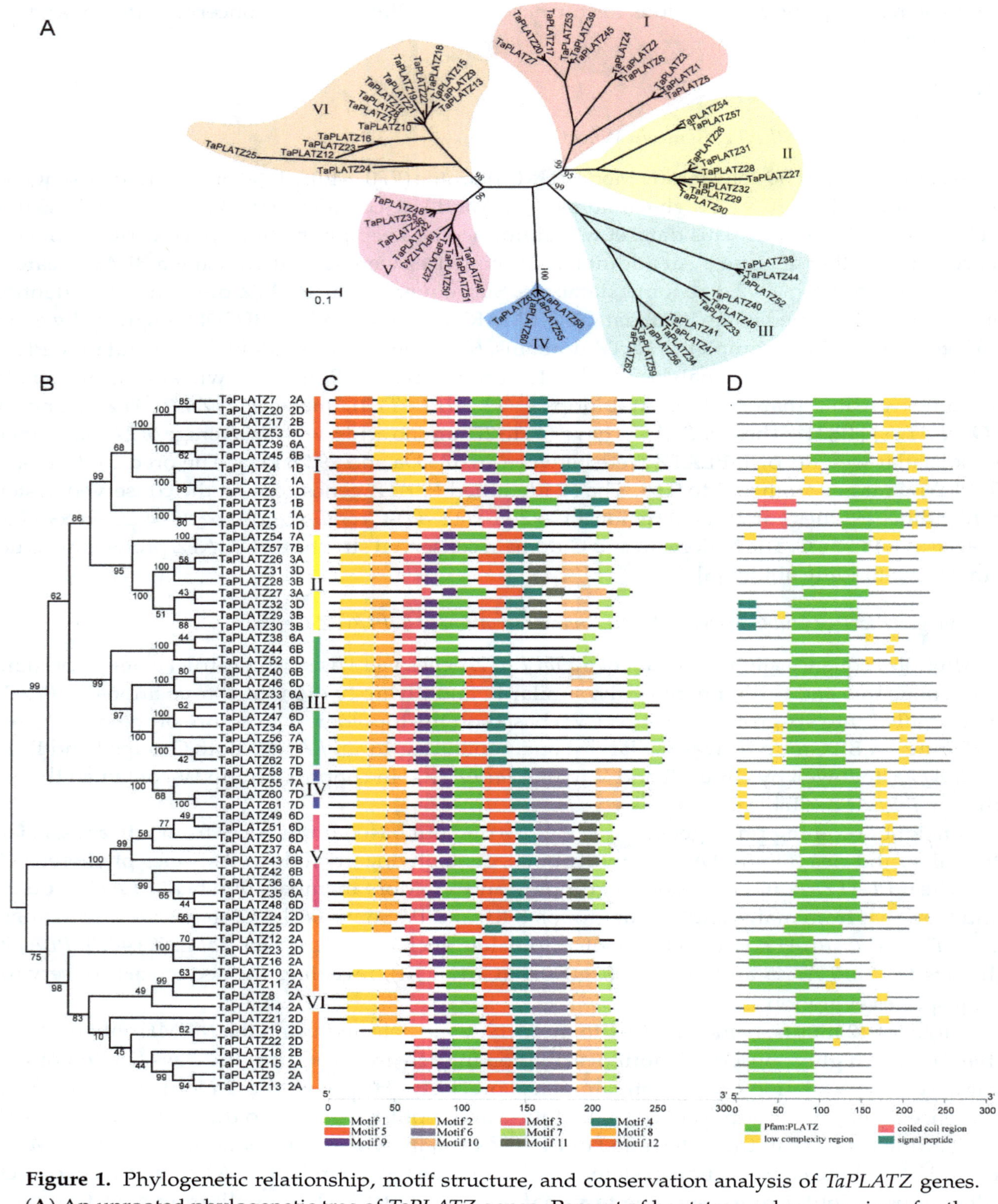


**Figure 1.** Phylogenetic relationship, motif structure, and conservation analysis of *TaPLATZ* genes. (**A**) An unrooted phylogenetic tree of *TaPLATZ* genes. Percent of bootstrap values are given for the main branches and support the classification in subfamilies; (**B**) The neighbor-joining tree of TaPLATZ proteins. I-VI: *TaPLATZs* were divided into six groups and are represented using different colors; (**C**) MEME motif structure shows the distinct divergence between groups; (**D**) Batch-smart analysis of PLATZ domain distribution of TaPLATZ proteins.

## 2.3. *Collinearity Analysis and Gene Duplication of TaPLATZs*

Collinear relationships between 62 *TaPLATZ* genes, 15 *OsPLATZ* genes, and 17 *ZmPLATZ* genes were analyzed for a better understanding of *PLATZ* genes evolution (Table S2, Figure 2). A total of thirty *TaPLATZ* genes showed collinear relationships with those in maize and rice, indicating that these orthologous pairs may already exist before the ancestral divergence. In addition, some *TaPLATZ* genes were associated with two orthologous genes located on different chromosomes, such as collinear gene pairs (*TaPLATZ54*, *TaPLATZ57*, *TaPLATZ60*, LOC_Os08g44620, and LOC_Os11g24130). Some collinear gene pairs were detected only in wheat and rice, such as *TaPLATZ7*, *TaPLATZ17*, *TaPLATZ20*.

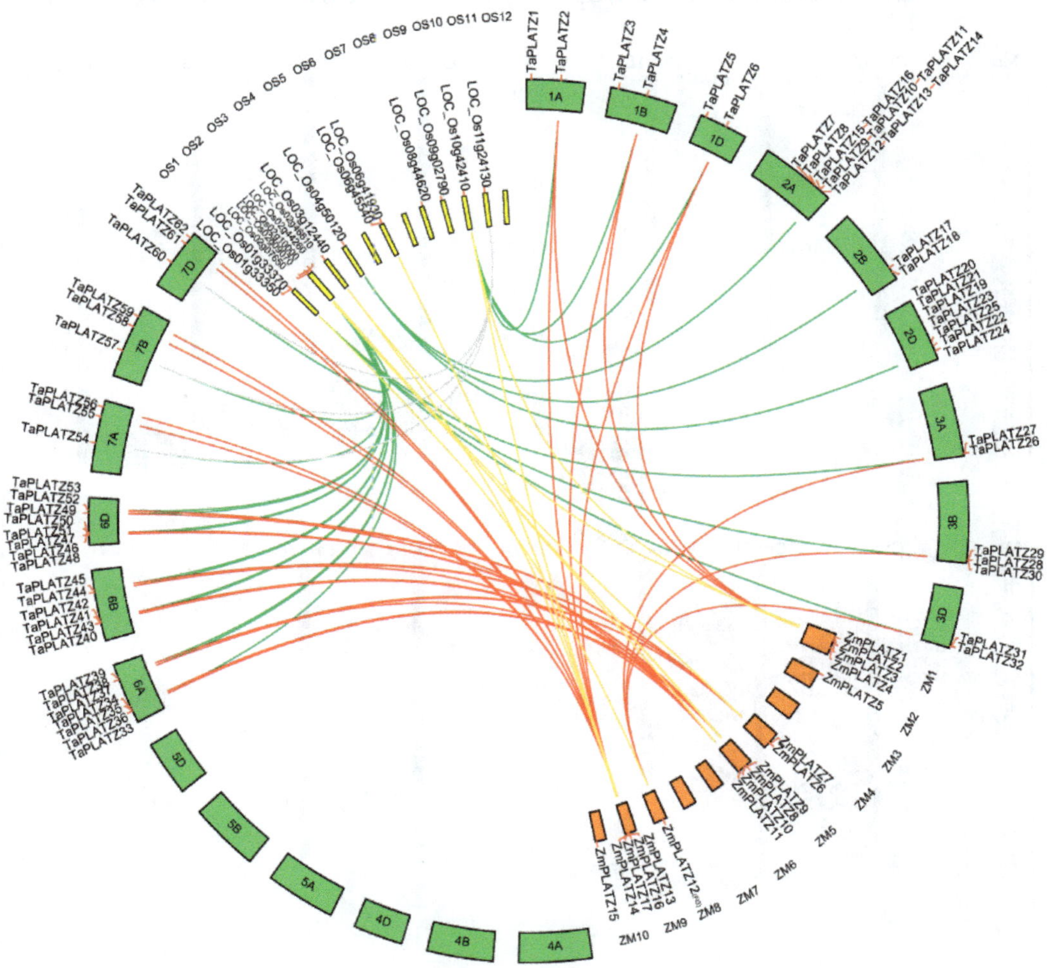


**Figure 2.** Collinear relationship analysis between orthologous *PLATZ* genes in wheat, rice, and maize. The circle atlas represents the orthologous pair position and collinearity. The red lines represent the orthologous *PLATZ* genes between wheat and maize. The green lines denote the orthologous *PLAYZ* genes in wheat and rice. The yellow lines represent the orthologous genes between maize and rice. Grey lines highlight the second orthologous pair of *TaPLATZ54*, *TaPLATZ57*, *TaPLATZ60* with rice. 1A–7D represent the twenty-one chromosomes of the wheat; OS1-OS12 represent the twelve chromosomes of rice; ZM1-ZM10 represent the ten chromosomes of maize.

In wheat, all 62 *TaPLATZs* were found to be unequally distributed on fifteen chromosomes, except for chromosomes 4A, 4B, 4D, 5A, 5B, and 5D. Most of the *TaPLATZs* (40/62) were located on chromosomes 2A, 2B, 2D, 6A, 6B, and 6D. Meanwhile, chromosomes 1A, 1B, and 1D contained only two *TaPLATZs*, respectively. It is indicated that duplication events tended to occur in chromosomes 2 and 6 during the evolution of gene families, which may be associated with gene functions. *OsPLATZs* were distributed unevenly among the nine rice chromosomes, as well. There were no *PLATZs* on chromosomes 5, 7, and 12. In addition, *ZmPLATZ* also exhibited uneven distribution. None of the *PLATZs* could be found on chromosomes 3, 6, 7, and 10 of maize.

Furthermore, the physical locations of the *TaPLATZ* genes were mapped on the corresponding chromosomes for a better understanding of the duplication events (Figure 3). During the process of evolution, gene duplication is essential for the generation of novel biological functions and the expansion of the gene family [23]. In wheat, most *TaPLATZs* had the corresponding homoeologous on the A, B, and D sub-genomes. Forty-six *TaPLATZ* genes exhibited a homology of 1:1:1 on the three sub-genomes with high identity, and can be referred to as triplets, indicating that wheat polyploidization was the main reason the expansion of the wheat PLATZ family. Apart from triplets, some genes contain several homologues due to complex gene-duplication events during wheat evolution.

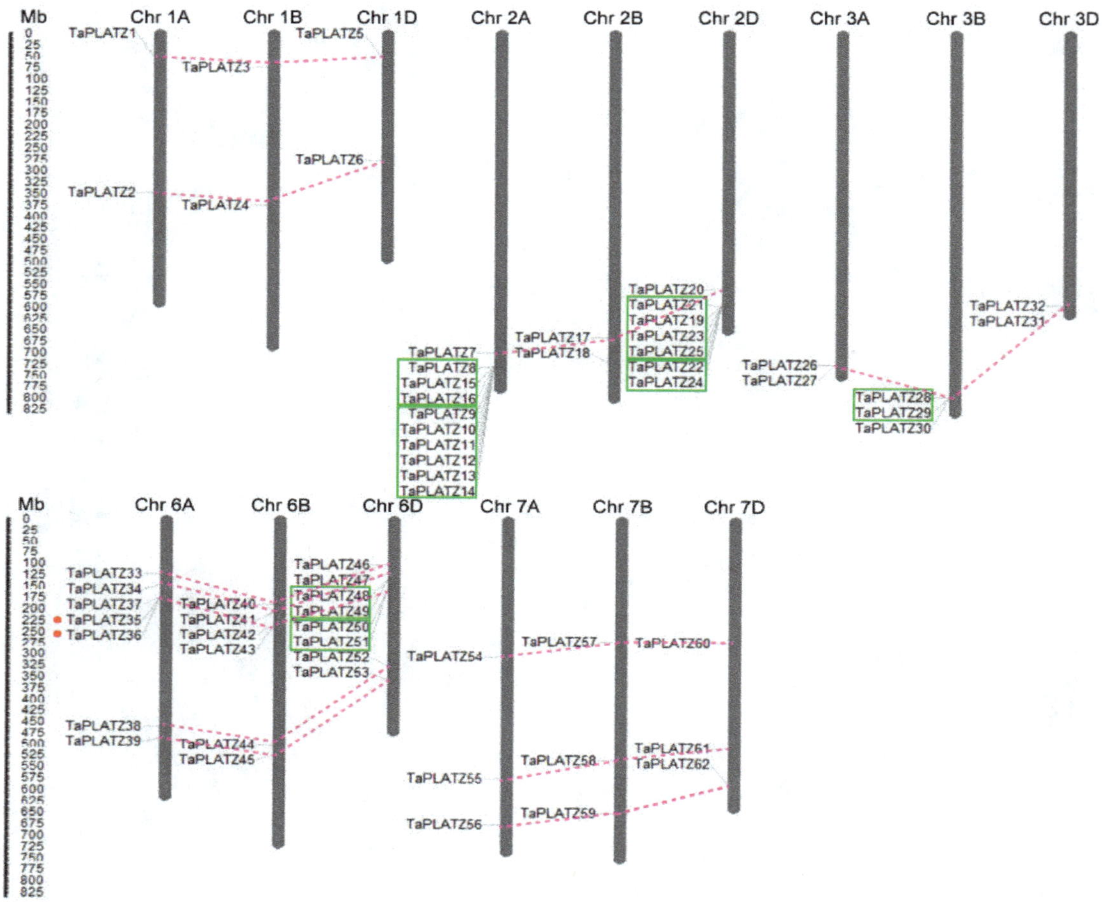


**Figure 3.** Distribution and duplication of *TaPLATZ* genes in *Triticum aestivum* chromosomes. The chromosome numbers are shown at the top of each bar. The length of chromosomes was their relative extent. The scale on the left is in megabases (Mb). Putative *TaPLATZ* homologous gene pairs were ligated with violet dotted lines. The tandem duplicated genes were marked with green boxes. Segmental duplicated genes were marked by red dots.

Duplication analysis revealed that obvious tandem duplication genes were found universally on chromosome 2A, 2D, 3B, and 6D (Figure 3) according to the criteria used in the analysis. Twenty-one out of 62 (33.9%) *TaPLATZ* genes were tandem duplicated, based on the similarity of their sequences and the proximity of position on the chromosome (Table S6). In addition, the *TaPLATZ* family has merely two segmental duplication genes (*TaPLATZ35* and *TaPLATZ36*), indicating that tandem duplication contributed more to the expansion of *TaPLATZ* family. Moreover, the majority of tandem duplicates were observed on subtelomeric distal regions of chromosomes result in more duplication events. On the other hand, genes located on the centromere proximal region of the chromosomes led to lower exchange frequency, such as *TaPLATZs* on chromosome 1A, 1B, and 1D.

## 2.4. *Variety of Cis-Acting Elements in Promoter Regions of TaPLATZs*

Studies of promoters that largely regulate gene expression at the transcriptional level are crucial for improving our basic understanding of gene regulation [24]. The 2.0-kb upstream sequences of translational initiation sites of *TaPLATZs* were used to predict cis-acting elements using the online database PlantCARE (Figure 4). Various cis-acting elements were found in the promoter of 62 *PLATZ* genes. Light responsive elements, including G-box (58/62), Sp1 (35/62), and Box 4 (32/62), were abundant in the 62 *TaPLATZ* genes. Drought-inducibility element MBS (35/62) accounted for a large part of the *TaPLATZ* genes, indicating the possibility of binding and regulation by the MYB transcription factor. Hormone responsive elements related to the response of gibberellin, salicylic acid (SA), abscisic acid (ABA), and methyl jasmonic acid (MEJA); as well as various abiotic stresses, such as ABRE (57/62), CGTCA-motif (52/62), and TGACG-motif (45/62), were identified in *TaPLATZ* genes.

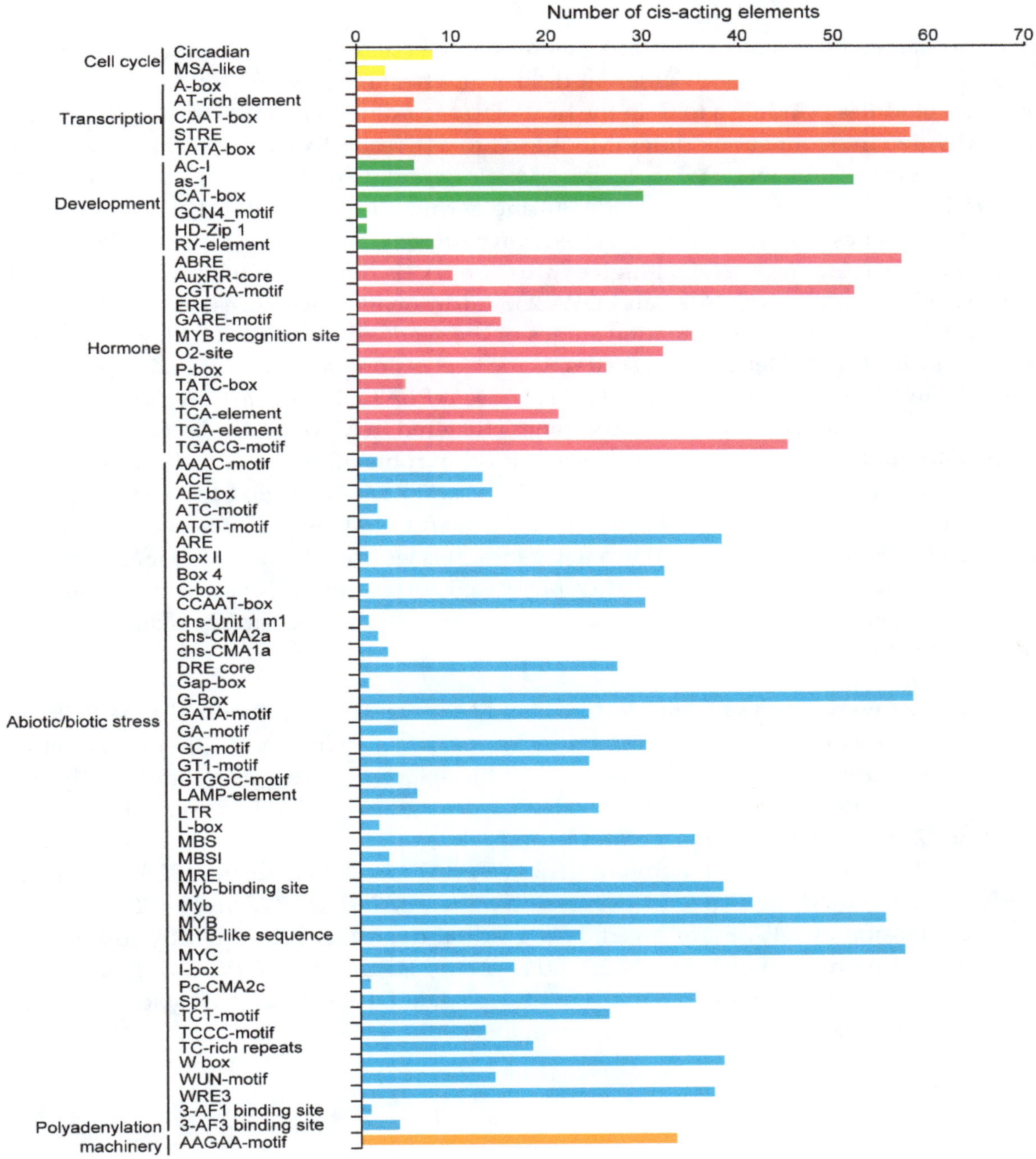


**Figure 4.** The number of cis-acting elements contained in the promoters of *TaPLATZ* genes. The cis-acting

element was identified with the online PlantCARE program using the 2k upstream from the transcription start site of *TaPLATZ* genes. The graph was generated based on the presence of cis-acting elements related to different conditions (*x*-axis) in 62 *TaPLATZs* (*y*-axis). The number of cis-acting elements involved in different regulatory pathways. Yellow columns refer to cell cycle, red columns represent transcription, green columns represent of the development process, blue columns refer to abiotic and biotic stress, orange columns represent the polyadenylation machinery.

The promoter-related elements CAAT-box and TATA-box were found in the promoter regions of all the 62 *TaPLATZ* genes, illustrating that most of the family members have transcriptional activity. Beyond that, eight *TaPLATZ* genes contained RY-element on promoters, such as *TaPLATZ54*, *TaPLATZ58*, and *TaPLATZ61*, giving rise to the possibility of regulating gene expression during late embryogenesis and seed development stage [25].

*2.5. Expression Patterns of TaPLATZs*

The expression patterns of 62 *TaPLATZs* were characterized using the transcriptomes (RNA-seq) data at different growth stages or in different tissues of wheat (Figure 5A). *TaPLATZs* were clustered into four main modules (Module a to d) in the heat map according to expression abundance. Module d was further divided into four submodules (Module d1 to d4), respectively. Eight *TaPLATZs* in Module a were highly expressed in roots, stems, leaves, developmental seeds at later stages, which refer to days-post-anthesis (DPA 20, DPA 25, and DPA 30) and germinating seeds after 12 and 24 h of imbibition (HAI 12, HAI 24). Genes with seed-specific expression patterns were clustered in Module b. Three genes' (*TaPLATZ26*, *TaPLATZ31*, *TaPLATZ32*) high expression levels remained in DPA 5 and DPA 10 seeds, which decreased gradually in DPA 20 and DPA 25. Additionally, the expression level of *TaPLATZ30* remained high in all of the developmental stages of the seeds. In Module d1, all genes (*TaPLATZ27*, *TaPLATZ28*, and *TaPLATZ29*) presented seed-specific expression patterns, but maintained a low level of expression. In Module d2, three genes (*TaPLATZ6*, *TaPLATZ2*, and *TaPLATZ4*) were exclusively expressed in roots and leaves. Another three genes clustered in Module d3 (*TaPLATZ33*, *TaPLATZ40*, and *TaPLATZ46*) had an obvious high expression level in roots of SHW-L1 and seeds at HAI 12 and HAI 24 of Chuanmai 32. Two genes (*TaPLATZ34*, *TaPLATZ47*) belonging to Module d4 were only expressed in DPA 30 of Fielder, and seeds at HAI 12, and HAI 24 of Chuanmai 32. In Module c, low expression levels were only detected in a few genes. It is worthy of noting that most genes clustered in the same subfamily of the phylogenetic tree possessed corresponding expression patterns. In other words, homoeologous genes possessed similar expression patterns and further ensured the functional stability in the process of evolution.

Ten *TaPLATZs* were selected from Module b and Module d1 and d2 for real-time PCR, and their expression patterns were further determined in leaves, roots, stems, and different developmental stages of seeds (Figure 5B). The qPCR results of all the selected genes were significantly correlated with the RNA-seq data. The expression levels of three genes (*TaPLATZ2*, *TaPLATZ4*, *TaPLATZ6*) from Module d2 were relatively high in roots and leaves. Three genes (*TaPLATZ27*, *TaPLATZ28*, *TaPLATZ29*) clustered in Module d1, and four genes (*TaPLATZ30*, *TaPLATZ26*, *TaPLATZ32*, *TaPLATZ31*) from Module b, presented seed-specific expression patterns. *TaPLATZ27* and *TaPLATZ28* exhibited the highest expression at DPA 8, followed by a gradual decrease, subsequently followed by seed development. The expression of *TaPLATZ26*, *TaPLATZ29*, *TaPLATZ30*, *TaPLATZ31*, and *TaPLATZ32* exhibited an up-down-up expression pattern, with a peak at DPA 8, followed by lower expression at DPA 12 and DPA 16, and then an increase again at DPA 20.

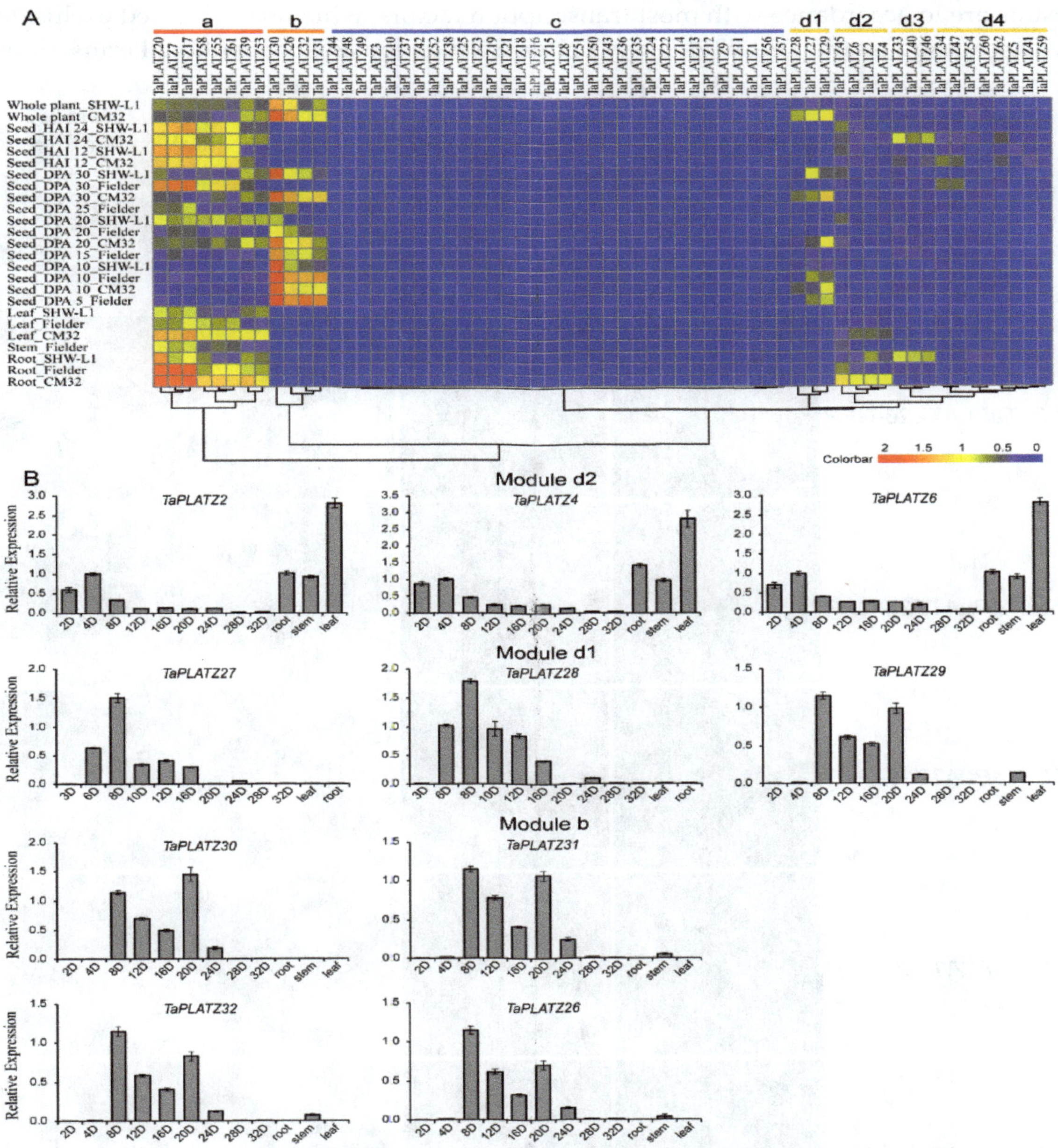

**Figure 5.** The expression profiles of *TaPLATZs* in wheat. (**A**) Heat map of *TaPLATZ* genes expression in different tissues and developmental stages. Colorbar represents the expression abundance of RNA-seq data. CM32: Chuanmai 32; HAI 12: seeds after 12 h of imbibition; HAI 24: seeds after 24 h of imbibition; DPA: days post-anthesis. (**B**) Real-time PCR data of ten selected *TaPLATZs* from Modules b, d1, and d2 in different developmental stages of seeds, root, stem, and leaves.

## *2.6. Subcellular Localization of TaPLATZ Proteins*

The predicted cellular localization by four different software programs showed that six TaPLATZ proteins (TaPLATZ26, TaPLATZ27, TaPLATZ29, TaPLATZ30, TaPLATZ31, and TaPLATZ32) were located in the nucleus (Table S1). Six TaPLATZ proteins were subsequently chosen to verify the subcellular localization by performing the transient expression. They were fused to a green fluorescent protein (GFP), and the constitutive 35S promoter drove all gene cassettes. The free GFP was used as the control. We transiently expressed the resulting constructs in tobacco leaves. Green fluorescence signals of all fusion proteins were localized in the nucleus (Figure 6), consistent with their predicted function as TFs, whereas the control 35S::GFP was detected both in the nucleus and cytoplasm ubiquitously.

These results are in accordance with most transcription factors, which were located exclusively at the nucleus, suggesting that the *TaPLATZs* likely perform a function as a conventional transcription factor.

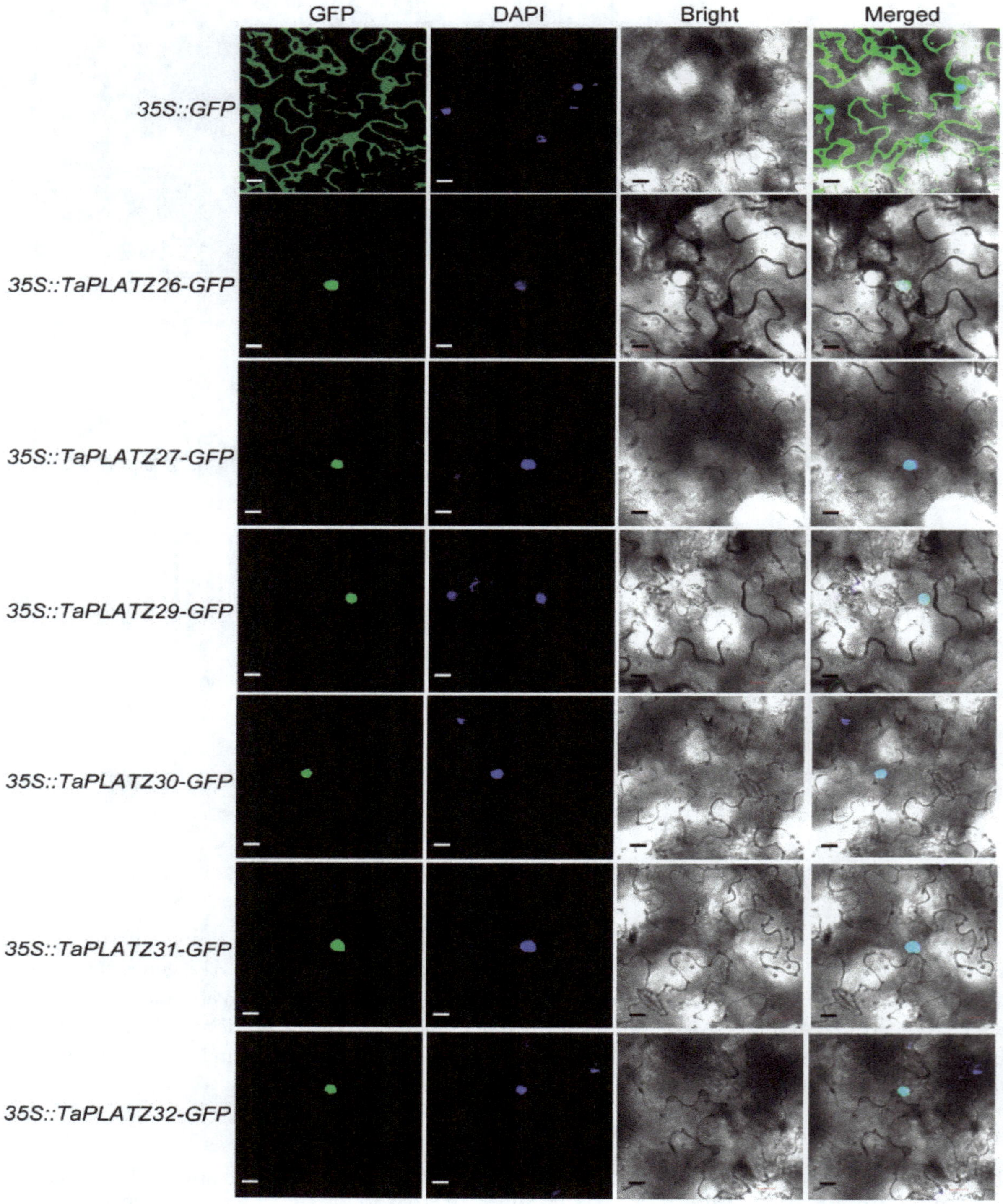


**Figure 6.** Subcellular localization of TaPLATZ proteins. The localization of the nucleus was detected by DAPI staining. GFP: Green fluorescence indicates the location of TaPLATZ proteins in *N. benthamiana*. DAPI: Blue fluorescence of DAPI indicated the location of the nucleus. Bright light: field of bright light; Merged: merge with the three former images. The scale bar = 20 μm.

## 3. Discussion

Ancient duplication events and a high rate of retention of extant pairs of duplicate genes have contributed to an abundance of duplicate genes in plant genomes [26–28]. A total of 62 *TaPLATZs* were characterized in wheat, which was almost threefold the numbers of *PLATZs* in Arabidopsis (13), rice (15), and maize (17). The phylogenetic tree of *PLATZ* genes associated with three species revealed that (Figure 1A), the ratio of *PLATZ* ortholog numbers in wheat to rice/maize was as expected—3:1 (wheat: rice and wheat: maize). However, the ratio within some groups was larger than expected. For example, the ratio was about 7:2 or 7:1 in Group II, and the ratio was 4:1 or 4:1 in Group IV. This large divergence suggests that the *PLATZ* gene family might have experienced multiple duplication events during the polyploidization of wheat. Gene duplications are considered to be among the primary driving forces in the evolution of genomes and genetic systems [29]. Duplicate genes provide raw materials for the evolution of mechanism novelties, in turn, facilitate the generation of new functions [29]. While in some of the other cases, the ratio was lower than expected. For instance, wheat orthologs of *ZmPLATZ2* and *ZmPLATZ14* could not be identified in Group III, indicating gene loss during the process of polyploidization of wheat.

Apart from this, among six groups in the phylogenetic tree of three PLATZ families (Figure 1A), genes clustered in Group V and VI have no ortholog gene in neither maize and rice PLATZ family. After sequence alignment, the amino acid sequence of *TaPLATZs* from Group V and VI showed the highest similarity of *TaPLATZs* in Group II. The large number of homoeologs can be inferred that most of *TaPLATZ* genes from Group V and VI may be retained after whole-genome duplications (WGDs).

The wheat (BBAADD) has a complex genome consisting of three related sub-genomes that were derived from three different diploid species [30]. The *TaPLATZs* were unequally distributed in sub-genomes A, B, and D, which included 24, 16, and 20 members, respectively (Table S1, Figure 2). This suggests that homologous genes on the B genome may be absent or become pseudogenes in the lineage leading to wheat [31]. During the evolutionary process of wheat, rapid alterations and sporadic changes in wheat genome took place due to hybridization, polyploidization, domestication, and mutation, resulting in some modifications and a high level of gene loss [32]. Previous reports have stated that the preferential retention of dosage-sensitive genes (e.g., regulatory genes such as transcription factors) and gene loss following WGDs played a significant role in the evolution of eukaryotes [33].

Through the analysis of the evolutionary mechanism, it has been determined that segmental duplication, tandem duplication, and transposition events, including retro-position and replication transposition, had an essential role in the expansion of the number of genes [34,35]. Among these patterns, segmental and tandem duplications are involved in the main patterns in plant gene family expansion [36]. The occurrence of segmental duplications in plant species was considered to be associated with plant polyploidization, followed by inter-chromosomal rearrangements [35].

Tandem duplications are identified by multiple members in one family occurring within the same intergenic region or in neighboring intergenic regions, results from unequal crossing-over and led to increasing or decreasing copy numbers in gene families [36]. Duplication analysis revealed that twenty-one *TaPLATZ* genes were identified as tandem duplicates (Figure 2, Table S6), which account for 37.9% of *TaPLATZ* duplicates, whereas segmental duplication genes account for 3.2%, indicating that tandem duplication pattern probably played a pivotal role in the expansion of *TaPLATZ* gene family. *TaPLATZ* genes are located in the subtelomeric region on chromosome leading to a high percentage of duplication events and results in the expansion of groups. Conversely, genes belonging to groups containing smaller members tended to be located close to the centromere of the chromosomes.

A variety of *TaPLATZ* gene expression patterns was shown in the heat map (Figure 5A). *TaPLATZs* genes in Module a displayed a continuous expression except in the early stage of seed development, and *TaPLATZs* genes in Module b presented a seed-specific expression pattern. Moreover, root-specific expression patterns can be found in Module d, indicating the functionally important and nonredundancy of *TaPLATZ* genes. *TaPLATZ* genes clustered in one subfamily could exhibit different expression patterns. For example, members in Group II were clustered into three modules, revealing that functional divergence and the biased expression of duplicated genes appear to be major factors promoting their retention in the genome [37,38]. It is noteworthy that the expression patterns of seven *TaPLATZs* (*TaPLATZ26* to *TaPLATZ32*) contained in Group II were highly similar to those of *ZmPLATZ12* (Fl3) and LOC_Os01g33350, LOC_Os01g33370, which displayed invariant seed-specific expression pattern in the early stage of seeds. These conserved expression patterns reflect that these *TaPLATZ*s may be involved in seed development and maturation, indicating the possibility of their participation in the common metabolic and or developmental processes of wheat [8,12].

Plants can respond and coordinate growth and stress tolerance to promote survival from abiotic and biotic stresses by modifying the production, distribution, or signal transduction of hormones. Zinc-finger transcription factors are a relatively large family of plant transcription factors (approximately 15% of the total), which regulate the expression of several genes in response to abiotic stress such as low temperature, salt, drought, osmotic stress, and oxidative stress [2,39]. AtPLATZ1 and AtPLATZ12 were identified as major nodes to positively regulate the acquisition of desiccation tolerance in Arabidopsis seeds and vegetative tissues [40]. The transcription level of *GhPLATZ1* (*Gossypium hirsutum* PLATZ1) was induced by abiotic and hormone stimuli in 20-day-old seedlings. Ectopic expression of *GhPLATZ1* in Arabidopsis resulted in enhanced insensitivity to osmotic stresses, ABA, and PAC [41]. Similarly, the RNA expression level of *GmPLATZ1* (*Glycine max* PLATZ1) dramatically increased when responding with exogenous ABA application on soybean plants, the leaves mRNA level of *GmPLATZ1* steadily increased after dealing with 24 h drought stress. Furthermore, ectopic expression of *GmPLATZ1* in Arabidopsis showed retarded germination during the early germination process with the addition of mannitol, ABA, and osmotic stress [7]. It is essential to investigate the RNA expression level of *TaPLATZs* under various abiotic stresses in further study, which will facilitate the identification of potential components to coordinate seedling growth during germination.

## 4. Materials and Methods

### 4.1. Plant Growth Conditions

The wheat landrace Chinese spring was planted in plant growth chambers in a 16-h-light/8-h-dark photoperiod at 20 °C in 2019. Tobacco (*Nicotiana benthamiana*) was grown in the plant growth chamber under a 16-h-light/8-h-dark photoperiod at a temperature of 20/25 °C.

### 4.2. Identification of TaPLATZ Family Members in the Wheat

The wheat protein sequence was obtained from the IWGSC database (https://urgi.versailles.inra.fr/download/iwgsc/IWGSC_RefSeq_Annotations/v1.1/). The Hidden Markov Model (HMM) analysis was carried out for the desired sequences search. The HMM profile of the PLATZ (PF04640) downloaded from the Pfam database (http://pfam.xfam.org/) was applied as a query using HMM search program (http://hmmer.janelia.org/) with an E-value cutoff of 1.0. The protein sequences containing complete or partial PLATZ domain, which may be pseudogenes, incomplete assemblies, sequencing errors, or mispredictions [42], were considered as putative *TaPLATZ*s. To eliminate the *TaPLATZ* sequences contained incomplete PLATZ domains, a BLASTP program was performed by using identified amino acid sequences as queries with an *e*-value $\leq 1 \times 10^{-3}$. Additionally, the amino acid sequences of maize and rice *PLATZ* genes were obtained from PlantTFDB (http://plntfdb.bio.uni-potsdam.de/v3.0/) and GrassTFDB (http://www.grassius.org/grasstfdb.php) databases, which were also used for blasting against wheat genomic reference in the IWGSC v2.0 database.

*TaPLATZ* protein sequences were reconfirmed by Pfam (http://pfam.xfam.org/), SMART (http://smart.embl-heidelberg.de/), and NCBI-CDD (https://www.ncbi.nlm.nih.gov/cdd/) to identify the conserved PLATZ domain. The sequences lacking the PLATZ domain were excluded. All of the non-redundant and high-confidence genes were named after their chromosomal positions on pseudomolecules. The putative TaPLATZ protein sequences were submitted to CDD (https://www.ncbi.nlm.nih.gov/Structure/bwrpsb/bwrpsb.cgi).

### 4.3. *Characterization of TaPLATZ: Conserved Motif, PLATZ Domain and Putative Cis-Acting Elements*

Conserved motifs of TaPLATZ protein were identified using the MEME website (http://meme-suite.org/tools/meme) [43] with the following parameters: distribution of motifs, 0 or 1 occurrence per sequence; maximum number of motifs, 12; minimum sites, 6; maximum width 50. Visualized of PLATZ domains on each TaPLATZ proteins were performed using TBtoolssoftware [44] (v1.046, Chen, C., GZ, China).

The 2.0-kb upstream of the transcription start site (−1) of *TaPLATZs* was extracted as a promoter to predict cis-acting elements using the PlantCARE (http://bioinformatics.psb.ugent.be/webtools/plantcare/html/) [45]. Then, statistics derived from hits of various cis-acting elements were constructed and displayed by the diagram. Theoretical pI/MW of TaPLATZs was calculated by the Compute pI/MW tool (http://web.expasy.org/compute_pi/).

### 4.4. *Phylogenetic Analysis, Collinear Relationships and Classification of PLATZ Genes in Wheat, Maize and Rice*

The amino acid sequences of PLATZ derived from maize and rice (Table S1), together with newly identified *TaPLATZs* were used for phylogenetic analysis. All of the amino acid sequences are first aligned by ClustalW with the default parameters. Subsequently, the Bayesian and Neighbor-joining phylogenetic trees were constructed using MEGA software with a bootstrap test of 1000 times [46] (v6.0, Tamura, K., Tokyo, Japan). The *TaPLATZs* were classified into different groups according to the topology of the phylogenetic tree and the classification in maize and rice from previous studies [12]. The collinear relationships of orthologous *PLATZ* genes in wheat, maize and rice were displayed using the Circos program [47] (Krzywinski, M., Vancouver, BC, Canada).

### 4.5. *Location of TaPLATZ Genes on the Chromosome; Identification of Duplication Genes*

The obtained chromosomal location information of *TaPLATZ* genes was visualized by performing MapInspect software (http://www.softsea.com/download/MapInspect.html) (R. van Berloo, Wageningen, The Netherlands)according to their chromosome locus and the length of each chromosome.

The duplication gene pairs in the *TaPLATZ* family were identified by BLASTP based on the criteria of the previous studies [48,49]: (a) the alignment covered >80% of the longer gene; (b) the aligned region had an identity > 80%.

Tandem duplicated PLATZ genes were defined as two or more adjacent homologous genes located physically on a single chromosome with an intergenic region less than 200 kb [50], while homologous genes with an interval greater than 200 kb, or between different chromosomes, were defined as segmentally duplicated genes [51].

### 4.6. *Expression Profiles of TaPLATZ*

A total of 25 RNA-seq (transcript) data files with wheat tissues (root, leaf, stem, grain and spike) at different developmental stages of the hexaploidy bread wheat (Chuanmai 32, SHW-L1 and Fielder) were obtained from the local database, the abbreviation of each tissue corresponding to the detailed description in Table S7.

The transcriptional results were sorted based on the phylogeny classification and visualized in a heatmap using the 'ggplot2' and 'heatmap' R-software (Ihaka R, Auckland, CA, USA)package with a normalization according to an individual gene. The bar graph represents the *TaPLATZs* expression in tissues. The blocks close to Red representing high expression levels, yellow indicating medium expression levels, and blue refers to almost no expression.

For real-time PCR, tissues including root, stem, the third leaf and seeds were obtained from at least three healthy seedlings and three spikelets after sowing. The roots, stems were isolated from the seedling after one-week after imbibition. The seeds were obtained from spikelets at 2, 3, 6, 8, 10, 12, 16, 20, 24, 28, and 32 days after pollination, respectively. RNA was extracted from wheat tissues using the RNA extraction kit (Magen, Beijing, China, Lot: R4165-02), refering to the manufacturer's instruction for specific steps, and then digested with RNase-free DNase I. The quantity and concentration of RNA were evaluated by Thermo Scientific™ NanoDrop™. The first-strand cDNA was generated using PrimeScript™ RT Reagent Kit (TaKaRa, Kyoto, Japan, Cat. # RR037B). Primer-BLAST (https://www.ncbi.nlm.nih.gov/tools/primer-blast/index.cgi?LINK_LOC=BlastHome) was used to design gene-specific primers. Wheat house-keeping genes 8 and 34 were used as the reference genes.

Real-time quantitative-PCR (qRT-PCR) was carried out using SYBR® Premix Ex Taq™ II (Tli RNaseH Plus) (TaKaRa, Kyoto, Japan, Cat. # RR820A) following the instructions in the manual with a Bio-Rad CFX96™ real-time PCR detection system (BioRad, Berkeley, CA, USA). For normalization, three reference genes were used as internal control genes: *TaGAPDH*, *Ta.7894.3.A1_at* and *Ta.14126.1.S1_at* [52,53]. Three biological replicates of tissues were applied for qPCR analyses. The relative gene expression levels were calculated using the $2^{-\Delta\Delta CT}$ method [54], and the primers used in qRT-PCR are listed in Table S3.

### 4.7. Subcellular Localization of TaPLATZ

The subcellular localization of TaPLATZ protein was predicted according to the results of CELLO Version 2.5 (http://cello.life.nctu.edu.tw/) [55], Plant-mPLoc (http://www.csbio.sjtu.edu.cn/bioinf/plant-multi/) [56], WoLF PSORT (https://wolfpsort.hgc.jp/) and SoftBerry (http://linux1.softberry.com/all.htm). Six seed-specifically expressed TaPLATZs were selected to verify the subcellular localization prediction. The similarity of homologous genes make application difficult, the full-length coding sequences of TaPLATZs without stop codon was synthesized by the company, subsequently inserted into pCAMBIA1300 plasmid driven by 35S promoter. The resulting vectors were *35S::TaPLATZ26-GFP*,*35S::TaPLATZ29-GFP*, *35S::TaPLATZ32-GFP*, *35S::TaPLATZ27-GFP*, *35S::TaPLATZ30-GFP*, and *35S::TaPLATZ31-GFP*, respectively. Agrobacterium-mediated transient transformation of *Nicotiana benthamiana* plants was conducted to check the subcellular locations of *TaPLATZs*. Agrobacterium tumefaciens strain GV3101(Weidi, Shanghai, China) carrying expression construct were grown in LB media with kanamycin and Rifampicin antibiotics, diluted in 1:6 and grown for 9 h at 28 °C, centrifugated for 4000 rpm 5 min, remove supernatant, pellets were resuspended in infiltration medium (10 mM $MgCl_2$, 10 mM MES-KOH, PH 5.7, 200 μM acetosyringone). The $OD_{600}$ was adjusted to 0.6–0.9. The resulting culture was infiltrated into 3-week-old *N. benthamiana* leaves using a 1 mL sterile syringe (no needle). At least three replicates were performed. Subcellular localization was observed after infiltration for 2 days. The GFP fluorescent signal was observed and imaged using a confocal laser scanning microscope (LSM 880, Karl Zeiss, Jena, Germany).

## 5. Conclusions

A total of 62 *PLATZ* genes were identified in the wheat genome database, and were distributed unevenly on 15 chromosomes. Chromosome 6 contained the most *TaPLATZ* genes. According to the phylogenetic tree, *TaPLATZ* genes could be classified into six subfamilies (Group I to VI), Group V and Group VI were without orthologues of other *PLATZ* genes in other species. Depending on the type of cis-acting elements, *TaPLATZ* genes may be regulated by a variety of hormones and environmental factors. Gene duplication events analyses suggested that tandem duplication events played a significant role in the expansion of *TaPLATZ* family. Tissue specificity in different developmental stages can be observed in RNA-seq data and qRT-PCR analysis, suggesting the potential role of *TaPLATZs* in tissue differentiation and seed development. This study provides valuable information for further understanding of the evolutionary mechanism and functional traits of the *PLATZ* genes family in wheat.

**Author Contributions:** Conceptualization, J.W.; methodology, Y.F.; data analysis, M.C. and X.G.; experimental validation, Y.F. and M.L.; formal analysis, Y.F.; investigation, Y.F.; resources, M.C.; data curation, M.C.; writing—original draft preparation, Y.F.; review and editing, J.W.; visualization, M.C.; supervision, Y.W. and J.W.; project administration, J.W.; funding acquisition, Y.W. and J.W. All authors have read and agreed to the published version of the manuscript.

**Acknowledgments:** We acknowledge the plant research community for sharing the relevant genomic data and making it publicly available.

## Abbreviations

| | |
|---|---|
| DPA | Days post-anthesis |
| HAI | Hours after imbibition |
| PLATZ | plant AT-rich protein and zinc-binding protein |
| NJ | Neighbor-Joining |
| TFs | Transcription factors |

## References

1. Ning, P.; Liu, C.C.; Kang, J.Q.; Lv, J.Y. Genome-wide analysis of WRKY transcription factors in wheat (*Triticum aestivum* L.) and differential expression under water deficit condition. *PeerJ* **2017**, *5*, e3232. [CrossRef] [PubMed]
2. Riechmann, J.L.; Heard, J.; Martin, G.; Reuber, L.; Jiang, C.-Z.; Keddie, J.; Adam, L.; Pineda, O.; Ratcliffe, O.J.; Samaha, R.R.; et al. Arabidopsis transcription factors: Genome-wide comparative analysis among eukaryotes. *Science* **2000**, *290*, 2105–2110. [CrossRef] [PubMed]
3. Eulgem, T.; Rushton, P.; Robatzek, S.; Somssich IEulgem, T.; Rushton, P.J.; Robatzek, S.; Somssich, I.E. The WRKY superfamily of plant transcription factors. *Trends Plant Sci.* **2000**, *5*, 199–206. [CrossRef]
4. Olsen, A.N.; Ernst, H.A.; Leggio, L.L.; Skriver, K. NAC transcription factors: Structurally distinct, functionally diverse. *Trends Plant Sci.* **2005**, *10*, 79–87. [CrossRef] [PubMed]
5. Kizis, D.; Lumbreras, V.; Pagès, M. Role of AP2/EREBP transcription factors in gene regulation during abiotic stress. *FEBS Lett.* **2001**, *498*, 187–189. [CrossRef]
6. Yukio, N.; Hirofumi, F.; Takehito, I.; Yukiko, S. A novel class of plant-specific zinc-dependent DNA-binding protein that binds to A/T-rich DNA sequences. *Nucleic Acids Res.* **2001**, *29*, 4097–4105. [CrossRef]
7. So, H.A.; Choi, S.J.; Chung, E.; Lee, J.H. Molecular characterization of stress-inducible PLATZ gene from soybean (*Glycine max* L.). *Plant Omics* **2015**, *8*, 479–484.
8. Li, Q.; Wang, J.; Ye, J.; Zheng, X.; Xiang, X.; Li, C.; Fu, M.; Wang, Q.; Zhang, Z.; Wu, Y. The Maize Imprinted Gene Floury3 Encodes a PLATZ Protein Required for tRNA and 5S rRNA Transcription through Interaction with RNA Polymerase III. *Plant Cell* **2017**, *29*, 2661–2675. [CrossRef]
9. Kim, J.H.; Kim, J.; Jun, S.E.; Park, S.; Timilsina, R.; Kwon, D.S.; Kim, Y.; Park, S.-J.; Hwang, J.Y.; Nam, H.G.; et al. ORESARA15, a PLATZ transcription factor, mediates leaf growth and senescence in Arabidopsis. *New Phytol.* **2018**, *220*, 609–623. [CrossRef]
10. Wang, A.; Hou, Q.; Si, L.; Huang, X.; Luo, J.; Lu, D.; Zhu, J.; Shangguan, Y.; Miao, J.; Xie, Y.; et al. The PLATZ Transcription Factor GL6 Affects Grain Length and Number in Rice. *Plant Physiol.* **2019**, *180*, 2077–2090. [CrossRef]
11. Zhou, S.R.; Xue, H.W. The rice PLATZ protein SHORT GRAIN6 determines grain size by regulating spikelet hull cell division. *J. Integr. Plant Biol.* **2019**, *62*, 847–864. [CrossRef] [PubMed]
12. Wang, J.; Ji, C.; Li, Q.; Zhou, Y.; Wu, Y. Genome-wide analysis of the plant-specific PLATZ proteins in maize and identification of their general role in interaction with RNA polymerase III complex. *BMC Plant Biol.* **2018**, *18*, 221. [CrossRef] [PubMed]
13. Heun, M.; SchaferPregl, R.; Klawan, D.; Castagna, R.; Accerbi, M.; Borghi, B.; Salamini, F. Site of einkorn wheat domestication identified by DNA fingerprinting. *Science* **1997**, *278*, 1312–1314. [CrossRef]
14. Marcussen, T.; Sandve, S.R.; Heier, L.; Spannagl, M.; Pfeifer, M.; Jakobsen, K.S.; Wulff, B.B.H.; Steuernagel, B.; Mayer, K.F.X.; Olsen, O.A. Ancient hybridizations among the ancestral genomes of bread wheat. *Science* **2014**, *345*, 286–292. [CrossRef] [PubMed]
15. Matsuoka, Y.; Nasuda, S. Durum wheat as a candidate for the unknown female progenitor of bread wheat: An empirical study with a highly fertile F1 hybrid with Aegilops tauschii Coss. *Theor. Appl. Genet.* **2004**, *109*, 1710–1717. [CrossRef]
16. Appels, R.; Eversole, K.; Stein, N.; Feuillet, C.; Keller, B.; Rogers, J.; Pozniak, C.J.; Choulet, F.; Distelfeld, A.; Poland, J.; et al. Shifting the limits in wheat research and breeding using a fully annotated reference genome. *Science* **2018**, *361*, 661–674. [CrossRef]

17. Avni, R.; Nave, M.; Barad, O.; Baruch, K.; Twardziok, S.O.; Gundlach, H.; Hale, I.; Mascher, M.; Spannagl, M.; Wiebe, K.; et al. Wild emmer genome architecture and diversity elucidate wheat evolution and domestication. *Science* **2017**, *357*, 93–97. [CrossRef]
18. Koenig, D.; Jimenez-Gomez, J.M.; Kimura, S.; Fulop, D.; Chitwood, D.H.; Headland, L.R.; Kumar, R.; Covington, M.F.; Devisetty, U.K.; Tat, A.V.; et al. Comparative transcriptomics reveals patterns of selection in domesticated and wild tomato. *Proc. Natl. Acad. Sci. USA* **2013**, *110*, E2655–E2662. [CrossRef]
19. Huang, X.; Kurata, N.; Wei, X.; Wang, Z.-X.; Wang, A.; Zhao, Q.; Zhao, Y.; Liu, K.; Lu, H.; Li, W.; et al. A map of rice genome variation reveals the origin of cultivated rice. *Nature* **2012**, *490*, 497–501. [CrossRef]
20. Uauy, C.; Wulff, B.B.H.; Dubcovsky, J. Combining Traditional Mutagenesis with New High-Throughput Sequencing and Genome Editing to Reveal Hidden Variation in Polyploid Wheat. *Annu. Rev. Genet.* **2017**, *51*, 435–454. [CrossRef]
21. Conte, M.G.; Gaillard, S.; Lanau, N.; Rouard, M.; Perin, C. GreenPhylDB: A database for plant comparative genomics. *Nucleic Acids Res.* **2007**, *36*, D991–D998. [CrossRef] [PubMed]
22. Schilling, S.; Kennedy, A.; Pan, S.; Jermiin, L.; Melzer, R. Genome-wide analysis of MIKC-type MADS-box genes in wheat: Pervasive duplications may have facilitated adaptation to different environmental conditions. *New Phytol.* **2019**, *225*, 511–529. [CrossRef] [PubMed]
23. Hurles, M. Gene duplication: The genomic trade in spare parts. *PLoS Biol.* **2004**, *2*, 900–904. [CrossRef] [PubMed]
24. Hernandez-Garcia, C.M.; Finer, J.J. Identification and validation of promoters and cis-acting regulatory elements. *Plant Sci.* **2014**, *109*, 217–218. [CrossRef] [PubMed]
25. Reidt, W.; Wohlfarth, T.; Ellerström, M.; Czihal, A.; Bäumlein, H. Gene regulation during late embryogenesis: The RY motif of maturation-specific gene promoters is a direct target of the FUS3 gene product. *Plant J.* **2000**, *21*, 401–408. [CrossRef]
26. Otto, S.P.; Whitton, J. Polyploid incidence and evolution. *Annu. Rev. Genet.* **2000**, *34*, 401–437. [CrossRef]
27. Blanc, G. A Recent Polyploidy Superimposed on Older Large-Scale Duplications in the Arabidopsis Genome. *Genome Res.* **2003**, *13*, 137–144. [CrossRef]
28. Panchy, N.; Lehti-Shiu, M.D.; Shiu, S.H. Evolution of Gene Duplication in Plants. *Plant Physiol.* **2016**, *171*, 2294–2316. [CrossRef]
29. Moore, R.C.; Purugganan, M.D. The early stages of duplicate gene evolution. *Proc. Natl. Acad. Sci. USA* **2003**, *100*, 15682–15687. [CrossRef]
30. Feldman, M.; Levy, A.A. Genome Evolution Due to Allopolyploidization in Wheat. *Genetics* **2012**, *192*, 763–774. [CrossRef]
31. Liu, Y.; Liu, N.N.; Deng, X.; Liu, D.M.; Li, M.F.; Cui, D.D.; Hu, Y.K.; Yan, Y.M. Genome-wide analysis of wheat DNA-binding with one finger (Dof) transcription factor genes: Evolutionary characteristics and diverse abiotic stress responses. *BMC Genom.* **2020**, *21*, 1–18. [CrossRef]
32. Rahman, S.; Islam, S.; Yu, Z.; She, M.; Nevo, E.; Ma, W. Current Progress in Understanding and Recovering the Wheat Genes Lost in Evolution and Domestication. *Int. J. Mol. Sci.* **2020**, *21*, 5836. [CrossRef] [PubMed]
33. Edger, P.P.; Pires, J.C. Gene and genome duplications: The impact of dosage-sensitivity on the fate of nuclear genes. *Chromosome Res.* **2009**, *17*, 699–717. [CrossRef] [PubMed]
34. Kong, H.; Landherr, L.L.; Frohlich, M.W.; Leebens-Mack, J.; Ma, H.; DePamphilis, C.W. Patterns of gene duplication in the plant SKP1 gene family in angiosperms: Evidence for multiple mechanisms of rapid gene birth. *Plant J.* **2007**, *50*, 873–885. [CrossRef] [PubMed]
35. Adams, K.L.; Wendel, J.F. Polyploidy and genome evolution in plants. *Curr. Opin. Plant Biol.* **2005**, *8*, 135–141. [CrossRef]
36. Cannon, S.B.; Mitra, A.; Baumgarten, A.; Young, N.D.; May, G. The roles of segmental and tandem gene duplication in the evolution of large gene families in Arabidopsis thaliana. *BMC Plant Biol.* **2004**, *4*, 10. [CrossRef]
37. Force, A.; Lynch, M.; Pickett, F.B.; Amores, A.; Yan, Y.L.; Postlethwait, J. Preservation of duplicate genes by complementary, degenerative mutations. *Genetics* **1999**, *151*, 1531–1545.
38. Lynch, M.; Force, A. The probability of duplicate gene preservation by subfunctionalization. *Genetics* **2000**, *154*, 459–473.
39. Kiełbowicz-Matuk, A. Involvement of plant $C_2H_2$-type zinc finger transcription factors in stress responses. *Plant Sci. Int. J. Exp. Plant Biol.* **2012**, *185–186*, 78–85. [CrossRef]

40. González-Morales, S.I.; Chávez-Montes, R.A.; Hayano-Kanashiro, C.; Alejo-Jacuinde, G.; Rico-Cambron, T.Y.; de Folter, S.; Herrera-Estrella, L. Regulatory network analysis reveals novel regulators of seed desiccation tolerance in Arabidopsis thaliana. *Proc. Natl. Acad. Sci. USA* **2016**, *113*, E5232. [CrossRef]
41. Zhang, S.; Yang, R.; Huo, Y.; Liu, S.; Yang, G.; Huang, J.; Zheng, C.; Wu, C. Expression of cotton PLATZ1 in transgenic Arabidopsis reduces sensitivity to osmotic and salt stress for germination and seedling establishment associated with modification of the abscisic acid, gibberellin, and ethylene signalling pathways. *BMC Plant Biol.* **2018**, *18*, 1–11. [CrossRef] [PubMed]
42. Rinerson, C.I.; Rabara, R.C.; Tripathi, P.; Shen, Q.J.; Rushton, P.J. The evolution of WRKY transcription factors. *BMC Plant Biol.* **2015**, *15*, 1–18. [CrossRef] [PubMed]
43. Bailey, T.L.; Boden, M.; Buske, F.A.; Frith, M.; Grant, C.E.; Clementi, L.; Ren, J.; Li, W.W.; Noble, W.S. MEME SUITE: Tools for motif discovery and searching. *Nucleic Acids Res.* **2009**, *37*, W202–W208. [CrossRef] [PubMed]
44. Chen, C.; Chen, H.; Zhang, Y.; Thomas, H.R.; Frank, M.H.; He, Y.; Xia, R. TBtools: An Integrative Toolkit Developed for Interactive Analyses of Big Biological Data. *Mol. Plant* **2020**, *13*, 1194–1202. [CrossRef] [PubMed]
45. Magali, L. PlantCARE, a database of plant cis-acting regulatory elements and a portal to tools for in silico analysis of promoter sequences. *Nucleic Acids Res.* **2002**, *1*, 325–327. [CrossRef]
46. Tamura, K.; Stecher, G.; Peterson, D.; Filipski, A.; Kumar, S. MEGA6: Molecular Evolutionary Genetics Analysis Version 6.0. *Mol. Biol. Evol.* **2013**, *30*, 2725–2729. [CrossRef]
47. Krzywinski, M.; Schein, J.; Birol, I.; Connors, J.; Gascoyne, R.; Horsman, D.; Jones, S.J.; Marra, M.A. Circos: An information aesthetic for comparative genomics. *Genome Res.* **2009**, *19*, 1639–1645. [CrossRef]
48. Wang, M.; Yue, H.; Feng, K.; Deng, P.; Song, W.; Nie, X. Genome-wide identification, phylogeny and expressional profiles of mitogen activated protein kinase kinase kinase (MAPKKK) gene family in bread wheat (*Triticum aestivum* L.). *BMC Genom.* **2016**, *17*, 668. [CrossRef]
49. Zhao, Y.; Ma, R.; Xu, D.; Bi, H.; Xia, Z.; Peng, H. Genome-Wide Identification and Analysis of the AP2 Transcription Factor Gene Family in Wheat (*Triticum aestivum* L.). *Front. Plant Sci.* **2019**, *10*, 1286. [CrossRef]
50. Holub, E.B. The arms race is ancient history in Arabidopsis, the wildflower. *Nat. Rev. Genet.* **2001**, *2*, 516–527. [CrossRef]
51. Bi, C.; Xu, Y.; Ye, Q.; Yin, T.; Ye, N. Genome-wide identification and characterization of WRKY gene family inSalix suchowensis. *PeerJ* **2016**, *4*, e2437. [CrossRef] [PubMed]
52. Jarosová, J.; Kundu, J.K. Validation of reference genes as internal control for studying viral infections in cereals by quantitative real-time RT-PCR. *BMC Plant Biol.* **2010**, *10*, 146. [CrossRef] [PubMed]
53. Long, X.-Y.; Wang, J.-R.; Ouellet, T.; Rocheleau, H.; Wei, Y.-M.; Pu, Z.-E.; Jiang, Q.-T.; Lan, X.-J.; Zheng, Y.-L. Genome-wide identification and evaluation of novel internal control genes for Q-PCR based transcript normalization in wheat. *Plant Mol. Biol.* **2010**, *74*, 307–311. [CrossRef]
54. Livak, K.J.; Schmittgen, T.D. Analysis of Relative Gene Expression Data Using Real-Time Quantitative PCR and the $2^{-\Delta\Delta CT}$ Method. *Methods* **2001**, *25*, 402–408. [CrossRef]
55. Yu, C.-S.; Chen, Y.-C.; Lu, C.-H.; Hwang, J.-K. Prediction of protein subcellular localization. *Proteins Struct. Funct. Bioinform.* **2006**, *64*, 643–651. [CrossRef]
56. Chou, K.-C.; Shen, H.-B. Plant-mPLoc: A Top-Down Strategy to Augment the Power for Predicting Plant Protein Subcellular Localization. *PLoS ONE* **2010**, *5*, e11335. [CrossRef]

2

# Exploring the Biochemical Origin of DNA Sequence Variation in Barley Plants Regenerated via In Vitro Anther Culture

**Piotr T. Bednarek [1,*], Jacek Zebrowski [2] and Renata Orłowska [1]**

[1] Department of Plant Physiology and Biochemistry, Plant Breeding and Acclimatization Institute—National Research Institute, Radzików, 05-870 Błonie, Poland; r.orlowska@ihar.edu.pl

[2] Institute of Biology and Biotechnology, University of Rzeszow, Al. Rejtana 16c A, 35-959 Rzeszow, Poland; jaze28@interia.pl

* Correspondence: p.bednarek@ihar.edu.pl.

**Abstract:** Tissue culture is an essential tool for the regeneration of uniform plant material. However, tissue culture conditions can be a source of abiotic stress for plants, leading to changes in the DNA sequence and methylation patterns. Despite the growing evidence on biochemical processes affected by abiotic stresses, how these altered biochemical processes affect DNA sequence and methylation patterns remains largely unknown. In this study, the methylation-sensitive Amplified Fragment Length Polymorphism (metAFLP) approach was used to investigate de novo methylation, demethylation, and sequence variation in barley regenerants derived by anther culture. Additionally, we used Attenuated Total Reflectance Fourier Transform Infrared (ATR-FTIR) spectroscopy to identify the spectral features of regenerants, which were then analyzed by mediation analysis. The infrared spectrum ranges (710–690 and 1010–940 $cm^{-1}$) identified as significant in the mediation analysis were most likely related to β-glucans, cellulose, and S-adenosyl-L-methionine (SAM). Additionally, the identified compounds participated as predictors in moderated mediation analysis, explaining the role of demethylation of CHG sites (CHG_DMV) in in vitro tissue culture-induced sequence variation, depending on the duration of tissue culture. The data demonstrate that ATR-FTIR spectroscopy is a useful tool for studying the biochemical compounds that may affect DNA methylation patterns and sequence variation, if combined with quantitative characteristics determined using metAFLP molecular markers and mediation analysis. The role of β-glucans, cellulose, and SAM in DNA methylation, and in cell wall, mitochondria, and signaling, are discussed to highlight the putative cellular mechanisms involved in sequence variation.

**Keywords:** ATR-FTIR spectroscopy; β-glucans; cellulose; DNA demethylation; S-adenosyl L-methionine; sequence variation; time of in vitro culture

## 1. Introduction

It is commonly believed that sequence variation during in vitro tissue culture is either caused by the activation of transposons [1], resulting in DNA demethylation [2] during cell reprogramming [3], or by the presence of modified cytosines [4], which undergo further modifications [5] but escape DNA repair mechanisms [6]. DNA demethylation is a widely used epigenetic marker in plants. DNA demethylation occurs in two sequence contexts: symmetric and asymmetric [7]. Symmetric DNA demethylation affects CG and CHG sequences (where H = A, T, and C) predominantly in gene-rich euchromatic regions, whereas asymmetric DNA demethylation affects CHH sequences mostly in heterochromatic regions. The maintenance of both types of DNA methylation patterns is regulated by distinct mechanisms [7].

Two mechanisms of DNA demethylation have been identified: passive and active. The passive mechanism assumes that DNA loses epigenetic marks during replication in the absence of the methylation of newly synthesized DNA strands by DNA (cytosine-5)-methyltransferase 1 (DNMT1), an enzyme that catalyzes the transfer of methyl groups to specific CpG sites during replication [8]. DNMT1 is responsible for the maintenance of DNA methylation patterns during replication. Passive demethylation has been reported in the male gametophyte and endosperm with reduced RNA-directed DNA methylation (RdDM) factors [9].

Active DNA demethylation utilizes 5-methylcytosine (5 mC) DNA glycosylases [10] and is a distinct feature of the maternal central cell. DEMETER-LIKE 2 (DML2) and DML3 function in nonsomatic and somatic tissues [11], and delete the methylated cytosines via the base excision repair (BER) pathway [12,13]. Active DNA demethylation depends on the REPRESSOR OF SILENCING 1 (ROS1) family, which comprises of 5 mC DNA glycosylases and apurinic/apyrimidinic lyases that antagonize de novo DNA methylation via the RdDM pathway to prevent DNA hypermethylation. ROS1 counteracts de novo DNA methylation in transgenes by actively promoting DNA demethylation. Other enzymes such as DEMETER (DME), DML2, and DML3 also participate in this process. DME regulates gene imprinting [14,15], whereas DML2 and DML3 are involved in the proper distribution of DNA methylation marks [13]. Recent studies show that active DNA demethylation is mediated by the increased DNA methylation (IDM) complex. Moreover, de novo DNA methylation and demethylation are under the control of a methylation-sensing genetic element [8]. In *Arabidopsis thaliana*, both de novo DNA methylation and demethylation processes are required for cellular reprogramming during in vitro tissue culture [16].

In active DNA demethylation, DNA glycosylases remove 5 mC and cleave the N-glycosidic (base–sugar) bond, generating apurinic or apyrimidinic sites, followed by gap-filling by a DNA polymerase and a DNA ligase [17]. In addition to the BER pathway, DNA demethylation is also accomplished by the removal of a DNA fragment, and the gap is filled using new nucleotides via the nucleotide excision (NER) or mismatch repair (MMR) mechanism. The NER and MMR mechanisms rely on the removal of 5 mC by further modifications such as oxidation of the methyl group, leading to the formation of 5-hydroxymethylcytosine (5 hmC) [5,18]. In active DNA demethylation, deamination of 5 mC generates thymine (T), which may result in C→T transition, as T is not recognized by the uracil-DNA glycosylase. Thus, mutations due to deamination are the most frequent, as they escape the cell repair system. Alternatively, oxidation of 5 mC may also contribute to sequence variation [19].

Recent studies demonstrate that abiotic stress conditions are responsible for DNA methylation-dependent stress memory in plants [20]. In some cases, changes in DNA methylation patterns are associated with plant stress responses [21]. Furthermore, abiotic stresses such as cold and drought are responsible for changes in DNA methylation patterns [22]. Additionally, studies on plant responses to in vitro tissue culture-induced stress show that tissue culture conditions promote changes in stress-induced DNA methylation and/or demethylation patterns [23–27]. Machczyńska and colleagues reported a relatively high level of DNA sequence variation between the donor plant and its regenerants [28]. However, evidence supporting a direct link between DNA demethylation/sequence variation and in vitro tissue culture conditions is lacking.

Sequence variation induced by in vitro tissue culture has been investigated using the methylation-sensitive Amplified Fragment Length Polymorphism (metAFLP) approach [28,29], which provides quantitative information about the DNA sequence. Furthermore, Attenuated Total Reflectance Fourier Transform Infrared (ATR-FTIR) spectroscopy has been used to analyze the extent to which nucleotide variation affects biochemical compounds in plants regenerated by an in vitro culture [30]. ATR-FTIR spectroscopy is a powerful technique that has been widely used to analyze the biochemical profiles of diverse biological materials, including plant tissues [31,32]. ATR-FTIR spectroscopy is a fast and simple measurement procedure that does not require special sample preparation, thus allowing the examination of a large number of samples. Although ATR-FTIR spectroscopy is not similarly precise in quantitative analysis of a specific analyte like time-consuming chromatography, it does allow for comprehensive survey of all essential compounds present in the material under investigation on the basis of a single short-time measurement. ATR-FTIR spectroscopy is generally used in combination with various mathematical and statistical tools that enhance its research potential, e.g., to discriminate between samples according to taxonomic categories [33,34] or to determine susceptibility to stresses [35–37]. Additionally, mediation analysis is often used as a method of choice to study the relationship between distinct compounds [38] as it does not involve expensive molecular experiments and may support valuable data on biological mechanisms.

Previously, we showed that in vitro tissue culture is moderated by copper ($Cu^{2+}$) and silver ($Ag^{+}$) ions present in the regeneration medium [38], depending on the duration of culture. Mediation analysis [38] showed that sequence variation in barley plants obtained by tissue culture was due to the presence of $Cu^{2+}$ and $Ag^{+}$ ions affecting CG sequences in different contexts, depending on the duration of tissue culture. Therefore, we hypothesized that sequence variation within other contexts might be induced by certain cellular compounds under the influence of abiotic stress, and this sequence variation may occur either throughout the duration of tissue culture or preferentially during specific intervals. To test this hypothesis, we investigated the putative origin of in vitro anther culture-induced sequence variation in barley regenerants using metAFLP markers, ATR-FTIR spectroscopy, and mediation analysis.

## 2. Results

Another culture of barley was performed in vitro under nine distinct treatments (M1–M9). Each treatment was conducted for a different duration and contained a different concentration of $Cu^{2+}$ and $Ag^{+}$ ions. A total of 35 barley regenerants were obtained. No morphological differences were detected among the regenerants, and all regenerants were in the type of an anther tissue donor plant. DNA isolated from fresh leaves of a donor and regenerated DH plants using a commercial kit yielded intact samples without detectable impurities.

Genotyping of the isolated DNAs using the metAFLP approach allowed the identification of 407 markers shared by the donor plant and regenerants [25]. Quantitative metAFLP characteristics of the DNAs are listed in Table 1.

**Table 1.** Quantitative metAFLP characteristics and corresponding infrared spectroscopy data of barley regenerants derived by in vitro tissue culture under nine different trials (M1–M9).

| Sample No. | Trial | Independent Variable [1] | | | | Mediator [1] | | Dependent Variable [1] | Moderator |
|---|---|---|---|---|---|---|---|---|---|
| | | F710.690 | F1010.940 | F710.690-F1010.940 | DMV | CG_DMV | CHG_DMV | SV | Time (Days) |
| 1 | M1 | 0.0012 | 0.0443 | 0.0455 | 1.6210 | 0.4636 | 1.1583 | 3.7058 | 21 |
| 2 | M1 | 0.0021 | 0.0267 | 0.0288 | 1.3900 | 0.2325 | 1.1588 | 2.5488 | 21 |
| 3 | M1 | 0.0018 | 0.0371 | 0.0389 | 1.3900 | 0.2315 | 1.1584 | 2.5479 | 21 |
| 4 | M1 | 0.0014 | 0.0317 | 0.0332 | 1.3900 | 0.2315 | 1.1584 | 2.7794 | 21 |
| 5 | M1 | 0.0011 | 0.0388 | 0.0399 | 1.6210 | 0.4630 | 1.1583 | 3.2423 | 21 |
| 6 | M2 | 0.0010 | 0.0283 | 0.0293 | 1.6220 | 0.4637 | 1.1585 | 2.5481 | 28 |
| 7 | M2 | 0.0015 | 0.0287 | 0.0302 | 1.6220 | 0.4650 | 1.1584 | 2.3171 | 28 |
| 8 | M2 | 0.0018 | 0.0322 | 0.0339 | 1.3900 | 0.4640 | 0.9268 | 2.5482 | 28 |
| 9 | M3 | 0.0014 | 0.0274 | 0.0288 | 1.3900 | 0.4640 | 0.9268 | 2.5482 | 35 |
| 10 | M3 | 0.0014 | 0.0264 | 0.0279 | 1.6210 | 0.4636 | 1.1583 | 3.7058 | 35 |
| 11 | M3 | 0.0008 | 0.0454 | 0.0461 | 1.6210 | 0.4637 | 1.1584 | 3.9377 | 35 |
| 12 | M3 | 0.0018 | 0.0263 | 0.0281 | 1.3900 | 0.4640 | 0.9268 | 2.5482 | 35 |
| 13 | M3 | 0.0019 | 0.0255 | 0.0275 | 1.3830 | 0.4616 | 0.9224 | 4.3802 | 35 |
| 14 | M4 | 0.0011 | 0.0282 | 0.0293 | 1.3900 | 0.4640 | 0.9268 | 2.5482 | 28 |
| 15 | M4 | 0.0013 | 0.0325 | 0.0338 | 1.3900 | 0.4640 | 0.9268 | 2.3166 | 28 |
| 16 | M5 | 0.0010 | 0.0329 | 0.0339 | 1.3900 | 0.4637 | 0.9268 | 2.5482 | 35 |
| 17 | M5 | 0.0009 | 0.0484 | 0.0493 | 1.3900 | 0.4630 | 0.9266 | 3.7055 | 35 |
| 18 | M5 | 0.0009 | 0.0483 | 0.0491 | 1.6220 | 0.6952 | 0.9271 | 3.4748 | 35 |

**Table 1.** *Cont.*

| Sample No. | Trial | Independent Variable [1] | | | Mediator [1] | | | Dependent Variable [1] | Moderator |
|---|---|---|---|---|---|---|---|---|---|
| | | F710.690 | F1010.940 | F710.690-F1010.940 | DMV | CG_DMV | CHG_DMV | SV | Time (Days) |
| 19 | M5 | 0.0018 | 0.0294 | 0.0312 | 1.8530 | 0.6954 | 1.1582 | 3.7057 | 35 |
| 20 | M6 | 0.0011 | 0.0353 | 0.0364 | 0.9270 | 0.0000 | 0.9271 | 2.5487 | 21 |
| 21 | M6 | 0.0013 | 0.0343 | 0.0356 | 1.3900 | 0.4640 | 0.9268 | 2.5482 | 21 |
| 22 | M6 | 0.0016 | 0.0238 | 0.0254 | 0.9270 | 0.0000 | 0.9268 | 2.5484 | 21 |
| 23 | M6 | 0.0018 | 0.0311 | 0.0329 | 1.1580 | 0.2323 | 0.9268 | 2.5483 | 21 |
| 24 | M6 | 0.0015 | 0.0331 | 0.0346 | 1.1580 | 0.2315 | 0.9271 | 2.3167 | 21 |
| 25 | M7 | 0.0014 | 0.0268 | 0.0283 | 1.8530 | 0.6944 | 1.1581 | 4.6315 | 35 |
| 26 | M7 | 0.0009 | 0.0482 | 0.0492 | 1.6210 | 0.4630 | 1.1584 | 2.5478 | 35 |
| 27 | M7 | 0.0010 | 0.0373 | 0.0383 | 1.3900 | 0.4630 | 0.9266 | 3.9370 | 35 |
| 28 | M8 | 0.0009 | 0.0440 | 0.0449 | 0.0000 | 0.0000 | 0.0000 | 13.5169 | 21 |
| 29 | M8 | 0.0004 | 0.0533 | 0.0537 | 0.0000 | 0.0000 | 0.0000 | 13.5170 | 21 |
| 30 | M8 | 0.0014 | 0.0426 | 0.0440 | 0.0000 | 0.0000 | 0.0000 | 14.0089 | 21 |
| 31 | M9 | 0.0013 | 0.0441 | 0.0454 | 2.0840 | 0.9267 | 1.1582 | 4.4006 | 28 |
| 32 | M9 | 0.0015 | 0.0361 | 0.0377 | 1.1580 | 0.2320 | 0.9268 | 2.7798 | 28 |
| 33 | M9 | 0.0013 | 0.0388 | 0.0400 | 1.1580 | 0.2320 | 0.9268 | 2.7798 | 28 |
| 34 | M9 | 0.0016 | 0.0335 | 0.0351 | 1.1580 | 0.2320 | 0.9268 | 2.5486 | 28 |
| 35 | M9 | 0.0019 | 0.0223 | 0.0243 | 2.0 940 | 0.9320 | 1.1651 | 3.4931 | 28 |

[1] DMV, demethylation; CG_DMV, demethylation of the CG context; CHG_DMV, demethylation of the CHG context; SV, sequence variation; F710.690, area under the spectral range from 710 to 690 $cm^{-1}$; F1010.940, area under the spectral range from 1010 to 940 $cm^{-1}$; F710.690-F1010.940, combination of F710.690 and F1010.940.

### *2.1. ATR-FTIR Spectroscopy*

ATR-FTIR spectroscopy was carried out to investigate the major biochemical features of the leaves of barley regenerants in a fast and straightforward manner. Figure 1A summarizes the variability in infrared data between all treatments. The spectra showed typical features of plants (Figure 1A, inset), with several usually complex broad peaks. Bands representing asymmetric and symmetric methylene ($CH_2$) stretching vibrations were located at 2925 and 2848 $cm^{-1}$, respectively, while the amide I band was centered at 1646 $cm^{-1}$. In turn, the fingerprint region was characterized by two dominant peaks: one at 1397 $cm^{-1}$, putatively assigned to the symmetric stretching vibrations of $COO^-$, $CH_2$ scissor, and CH bending [39,40], and the other at 1058 $cm^{-1}$, assigned to the C-O, C-O-C stretch, and ring vibrations, attributed mainly to polysaccharides [41,42]. The spectral range between 1200 and 600 $cm^{-1}$, which was the focus of the present study, is depicted in Figure 1A. The local peak at 1153 $cm^{-1}$ was assigned mostly to C-O-C asymmetric stretch vibrations, which are characteristic of various cell wall polysaccharides and therefore are not distinctive. The local peak at 1104 $cm^{-1}$ representing C-O, C-O-C stretch, and ring vibrations is attributed mostly to cellulose [42,43]. To resolve the broad complex band centered at 1058 $cm^{-1}$, we performed Gaussian deconvolution of the wavenumber range going from 1200 to 600 $cm^{-1}$, achieving a very close fit to the original contour. The component peaks are presented in Figure 1B. The band at 1069 $cm^{-1}$ was assigned mostly to C–O and C–C stretching vibrations of β-(1→3, 1→4) glucan [44], whereas the band at 1036 $cm^{-1}$ was attributed to the C-O-C skeletal vibration of the polysaccharide ring of both cellulose and β-glucan [42]. Two small peaks were also generated at 991 and 960 $cm^{-1}$, which seemed to contribute to the shape of the band shoulder; the former peak putatively reflects the β-glycosidic linkage, whereas the latter peak does not seem to be related to cell wall polysaccharides. The band at 960 $cm^{-1}$ was attributed to C–O and C–C stretching in deoxyribose [45].

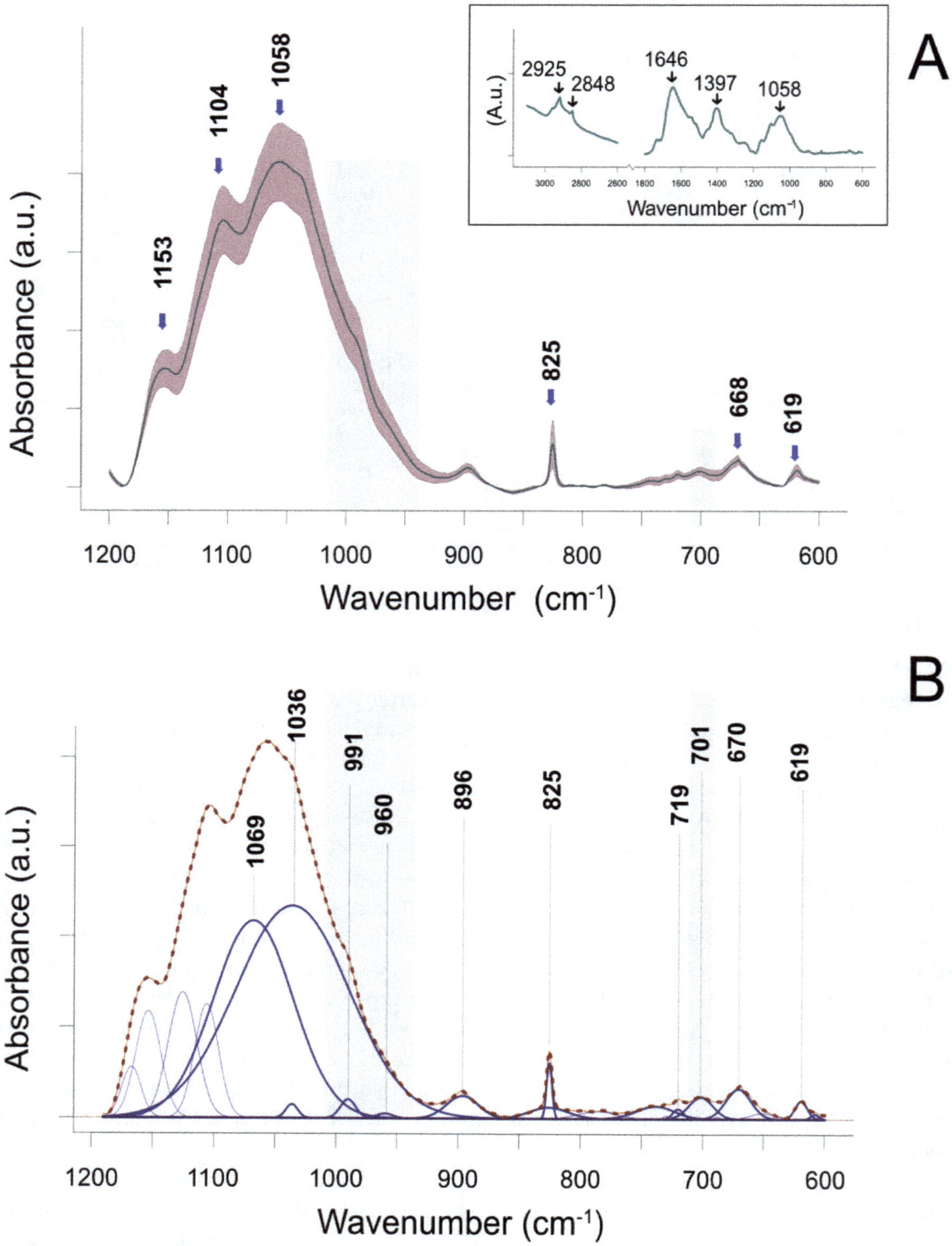


**Figure 1.** Attenuated Total Reflectance Fourier Transform Infrared (ATR-FTIR) spectrum of the leaves of barley plants regenerated by an in vitro anther culture. (**A**) ATR-FTIR spectra of the region between 1200 and 600 $cm^{-1}$. The most prominent peaks are marked. The inset presents an extended range of the spectrum between 3100 and 600 $cm^{-1}$. Data represent mean ± standard deviation (SD). (**B**) Numerical (Gaussian) deconvolution of overlapping peaks. Data represent the sum of fitted curves (dashed line) closely overlapping the experimental data (solid red line). The centered frequencies of major resolved peaks are shown in blue. Regions shaded in gray represent spectral regions relevant to mediation analysis.

The band at 896 $cm^{-1}$ was assigned to β-glycosidic (C1-H) deformation [42]. A narrow, moderately intense peak at 825 $cm^{-1}$ was probably caused by phenolic compounds containing a specifically substituted benzene ring; however, this peak was not considered essential in the present study. The resolved peak in the low-frequency region showed compounds at 719, 710, and 670 $cm^{-1}$ [46]. Two bands at 701 and 670 $cm^{-1}$ typically represent, in the case of cell wall material, H-bonded O-H out-of-plane bending vibrations of cellulose [47,48]. However, these bands may be associated with vibrations unrelated to cell wall compounds, e.g., C-S stretching vibrations present in thiols or other compounds with C-S-C bonds [49–51].

### 2.2. Mediation Analysis

The analysis consisted of a series of regression mediation analyses that aimed to establish whether the DNA demethylation affecting CHG contexts (CHG_DMV) was a mediator in the relationship among F710.690, F1010.940, and F710.690-F1010.940 variables and sequence variation (SV). We have also tested the model of relationships between F710.690-F1010.940 and SV mediated by CHG_DMV and moderated by the time of in vitro tissue cultures.

The F710.690 mediation on sequence variation through CHG_DMV (mediator) was assumed to be insignificant based on indirect and partial indirect effects (not shown). Based on the complete standardized indirect effect, the mediation was nearly significant (Table 2). Moreover, based on the casual mediation approach [52] coefficient relating F710.690 and SV (path c) was significant, coefficient relating F710.690 and CHG_DMV (*a*) was significant. Furthermore, the CHG_DMV (mediator) was related to the SV (DV) when both F710.690 and CHG_DMV are predictors of the dependent variable ($c'$), the coefficient relating the F710.690 to the SV ($c$) was larger (in absolute value) than the coefficient relating the F710.690 to the SV in the regression model with both the F710.690 and the CHG_DMV ($c'$) predicting the dependent variable and indicating mediation. There were significant mediation effects of F1010.940 and F710.690-F1010.940 (Figure 2) as independent variables (IDs) on sequence variation through CHG_DMV (mediator). Table 2 shows the outcomes of simple mediation analyses from IDs to SV assessing the indirect effects of CHG_DMV (indicated by a 95% confidence interval). All indirect effects except for the first model were positive. Mediation analysis showed that F1010.940 ATR-FTIR spectrum was a predictor of mediation, based on the indirect effect (IE) bootstrap value. Similarly, the indirect effect of the F710.690-F1010.940 variable on independent SV variables via the CHG_DMV (mediator) was significantly different from zero. The significance of mediations was assessed based on the Goodman test (F710.690: $Z = 2.07$, $p = 0.04$; F1010.940: $Z = 2.47$, $p = 0.01$; F710.690-F1010.940-: $Z = 2.45$, $p = 0.01$). The mediator variable's value accounted for (VAF) 84.5, 71.4, and 71% of the variance when the F710.690, F1010.940, and F710.690-F1010.940 variables were used respectively.

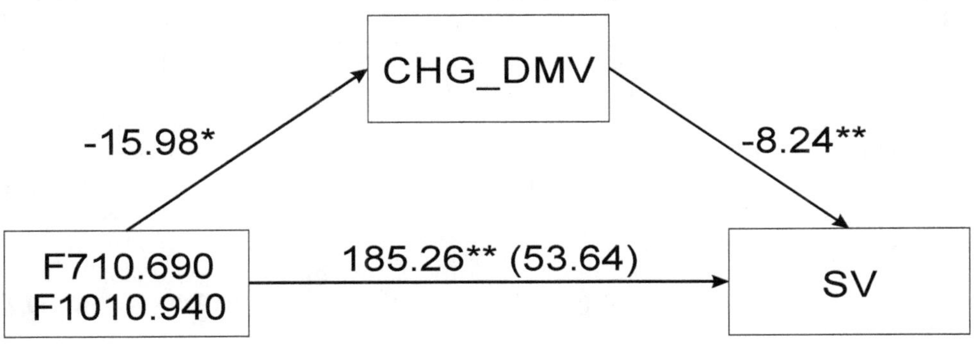


**Figure 2.** Mediation analysis of compounds associated with infrared spectral regions of 710–690 and 1010–940 $cm^{-1}$ to demonstrate their effect on sequence variation (SV) via the demethylation of the CHG sequence context (CHG_DMV). * $p < 0.05$, ** $p < 0.01$

**Table 2.** Outcomes of mediation analyses from F710.690, F1010.940, and F710.690-F1010.940 to sequence variation assessing indirect effects. The unstandardized coefficients are given and explained in the table.

| Model | Effects | | | | | | 95% CI | |
|---|---|---|---|---|---|---|---|---|
| | $R^2$ | $c'$ | $a$ | $b$ | $c$ | IE | L | U |
| F710.690- > CHG_DMV- > SV | 0.79 | −436.37 | 274.82 * | −8.59 *** | −2797.49 * | −0.299 † | −0.56 † | 0.06 † |
| F1010.940- > CHG_DMV- > SV | 0.80 | 51.35 | −15.57 * | −8.24 *** | 179.63 ** | 128.28 | 0.18 | 286.98 |
| F710.690-F1010.940- > CHG_DMV- > SV | 0.80 | 53.64 | −15.98 * | −8.24 *** | 185.26 *** | 131.62 | 0.18 | 298.27 |

F710.690, F1010.940, and F710.690-F1010.940 are the ATR-FTIR; CHG_DMV (mediator) states for DNA demethylation of the CHG contexts evaluated by metAFLP approach; SV is sequence variation also evaluated by the metAFLP approach and shared among anther culture derived regenerants of barley; $c'$ = direct effect of predictor (IDs) on outcome (SV) while controlling for the mediator (CHG_DMV); $a$ = effect of the predictor (ID) on the mediator (CHG_DMV); $b$ = effect of the mediator (CHG_DMV) on the outcome (SV); $c$ = total effect focal predictor (F710.690, F1010.940, and F710.690-F1010.690) on the outcome (SV); IE = indirect effect of predictor (ID) on outcome (SV) through the mediator (CHG_DMV); $R^2$—the amount of variance explained by the model; * $p < 0.05$, ** $p < 0.01$; *** $p < 0.001$; L and U are lower and upper values of 95% confidence interval (CI). †—complete standardized indirect effect (C IE).

When the analysis tested relationship between F710.690-F1010.940 and SV mediated by CHG_DMV and moderated by the time of in vitro tissue cultures (Table 3) the model (Figure 3) was significant ($F(4, 30) = 66.25$, $p < 0.0001$) and explained 90% of the variance. The interaction between F710.690-F1010.940 and time as well as between CHG_DMV and time was significant based on bootstrap confidence intervals, indicating that mediation was moderated by time. Test of the highest order of unconditional interaction between F710.690-F1010.940 and time ($F(1, 31) = 9.89$, $R^2_{chang} = 0.1848$, $p < 0.01$) as well as between CHG_DMV and SV ($F(1, 30) = 28.32$, $R^2_{chang} = 0.096$, $p < 0.0001$) were also significant and explained 18.48% and 9.6% of variance, respectively. Johnson–Neuman's statistics indicated that moderation of F710.690-F1010.940 by time was significant through 30 days (Figure 4), and CHG_DMV through 28 days of in vitro tissue culture (Figure 5; data not shown). The indirect effect of moderated mediation was significant through 21 days of in vitro culture (Table 3). The indirect and direct effects of the moderated mediation calculated as $\beta a \times \beta b$ and $\beta a \times (\beta b + c')$ equaled 2804.153 and 2844.715, respectively. Thus, the VAF values of the mediator variables accounted for 98.57% of the variance. The significance of the moderated mediator was also confirmed by the Goodman test ($Z = 3.29$, $p < 0.001$).

**Table 3.** The outcomes (unstandardized regression coefficients) of mediation analyses from F710.690-F1010.940 to SV assessing the indirect effects of CHG_DMV (indicated by 95% confidence interval) moderated by the time of in vitro tissue culture. For further explanation, see Table 2.

| Model: Moderated Mediation | *B* | *SE* | 95% CI L | 95% CI U |
|---|---|---|---|---|
| $c'$ | 40.5613 | 25.167 | −10.8377 | 91.9603 |
| $a$ | −94.5994 *** | 25.8835 | −147.3903 | −41.8085 |
| Time (moderator) | −0.0887 * | 0.0335 | −0.1571 | −0.0204 |
| Moderation: F710.690-F1010.940 × Time | 2.7594 ** | 0.8776 | 0.9696 | 4.5493 |
| $b$ | −29.6424 **** | 4.0666 | −37.9477 | −21.3371 |
| Time (moderator) | −0.9608 **** | 0.1848 | −1.3382 | −0.5834 |
| Moderation: CHG_DMV × Time | 0.9633 **** | 0.181 | 0.5936 | 1.333 |
| $c$ | 40.5613 | 25.167 | −10.8377 | 91.9603 |
| Conditional indirect effect of F710.690-F1010.940 on SV (F710.690-F1010.940 –> CHG_DMV –> SV) | | | | |
| Time | *B* | *SE* | L | U |
| 21 | 344.9937 | 154.151 | 9.9749 | 633.6907 |
| 28 | 46.2799 | 114.4689 | −7.6518 | 113.7094 |
| 35 | 8.0692 | 53.3305 | −43.0736 | 72.1826 |

* $p < 0.05$, ** $p < 0.01$; *** $p < 0.001$; **** $p < 0.0001$.

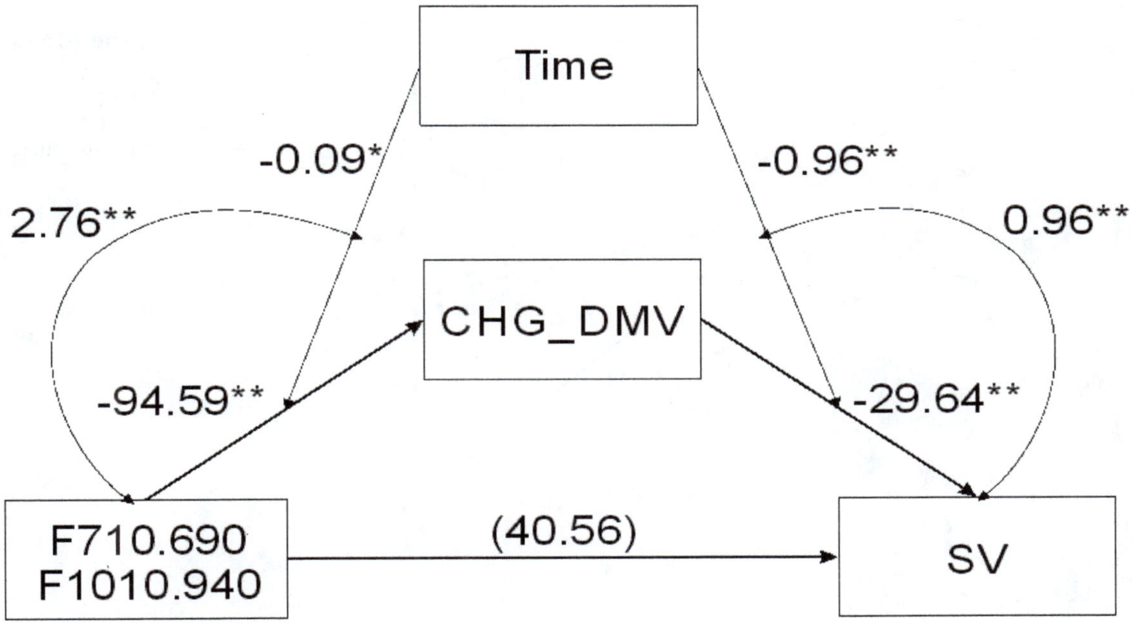


**Figure 3.** Mediation analysis of the compounds associated with infrared spectral regions of 710–690 and 1010–940 cm$^{-1}$ to demonstrate their effect on sequence variation (SV) via demethylation of the CHG sequence context (CHG_DMV) moderated by time of tissue culture. * $p < 0.05$, ** $p < 0.01$.

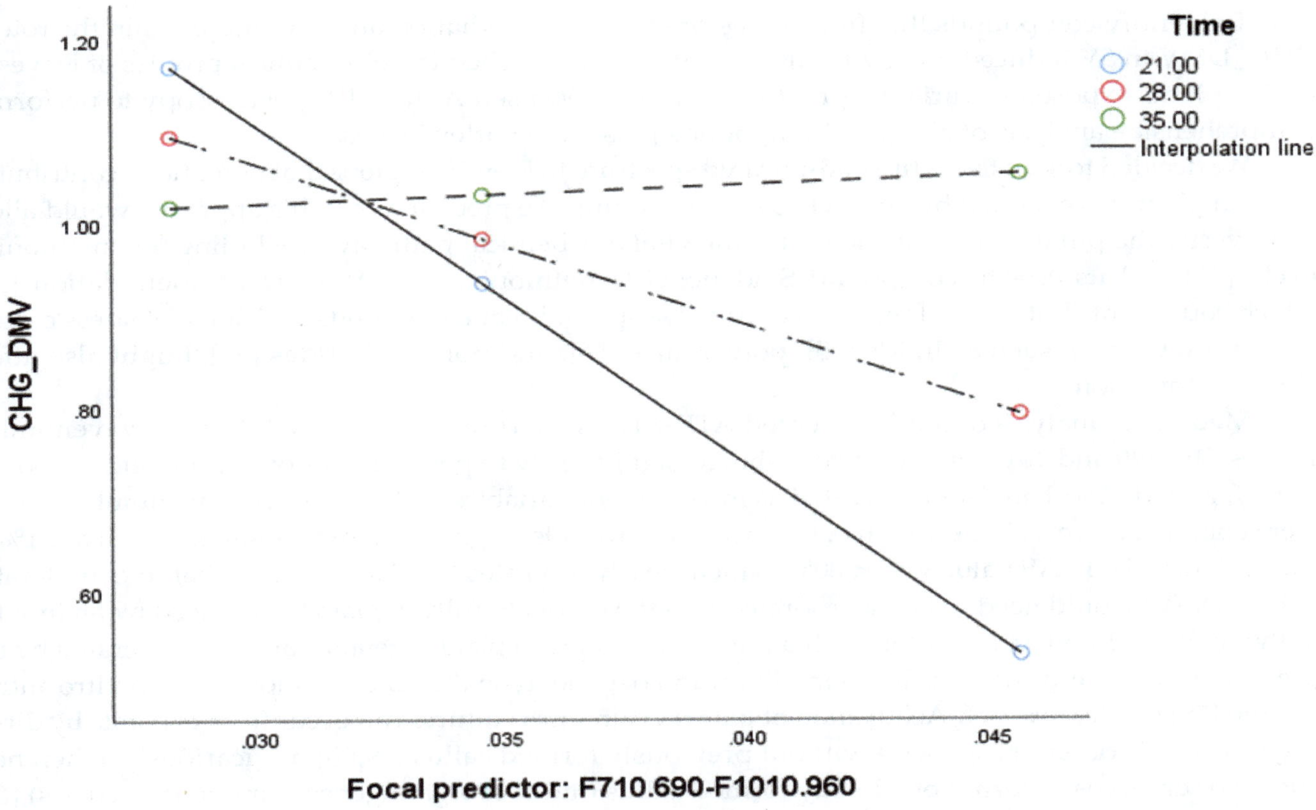


**Figure 4.** Conditional effect of the integrated absorbance area for regions 710–690 and 1010–940 cm$^{-1}$ (F710.690-F1010.940; predictor) under different durations of in vitro anther culture.

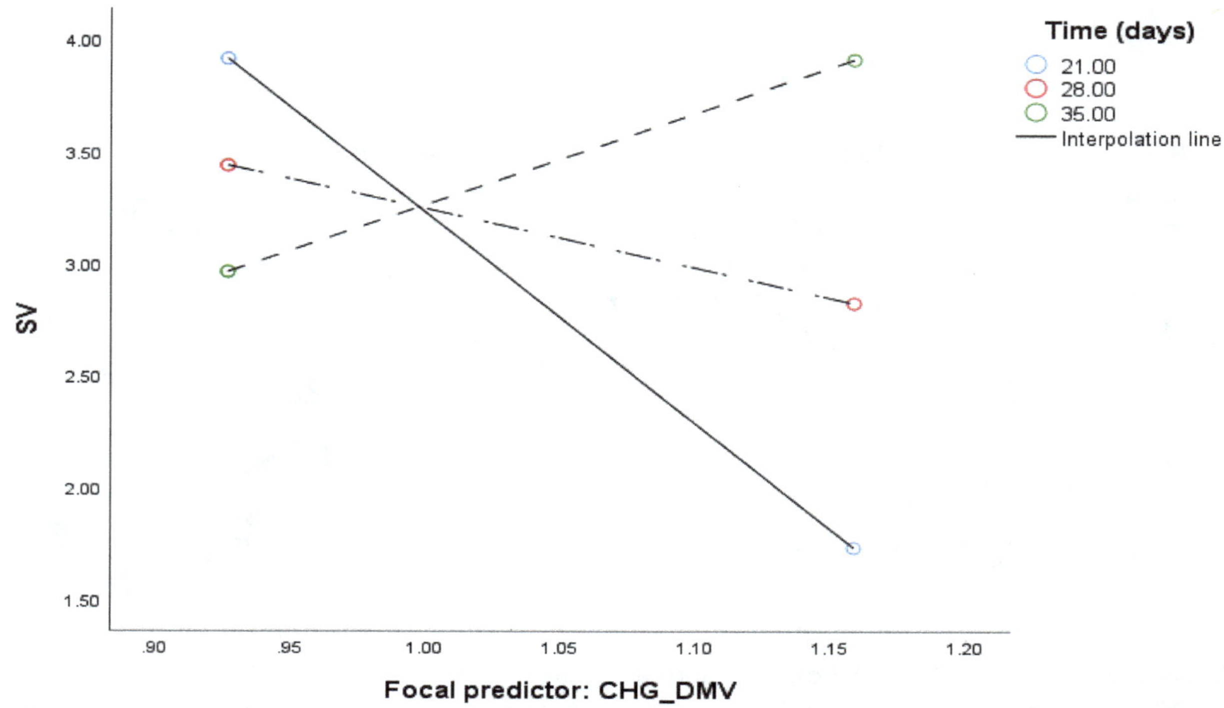


**Figure 5.** Conditional effect of demethylation of the CHG sequence context (CHG_DMV; predictor) under different durations of in vitro anther culture.

## 3. Discussion

To identify compounds that function as predictors in mediation analysis and explain the role of CHG_DMV in SV induced by in vitro tissue culture, we searched the biochemical profiles of leaves of barley plants exposed to variable growth conditions. We used ATR-FTIR spectroscopy to perform a comprehensive analysis of all major compounds present in barley leaves.

We decided to scan the entire mid-infrared spectrum to identify regions that potentially contributed to SV in plants regenerated by an in vitro anther culture. We predicted that this approach would allow us to verify the putative involvement of some vital biochemical pathways, including the methionine cycle [53] and its crucial compound S-adenosyl-L-methionine (SAM), in DNA methylation [54], which could contribute to SV [19]. Moreover, we suspected that compounds indicative of stress caused by in vitro culture, such as lipids [55], polyamines [56], and polysaccharides [57], might also affect DNA methylation.

Mediation analysis of the fragmented ATR-FTIR spectrum demonstrated that the wavenumber ranges 710–690 and 1010–940 $cm^{-1}$ contributed significantly to performance of the statistical model if the integrated absorbances were treated as independent variables. When the two wavenumber ranges were combined into a single variable, the mediation was also significant, explaining as much as 71% of the variance. The VAF value was $< 80\%$, which clearly indicates that factors other than the moderator (CHG_DMV) would need to be incorporated into the model to fully explain SV induced by an in vitro anther culture. Previously, we found that the process of green plant regeneration was moderated by the presence of $Cu^{2+}$ and $Ag^{+}$ ions in the medium, and depended on the time duration of the in vitro anther culture [58] (under review). Acquisition of plants via in vitro cultures can occur from explants by direct organogenesis or embryogenesis without previously formed callus [59,60] or vicariously, when new shoots or embryos occurred on already formed callus (indirect organogenesis or embryogenesis) [61]. It is being suggested that under tissue culture conditions direct embryogenesis may proceed an indirect one. Thus, the role of time could be related to a moment when plant regeneration via indirect exceeds direct embryogenesis what might be linked to the utilization of β-glucans as a source of glucose under darkness [62]. Moreover, it was also demonstrated by us [38] that time was influential in the case of mediation between DNA demethylation of the CG sites and SV due to the presence of copper and silver ions in the medium. In this study, the time duration of in vitro culture was also a significant moderator. When these elements were incorporated into the model of moderated mediation, the VAF value increased to nearly 100%, indicating that most of the essential factors were included in the model.

The 710–690 $cm^{-1}$ band region is putatively associated with cellulose, whereas another band at approximately 665 $cm^{-1}$ was also detected in the spectra of the examined material [47,48]. However, mediation analysis did not show a significant effect of the 665 $cm^{-1}$ band, possibly because of the presence of cellulose in dissimilar forms and/or at different crystallinity levels, generating separate peaks. This contradiction may result from diverse patterns of intermolecular bonds and high sensitivity of the O-H out-of-plane bending vibrations due to hydrogen bonding. Variation in the position of the peak attributed to cellulose, independent of the hydrogen bonding pattern and crystallinity level, has been reported previously [47,48]. The band region at 700 $cm^{-1}$ may be associated with a more amorphic fraction of cellulose compared with that at 665 $cm^{-1}$, and consequently only this fraction may be relevant for mediation analysis. Abiotic stresses during the regeneration process may be more influential on the amorphic fraction of cellulose than on the crystalline fraction. It is possible that stress due to the supply of $Cu^{2+}$ and $Ag^{+}$ ions in the regeneration medium affected the status of cellulose primarily in a relatively more amorphic phase. In *Miscanthus giganteum*, a slight shift in the cellulose peak in this region of frequency has been observed as a result of mild chilling stress [63], which was associated with the formation of a more compact cellulose structure. The band at 700 $cm^{-1}$ may be assigned not only to cellulose but also to the C-S stretching vibrations, which may be attributed to thiol or thioester functional groups. In the context of the apparent involvement of spectral data in the methylation process, an indication of SAM as a compound associated with this spectral region seems to be the most rational. The C-S stretch vibrations in the spectral region displayed a relatively strong signal, and the integrated absorbance may well explain the observed band even at a reasonably low concentration of the compound. Concerning the 710–690 $cm^{-1}$ band region it should be stressed that mediation was not significant based on the indirect effect and nearly significant based on the completely unstandardized indirect effect. We tend to think that such a result is explained by a relatively small sample size rather than by a lack of mediation. The notion is supported by the fact that the compounds reflecting the 710–690 $cm^{-1}$ spectrum region are documented in processes that may affect sequence variation.

Other spectra that acted as predictors in the mediation model included absorbances between 1010 and 940 $cm^{-1}$. This spectral range constitutes a shoulder of intricate broadband, with a maximum at 1058 $cm^{-1}$. The deconvolution of the band region generated two main components (located at 1069 and 1036 $cm^{-1}$) and two much smaller peaks (at 991 and 960 $cm^{-1}$), which could modulate the data associated with the F1010.940 variable. The broad peak with a maximum at 1058 $cm^{-1}$ is greatly influenced by cellulose and hemicellulose [42]. In the hemicellulose fraction, β-glucans seemed to have the most substantial effect as signals, as pectins and other putative polysaccharides did not show clear specific peaks.

Since the contribution of β-glucans to the band region is marked, a similarly essential effect may be attributed to β-glucans in the 1010–940 $cm^{-1}$ region. Therefore, both β-glucans and cellulose may contribute to mediation, and variability in the proportion of contribution between these two compounds may affect the model in mediation analysis.

The ATR-FTIR analysis of the 710–690 and 1010–940 $cm^{-1}$ spectrum ranges suggests that these regions are associated with both cellulose and β-glucans, but the assignment of the 710–690 bands to cellulose was much more apparent than to β-glucans. Despite that the 710–690 $cm^{-1}$ band is very weak, we are convinced that it reflects biological differentiation, due to both plant individual variability and the applied treatments. The technical repetitions (data not shown) were almost identical and did not infer observed data. We also exclude the randomness effect because the used spectrometer provided the spectra with the signal-to-noise ratio above 35,000:1. The absorbance in the 1010–940 $cm^{-1}$ range showed an equal contribution of β-glucans and cellulose, although it was difficult to split the share.

Thus, both β-glucans and cellulose can be considered as mediators, but the variability in the glucan content may be crucial for the model.

At a first glance, the role of β-glucans as a putative predictor of mediation seems suspicious. However, studies on cell wall signaling suggest that mitochondria and the cell wall cooperate to mediate stress responses [64]. Stresses can also affect the rate of respiratory electron transport, and mitochondria act as hubs for signaling to other cellular organelles [65]. β-glucans are a major component of the cell wall [66]. In vitro anther cultures are maintained in the dark [25]. Under such conditions, β-glucans are a source of metabolizable glucose [67]. Glucose from the cell wall in the form of (1,3;1,4)-β-glucans is mobilized quicker than glucose from starch (which is not available in tissue culture carried out in the dark). The process of β-glucan depolymerization requires only two enzymes: (1,3;1,4)-β-glucan endohydrolase and the exo-acting β-glucan glucohydrolase [68], biosynthesis of lipids [69] that may act as signaling compounds [70,71] affecting gene expression [72,73]. Interestingly, β-1,3-linked D-glucose in the cell wall also protects plants from external stresses [74]. Other pathways linked to glycolysis and the tricarboxylic acid cycle (TCA) include the methionine cycle and ethylene biosynthesis pathway.

It is known that the ethylene biosynthesis pathway involving SAM [75–77] contributes to DNA methylation [78]. At low SAM concentration, the expression of genes involved in DNA repair and lipid and glucose metabolism is altered [79]. The decarboxylated form of SAM (dcAdoMet) is reported to block cytosine methylation [80]. Decarboxylated SAM may block DNA methylation by SAM and is also involved in polyamine synthesis [80]. Polyamines participate in the regulation of transcription, translation, cell growth, and apoptosis [80]. Thus, either changes in the expression of genes involved in DNA repair or the presence of mutations (such as C→T transitions) not recognized by DNA repair systems may act as a link between DNA methylation and SV induced by in vitro tissue culture [81]. Alternatively, SAM-e could be cleaved reductively by an iron-sulfur cluster-containing enzymes, producing an intermediate called 5′-deoxyadenosyl radical [82]. Such radicals may modify tRNA [83], causing rearrangements and other post-transcriptional modifications [84]. The 5′-deoxyadenosyl radical removes a hydrogen from proteins, RNA, or DNA (substrate) initiating a radical mechanism [80]. Radicals may induce DNA modifications leading to DNA mutations [85]. SAM is a key player in the methionine cycle, leading to ethylene biosynthesis. The level of ethylene is dependent on light conditions, and its regulation is dependent on enzymes participating in the ethylene pathway [86]. Since ethylene is a phytohormone that may affect gene expression and may depend on light conditions it may potentially contribute to SV. Thus, SAM and its derivatives may play a role in DNA mutations.

In the ATR-FTIR spectrum, the 1010–940 $cm^{-1}$ range was linked to cellulose and β-glucans. Mediation analysis of the 1010–940 $cm^{-1}$ range explained less variance, and mediation was not classified as full. This suggests that either cellulose is less critical for mediation and only influences the signal from β-glucans, or that cellulose participates in different mechanisms affecting SV. One possibility is that the level of cellulose affects $Cu^{2+}$ and $Ag^{+}$ ions present in the regeneration medium and delivered to the cell via the cell wall. If this is true, then insufficient concentration of $Cu^{2+}$ and Ag+ ions in the medium should alter the biochemical pathways that require these ions. However, in this study, $Cu^{2+}$ and $Ag^{+}$ ions were not incorporated in the model tested, as they were statistically insignificant (data not shown). The other possibility concerns the putative signaling pathway that links cellulose and mitochondria. Two pentatricopeptide repeat proteins, cell wall maintainer1 (cwm1) and cwm2, are involved in editing mitochondrial transcripts that encode subunits of the respiratory complex III, which is linked to the maintenance of cell wall integrity under stress and activation of retrograde mitochondrial signaling via ANAC017, a transcription factor involved in retrograde signaling to the nucleus upon mitochondrial dysfunction [87]. A complex hierarchy of transcription factors exists downstream of ANAC017, including ANAC and WRKY transcription factors associated with organellar signaling and senescence, and ethylene- and gibberellic acid-related transcription factors involved in stress responses [88].

Recently, a study showed that mitochondrial nicotinamide adenine dinucleotide (NAD) reduced oxidation links the TCA with methionine metabolism and nuclear DNA methylation [89].

The methionine cycle is a vital pathway in DNA methylation [90]. Studies show that 1-aminocyclopropane-1-carboxylic acid (ACC), the immediate precursor of ethylene, is involved in the regulation of the cell wall function and response to cell wall damage [91]. Thus, we believe that the signal from cellulose coincides with that from β-glucans using, at least to some extent, the same pathways that lead to DNA methylation.

Analysis of the moderated mediation showed that the model was highly influential, and inclusion of the time duration factor in the in vitro culture resulted in 98.57% VAF. Thus, the model explained nearly all variance. This is not surprising, as many cellular processes, including reprogramming, depend on time. The moderation effect of time was also observed when analyzing moderated moderation, where the moderation was conditional on the time of the entire in vitro culture period (21–30.1 days when relations between β-glucans and CHG_DMV were assumed, and until 28 days when CHG_DMV and SV were analyzed). Previously, we proposed that the moment when β-glucans become the predominant source of glucose in the cell coincides with the transition from direct to indirect embryogenesis [38]. However, in this study, the time of in vitro tissue culture was a non-significant moderator, possibly because of a different reason. It is possible that after 28 days of tissue culture, demethylation of the CHG context results in such a low level of methylated cytosines that the cell repair systems can easily cope with putative mutagenic modifications of cytosines. If this hypothesis is correct, the level of mutations after 28 days of tissue culture should be stabilized. Our results may also be interpreted in terms of the cell reprogramming process [12] affecting DNA demethylation change. Further studies are needed to confirm or reject this hypothesis.

Finally, it should be emphasized that the moderated mediation model presented in this study failed to explain the demethylation of CG sites, although the reason for this result is not apparent. However, we interpret the results in terms of how the process of DNA demethylation and its maintenance differ between the two sequence contexts [7,92]. Epigenetic mechanisms involved in demethylation of the CHG context [93] or maintenance of methylation [94] are most likely crucial here. We cannot exclude the possibility that differences between the two contexts are due to the fact that CG is predominant in euchromatic regions, whereas CHG and CHH sequences are more abundant in heterochromatic regions, and distinct mechanisms regulate the maintenance of the methylation of the two contexts [7].

## 4. Materials and Methods

### 4.1. Plant Material

The experiment was carried out on barley (*Hordeum vulgare* L.) of the NAD2 spring genotype obtained from Poznan Plant Breeding Ltd. (Nagradowice, Poland). The presented research was conducted on donor plants and their regenerants. The donor plants were the generative offspring of regenerants obtained by androgenesis in anther cultures. To obtain donor plants, seeds of the NAD2 genotype were sown and grown under controlled conditions: 16 h light/8 h dark photoperiod, 16 °C/12 °C during day/night, and approximately 190 μE $m^{-2}$ $s^{-1}$ light intensity using sodium lamps. Spikes from these plants were harvested when microspores were in the mid- to the late-uninucleate stage. To induce androgenesis, the spikes were kept in the dark at 4 °C for 21 days. After this time, the anthers were removed and placed on the N6L induction medium containing macro- and microelements [95] supplemented with 2 mg $L^{-1}$ 2,4-dichlorophenoxyacetic acid, 0.5 mg $L^{-1}$ naphthaleneacetic acid, and 0.5 mg $L^{-1}$ kinetin. The anther plates prepared in this way were incubated in the dark at 26 °C. First androgenic structures were transferred onto regeneration medium K4NB [96] supplemented with 0.225 mg $L^{-1}$ 6-benzylaminopurine. The incubation of the callus on the regeneration medium was carried out at temperature 26 °C under the following photoperiod: 16 h light/8 h dark. Appearing green regenerants were transferred to flasks on the N6I rooting medium [95] supplemented with 2 mg $L^{-1}$ indole-3-acetic acid and then to pots and grown and greenhouses. The randomly chosen spikes of doubled haploid regenerants were self-pollinated, and finally, seeds were collected. Twenty four offspring obtained from doubled haploids were grown as donor plants to get regenerants for this experiment. The acquisition of regenerants for the present investigation was carried out according to the procedure described above for the donor plants. The difference was the use of nine variants of induction media, the base composition of which was the same. Still, the supplementation of copper and silver ions and the incubation time differed. Detailed supplementation of the induction media is included in Table 4. Each variant was treated as separate in vitro culture conditions and described as trial M1–M9. Trail M1, in which there are no Ag ions, and Cu ions are at the standard medium level, is a control sample, while trial M2–M9 are test samples. From the pool of 24 donor plants and their regenerants, it was possible to obtain regenerants in all the tested trials (M1–M9) for only one donor plant. This donor plant and its regenerants were taken for analysis. Hence, nine media with different concentrations of $CuSO_4$ and $AgNO_3$ and the different in vitro tissue culture duration were tested in terms of the influence of these conditions on the efficiency of green regenerant production and observed genetic/epigenetic variability and the biochemical compounds that may affect DNA methylation patterns and sequence variation.

**Table 4.** Concentrations of ions and number of days applied in trials (M1–M9).

| Trial | $Cu^{2+}$ (μM) | $Ag^{+}$ (μM) | Time (Days) |
|---|---|---|---|
| M1 | 0.1 | 0 | 21 |
| M2 | 0.1 | 10 | 28 |
| M3 | 0.1 | 60 | 35 |
| M4 | 5 | 60 | 28 |
| M5 | 5 | 0 | 35 |
| M6 | 5 | 10 | 21 |
| M7 | 10 | 10 | 35 |
| M8 | 10 | 60 | 21 |
| M9 | 10 | 0 | 28 |

### *4.2. DNA Isolation and Analysis*

DNA was extracted from the fresh leaves of a donor plant and its regenerants using the DNeasy MiniPrep Kit (Qiagen, Hilden, Germany). MetAFLP was performed as described previously [25,28]. Quantitative metAFLP characteristics of demethylated CG and CHG sequences were evaluated as described previously [25].

To conduct ATR-FTIR spectroscopy, leaves were collected from barley regenerants and lyophilized using a speed-vac. The lyophilized leaves were homogenized to a fine powder using an agate mortar and pestle, and immediately used for ATR-FTIR analysis. Mid-infrared spectra were acquired using the iZ10 FTIR module of the Nicolet iN10 MX infrared microscope (Thermo Nicolet Corporation, Madison, WI, USA), equipped with a deuterated triglycine sulfate detector and potassium bromide (KBr) beam splitter. Measurements were recorded in the ATR mode using an ATR accessory (Smart Orbit, Thermo Fisher Scientific) equipped with a single-bounce diamond crystal. A pinch of the homogenized tissue was deposited on the crystal and pressurized against its surface using an attached pressure clamp. Sixty-four interferograms were collected and co-added before Fourier transformation at a wavelength ranging from 4000 to 400 $cm^{-1}$ and at a resolution of 4 $cm^{-1}$. The ATR crystal was carefully cleaned using ethanol before each measurement to remove any residual trace of the previous sample. The spectra were ATR, water vapor and baseline corrected using the OMNIC software (v. 9, Thermo Fischer Scientific). Spectra area normalization (1800–900 $cm^{-1}$) and band integrated area calculations were performed using ChemoSpec [97], while spectra plots were generated using hyperSpec [98], the packages in R programming language [99]. To resolve overlapping peaks, deconvolution was performed using the Gaussian function and nonlinear least-squares fitting [100]. The whole spectrum was initially divided into sections of 10 $cm^{-1}$, and each fragmented spectrum region was then subjected to a mediation analysis [101].

### 4.3. Mediation Analysis

The mediation analysis was performed in the SPSS software v. 26 (https://www.ibm.com/support/pages/node/874712) using the A. F. Hayes Process v. 3.4 macro [101]. Model number 4 and 59 were exploited. Demethylation (DMV), demethylation of the CG contexts (CG_DMV), and demethylation of the CHG contexts (CHG_DMV) were used as putative mediators of sequence variation (SV). The infrared spectrum range (940–1010) was tested as the putative predictor of mediation. The mediation strength was measured using the variance accounted for (VAF) [102] value of the mediator (VAF = indirect effect (IE)/total effect (TE)), indirect, partial, and complete standardized indirect effects of X on Y [103], and the Goodman test [104].

The minimum population size was calculated using the G-Power software [105]. Squared multiple correlation ($r^2$) was set to 0.1 to calculate the effect size ($f^2$) at $\alpha = 0.05$ with three variables and a fixed value of power (1-$\beta$ error probability = 0.28).

**Author Contributions:** P.T.B. and R.O. conceived and designed the research. R.O. conducted the experiments. J.Z. conducted and designed the ATR-FTIR experiments. P.T.B. performed statistical analyzes. P.T.B., R.O., and J.Z. analyzed the generated data and wrote the manuscript. All authors have read and agreed to the published version of the manuscript.

## Abbreviations

| | |
|---|---|
| 5 hmC | 5-hydroxymethylcytosine |
| 5 mC | 5-methylcytosine |
| ATR-FTIR | Attenuated Total Reflectance Fourier Transform Infrared |
| CG_DMV | demethylation of the CG contexts |
| CHG_DMV | demethylation of the CHG contexts |
| dcAdoMet | decarboxylated form of SAM |
| DMV | demethylation |
| IDM | increased DNA methylation complex |
| IF | indirect effect |
| metAFLP | methylation-sensitive Amplified Fragment Length Polymorphism |
| MMR | mismatch repair |
| NER | nucleotide excision repair |
| RdDM | RNA-directed DNA methylation |
| ROS1 | REPRESSOR OF SILENCING 1 |
| SAM | S-adenosyl-L-methionine, |
| SV | sequence variation |
| TCA | tricarboxylic acid cycle |
| TE | total effect |
| VAF | variance accounted for |

## References

1. Naito, K.; Zhang, F.; Tsukiyama, T.; Saito, H.; Hancock, C.N.; Richardson, A.O.; Okumoto, Y.; Tanisaka, T.; Wessler, S.R. Unexpected consequences of a sudden and massive transposon amplification on rice gene expression. *Nature* **2009**, *461*, 1130. [CrossRef] [PubMed]
2. Tanurdzic, M.; Vaughn, M.W.; Jiang, H.; Lee, T.-J.; Slotkin, R.K.; Sosinski, B.; Thompson, W.F.; Doerge, R.W.; Martienssen, R.A. Epigenomic consequences of immortalized plant cell suspension culture. *PLoS Biol.* **2008**, *6*, e302. [CrossRef] [PubMed]
3. Birnbaum, K.D.; Roudier, F. Epigenetic memory and cell fate reprogramming in plants. *Regeneration* **2017**, *4*, 15–20. [CrossRef] [PubMed]

4. Maki, H. Origins of Spontaneous Mutations: Specificity and Directionality of Base-Substitution, Frameshift, and Sequence-Substitution Mutageneses. *Annu. Rev. Genet.* **2002**, *36*, 279–303. [CrossRef]
5. Ito, S.; Shen, L.; Dai, Q.; Wu, S.C.; Collins, L.B.; Swenberg, J.A.; He, C.; Zhang, Y. Tet proteins can convert 5-methylcytosine to 5-formylcytosine and 5-carboxylcytosine. *Science* **2011**, *333*, 1300–1303. [CrossRef]
6. Fujii, S.; Akiyama, M.; Aoki, K.; Sugaya, Y.; Higuchi, K.; Hiraoka, M.; Miki, Y.; Saitoh, N.; Yoshiyama, K.; Ihara, K.; et al. DNA Replication Errors Produced by the Replicative Apparatus of Escherichia coli. *J. Mol. Biol.* **1999**, *289*, 835–850. [CrossRef]
7. Law, J.A.; Jacobsen, S.E. Establishing, maintaining and modifying DNA methylation patterns in plants and animals. *Nat. Rev. Genet.* **2010**, *11*, 204. [CrossRef]
8. Zhang, H.; Lang, Z.; Zhu, J.-K. Dynamics and function of DNA methylation in plants. *Nat. Rev. Mol. Cell Biol.* **2018**, *19*, 489–506. [CrossRef]
9. Slotkin, R.K.; Vaughn, M.; Borges, F.; Tanurdzic, M.; Becker, J.D.; Feijo, J.A.; Martienssen, R.A. Epigenetic reprogramming and small RNA silencing of transposable elements in pollen. *Cell* **2009**, *136*, 461–472. [CrossRef]
10. Zhang, H.; Zhu, J.K. Active DNA demethylation in plants and animals. *Cold Spring Harb. Symp. Quant. Biol.* **2012**, *77*, 161–173. [CrossRef]
11. Ooi, S.K.; Bestor, T.H. The colorful history of active DNA demethylation. *Cell* **2008**, *133*, 1145–1148. [CrossRef]
12. Li, Y.; Kumar, S.; Qian, W. Active DNA demethylation: Mechanism and role in plant development. *Plant Cell Rep.* **2018**, *37*, 77–85. [CrossRef] [PubMed]
13. Ortega-Galisteo, A.P.; Morales-Ruiz, T.; Ariza, R.R.; Roldán-Arjona, T. Arabidopsis DEMETER-LIKE proteins DML2 and DML3 are required for appropriate distribution of DNA methylation marks. *Plant Mol. Biol.* **2008**, *67*, 671–681. [CrossRef] [PubMed]
14. Gehring, M.; Bubb, K.L.; Henikoff, S. Extensive demethylation of repetitive elements during seed development underlies gene imprinting. *Science* **2009**, *324*, 1447–1451. [CrossRef] [PubMed]
15. Hsieh, T.F.; Ibarra, C.A.; Silva, P.; Zemach, A.; Eshed-Williams, L.; Fischer, R.L.; Zilberman, D. Genome-wide demethylation of Arabidopsis endosperm. *Science* **2009**, *324*, 1451–1454. [CrossRef]
16. Ausin, I.; Greenberg, M.V.C.; Simanshu, D.K.; Hale, C.J.; Vashisht, A.A.; Simon, S.A.; Lee, T.-F.; Feng, S.; Española, S.D.; Meyers, B.C.; et al. INVOLVED IN DE NOVO 2-containing complex involved in RNA-directed DNA methylation in *Arabidopsis*. *Proc. Natl. Acad. Sci. USA* **2012**, *109*, 8374–8381. [CrossRef]
17. He, X.-J.; Chen, T.; Zhu, J.-K. Regulation and function of DNA methylation in plants and animals. *Cell Res.* **2011**, *21*, 442–465. [CrossRef]
18. Globisch, D.; Münzel, M.; Müller, M.; Michalakis, S.; Wagner, M.; Koch, S.; Brückl, T.; Biel, M.; Carell, T. Tissue Distribution of 5-Hydroxymethylcytosine and Search for Active Demethylation Intermediates. *PLoS ONE* **2010**, *5*, e15367. [CrossRef]
19. Nabel, C.S.; Manning, S.A.; Kohli, R.M. The Curious Chemical Biology of Cytosine: Deamination, Methylation, and Oxidation as Modulators of Genomic Potential. *ACS Chem. Biol.* **2012**, *7*, 20–30. [CrossRef]
20. Wibowo, A.; Becker, C.; Marconi, G.; Durr, J.; Price, J.; Hagmann, J.; Papareddy, R.; Putra, H.; Kageyama, J.; Becker, J.; et al. Hyperosmotic stress memory in Arabidopsis is mediated by distinct epigenetically labile sites in the genome and is restricted in the male germline by DNA glycosylase activity. *Elife* **2016**, *5*, e13546. [CrossRef]
21. Yong-Villalobos, L.; González-Morales, S.I.; Wrobel, K.; Gutiérrez-Alanis, D.; Cervantes-Peréz, S.A.; Hayano-Kanashiro, C.; Oropeza-Aburto, A.; Cruz-Ramírez, A.; Martínez, O.; Herrera-Estrella, L. Methylome analysis reveals an important role for epigenetic changes in the regulation of the Arabidopsis response to phosphate starvation. *Proc. Natl. Acad. Sci. USA* **2015**, *112*, E7293–E7302. [CrossRef] [PubMed]
22. Ito, H.; Gaubert, H.; Bucher, E.; Mirouze, M.; Vaillant, I.; Paszkowski, J. An siRNA pathway prevents transgenerational retrotransposition in plants subjected to stress. *Nature* **2011**, *472*, 115–119. [CrossRef] [PubMed]
23. Coronel, C.J.; González, A.I.; Ruiz, M.L.; Polanco, C. Analysis of somaclonal variation in transgenic and regenerated plants of Arabidopsis thaliana using methylation related metAFLP and TMD markers. *Plant Cell Rep.* **2018**, *37*, 137–152. [CrossRef] [PubMed]
24. Machczyńska, J.; Zimny, J.; Bednarek, P. Tissue culture-induced genetic and epigenetic variation in triticale (× *Triticosecale* spp. Wittmack ex A. Camus 1927) regenerants. *Plant Mol. Biol.* **2015**, *89*, 279–292. [CrossRef]

25. Orłowska, R.; Bednarek, P.T. Precise evaluation of tissue culture-induced variation during optimisation of in vitro regeneration regime in barley. *Plant Mol. Biol.* **2020**. [CrossRef]
26. Polanco, C.; Ruiz, M.L. AFLP analysis of somaclonal variation in *Arabidopsis thaliana* regenerated plants. *Plant Sci.* **2002**, *162*, 817–824. [CrossRef]
27. Bednarek, P.T.; Orłowska, R. Plant tissue culture environment as a switch-key of (epi) genetic changes. *Plant Cell Tissue Organ Cult. (PCTOC)* **2020**, *140*, 245–257. [CrossRef]
28. Machczyńska, J.; Orłowska, R.; Zimny, J.; Bednarek, P.T. Extended metAFLP approach in studies of the tissue culture induced variation (TCIV) in case of triticale. *Mol. Breed.* **2014**, *34*, 845–854. [CrossRef]
29. Bednarek, P.T.; Orłowska, R.; Koebner, R.M.D.; Zimny, J. Quantification of the tissue-culture induced variation in barley (*Hordeum vulgare* L.). *BMC Plant Biol.* **2007**, *7*, 10. [CrossRef]
30. Kazarian, S.G.; Chan, K.L.A. ATR-FTIR spectroscopic imaging: Recent advances and applications to biological systems. *Analyst* **2013**, *138*, 1940–1951. [CrossRef]
31. Baker, M.J.; Trevisan, J.; Bassan, P.; Bhargava, R.; Butler, H.J.; Dorling, K.M.; Fielden, P.R.; Fogarty, S.W.; Fullwood, N.J.; Heys, K.A.; et al. Using Fourier transform IR spectroscopy to analyze biological materials. *Nat. Protoc.* **2014**, *9*, 1771–1791. [CrossRef] [PubMed]
32. Kumar, S.; Lahlali, R.; Liu, X.; Karunakaran, C. Infrared spectroscopy combined with imaging: A new developing analytical tool in health and plant science. *Appl. Spectrosc. Rev.* **2016**, *51*, 466–483. [CrossRef]
33. Gorgulu, S.T.; Dogan, M.; Severcan, F. The Characterization and Differentiation of Higher Plants by Fourier Transform Infrared Spectroscopy. *Appl. Spectrosc.* **2007**, *61*, 300–308. [CrossRef]
34. Lu, H.-F.; Shen, J.-B.; Lin, X.-Y.; Fu, J.-L. Relevance of Fourier transform infrared spectroscopy and leaf anatomy for species classification in Camellia (Theaceae). *TAXON* **2008**, *57*, 1274–1278. [CrossRef]
35. Alonso-Simón, A.; García-Angulo, P.; Mélida, H.; Encina, A.; Álvarez, J.M.; Acebes, J.L. The use of FTIR spectroscopy to monitor modifications in plant cell wall architecture caused by cellulose biosynthesis inhibitors. *Plant Signal. Behav.* **2011**, *6*, 1104–1110. [CrossRef]
36. Westworth, S.; Ashwath, N.; Cozzolino, D. Application of FTIR-ATR spectroscopy to detect salinity response in Beauty Leaf Tree (Calophyllum inophyllum L). *Energy Procedia* **2019**, *160*, 761–768. [CrossRef]
37. Willick, I.R.; Lahlali, R.; Vijayan, P.; Muir, D.; Karunakaran, C.; Tanino, K.K. Wheat flag leaf epicuticular wax morphology and composition in response to moderate drought stress are revealed by SEM, FTIR-ATR and synchrotron X-ray spectroscopy. *Physiol. Plant.* **2018**, *162*, 316–332. [CrossRef]
38. Bednarek, P.T.; Orłowska, R. CG Demethylation Leads to Sequence Mutations in an Anther Culture of Barley Due to the Presence of Cu, Ag Ions in the Medium and Culture Time. *Int. J. Mol. Sci.* **2020**, *21*, 4401. [CrossRef]
39. Meyer, C.-L.; Juraniec, M.; Huguet, S.; Chaves-Rodriguez, E.; Salis, P.; Isaure, M.-P.; Goormaghtigh, E.; Verbruggen, N. Intraspecific variability of cadmium tolerance and accumulation, and cadmium-induced cell wall modifications in the metal hyperaccumulator Arabidopsis halleri. *J. Exp. Bot.* **2015**, *66*, 3215–3227. [CrossRef]
40. Stuart, B.H. *Biological Applications of Infrared Spectroscopy*; John WIley & Sons, Ltd.: Chichester, UK, 1997.
41. Fan, M.; Dai, D.; Huang, B. Fourier Transform Infrared Spectroscopy for Natural Fibres. In *Fourier Transform—Materials Analysis*; Salih, S., Ed.; InTech: Rijeka, Croatia, 2012; pp. 45–68.
42. Kacuráková, M.; Capek, P.; Sasinková, V.; Wellner, N.; Ebringerová, A. FT-IR study of plant cell wall model compounds: Pectic polysaccharides and hemicelluloses. *Carbohydr. Polym.* **2000**, *43*, 195–203. [CrossRef]
43. Mathlouthi, M.; Koenig, J.L. Vibrational Spectra of Carbohydrates. In *Advances in Carbohydrate Chemistry and Biochemistry*; Tipson, R.S., Horton, D., Eds.; Academic Press: Cambridge, MA, USA, 1987; Volume 44, pp. 7–89.
44. Šandula, J.; Kogan, G.; Kačuráková, M.; Machová, E. Microbial (1→3)-β-d-glucans, their preparation, physico-chemical characterization and immunomodulatory activity. *Carbohydr. Polym.* **1999**, *38*, 247–253. [CrossRef]
45. Dovbeshko, G.I.; Gridina, N.Y.; Kruglova, E.B.; Pashchuk, O.P. FTIR spectroscopy studies of nucleic acid damage. *Talanta* **1997**, *53*, 233–246. [CrossRef]
46. Stuart, B.H. *Infrared Spectroscopy: Fundamentals and Applications*; John Wiley & Sons: Chichester, UK, 2004.
47. Ilharco, L.M.; Garcia, A.R.; Lopes da Silva, J.; Vieira Ferreira, L.F. Infrared Approach to the Study of Adsorption on Cellulose: Influence of Cellulose Crystallinity on the Adsorption of Benzophenone. *Langmuir* **1997**, *13*, 4126–4132. [CrossRef]

48. Kondo, T.; Sawatari, C. A Fourier transform infra-red spectroscopic analysis of the character of hydrogen bonds in amorphous cellulose. *Polymer* **1996**, *37*, 393–399. [CrossRef]
49. Cardoso, P.B.; Machado, T.O.; Feuser, P.E.; Sayer, C.; Meier, M.A.R.; Araújo, P.H.H. Biocompatible Polymeric Nanoparticles From Castor Oil Derivatives via Thiol-Ene Miniemulsion Polymerization. *Eur. J. Lipid Sci. Technol.* **2018**, *120*, 1700212. [CrossRef]
50. Hostetler, M.J.; Stokes, J.J.; Murray, R.W. Infrared Spectroscopy of Three-Dimensional Self-Assembled Monolayers: N-Alkanethiolate Monolayers on Gold Cluster Compounds. *Langmuir* **1996**, *12*, 3604–3612. [CrossRef]
51. Nyquist, R.A. Thiols, Sulfides and Disulfides, Alkanethiols, and Alkanedithiols (S-H stretching). In *Interpreting Infrared, Raman, and Nuclear Magnetic Resonance Spectra*; Academic Press: Cambridge, MA, USA, 2001; pp. 65–83.
52. MacKinnon, D.P.; Fairchild, A.J.; Fritz, M.S. Mediation Analysis. *Annu. Rev. Psychol.* **2007**, *58*, 593–614. [CrossRef]
53. Sauter, M.; Moffatt, B.; Saechao, M.C.; Hell, R.; Wirtz, M. Methionine salvage and S-adenosylmethionine: Essential links between sulfur, ethylene and polyamine biosynthesis. *Biochem. J.* **2013**, *451*, 145–154. [CrossRef]
54. Chiang, P.K.; Gordon, R.K.; Tai, J.; Zeng, G.C.; Doctor, B.P.; Pardhasaradhi, K.; McCann, P.P. S-Adenosylmethionine and methylation. *FASEB J.* **1996**, *10*, 471–480. [CrossRef]
55. Cui, X.-H.; Murthy, H.N.; Wu, C.-H.; Paek, K.-Y. Sucrose-induced osmotic stress affects biomass, metabolite, and antioxidant levels in root suspension cultures of Hypericum perforatum L. *Plant Cell Tissue Organ Cult. (PCTOC)* **2010**, *103*, 7–14. [CrossRef]
56. Carone, S.B.; Santa-Catarina, C.; Silveira, V.; Floh, E.I.S. Polyamine patterns in haploid and diploid tobacco tissues and in vitro cultures. *Braz. Arch. Biol. Technol.* **2010**, *53*, 409–417. [CrossRef]
57. Bishimbayeva, N.; Murtazina, A.; McDougall, G. Influence of Phytohormones on Monosaccharide Composition of Polysaccharides from Wheat Suspension Culture. *Eurasian Chem. Technol. J.* **2017**, *19*, 237–321. [CrossRef]
58. Bednarek, P.T.; Orłowska, R. Time of in vitro anther culture may moderate action of copper and silver ions that affect the relationship between DNA methylation change and the yield of barley green regenerants. *Plants* **2020**, *9*, unpublished work.
59. Ganeshan, S.; Chodaparambil, S.V.; Båga, M.; Fowler, D.B.; Hucl, P.; Rossnagel, B.G.; Chibbar, R.N. In vitro regeneration of cereals based on multiple shoot induction from mature embryos in response to thidiazuron. *Plant Cell Tissue Organ Cult.* **2006**, *85*, 63–73. [CrossRef]
60. Miroshnichenko, D.; Chaban, I.; Chernobrovkina, M.; Dolgov, S. Protocol for efficient regulation of in vitro morphogenesis in einkorn (Triticum monococcum L.), a recalcitrant diploid wheat species. *PLoS ONE* **2017**, *12*, e0173533. [CrossRef]
61. Biesaga-Kościelniak, J.; Kościelniak, J.; Janeczko, A. The impact of zearalenone and thidiazuron on indirect plant regeneration of oilseed rape and wheat. *Acta Physiol. Plant.* **2010**, *32*, 1047–1053. [CrossRef]
62. Burton, R.A.; Fincher, G.B. (1,3;1,4)-β-D-Glucans in Cell Walls of the Poaceae, Lower Plants, and Fungi: A Tale of Two Linkages. *Mol. Plant* **2009**, *2*, 873–882. [CrossRef]
63. Bilska-Kos, A.; Mytych, J.; Suski, S.; Magoń, J.; Ochodzki, P.; Zebrowski, J. Sucrose phosphate synthase (SPS), sucrose synthase (SUS) and their products in the leaves of *Miscanthus* × *giganteus* and *Zea mays* at low temperature. *Planta* **2020**, *252*, 23. [CrossRef]
64. Hofmann, N.R. A Functional Link between Mitochondria and the Cell Wall in Stress Responses. *Plant Cell* **2016**, *28*, 1996. [CrossRef]
65. Huang, S.; Van Aken, O.; Schwarzländer, M.; Belt, K.; Millar, A.H. The Roles of Mitochondrial Reactive Oxygen Species in Cellular Signaling and Stress Response in Plants. *Plant Physiol.* **2016**, *171*, 1551–1559. [CrossRef]
66. Bulone, V.; Schwerdt, J.G.; Fincher, G.B. Co-evolution of Enzymes Involved in Plant Cell Wall Metabolism in the Grasses. *Front. Plant Sci.* **2019**, *10*. [CrossRef] [PubMed]
67. Roulin, S.; Buchala, A.J.; Fincher, G.B. Induction of (1→3,1→4)-β-D-glucan hydrolases in leaves of dark-incubated barley seedlings. *Planta* **2002**, *215*, 51–59. [CrossRef] [PubMed]

68. Hrmova, M.; Fincher, G.B. Dissecting the catalytic mechanism of a plant β-d-glucan glucohydrolase through structural biology using inhibitors and substrate analogues. *Carbohydr. Res.* **2007**, *342*, 1613–1623. [CrossRef] [PubMed]
69. Zhai, Z.; Liu, H.; Xu, C.; Shanklin, J. Sugar Potentiation of Fatty Acid and Triacylglycerol Accumulation. *Plant Physiol.* **2017**, *175*, 696–707. [CrossRef] [PubMed]
70. Funck, D.; Clauß, K.; Frommer, W.; Hellmann, H. The Arabidopsis CstF64-Like RSR1/ESP1 Protein Participates in Glucose Signaling and Flowering Time Control. *Front. Plant Sci.* **2012**, *3*, 80. [CrossRef]
71. Price, J.; Laxmi, A.; St Martin, S.K.; Jang, J.-C. Global transcription profiling reveals multiple sugar signal transduction mechanisms in Arabidopsis. *Plant Cell* **2004**, *16*, 2128–2150. [CrossRef]
72. Jang, J.C.; Sheen, J. Sugar sensing in higher plants. *Plant Cell* **1994**, *6*, 1665–1679. [CrossRef]
73. Pego, J.V.; Kortstee, A.J.; Huijser, C.; Smeekens, S.C.M. Photosynthesis, sugars and the regulation of gene expression. *J. Exp. Bot.* **2000**, *51*, 407–416. [CrossRef]
74. Kesten, C.; Menna, A.; Sánchez-Rodríguez, C. Regulation of cellulose synthesis in response to stress. *Curr. Opin. Plant Biol.* **2017**, *40*, 106–113. [CrossRef]
75. Ding, W.; Smulan, L.J.; Hou, N.S.; Taubert, S.; Watts, J.L.; Walker, A.K. s-Adenosylmethionine Levels Govern Innate Immunity through Distinct Methylation-Dependent Pathways. *Cell Metab.* **2015**, *22*, 633–645. [CrossRef]
76. Roje, S. S-Adenosyl-L-methionine: Beyond the universal methyl group donor. *Phytochemistry* **2006**, *67*, 1686–1698. [CrossRef] [PubMed]
77. Wang, K.L.-C.; Li, H.; Ecker, J.R. Ethylene Biosynthesis and Signaling Networks. *Plant Cell* **2002**, *14*, S131–S151. [CrossRef] [PubMed]
78. Zuo, J.; Wang, Y.; Zhu, B.; Luo, Y.; Wang, Q.; Gao, L. Comparative Analysis of DNA Methylation Reveals Specific Regulations on Ethylene Pathway in Tomato Fruit. *Genes* **2018**, *9*, 266. [CrossRef] [PubMed]
79. Yu, J. *Obesity, Fatty Liver and Liver Cancer*; Yu, J., Ed.; Springer: Singapore, 2018; p. 157.
80. Lu, S.C.; Mato, J.M. S-adenosylmethionine in Liver Health, Injury, and Cancer. *Physiol. Rev.* **2012**, *92*, 1515–1542. [CrossRef] [PubMed]
81. Modrich, P. DNA MISMATCH CORRECTION. *Annu. Rev. Biochem.* **1987**, *56*, 435–466. [CrossRef] [PubMed]
82. Booker, S.J.; Grove, T.L. Mechanistic and functional versatility of radical SAM enzymes. *F1000 Biol. Rep.* **2010**, *2*, 52. [CrossRef]
83. Zhang, B.; Arcinas, A.J.; Radle, M.I.; Silakov, A.; Booker, S.J.; Krebs, C. First Step in Catalysis of the Radical S-Adenosylmethionine Methylthiotransferase MiaB Yields an Intermediate with a [3Fe-4S]0-Like Auxiliary Cluster. *J. Am. Chem. Soc.* **2020**, *142*, 1911–1924. [CrossRef]
84. Broderick, J.B.; Duffus, B.R.; Duschene, K.S.; Shepard, E.M. Radical S-Adenosylmethionine Enzymes. *Chem. Rev.* **2014**, *114*, 4229–4317. [CrossRef]
85. Manova, V.; Gruszka, D. DNA damage and repair in plants—from models to crops. *Front. Plant Sci.* **2015**, *6*, 885. [CrossRef]
86. Harkey, A.F.; Yoon, G.M.; Seo, D.H.; DeLong, A.; Muday, G.K. Light Modulates Ethylene Synthesis, Signaling, and Downstream Transcriptional Networks to Control Plant Development. *Front. Plant Sci.* **2019**, *10*. [CrossRef]
87. Hu, Z.; Vanderhaeghen, R.; Cools, T.; Wang, Y.; De Clercq, I.; Leroux, O.; Nguyen, L.; Belt, K.; Millar, A.H.; Audenaert, D.; et al. Mitochondrial Defects Confer Tolerance against Cellulose Deficiency. *Plant Cell* **2016**, *28*, 2276–2290. [CrossRef] [PubMed]
88. Meng, X.; Li, L.; De Clercq, I.; Narsai, R.; Xu, Y.; Hartmann, A.; Lozano Claros, D.; Custovic, E.; Lewsey, M.G.; Whelan, J.; et al. ANAC017 coordinates organellar functions and stress responses by reprogramming retrograde signaling. *Plant Physiol.* **2019**. [CrossRef]
89. Lozoya, O.A.; Martinez-Reyes, I.; Wang, T.; Grenet, D.; Bushel, P.; Li, J.; Chandel, N.; Woychik, R.P.; Santos, J.H. Mitochondrial nicotinamide adenine dinucleotide reduced (NADH) oxidation links the tricarboxylic acid (TCA) cycle with methionine metabolism and nuclear DNA methylation. *PLoS Biol.* **2018**, *16*, e2005707. [CrossRef] [PubMed]
90. Shen, W.; Gao, C.; Cueto, R.; Liu, L.; Fu, H.; Shao, Y.; Yang, W.Y.; Fang, P.; Choi, E.T.; Wu, Q.; et al. Homocysteine-methionine cycle is a metabolic sensor system controlling methylation-regulated pathological signaling. *Redox Biol.* **2020**, *28*, 101322. [CrossRef] [PubMed]

91. Wolf, S.; Hématy, K.; Höfte, H. Growth Control and Cell Wall Signaling in Plants. *Annu. Rev. Plant Biol.* **2012**, *63*, 381–407. [CrossRef] [PubMed]
92. Parrilla-Doblas, J.T.; Roldán-Arjona, T.; Ariza, R.R.; Córdoba-Cañero, D. Active DNA Demethylation in Plants. *Int. J. Mol. Sci.* **2019**, *20*, 4683. [CrossRef] [PubMed]
93. Liu, R.; Lang, Z. The mechanism and function of active DNA demethylation in plants. *J. Integr. Plant Biol.* **2020**, *62*, 148–159. [CrossRef]
94. Bartels, A.; Han, Q.; Nair, P.; Stacey, L.; Gaynier, H.; Mosley, M.; Huang, Q.Q.; Pearson, J.K.; Hsieh, T.-F.; An, Y.-Q.C.; et al. Dynamic DNA Methylation in Plant Growth and Development. *Int. J. Mol. Sci.* **2018**, *19*, 2144. [CrossRef]
95. Chu, C.C. The N6 medium and its applications to anther culture of cereal crops. *Proc. Symp. Plant Tissue Cult. Sci. Press Pekin.* **1978**, 43–50.
96. Kumlehn, J.; Serazetdinova, L.; Hensel, G.; Becker, D.; Loerz, H. Genetic transformation of barley (*Hordeum vulgare* L.) via infection of androgenetic pollen cultures with *Agrobacterium tumefaciens*. *Plant Biotechnol. J.* **2006**, *4*, 251–261. [CrossRef]
97. Hanson, B.A. ChemoSpec: Exploratory Chemometrics for Spectroscopy. R Package Version 4.4.97. 2017. Available online: https://CRAN.R-project.org/package=ChemoSpec (accessed on 12 December 2017).
98. Beleites, C.; Sergo, V. hyperspec: A Package to Handle Hyperspectral Data Sets in R. R Package Version 0.98-20161118. 2016. Available online: http://hyperspec.R-Forge.R-project.org/ (accessed on 19 November 2016).
99. R Core Team. R: A Language and Environment for Statistical Computing. R Foundation for Statistical Computing, Vienna, Austria. 2018. Available online: http://www.R-project.org/ (accessed on 20 December 2018).
100. Wojdyr, M. Fityk: A general-purpose peak fitting program. *J. Appl. Crystallogr.* **2010**, *43*, 1126–1128. [CrossRef]
101. Hayes, A.F. Introduction to Mediation, Moderation, and Conditional Process Analysis. In *A Regression Bases Approach*; A Division of Guilford Publications, Inc.: New York, NY, USA, 2018; p. 507.
102. Nitzl, C.; Roldan Jose, L.; Cepeda, G. Mediation analysis in partial least squares path modeling: Helping researchers discuss more sophisticated models. *Ind. Manag. Data Syst.* **2016**, *116*, 1849–1864. [CrossRef]
103. Preacher, K.J.; Kelley, K. Effect size measures for mediation models: Quantitative strategies for communicating indirect effects. *Psychol. Methods* **2011**, *16*, 93–115. [CrossRef] [PubMed]
104. Preacher, K.J.; Leonardelli, G.J. Calculation for the Sobel Test: An Interactive Calculation Tool for Mediation Tests. 2001. Available online: http://quantpsy.org/sobel/sobel.htm (accessed on 20 January 2009).
105. Faul, F.; Erdfelder, E.; Lang, A.-G.; Buchner, A. G*Power 3: A flexible statistical power analysis program for the social, behavioral, and biomedical sciences. *Behav. Res. Methods* **2007**, *39*, 175–191. [CrossRef] [PubMed]

3

# Resistance to Cereal Cyst Nematodes in Wheat and Barley: An Emphasis on Classical and Modern Approaches

**Muhammad Amjad Ali [1,*], Mahpara Shahzadi [1], Adil Zahoor [1], Abdelfattah A. Dababat [2], Halil Toktay [3], Allah Bakhsh [4], Muhammad Azher Nawaz [5] and Hongjie Li [6,*]**

1 Department of Plant Pathology, Faculty of Agriculture, University of Agriculture, Faisalabad 38040, Pakistan; mahpara015@gmail.com (M.S.); adilzahoor3253@gmail.com (A.Z.)
2 International Maize and Wheat Improvement Center (CIMMYT), Ankara 06511, Turkey; A.Dababat@cgiar.org
3 Department of Plant Production and Technologies, Faculty of Agricultural Sciences and Technologies, Nigde Omer Halisdemir University, Nigde 51240, Turkey; toktay@yahoo.com
4 Department of Agricultural Genetic Engineering, Faculty of Agricultural Sciences and Technologies, Nigde Omer Halisdemir University, Nigde 51240, Turkey; abthebest@gmail.com
5 College of Agriculture, University of Sargodha, Sargodha 40100, Pakistan; azher490@hotmail.com
6 National Engineering Laboratory for Crop Molecular Breeding, Institute of Crop Sciences, Chinese Academy of Agricultural Sciences, Beijing 100081, China
* Correspondence: amjad.ali@uaf.edu.pk (M.A.A.); lihongjie@caas.cn (H.L.).

**Abstract:** Cereal cyst nematodes (CCNs) are among the most important nematode pests that limit production of small grain cereals like wheat and barley. These nematodes alone are estimated to reduce production of crops by 10% globally. This necessitates a huge enhancement of nematode resistance in cereal crops against CCNs. Nematode resistance in wheat and barley in combination with higher grain yields has been a preferential research area for cereal nematologists. This usually involved the targeted genetic exploitations through natural means of classical selection breeding of resistant genotypes and finding quantitative trait luci (QTLs) associated with resistance genes. These improvements were based on available genetic diversity among the crop plants. Recently, genome-wide association studies have widely been exploited to associate nematode resistance or susceptibility with particular regions of the genome. Use of biotechnological tools through the application of various transgenic strategies for enhancement of nematode resistance in various crop plants including wheat and barley had also been an important area of research. These modern approaches primarily include the use of gene silencing, exploitation of nematode effector genes, proteinase inhibitors, chemodisruptive peptides and a combination of one or more of these approaches. Furthermore, the perspective genome editing technologies including CRISPR-Cas9 could also be helpful for improving CCN resistance in wheat and barley. The information provided in this review will be helpful to enhance resistance against CCNs and will attract the attention of the scientific community towards this neglected area.

**Keywords:** cereal cyst nematodes; wheat; barely; breeding; biotechnology; resistance

---

## 1. Introduction

Small grain cereals have served as the basis for staple foods, beverages, and animal feed for thousands of years [1–3]. Wheat (*Ttriticum aestivum* L.), barley (*Hordeum vulgare* L.), oats (*Avena sativa* L.), rye (*Secale cereale* L.), triticale (*X Triticicosecale* Wittm.), rice (*Oryza sativa* L.), and some other cereals are rich in calories, proteins, carbohydrates, vitamins, and minerals. These cereals supply around 20% of the calories consumed by people worldwide and are therefore a primary source of energy for humans. Global production of small grains increased exponentially from 1960 to 2005, and then began to decline [4]. Further decline in production is projected to continue through 2050 [5], while global demand for these grains is projected to increase by 1% per annum [4].

Currently, wheat, barley, and oat production exceeds consumption in developed countries, while in developing countries the consumption rate is higher than production [5]. Current production levels and trends will not be sufficient to fulfill the projected global demand spawned by the increasing population pressure. For wheat, global production will need to be increased by 60% to fulfill the estimated demand in 2050 [6]. Until recently, global wheat production increased mostly in response to development of improved cultivars and farming practices and technologies. Production is now limited by biotic and abiotic constraints, including diseases, nematodes, insect pests, weeds, and climate. Among these constraints, plant-parasitic nematodes (PPNs) alone are estimated to reduce production of all world crops by 10% [7]. Nematodes are the second biggest group of animals after insects and are present everywhere on the earth ranging from the Polar Regions to the bottom of the oceans [8]. They are present in the ecosystem as free living and as saprophytes, bacteriovores, fungivores, algaevores and as parasites of human, animals and plants. PPNs are 7% of total species of the phylum Nematoda belonging to 4300 species and 197 genera and infect a huge range of economically important crop plants, including wheat and barley [9,10]. The most dangerous species of PPNs belong to Heteroderidae, which also exhibits the cereal cyst nematodes (CCNs). These nematodes are obligate sedentary endo-parasites and are among the important pests that limit the production of small grain cereals. Heavily invaded young cereal plants are stunted and their lower leaves are often chlorotic, forming pale green patches in the field. Mature plants are also stunted and have a reduced number of tillers, and the roots are shallow and abbreviated and have a "bushy-knotted" appearance [11,12]. CCNs comprise a number of closely-related species and are found in most regions where cereals are produced [13–17]. Several efforts have been made to enhance resistance against CCNs in wheat and barley. A major proportion of these efforts are screening and selection of suitable parents for breeding programs followed by use of molecular markers associated with this resistance. Similarly, some attempts have also been made to silence certain nematode parasitism genes and to develop plants with increased resistance using biotechnological approaches. This review is an update on the use of various classical and modern approaches to induce resistance against CCNs in wheat and barley.

## 2. Economic Impact of CCNs in Wheat and Barely

CCNs have reduced yields in individual research trials or fields by as much as 20% in Pakistan, 50% in Australia, 50% in Turkey, and 90% in Saudi Arabia [14,18]. More than half the fields are reported to be infested by CCNs in selected cereal-producing regions of Turkey [19,20], Iran [21], the U.S.A. [22], and Europe [23]. Wheat fields are infested by CCNs in at least 16 provinces of China [24]. A bibliography of 123 CCN publications relating to all aspects of CCN biology and management in China, from 1991 to 2014, was published by Riley and Qi [25]. Reports of crop losses at the magnitudes shown above do not accurately depict the magnitude of economic losses at the regional or national level because documentation was based mostly on research plots located in infested areas of fields. Since the nematode density varies greatly across most fields, published estimates nearly always fail to represent field-wide yield reductions, which are rarely documented. A further complication is that some reports initially attributed to yield reduction by *H. avenae* are now known or assumed to have been attributable to species recently reclassified as *H. australis*, *H. filipjevi*, *H. latipons*, or *H. sturhani*. Nevertheless, several reports of regional or national crop losses caused by CCNs are available. In Australia, annual yield losses due to Australian populations of *H. avenae* were estimated at 300 thousand tonnes [26]. Losses in Australia were at one time much higher but have been reduced greatly by deploying resistant varieties because only one biotype is present [18,27]. Yield losses in three provinces of China, caused by *H. avenae*, were estimated at 1.2 Mt, assuming that 22% of the production area was infested and that the overall yield reduction was 10% in those areas [24]. National production of cereals in Norway was estimated to be reduced by between 1% and 5% by multiple CCN species [28]. Losses from *H. avenae* and *H. filipjevi* in four northwestern states of the U.S.A. are estimated at 22 thousand tons, assuming that 0.04% of the wheat and barley fields are infested and the average field-wide yield reduction in infested fields is 10% [29]. Economic losses in other infested regions of the western U.S.A. and globally are poorly documented [14,30,31].

## 3. Factors Affecting Yield Losses in CCN-Disease

Nematode population has a strong correlation with yield reduction and this relationship is primarily influenced by crop variety, prevailing climatic conditions, cultural techniques, soil quality, and the distribution and density of nematode at a particular location. Similarly, improper use of fertilizer, drought conditions, weakness of root proliferation in soil and unfavorable temperature affected plants are also important determinants of nematode density at the planting time. The varieties have the capability to replace the damaged roots, and thus it is considered to have a better potential among varieties. Nevertheless, associations have been made throughout the history of CCN research to demonstrate a generally linear relationship between initial population density and potential for reduced grain yield, and reductions of other growth and yield components. As an example of the relationship between CCN density and grain yield, *H. filipjevi* in Iran reduced the yield of rainfed winter wheat at all densities ranging from 2.5 to 20 eggs plus J2s/g of soil, with the lowest and highest densities causing 11 and 48% reductions in yield, respectively [32]. Andersen [33] reported that the numbers of *H. avenae* cysts produced on barley were 12, 24, 26, and 33 cysts/plant at initial densities of 1, 2.5, 5, and 10 eggs plus J2s/g of soil. In the northwestern U.S.A., rainfed wheat yields are generally reduced when the number of *H. avenae* eggs plus J2s from extracted cysts, present within the soil, exceeds 3/g of soil [29]. Densities of five *H. avenae* or *H. filipjevi* J2s/g of soil are capable of causing economic damage to irrigated wheat in India [34]. This demonstrates that the initial population pressure is a key player involved in the reduction of grain yield in addition to various other biotic, abiotic and edaphic factors associated with the plants and nematodes.

## 4. Resistance and Tolerance Responses against CCNs

Use of host-plant resistance and tolerance against CCNs is one of the most cost effective and prominent approaches to minimize crop losses below the threshold level [35]. Continuous cultivation of resistant wheat or barley crops has successfully reduced CCN densities to negligible levels in several countries around the globe (Reviewed by Smiley et al. [29]). Cook and Evans [36] and Cui et al. [37] documented the suppression of nematode reproduction due to the plantation of resistant cultivars. It is established that most of the resistances reported against CCNs in commercial cultivars have been based on introgressions of single dominant genes [12,15,38–41]. However, the resistance response must be combined with tolerance to attain yield stability [36,42]. Similarly, Cook and Evans [36] reported that the ability to get better crop yield from tolerant cultivars is possible, which is normally coupled with a substantial degree of nematode control. However, generally tolerance is tested under field conditions by comparing the yield of a control treated with a nematocide, i.e., aldicarb, and untreated infested soil that is also used to assess the effect of pre-planting nematode population density [16,42,43].

Roots of resistant and susceptible cultivars are initially attacked by J2s, which may result in an intolerant reaction prior to the expression of resistance in a resistant cultivar [44–48]. The tolerance character is primarily attributed to particular characteristics of root growth and physiological response of the plants to nematode invasion [49,50]. During the course of establishment of syncytia, root development is highly compromised due to root abbreviation and sometimes proliferated adventitious roots. The growth of infected roots, in the increasing depth of the soil, mostly failed. Nematode resistance is mostly negatively associated with grain yield and susceptible cultivars without nematode infestation showed a higher grain yield as compared to that of resistant cultivars [51]. Due to this reason, farmers mostly show unwillingness to use resistant cultivars, which presents a lower yield than susceptible in non-infested soils [23].

Andersson [44] and O'Brien and Fisher [45] concluded that barley normally shows more tolerance against *H. avenae* as compared to oats or wheat. Andersen [33] reported comparative damage thresholds of 5, 1, and 0.2 eggs plus J2s per gram of soil for these crops, respectively. Those results were further confirmed in selected cultivars of barley and wheat in the U.S.A. [43]. The mechanism behind this enhanced tolerance in barley is generally associated with earlier development of crown roots in barley seedlings than in wheat seedlings, which enables the crown roots to compensate more rapidly for early damage on seminal roots [52].

On the other hand, a combination of resistance and tolerance responses in wheat and barley cultivars leads to improved profitability and production efficiency [29]. However, most wheat breeding programs in the U.S.A. and other regions with established populations of CCNs are not yet breeding for resistance or selecting for tolerance against CCNs. Less financial support, technical and institutional support, uniformity of infestation and a lack of field testing sites could be the possible reasons behind this. Nonetheless, an understanding of accurate resistance mechanisms would be the key to enhance nematode resistant and/or tolerant wheat and barley cultivars [53–56]. Various isolates of *Heterodera* from different regions of the world differ in their behavior against the resistance response in different cereal crops. The resistant cultivars of wheat, barley, or oat to the populations of *Heterodera* in one region are susceptible to the CCN populations in the other region [29]. Oat cultivars exhibiting resistance and tolerance to *H. avenae* in Australia were susceptible to *H. avenae* in Britain [57].

## 5. Marker-Assisted Breeding and QTL Mapping for Nematode Resistance

The first resistance gene against *H. avenae* was discovered in barley during 1920 in Sweden; however, it was not characterized until 1961 [33]. After that, a number of scientists in several countries with five decades focused on the improvement of resistance in barley against CCN (Reviewed by Smiley et al. [29]). The source of resistant to *H. avenae* was found on chromosome 2H of barley, which was mapped as the *Ha2* locus [58] through restriction fragment length polymorphism (RFLP) molecular markers. Similarly, Barr et al. [59] mapped *Ha4* locus on chromosome 5H in barley using RFLP.

It is well established that most of the resistant sources against various diseases in common wheat were obtained from wild wheat relatives through breeding programs [46]. Barloy et al. [60] reported that 9 resistance genes were transferred into common wheat from its wild relatives like *Aegilops* and other *Triticum* spp. to enhance resistance against *H. avenae*. These genes include *Cre1* and *Cre8* from *T. aestivum*; *Cre3* and *Cre4* from *Ae. tauschii* Coss.; *Cre2*, *Cre5* and *Cre6* from *Ae. ventricosa* (Zhuk.); *Cre7* from *Ae. triuncialis* L.; *CreR* from rye and *CreV* from *Dasypium villosum* L. Can. [61]. Recently, Baloch et al. [62] have reviewed and given detailed information regarding the transfer of these genes and their associated QTL in common wheat from its wild relatives.

Furthermore, some other resistance loci, i.e., *CreX* and *CreY* from *Ae. variabilis* Eig. [60], were documented, while their location on the chromosomes and mode of inheritance is still uncharacterized. Similarly, in the hexaploid wheat, many resistance genes are introgressed from its progenitors. Mokabli et al. [63] and Rivoal et al. [41] reported that *Cre1* gene is highly responsive against *H. avenae* population in north Africa, Europe, and North America, while in Asia and Australia it is less effective or ineffective against the population of CCNs [41,63]. However, the virulence to *Cre1* gene, compared with *H. avenae*, differs for the population of *H. latipons* in Syria and in India for *H. filipjevi* [63]. Gene *Cre1* was effective against *H. filipjevi* but *Cre3* was not effective against the Turkish population of *H. filipjevi* [29]. *Cre3* is mostly effective against the population of *H. avenae* in Australia [64], while not being effective against the European populations of *H. avenae* [65,66]. Nicol et al. [67] reported that *Cre2* and *Cre4* genes from *Aegilops* spp. along with an unidentified gene from wheat line AUS4930 have broad spectrum resistance against several species of genus *Heterodera* and its pathotypes. The coordination of CIMMYT established these loci in wheat in many regions of the world, which included *Cre1* to *Cre7* with a substantial degree of resistance against CCNs. The CCN resistance QTLs are mapped on chromosomes 1A, 1D, 4D, 5A, 5B, 5D, 6A, 6B, 7A, and 7D of hexaploid wheat [68,69]. Additionally, 11 DArT markers associated with resistance against CCNs have also been reported in wheat and could lead to the identification of new resistance loci and tools that may become useful in wheat breeding programs [68]. Smiley et al. [29] reviewed the findings of previous scientists that a single resistance gene to *H. avenae*, which has been used for the identification of new pathotypes for a long period of time in wheat, oat, and barley.

Some researchers have developed the molecular markers linked to the resistance against *H. avenae* in wheat and barley (Reviewed by Smiley et al. [29]). Marker-assisted selection (MAS) and backcrossing to improve genetic resistance is being applied, but effective resistance genes for CCNs are not yet available for all crops and are not effective against all pathotypes. Large scale tube, pot, or field test screening to identify lines of wheat, barley, oats, and triticale resistant to the Australian populations of *H. avenae* have been undertaken in Australia for more than 30 years [27]. Initially, the pot test was the method of choice, with resistance determined by white cysts counted on the surface of root balls enabling up to 600 pots to be evaluated each day. Up to 130,000 plants per annum have been screened in this way, resulting in the release of many cultivars resistant or moderately resistant to the Australian populations [27]. However, this approach is labor intensive and time consuming, taking a full growing season to complete. With the development and validation of codominant molecular markers linked to resistance to the Australian populations of *H. avenae*, selection can be applied to leaf samples from small seedlings, and the tests can be automated to determine the presence of resistance genes in 1 to 2 days, with substantial savings in costs and time. As a result, MAS for resistance to CCNs in wheat is now used routinely in Australia to identify resistant germplasm in breeding programs [29].

The combination of pot tests and MAS has been used very successfully to reduce infestation levels and losses caused by the CCN in Australia [70]. The strategy followed for MAS involves two phases: prebreeding, to identify and characterize resistant sources and the development of linked markers, followed by their incorporation by backcrossing into advanced breeding lines. This includes pyramiding of resistance genes from different sources (e.g., *Cre1* on chromosome arm 2BL, *Cre3* on chromosome arm 2DL, and *Cre8* on chromosome arm 6BL), using specific linked PCR-based molecular markers to follow each gene [70]. Deployment of resistant cultivars, starting in about 1975, was also responsible for a strong decrease in damage caused by *H. avenae* in Sweden [44].

Transgenic expression of resistance genes results in the induction of a variety of defense responses, including the up-regulation of pathogenesis related (PR) proteins to establish nematode resistance in plants [71]. Uehara et al. [72] reported that induction of PR-1(P4) involved in the regulation of resistance conferred by *HeroA* R gene against potato cyst nematodes [PCNs, *Globodera rostochiensis* (Wollenweber, 1923) and *G. pallida* (Stone, 1973)] through salicylic acid (SA) signaling and nematode infection resulted in the inhibition of the SA signaling pathway in the susceptible cultivars. Similar effects were found in resistant line of hexaploid wheat carrying *Cre2* gene, which showed upregulation of ascorbate peroxidase coding gene in response to cereal cyst nematode (*H. avenae*) when compared with the expression in the susceptible lines [73].

## 6. Genome-Wide Association Studies for CCN Resistance

Different molecular markers linked to genes for nematode resistance have been identified and reported by developing and utilizing bi-parental populations acquired by crossing contrasting parents, i.e., resistant and susceptible to particular isolates of CCNs. This provides the basis of QTL mapping for nematode resistance which normally involves at least a couple years to develop mapping populations and gene discovery based on the genetic background of only two parental genotypes. Recently, 3 QTL linked to the genomic regions associated with resistance against *Pratylechus thorni* have been identified on chromosomes 2B and 6D in wheat, validating the robustness of these QTL as useful sources of resistance in different genetic backgrounds [74]. Wheat genotypes resistant to *H. avenae* in one region of a same country could be highly susceptible to isolates of *H. avenae* from other regions. This was confirmed by Imren et al. [75] who assessed various landraces and national wheat cultivars from Turkey that wheat genotypes and isolates from different regions in the same country behaved differently. Moreover, QTL recognized from a particular genetic background and working in a particular environment may not be similarly responsive in the other. This shows the need for the development and use of other alternative strategies like association mapping (AM) for the identification of molecular markers which are tightly associated with nematode resistance in crop plants.

Association mapping makes the use of natural plant populations instead of developing the bi-parental mapping populations. However, substantial diversity is required in AM for the detection of QTL with enhanced resolution because AM employs linkage disequilibrium (LD) between alleles within diverse populations to identify markers associated with particular traits. This technique primarily involves genotyping of crop populations, i.e., landraces, cultivars, and advanced lines etc., using particular molecular markers like simple sequence repeats (SSRs) and single-nucleotide polymorphisms (SNPs) and simultaneously phenotying this particular collection of germplasm against a specific biotic stress. For instance, Yang et al. [76] used this approach to map genomic regions associated with stripe rust resistance in wheat. Likewise, recent approaches, such as genome-wide association mapping, utilize the adapted germplasm for identification of QTL associated with a particular plant trait. Moreover, comparing with bi-parental mapping populations, in addition to higher mapping resolution, genome-wide association mapping leads to an increased number of identified alleles and immediate application of the mapping results in a breeding program [62].

Recently, genome-wide association studies (GWAS) have been largely exploited to map the genomic locations associated with a specific plant trait. This approach comprises the high-throughput genotyping of germplasm accessions by sequencing (GBS) through PCR based and next generation sequencing technologies followed by the phenotying. This genotyping results in high-density SNP data that could be utilized to detect marker-character associations in the mapping experiments [77–79]. Advancement in next-generation sequencing technologies has considerably enabled the discovery of SNPs by whole genome sequencing [80–82], transcriptome sequencing [83,84], or reduced-representation sequencing in diverse populations of individuals [85]. GWAS could also be an effective way to identity resistance QTL associated with nematode resistance. For instance, very recently, 13 SNPs associated with resistance to soybean cyst nematode (*Heterodera glycines* Ichinohe, 1952) were recognized, out of which 3 SNPs were associated with already reported QTLs *Rhg1* and *Rhg4* [86].

Similarly, Pariyar et al. [87] performed GWAS of 161 winter wheat accessions using 90K iSelect SNP chip and identified 11 QTL on chromosomes 1AL, 2AS, 2BL, 3AL, 3BL, 4AS, 4AL, 5BL, and 7BL associated with resistance against *H. filipjevi*. They further reported that 8 of these 11 QTL present on chromosome arms 1AL, 2AS, 2BL, 3AL, 4AL, and 5BL were tightly associated with putative genes important for resistant response during plant-pathogen interactions. For instance, methyl transferase 1-associated protein 1 (DMAP1) gene and a putative a RING/FYVE/PHD-type Zinc finger gene linked to chromosome arms 1AL and 2AS are involved in programmed cell death and resistant response against pests. According to the authors, it was the first report of GWAS to map the resistance against CCNs in wheat.

GWAS was recently used to identify Diversity Array Technology (DArT) markers associated with resistance against *H. avenae*, *P. neglectus*, and *P. thornei* from 126 CIMMYT advanced lines of spring wheat adapted to semi-arid conditions [68]. The results demonstrated association of 11 markers with resistance against Ha21 pathotype of *H. avenae*, 25 markers with resistance against *P. neglectus*, and 9 significant markers linked to resistance against *P. thornei*. In that work, chromosome 4A (~90–105 cM) proved to be a source of resistance to *P. thornei*, as has been recently reported. All the QTL associated with resistance against *H. avenae* were mapped on chromosomes 5A, 6A, and 7A of wheat.

## 7. Nematode Resistance through Modern Approaches

Conventional breeding for pest resistance requires either large scale screening of genotypes to identify suitable resistance genes and multiple generations of plants to incorporate the resistance, or more recently, the application of MAS to combine validated resistance genes. Transgenic plants provide an alternative strategy to develop new forms of resistance by homologous or heterologous transfer of functional resistance genes directly, or by developing 'synthetic' resistance genes that interfere with vital processes in the nematode-plant interaction. We have recently reviewed various transgenic strategies to enhance nematode resistance in plants [10]. Here, some of these important approaches are discussed to develop plants that are resistant against CCNs.

### *7.1. Gene Silencing for the Enhancement of Nematode Resistance*

PPNs can take up macromolecules from the cytoplasm of plant cell during the feeding process. Initially it was thought that J2s would not ingest external solution when outside a host plant. It is now evident that J2s can take up external solutions containing dsRNA when they are 'soaked' or incubated in a solution containing dsRNA [88], with some studies showing that the addition of neurostimulants like octopamine to the soaking solution can enhance the uptake of solution and thus dsRNA by PPNs. This finding enabled the study of gene silencing through in vitro soaking or feeding of dsRNA to the infective J2s. As a result, many effectors and other genes involved in biochemical and developmental processes have been targeted in functional studies to determine how their silencing can affect nematode parasitism, reproduction, viability and ability to establish feeding sites [10]. In general, in vitro studies involving the soaking of J2s in dsRNA solutions for up to 24 h have been used to assess the phenotypic effects (e.g., reduced motility or rigidity, aberrant behavior, reduced attraction to roots, reduction in migration in roots), and subsequent quantification of changes in the expression of the targeted gene [89,90].

Such studies target genes involved in nematode parasitism, development, locomotion, invasion and important biological pathways. To date, there have been relatively few studies on gene silencing for CCNs, partly because these species are relatively difficult to handle and culture, and because genomic or transcriptomic data were rather limited. Recently, Gantasala et al. [91] used in vitro silencing of 4 genes from *H. avenae* viz., nuclear hormone receptor, ployadenalyte binding protein, intron binding protein and epsin, through siRNA soaking. They reported that silencing of these genes resulted in a 71%, 26%, and 60% reduction in females and eggs due to the silencing of epsin, intron binding protein and ployadenalyte binding protein, respectively. Conversely, a 25% increase in females and eggs was reported due to the suppression of the nuclear hormone receptor [91].

Host-induced gene silencing (HIGS) is another transgenic approach that has been widely used by the scientists to silence nematode genes involved in the invasion, virulence and establishment of feeding sites. This technique involves the cloning and transformation of dsRNA of nematode gene into the host plants from where it is taken up by the nematodes through their sylet in the form of siRNAs. These siRNAs interfere with expression of nematode genes at the transcript level to induce silencing of that particular gene. A diagrammatic representation of HIGS in provided in Figure 1. We have recently reviewed gene silencing studies that have been undertaken for cyst nematodes [10,29,92].

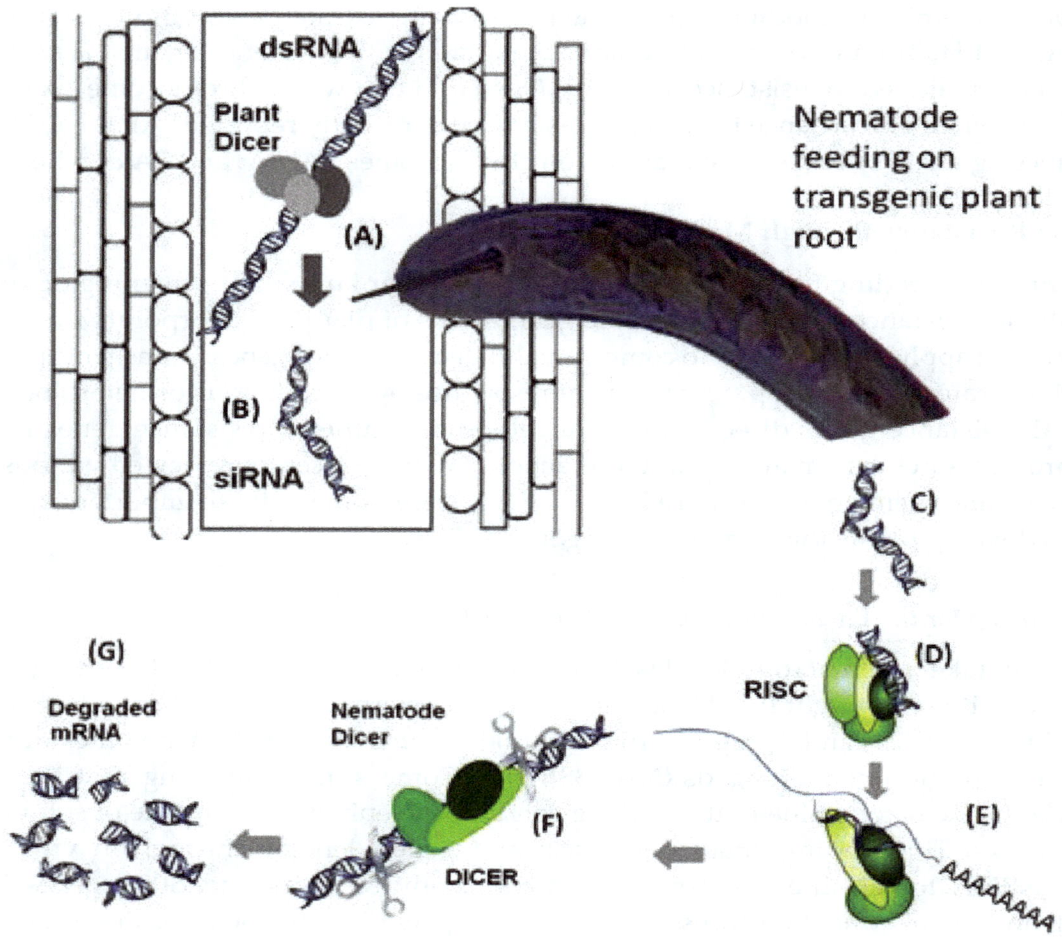


**Figure 1.** Host induced gene silencing for nematode resistance in plants. (**A**) and (**B**): Double stranded RNA (dsRNA) sequence of the target nematode effector gene is transformed into the plant which is cut by the plant DICER enzyme into small interfering RNAs (siRNAs). (**C**), (**D**) and (**E**): These siRNAs are taken up by the nematodes through their stylets which are detected by the RISC complex that binds to mRNA (complementary to the siRNA sequences) of the target effector gene. (**F**) and (**G**): This is followed by the activation of nematode DICER to cut the double stranded RNA to degrade the mRNA of the target nematode effector gene. Reproduced from Ali et al. [10].

In most of the HIGS studies, different goals set by the researchers have been achieved by knocking down the vital genes from nematodes to reduce their ability to parasitize the plant, or by interfering with chemotaxis towards and invasion into host roots, migration into the root tissues, development of nematode feeding structures, or reproduction (reviewed by references [8,10,93]). For example, some data indicates that more than a 90% reduction can be achieved in cyst development for *H. schachtii* on *Arabidopsis*, and a reduction in cyst formation of up to 94% was found after silencing the synaptobrevin (*snb-1*) gene in soybean cyst nematode through HIGS [94]. It is important to perform bioinformatic analyses of the target gene sequences before using them in the gene silencing studies to get rid of off-target effects. In addition, this approach can be used to silence genes that are induced in the host plants and interact with nematode-secreted effector proteins for compatible plant-nematode interactions. For example, the CLE-like nematode effectors are recognized by plant CLE receptors that are required for syncytium development in plants [95,96]. Knockdown of these receptors in soybean roots resulted in reduced soybean cyst nematode infestations [97]. However, reducing expression of plant genes involved in the nematode-plant interaction is likely to be detrimental to plant crops, since the plant genes will have a functional role in the plant, and interfering with this could reduce field performance. To overcome this problem, we have recently reported the site specific delivery of nematode resistance genes and miRNAs of plant genes involved in the establishment of nematode feeding sites using syncytia specific promoters [98,99].

Some of the nematode effector proteins are localized in the nucleus of the host plant cells, which act as transcription factors to regulate the expression of plant genes [92]. It has been shown recently that annexin like protein (Ha-annexin) from *H. avenae* is localized in the nuclei of host plant cells and is involved in the suppression of basal plant defense responses [100]. A transgenic wheat line containing a HIGS construct of Ha-annexin revealed compromised nematode establishment on the plants. Similarly, the transient expression of Ha-annexin led to the down-regulation of the host hypersensitive (HR) response induced by BAX protein and different pathogen associated molecular patterns (PAMPs) such as flagellin [100]. Very recently two venom allergen-like effector proteins (HaVAP1 and HaVAP2) from *H. avenae* have been characterized through gene silencing [101]. The results indicated that both of them are involved in the suppression of programmed cell death induced by BAX in *Nicotiana benthamiana* leaves. Kumar et al. [102] have reported the de novo transcriptome of *H. avenae* which may lead to the identification of various candidate target genes for HIGS to enhance CCN resistance in wheat.

However, most of the times, HIGS using RNAi does not confer 100% resistance to nematodes. There are a number of possible explanations for this. For instance, the first reason could be the specific target gene chosen with several questions like, is its expression vital, is it unique, or does it belong to a multigene family? Similarly, other factors that affect the extent of gene silencing include where and when the target gene is expressed, the specific dsRNA sequence chosen, the stability of target mRNA or encoded protein, the presence of a 'recovery' phenomenon, and experimental variables such as the vigor of the nematodes treated, and differences in the RNAi machinery (e.g., for systemic spread of the silencing signal) between genera or species [103]. This shows the need for other appropriate biotechnologies to increase nematode resistance in plants.

### 7.2. *Utilization of Proteinase Inhibitors and Chemosensory Disruptive Peptides*

RNAi is not the only transgenic approach to confer resistance to nematode pests. Quite a number of different protease inhibitors (PIs) such as cystatins, cowpea trypsin inhibitor (CpTI) and serine proteinase inhibitors have been documented to be successful in producing nematode resistant plants [104]. Heterologous expression of a serine proteinase inhibitor, PIN2 from potato (*Solanum tuberosum* L.) into durum wheat (*Triticum durum* Desf.) led to increased resistance in wheat against *H. avenae* [105]. The main digestive enzymes of many nematodes are cysteine proteinases which have been tackled with the transgenic expression of small proteinase inhibitors, i.e., cystatins, in several plants species to enhance nematode resistance. This strategy has been successfully utilized in a number of crop plants like tomato (*Lycopersicon esculentum* Mill.), rice (*Oryza sativa* L.), potato, banana (*Musa saplentum* L.), and plantain [10]. The human diet normal part is cystatin and is quickly sullied by gastric juice and also it is not allergenic, indicating there is no biosafety issue: the peptide is too small to be allergenic and is degraded in the human small intestine. Similarly, there is no evidence for environmental safety concerns.

In addition to utilization of anti-feedant cysteine proteinase inhibitors to interfere with nematode digestion, anti-root-invasion non-lethal synthetic peptides have been used to hinder the invasion of the nematodes in host plants. These peptides are able to disrupt the chemosensory ability of the nematodes, which make them unable to sense the presence of the plants and the process of nematode chemotaxis towards plants is greatly affected. Winter et al. [106] first reported that synthetic peptides can interfere with the chemoreceptive ability of nematodes to chemically signal in very small concentrations. It was reported that expression of a synthetic peptide nAChRbp in potato resulted in the inhibition of nematode acetylcholinesterase (*AChE*) gene, leading to disorientation of invading J2s of PCN (*G. pallida*) that led to a 52% reduction of female nematodes established on potato roots [107]. Moreover, Costa et al. [108] revealed that *AChE* is highly expressed in chemo- and mechano-sensory neurons of *C. elegan*, further supporting the hypothesis that inhibition of this gene results in chemodisrution of the invading J2s of PPNs [109]. These reports suggest a potential use of chemoreceptive nematode repellent peptides to induce transgenic resistance in cereal crops against various species of CCNs.

*7.3. Coupling of Various Resistance Strategies to Augment Nematode Resistance*

Sometimes, if the HIGS does not work efficiently, it could be coupled with some other transgenic approaches to augment nematode resistance in plants. One way to improve transgenic resistance to nematodes is to combine two different modes of resistance, such as a cystatin and anti-invasion peptide [110]. The coupling of these approaches in potato has resulted in high degree of resistance against PCN without effecting soil quality [111]. The same combination approach has been utilized by the scientists at International Institute of Tropical Agriculture (IITA, Ibadan, Nigeria), in collaboration with researchers at University of Leeds, UK using maize (*Zea mays* L.) cystatin and synthetic nematode repellent peptide, nAChRbp to developed transgenic plantain for resistance against three nematode species, i.e., *R. similis*, *H. multicinctus* and *Meloidogyne* spp. [112,113]. Furthermore, pyramiding of cystatins and chemodisruptive peptide into different crop plants has shown a high degree of nematode resistance and enhanced crop yields [114–116]. These approaches could be further combined with RNAi to give more effective and more durable transgenic resistance. Durability of resistance to pests and diseases is an important consideration, and where RNAi-based traits have already been deployed commercially, the expression of the trait seems to be consistent in the following generations [117,118]. This emphasizes the potential of combining all these approaches to multiply the degree of resistance in cereal crops against various CCN species.

## 8. Genome Editing Technologies: A Potential Perspective for Nematode Resistance in Plants

To date, conventional breeding practices, due to their laborious and time consuming nature, are being replaced with genome editing and other advanced molecular techniques. However, genome editing is being integrated with plant breeding for the development of crop cultivars with improved resistance against pests and other diseases. Until now, different genome editing technologies have been practiced to enhance disease resistance in different crop plants. However, one of the recent and breakthrough in genome editing has been accomplished by the CRISPR (Clustered Regular Interspaced Palindromic Repeats) Cas9 (CRISPR-associated protein) system that is the key player of bacterial immune system.

CRISPR/Cas9 is RNA-guided machinery for effective and precise editing of genomes as compared to any other gene editing technique like Transcription Activator Like Effector Nucleases (TALENS) and Zinc Finger Nucleases (ZFNs) [119,120]. Being precise, highly specific, a multi-gene editor and highly efficient, CRISPR/Cas9 is regarded as promising gene editing system for crop plants [119]. Because CRISPR/Cas9 technology is sequence specific nuclease, it could be employed to exploit defense related mechanisms in plants against invading pathogens. For instance, Ali et al. [119] and Baltes et al. [120] described the implementation of the CRISPR/Cas9 technique to develop resistant plants against geminiviruses and exhibit great potential for enhancement of CCN resistance in cereal crops.

Genome editing using this system is an exciting and powerful alternative to RNAi for gene silencing. It is essentially targeted mutagenesis, in which mutations can be induced in target genes in a nontransgenic manner termed nonhomologous end joining (NHEJ). Alternatively, by insertion of oligonucleotide sequences with ends homologous to each side of the cut site, specific additions to a sequence can be made, known as homologous end joining (HEJ). The technology has passed through a number of iterations, with the use of a 'guide' RNA sequence directing a dsDNAse enzyme (CRISPR-Cas9) to cut a target sequence at a specific site [121]. In NHEJ, the cell repair enzymes frequently make a mistake in joining the ends, resulting in a targeted mutation or total inactivation of the gene. When a cassette consisting of a selectable marker gene with CRISPR and Cas 9 is used to select edited cells for regeneration to plants, the site of gene editing will be elsewhere in the genome from the editing cassette. Hence, for cereals, it is possible to generate edited genotypes that no longer contain an editing cassette. This can be achieved by making a cross with the original (or another) genotype, and identifying genotypes with the edited gene but without the introduced cassette. A plant with a targeted mutation but lacking any introduced DNA may well be regulated as non-Genetically Modified (Non-GM) [29]. However, since the silencing trigger is delivered from the plant to the pest in HIGS, it is not possible to deliver a (nontransgenic) genome editing signal in this way. Application of this technology for nematode management necessitates the identification of a nonvital host plant gene whose expression is very much needed for nematode parasitism to edit it to be nonfunctional.

Recently, genome editing using CRISPR/Cas 9 system has been established in free living nematode, *Caenorhabditis elegans* [122,123]. This development would lead to the characterization of several important genes involved in different physiological processes of nematodes. Nonetheless, there are very few reports available on the application of CRISPR/Cas 9 system to study the resistance responses of the plants against nematodes. Kang [124] recently used soybean hairy roots to study the resistance response knockouts of two serine hydroxymethyltransferase genes, *GmSHMT08* and *GmSHMT05*, generated through this system in soybean-*H. glycines* model organisms. However, to date no report is available regarding the use of this technology in CCN-plant interactions. The recent availability of genome sequences for hosts of CCNs could well lead to the development of nontransgenic cereals genome edited for resistance [125,126].

## 9. Conclusions and Future Perspectives

Current production levels and trends of wheat and barley will not be sufficient to fulfill the projected global demand created by increased populations. For wheat, global production will need to be increased by 60% to fulfill the estimated demand in 2050. Recently, the global wheat production has increased mostly in response to the development of improved cultivars and farming practices and technologies. However, its production is still limited by biotic and abiotic constraints, including diseases, nematodes, insect pests, weeds, and changing climate. Since domestication, improvements in crop plants regarding pest resistance and higher yields are preferential research area of all times in plant sciences, and usually involve targeted genetic exploitations through natural means as well as transgenic means.

Genotyping by sequencing (GBS) followed by association mapping of resistance genes in crop plants could be an important strategy. Similarly, whole genome sequence availability of various CCNs could provide a great platform regarding the vital effector genes that could be manipulated via different transgenic technologies [126,127]. Moreover, full sequence and annotation of wheat genome has been carried out recently containing huge data that could be helpful in understanding nematode-wheat interactions at the molecular level, which could be used for the enhancement of resistance [127]. Furthermore, wheat and barley transcriptome studies in response to CCN infection could be used to manipulate different up- and down-regulated genes. The genes involved in nematode establishment could be down-regulated and suppressed defense genes in syncytia could be overexpressed using syncytia specific genes, as has been demonstrated in Arabidopsis [10,97,98]. Recently, the role of silicon derived resistance has been described in plants against a variety of phytopathogens [128]. This silicon based resistance could be employed for managing CCNs.

Although about 10% of the world's crops are transgenic at present, the costs and issues that must be overcome to deploy any form of transgenic crop resistance to nematodes, and to CCN in particular, are not insubstantial. Hence, there is current research aimed at delivering dsRNA in spray form (ectopic delivery) rather than by transgenic plants. This strategy requires low cost production of dsRNA sequences, methods to stabilize them for field delivery, uptake of dsRNA by leaves, its systemic basipetal movement through plants to roots, and uptake by nematode on feeding [129,130]. If the technical aspects of ectopic delivery of dsRNA can be overcome in a cost-effective manner, this could bypass the issues of RNAi-based transgenic nematode control. In addition to ectopic delivery of RNAi, gene silencing technology can be used to determine targets for new nematicides. This process involves genome-enabled novel chemical nematicides [129]. Very recently, it has been reported that the tryptophan decarboxylase 1 (*AeVTDC1*) gene from a wide relative of wheat *Aegilops variabilis* regulates the resistance against *H. avenae* by altering the downstream secondary metabolite contents rather than auxin synthesis [131]. This shows the potential use of genetic resources from wide relatives for the enhancement of CCN resistance in cereal crops.

## References

1. Breiman, A.; Graur, D. Wheat evolution. *Israel J. Plant Sci.* **1995**, *43*, 85–98. [CrossRef]
2. Gustafson, P.; Raskina, O.; Ma, X.-F.; Nevo, E. Wheat evolution, domestication, and improvement. In *Wheat Science and Trade*; Carver, B., Ed.; Wiley-Blackwell: Ames, IA, USA, 2009; pp. 5–30.
3. Newman, R.K.; Newman, C.W. Barley history: Relationship of humans and barley through the ages. In *Barley for Food and Health: Science, Technology, and Products*; Newman, R.K., Walter Newman, C., Eds.; John Wiley and Sons: Hoboken, NJ, USA, 2008; pp. 1–17.
4. Alexandratos, N.; Bruinsma, J. *World Agriculture towards 2030/2050: The 2012 Revision*; Division, A.D.E., Ed.; Food and Agriculture Organization of the United Nations: New York, NY, USA, 2012.
5. OECD-FAO. Agricultural Outlook Organisation for Economic Cooperation and Development (OECD) and Food and Agricultural Organization of the United Nations (FAO). OECD Agricultural Statistics (Database). Available online: http://dx.doi.org/10.1787/agr-data-en (accessed on 7 December 2018).
6. Ackerman, F.; Stanton, E.A. Can climate change save lives? A comment on economy-wide estimates of the implications of climate change: Human health. *Ecol. Econ.* **2008**, *66*, 8–13. [CrossRef]
7. Whitehead, A. *Plant Nematode Control*; CAB International: Oxon, UK; New York, NY, USA, 1998; Volume VIII.
8. Ali, M.A.; Abbas, A.; Azeem, F.; Javed, N.; Bohlmann, H. Plant-nematode Interactions: From genomics to metabolomics. *Int. J. Agric. Biol.* **2015**, *17*, 1071–1082.
9. Decraemer, W.; Hunt, D.J. Structure and classification. In *Plant Nematology*; Perry, R.N., Moens, M., Eds.; CABI1: Oxfordshire, UK, 2006; pp. 4–32.
10. Ali, M.A.; Azeem, F.; Abbas, A.; Joyia, F.A.; Li, H.J.; Dababat, A.A. Transgenic strategies for enhancement of nematode resistance in plants. *Front. Plant Sci.* **2017**, *8*, 750. [CrossRef]
11. Nicol, J.M.; Turner, S.J.; Coyne, D.L.; den Nijs, L.; Hockland, S.; Tahna Maafi, Z. Current nematode threats to world agriculture. In *Genomics and Molecular Genetics of Plant-Nematode Interactions*; Jones, J., Gheysen, G., Fenoll, C., Eds.; Springer: Dordrecht, The Netherlands, 2011; pp. 21–43.
12. Smiley, R.W.; Nicol, J.M. Nematodes which challenge global wheat production. In *Wheat Science and Trade*; Carver, B.F., Ed.; Wiley-Blackwell: Ames, IA, USA, 2009; pp. 171–187.
13. Dababat, A.A.; Pariyar, S.; Nicol, J.; Erginbasx-Orakçi, G.; Wartin, C.; Klix, M.; Bolat, N.; Braun, H.; Sikora, R. Influence of fungicide seed treatment on the integrated control of *Heterodera filipjevi* on six wheat germplasm with different levels of genetic resistance under controlled conditions. *Nematropica* **2014**, *44*, 25–30.
14. Dababat, A.A.; Muminjanov, H.; Smiley, R.W. *Nematodes of Small Grain Cereals: Current Status and Research*; FAO: Ankara, Turkey, 2015.
15. Nicol, J.M.; Rivoal, R. Global knowledge and its application for the integrated control and management of nematodes on wheat. In *Integrated Management and Biocontrol of Vegetable and Grain Crops Nematodes*; Ciancio, A., Mukerji, K.G., Eds.; Springer: Dordrecht, The Netherlands, 2008; pp. 243–287.
16. Smiley, R.W. Occurrence, distribution and control of *Heterodera avenae* and *H. filipjevi* in the western USA. In *Cereal Cyst Nematodes: Status, Research and Outlook*; Riley, I.T., Nicol, J.M., Dababat, A.A., Eds.; CIMMYT: Ankara, Turkey, 2009; pp. 35–40.
17. Subbotin, S.A.; Sergei, A.; Baldwin, J. *Systematics of Cyst Nematodes (Nematoda: Heteroderinae)*; Brill: Leiden, The Netherlands, 2010; Volume 8A.
18. Riley, I.T.; McKay, A. Cereal cyst nematode in Australia: Biography of a biological invader. In *Cereal Cyst Nematodes: Status, Research and Outlook*; Riley, I.T., Nicol, J.M., Dababat, A.A., Eds.; CIMMYT: Ankara, Turkey, 2009; pp. 23–28.

19. Abidou, H.; El-Ahmed, A.; Nicol, J.M.; Bolat, N.; Rivoal, R.; Yahyaoui, A. Occurrence and distribution of species of the *Heterodera avenae* group in Syria and Turkey. *Nematol. Mediterr.* **2005**, *33*, 195–201.
20. Elekçioğlu, İ.H.; Nicol, J.; Bolat, N.; Sahin, E.; Yorgancılar, A.; Braun, H.; Yorgancılar, O.; Yılddırım, A.; Kılınç, A.; Toktay, H.; Çalisxkan, M. Longterm studies on the cereal cyst nematode *Heterodera filipjevi* in Turkey: International collaboration with regional implications. In *Cereal Cyst Nematodes: Status, Research and Outlook*; Riley, I.T., Nicol, J.M., Dababat, A.A., Eds.; CIMMYT: Ankara, Turkey, 2009; pp. 11–16.
21. Tanha Maafi, Z.; Nicol, J.M.; Kazemi, H.; Ebrahimi, N.; Gitty, M.; Ghalandar, M.; Mohammadi-Pour, M.; Khoshkhabar, Z. Cereal cyst nematodes, root rot pathogens and root lesion nematodes affecting cereal production in Iran. In *Cereal Cyst Nematodes: Status, Research and Outlook*; Riley, I.T., Nicol, J.M., Dababat, A.A., Eds.; CIMMYT: Ankara, Turkey, 2009; pp. 51–55.
22. Smiley, R.W.; Ingham, R.E.; Uddin, W.; Cook, G.H. Crop sequences for managing cereal cyst nematode and fungal populations of winter wheat. *Plant Dis.* **1994**, *78*, 1142–1149. [CrossRef]
23. Rivoal, R.; Cook, R. Nematode pests of cereals. In *Plant Parasitic Nematodes in Temperate Agriculture*; Evans, K., Trudgill, D.L., Webster, J.M., Eds.; CAB International: Wallingford, UK, 1993; pp. 259–303.
24. Peng, D.L.; Nicol, J.; Li, H.M.; Hou, S.Y.; Li, H.X.; Chen, S.L.; Ma, P.; Li, H.L.; Riley, I. Current knowledge of cereal cyst nematode (*Heterodera avenae*) on wheat in China. In *Cereal Cyst Nematodes: Status, Research and Outlook*; Riley, I.T., Nicol, J.M., Dababat, A.A., Eds.; CIMMYT: Ankara, Turkey, 2009; pp. 29–34.
25. Riley, I.T.; Qi, R.D. Annotated bibliography of cereal cyst nematodes (*Heterodera avenae* and *H. filipjevi*) in China, 1991 to 2014. *Australas. Nematol. Newsl.* **2015**, *26*, 1–46.
26. Murray, G.; Brennan, J. Estimating disease losses to the Australian wheat industry. *Australas. Plant Pathol.* **2009**, *38*, 558–570. [CrossRef]
27. Lewis, J.M.; Matic, M.; McKay, A.C. Success of cereal cyst nematode resistance in Australia: History and status of resistance screening systems. In *Cereal Cyst Nematodes: Status, Research and Outlook*; Riley, I.T., Nicol, J.M., Dababat, A.A., Eds.; CIMMYT: Ankara, Turkey, 2009; pp. 137–142.
28. Holgado, R.; Støen, M.; Magnusson, C.; Hammeraas, B. The occurrence and hosts of cereal cyst nematodes (*Heterodera* spp.) in Norway. *Int. J. Nematol. Mediterr.* **2003**, *13*, 1–19.
29. Smiley, R.W.; Dababat, A.A.; Iqbal, S.; Jones, M.G.K.; Maafi, Z.T.; Peng, D.L.; Subbotin, S.A.; Waeyenberge, L. Cereal cyst nematodes: A complex and destructive group of *Heterodera* species. *Plant Dis.* **2017**, *101*, 1692–1720. [CrossRef]
30. Riley, I.T.; Nicol, J.M.; Dababat, A.A. *Cereal Cyst Nematodes: Status, Research and Outlook*; CIMMYT: Ankara, Turkey, 2009.
31. Dababat, A.; Mokrini, F.; Smiley, R.W. *Proceedings of the Sixth International Cereal Nematodes Symposium (Agadir, Morocco)*; CIMMYT: Ankara, Turkey, 2017.
32. Hajihasani, A.; Tanha Maafi, Z.; Nicol, J.M.; Rezaee, S. Effect of the cereal cyst nematode, *Heterodera filipjevi*, on wheat in microplot trials. *Nematology* **2010**, *12*, 357–363.
33. Andersen, S. *Resistens mod Havreål Heterodera avenae*; I Kommission hos Dansk Videnskabs Forlag: Copenhagen, Denmark, 1961.
34. Singh, A.; Sharma, A.K.; Shoran, J. Heterodera avenae and its management in India. In *Cereal Cyst Nematodes: Status, Research and Outlook*; Riley, I.T., Nicol, J.M., Dababat, A.A., Eds.; CIMMYT: Ankara, Turkey, 2009; pp. 17–22.
35. Trudgill, D.L. Resistance to and tolerance of plant parasitic nematodes in plants. *Annu. Rev. Phytopathol.* **1991**, *29*, 167–192. [CrossRef]
36. Cook, R.; Evans, K. Resistance and tolerance. In *Principles and Practice of Nematode Control in Crops*; Brown, R.H., Kerry, B.R., Eds.; Academic Press: Sydney, Australia, 1987; pp. 179–231.
37. Cui, L.; Sun, L.; Gao, X.; Song, W.; Wang, X.M.; Li, H.L.; Liu, Z.Y.; Tang, W.H.; Li, H.J. The impact of resistant and susceptible wheat cultivars on the multiplication of *Heterodera filipjevi* and *H. avenae* in parasite-infected soil. *Plant Pathol.* **2016**, *65*, 1192–1199. [CrossRef]
38. McDonald, A.; Nicol, J. Nematode parasites of cereals. In *Plant Parasitic Nematodes in Subtropical and Tropical Agriculture*; Luc, M., Sikora, R.A., Bridge, J., Eds.; CAB Int.: Wallingford, UK, 2005; pp. 131–191.
39. Nicol, J.M. Important nematode pests of cereals. In *Bread Wheat: Improvement and Production*; Curtis, B.C., Rajaram, S., Gómez, M., Eds.; FAO: Rome, Italy, 2002; pp. 345–366.

40. Nicol, J.M.; Rivoal, R.; Taylor, S.; Zaharieva, M. Global importance of cyst (*Heterodera* spp.) and lesion nematodes (*Pratylenchus* spp.) on cereals: Distribution, yield loss, use of host resistance and integration of molecular tools. *Nematol. Monogr. Perspect.* **2003**, *2*, 1–19.
41. Rivoal, R.; Bekal, S.; Valette, S.; Gauthier, J.P.; Fradj, M.B.; Mokabli, A.; Jahier, J.; Nicol, J.M.; Yahyaoui, A. Variation in reproductive capacity and virulence on different genotypes and resistance genes of Triticeae, in the cereal cyst nematode species complex. *Nematology* **2001**, *3*, 581–590.
42. Brown, R.H. Control strategies in low-value crops. In *Principles and Practice of Nematode Control in Crops*; Brown, R.H., Kerry, B.R., Eds.; Academic Press: Sydney, Australia, 1987; pp. 351–387.
43. Smiley, R.W.; Marshall, J.M. Detection of dual *Heterodera avenae* resistance plus tolerance traits in spring wheat. *Plant Dis.* **2016**, *100*, 1677–1685. [CrossRef]
44. Andersson, S. Population dynamics and control of *Heterodera avenae*—A review with some original results. *EPPO Bull.* **1982**, *12*, 463–475. [CrossRef]
45. O'Brien, P.C.; Fisher, J.M. Development of *Heterodera avenae* on resistant wheat and barley cultivars. *Nematologica* **1977**, *23*, 390–397. [CrossRef]
46. Ogbonnaya, F.C.; Seah, S.; Delibes, A.; Jahier, J.; Lopez-Braña, I.; Eastwood, R.F.; Lagudah, E.S. Molecular genetic characterisation of a new nematode resistance gene in wheat. *Theor. Appl. Genet.* **2001**, *102*, 623–629. [CrossRef]
47. Oka, Y.; Chet, I.; Speigel, Y. Accumulation of lectins in cereal roots invaded by the cereal cyst nematode *Heterodera avenae*. *Physiol. Mol. Plant Pathol.* **1997**, *51*, 333–345. [CrossRef]
48. Cui, L.; Gao, X.; Wang, X.M.; Jian, H.; Tang, W.H.; Li, H.L.; Li, H.J. Characterization of interaction between wheat roots with different resistance and *Heterodera filipjevi*. *Acta Agron. Sin.* **2012**, *38*, 1009–1017. [CrossRef]
49. Stanton, J.M.; Fisher, J.M. Factors of early growth associated with tolerance of wheat to *Heterodera avenae*. *Nematologica* **1988**, *34*, 188–197. [CrossRef]
50. Volkmar, K.M. The cereal cyst nematode (*Heterodera avenae*) on oats. I. Identification of attributes useful in early screening for tolerance to *H. avenae*. *Aust. J. Agric. Res.* **1990**, *41*, 39–49. [CrossRef]
51. Wilson, R.E.; Hollamby, G.J.; Bayraktar, A. Selecting for high yield potential in wheat with tolerance to cereal cyst nematode. *Aust. Field Crops Newsl.* **1983**, *18*, 21–25.
52. Williams, T.D.; Salt, G.A. The effects of soil sterilants on the cereal cyst-nematode (*Heterodera avenae* Woll.), take-all (*Ophiobolus graminis* Sacc.) and yields of spring wheat and barley. *Ann. Appl. Biol.* **1970**, *66*, 329–338. [CrossRef]
53. Andres, M.F.; Melillo, M.T.; Delibes, A.; Romero, M.D.; Bleve-Zacheo, T. Changes in wheat root enzymes correlated with resistance to cereal cyst nematodes. *New Phytol.* **2001**, *152*, 343–354. [CrossRef]
54. Montes, M.J.; López-Braña, I.; Delibes, A. Root enzyme activities associated with resistance to *Heterodera avenae* conferred by gene *Cre7* in a wheat/*Aegilops triuncalis* introgression line. *J. Plant Physiol.* **2004**, *161*, 493–495. [CrossRef]
55. Montes, M.J.; López-Braña, I.; Romero, M.D.; Sin, E.; Andrés, M.F.; Martín-Sánchez, J.; Delibes, A. Biochemical and genetic studies of two *Heterodera avenae* resistance genes transferred from *Aegilops ventricosa* to wheat. *Theor. Appl. Genet.* **2003**, *107*, 611–618. [CrossRef]
56. Seah, S.; Miller, C.; Sivasithamparam, K.; Lagudah, E.S. Root responses to cereal cyst nematode (*Heterodera avenae*) in hosts with different resistance genes. *New Phytol.* **2000**, *146*, 527–533. [CrossRef]
57. Cook, R.; York, P.A. Reaction of some European and Australian oat cultivars to cereal cyst nematode. *Ann. Appl. Biol.* **1988**, *122*, 84–85.
58. Kretschmer, J.M.; Chalmers, K.J.; Manning, S.; Karakousis, A.; Barr, A.R.; Islam, A.K.M.R.; Logue, S.J.; Choe, Y.W.; Barker, S.J.; Lance, R.C.M.; et al. RFLP mapping of the *Ha2* cereal cyst nematode resistance gene in barley. *Theor. Appl. Genet.* **1997**, *94*, 1060–1064. [CrossRef]
59. Barr, A.R.; Chalmers, K.J.; Karakousis, A.; Kretschmer, J.M.; Manning, S.; Lance, R.C.M.; Lewis, J.; Jeffries, S.P.; Langridge, P. RFLP mapping of a new cereal cyst nematode resistance locus in barley. *Plant Breed* **1998**, *117*, 185–187. [CrossRef]
60. Barloy, D.; Lemoine, J.; Abelard, P.; Tanguy, A.M.; Rivoal, R.; Jahier, J. Marker-assisted pyramiding of two cereal cyst nematode resistance genes from *Aegilops variabilis* in wheat. *Mol. Breed* **2007**, *20*, 31–40. [CrossRef]
61. Zhang, R.Q.; Feng, Y.X.; Li, H.F.; Yuan, H.X.; Dai, J.L.; Cao, A.Z.; Xing, X.P.; Li, H.L. Cereal cyst nematode resistance gene *CreV* effective against *Heterodera filipjevi* transferred from chromosome 6VL of *Dasypyrum villosum* to bread wheat. *Mol. Breed.* **2016**, *36*, 122. [CrossRef]

62. Baloch, F.S.; Cömertbay, G.; Özkan, H. DNA molecular markers for disease resistance in plant breeding with example in wheat. In *Nematodes of Small Grain Cereals: Current Status and Research*; Dababat, A., Muminjanov, H., Smiley, R.W., Eds.; FAO: Ankara, Turkey, 2015; pp. 159–166.
63. Mokabli, A.; Rivoal, R.; Gauthier, J.P.; Valette, S. Variation in virulence of cereal cyst nematode populations from North Africa and Asia. *Nematology* **2002**, *4*, 521–525.
64. Vanstone, V.A.; Hollaway, G.J.; Stirling, G.R. Managing nematode pests in the southern and western regions of the Australian cereal industry: Continuing progress in a challenging environment. *Australas. Plant Pathol.* **2008**, *37*, 220–234. [CrossRef]
65. De Majnik, J.; Ogbonnaya, F.C.; Moullet, O.; Lagudah, E.S. The *Cre1* and *Cre3* nematode resistance genes are located at homeologous loci in the wheat genome. *Mol. Plant-Microbe Interact.* **2003**, *16*, 1129–1134. [CrossRef] [PubMed]
66. Safari, E.; Gororo, N.N.; Eastwood, R.F.; Lewis, J.; Eagles, H.A.; Ogbonnaya, F.C. Impact of *Cre1*, *Cre8* and *Cre3* genes on cereal cyst nematode resistance in wheat. *Theor. Appl. Genet.* **2005**, *110*, 567–572. [CrossRef] [PubMed]
67. Nicol, J.M.; Rivoal, R.; Trethowan, R.M.; van Ginkel, M.; Mergoum, M.; Singh, R.P. CIMMYT's approach to identify and use resistance to nematodes and soil-borne fungi, in developing superior wheat germplasm. In *Wheat in a Global Environment*; Bedo, Z., Lang, L., Eds.; Kluwer Academic Publishers: Dordrecht, The Netherlands, 2001; pp. 381–389.
68. Dababat, A.A.; Ferney, G.-B.H.; Erginbas-Orakci, G.; Dreisigacker, S.; Imren, M.; Toktay, H.; Elekçioğlu, İ.H.; Mekete, T.; Nicol, J.M.; Ansari, O.; et al. Association analysis of resistance to cereal cyst nematodes (*Heterodera avenae*) and root lesion nematodes (*Pratylenchus neglectus* and *P. thornei*) in CIMMYT advanced spring wheat lines for semi-arid conditions. *Breed. Sci.* **2016**, *66*, 692–702. [CrossRef] [PubMed]
69. Mulki, M.A.; Jighly, A.; Ye, G.; Emebiri, L.C.; Moody, D.; Ansari, O.; Ogbonnaya, F.C. Association mapping for soilborne pathogen resistance in synthetic hexaploid wheat. *Mol. Breed.* **2013**, *31*, 299–311. [CrossRef]
70. Ogbonnaya, F.C.; Eastwood, R.F.; Lagudah, E. Identification and utilisation of genes for cereal cyst nematode (*Heterodera avenae*) resistance in wheat: The Australian experience. In *Cereal Cyst Nematodes: Status, Research and Outlook*; Riley, I.T., Nicol, M.J., Dababat, A.A., Eds.; CIMMYT: Ankara, Turkey, 2009; pp. 166–171.
71. Ali, M.; Anjam, M.; Nawaz, M.; Lam, H.M.; Chung, G. Signal transduction in plant–nematode interactions. *Int. J. Mol. Sci.* **2018**, *19*, 1648. [CrossRef]
72. Uehara, T.; Sugiyama, S.; Matsuura, H.; Arie, T.; Masuta, C. Resistant and susceptible responses in tomato to cyst nematode are differentially regulated by salicylic acid. *Plant Cell Physiol.* **2010**, *51*, 1524–1536. [CrossRef]
73. Simonetti, E.; Alba, E.; Montes, M.J.; Delibes, A.; López-Braña, I. Analysis of ascorbate peroxidase genes expressed in resistant and susceptible wheat lines infected by the cereal cyst nematode, *Heterodera avenae*. *Plant Cell Rep.* **2010**, *29*, 1169–1178. [CrossRef]
74. Linsell, K.J.; Rahman, M.S.; Taylor, J.D.; Davey, R.S.; Gogel, B.J.; Wallwork, H.; Forrest, K.L.; Hayden, M.J.; Taylor, S.P.; Oldach, K.H. QTL for resistance to root lesion nematode (*Pratylenchus thornei*) from a synthetic hexaploid wheat source. *Theor. Appl. Genet.* **2014**, *127*, 1409–1421. [CrossRef]
75. Imren, M.; Toktay, H.; Bozbuğa, R.; Dababat, A.; Elekçioğlu, İ.H. Pathotype determination of the cereal cyst nematode, *Heterodera avenae* (Wollenweber, 1924) in the Eastern Mediterranean Region in Turkey. *Türk. J. Entomol.* **2013**, *37*, 13–19.
76. Yang, T.M.; Xie, C.J.; Sun, Q.X. Situation of the sources of stripe rust resistance of wheat in the post-CYR32 era in China. *Acta Agron. Sin.* **2003**, *29*, 161–168. [CrossRef]
77. Jia, J.Z.; Zhao, S.C.; Kong, X.Y.; Li, Y.R.; Zhao, G.Y.; He, W.M.; Appels, R.; Pfeifer, M.; Tao, Y.; Zhang, X.Y.; et al. International Wheat Genome Sequencing Consortium, Yang, H.M.; Liu, X.; He, Z.H.; Mao, L.; Wang, J. *Aegilops tauschii* draft genome sequence reveals a gene repertoire for wheat adaptation. *Nature* **2013**, *496*, 91–95. [CrossRef]
78. Tian, F.; Bradbury, P.J.; Brown, P.J.; Hung, H.; Sun, Q.; Flint-Garcia, S.; Rocheford, T.R.; McMullen, M.D.; Holland, J.B.; Buckler, E.S. Genome-wide association study of leaf architecture in the maize nested association mapping population. *Nat. Genet.* **2011**, *43*, 159–162. [CrossRef]
79. Zhao, K.; Tung, C.-W.; Eizenga, G.C.; Wright, M.H.; Ali, M.L.; Price, A.H.; Norton, G.J.; Islam, M.R.; Reynolds, A.; Mezey, J.; et al. Genome-wide association mapping reveals a rich genetic architecture of complex traits in *Oryza sativa*. *Nat. Commun.* **2011**, *2*, 467. [CrossRef]

80. Berkman, P.J.; Lai, K.; Lorenc, M.T.; Edwards, D. Next-generation sequencing applications for wheat crop improvement. *Am. J. Bot.* **2012**, *99*, 365–371. [CrossRef]
81. Chia, J.-M.; Song, C.; Bradbury, P.J.; Costich, D.; de Leon, N.; Doebley, J.; Elshire, R.J.; Gaut, B.; Geller, L.; Glaubitz, J.C.; et al. Maize HapMap2 identifies extant variation from a genome in flux. *Nat. Genet.* **2012**, *44*, 803–807. [CrossRef]
82. Xu, X.; Liu, X.; Ge, S.; Jensen, J.D.; Hu, F.Y.; Li, X.; Dong, Y.; Gutenkunst, R.N.; Fang, L.; Huang, L.; et al. Resequencing 50 accessions of cultivated and wild rice yields markers for identifying agronomically important genes. *Nat. Biotechnol.* **2011**, *30*, 105–111. [CrossRef]
83. Allen, A.M.; Barker, G.L.A.; Berry, S.T.; Coghill, J.A.; Gwilliam, R.; Kirby, S.; Robinson, P.; Brenchley, R.C.; D'Amore, R.; McKenzie, N.; et al. Transcript-specific, single-nucleotide polymorphism discovery and linkage analysis in hexaploid bread wheat (*Triticum aestivum* L.). *Plant Biotechnol. J.* **2011**, *9*, 1086–1099. [CrossRef]
84. Cavanagh, C.R.; Chao, S.; Wang, S.; Huang, B.E.; Stephen, S.; Kiani, S.; Forrest, K.; Saintenac, C.; Brown-Guedira, G.L.; Akhunova, A.; et al. Genome-wide comparative diversity uncovers multiple targets of selection for improvement in hexaploid wheat landraces and cultivars. *Proc. Natl. Acad. Sci. USA* **2013**, *110*, 8057–8062. [CrossRef]
85. Elshire, R.J.; Glaubitz, J.C.; Sun, Q.; Poland, J.A.; Kawamoto, K.; Buckler, E.S.; Mitchell, S.E. A robust, simple genotyping-by-sequencing (GBS) approach for high diversity species. *PLoS ONE* **2011**, *6*, e19379. [CrossRef]
86. Zhang, J.; Wen, Z.X.; Li, W.; Zhang, Y.W.; Zhang, L.F.; Dai, H.Y.; Wang, D.C.; Xu, R. Genome-wide association study for soybean cyst nematode resistance in Chinese elite soybean cultivars. *Mol. Breed.* **2017**, *37*, 60. [CrossRef]
87. Pariyar, S.R.; Dababat, A.A.; Sannemann, W.; Erginbas-Orakci, G.; Elashry, A.; Siddique, S.; Morgounov, A.; Leon, J.; Grundler, F.M.W. Genome-wide association study in wheat identifies resistance to the cereal cyst nematode *Heterodera filipjevi*. *Phytopathology* **2016**, *106*, 1128–1138. [CrossRef]
88. Urwin, P.E.; Lilley, C.J.; Atkinson, H.J. Ingestion of double-stranded RNA by preparasitic juvenile cyst nematodes leads to RNA interference. *Mol. Plant-Microbe Interact.* **2002**, *15*, 747–752. [CrossRef]
89. Fosu-Nyarko, J.; Jones, M.G.K. Advances in understanding the molecular mechanisms of root lesion nematode host interactions. *Annu. Rev. Phytopathol.* **2016**, *54*, 253–278. [CrossRef]
90. Tan, J.A.C.H.; Jones, M.G.K.; Fosu-Nyarko, J. Gene silencing in root lesion nematodes (*Pratylenchus* spp.) significantly reduces reproduction in a plant host. *Exp. Parasitol.* **2013**, *133*, 166–178. [CrossRef]
91. Gantasala, N.P.; Kumar, M.; Banakar, P.; Thakur, P.K.; Rao, U. Functional validation of genes in cereal cyst nematode, *Heterodera avenae*, using siRNA gene silencing. In *Nematodes of Small Grain Cereals: Current Status and Research*; Dababat, A., Muminjanov, H., Smiley, R.W., Eds.; FAO: Ankara, Turkey, 2015; pp. 353–356.
92. Ali, M.A.; Azeem, F.; Li, H.J.; Bohlmann, H. Smart parasitic nematodes use multifaceted strategies to parasitize plants. *Front. Plant Sci.* **2017**, *8*, 1699. [CrossRef]
93. Dutta, T.K.; Banakar, P.; Rao, U. The status of RNAi-based transgenic research in plant nematology. *Front. Microbiol.* **2015**, *5*, 760. [CrossRef]
94. Klink, V.P.; Kim, K.-H.; Martins, V.; MacDonald, M.H.; Beard, H.S.; Alkharouf, N.W.; Lee, S.-K.; Park, S.-C.; Matthews, B.F. A correlation between host-mediated expression of parasite genes as tandem inverted repeats and abrogation of development of female *Heterodera glycines* cyst formation during infection of *Glycine max*. *Planta* **2009**, *230*, 53–71. [CrossRef]
95. Replogle, A.; Wang, J.; Bleckmann, A.; Hussey, R.S.; Baum, T.J.; Sawa, S.; Davis, E.L.; Wang, X.; Simon, R.; Mitchum, M.G. Nematode CLE signaling in Arabidopsis requires CLAVATA2 and CORYNE. *Plant J.* **2011**, *65*, 430–440. [CrossRef]
96. Replogle, A.; Wang, J.; Paolillo, V.; Smeda, J.; Kinoshita, A.; Durbak, A.; Tax, F.E.; Wang, X.; Sawa, S.; Mitchum, M.G. Synergistic interaction of CLAVATA1, CLAVATA2, and RECEPTOR-LIKE PROTEIN KINASE 2 in cyst nematode parasitism of Arabidopsis. *Mol. Plant-Microbe Interact.* **2013**, *26*, 87–96. [CrossRef]
97. Guo, X.; Chronis, D.; De La Torre, C.M.; Smeda, J.; Wang, X.; Mitchum, M.G. Enhanced resistance to soybean cyst nematode *Heterodera glycines* in transgenic soybean by silencing putative CLE receptors. *Plant Biotechnol. J.* **2015**, *13*, 801–810. [CrossRef]
98. Ali, M.A.; Plattner, S.; Radakovic, Z.; Wieczorek, K.; Elashry, A.; Grundler, F.M.; Ammelburg, M.; Siddique, S.; Bohlmann, H. An Arabidopsis ATPase gene involved in nematode-induced syncytium development and abiotic stress responses. *Plant J.* **2013**, *74*, 852–866. [CrossRef]

99. Ali, M.A.; Wieczorek, K.; Kreil, D.P.; Bohlmann, H. The beet cyst nematode *Heterodera schachtii* modulates the expression of WRKY transcription factors in syncytia to favour its development in Arabidopsis roots. *PLoS ONE* **2014**, *9*, e102360. [CrossRef]
100. Chen, C.L.; Liu, S.S.; Liu, Q.; Niu, J.H.; Liu, P.; Zhao, J.L.; Jian, H. An ANNEXIN-like protein from the cereal cyst nematode *Heterodera avenae* suppresses plant defense. *PLoS ONE* **2015**, *10*, e0122256. [CrossRef]
101. Luo, S.J.; Liu, S.M.; Kong, L.A.; Peng, H.; Huang, W.K.; Jian, H.; Peng, D.L. Two venom allergen-like proteins, HaVAP1 and HaVAP2, are involved in the parasitism of *Heterodera avenae*. *Mol. Plant Pathol.* **2018**, in press. [CrossRef]
102. Kumar, M.; Gantasala, N.P.; Roychowdhury, T.; Thakur, P.K.; Banakar, P.; Shukla, R.N.; Jones, M.G.K.; Rao, U. *De novo* transcriptome sequencing and analysis of the cereal cyst nematode, *Heterodera avenae*. *PLoS ONE* **2014**, *9*, e96311. [CrossRef]
103. Jones, M.G.K.; Fosu-Nyarko, J. Molecular biology of root lesion nematodes (*Pratylenchus* spp.) and their interaction with host plants. *Ann. Appl. Biol.* **2014**, *164*, 163–181. [CrossRef]
104. Lilley, C.J.; Devlin, F.; Urwin, P.E.; Atkinson, H.J. Parasitic nematodes, proteinases and transgenic plants. *Parasitol. Today* **1999**, *15*, 414–417. [CrossRef]
105. Vishnudasan, D.; Tripathi, M.N.; Rao, U.; Khurana, P. Assessment of nematode resistance in wheat transgenic plants expressing potato proteinase inhibitor (*PIN2*) gene. *Transgenic Res.* **2005**, *14*, 665–675. [CrossRef]
106. Winter, M.D.; McPherson, M.J.; Atkinson, H.J. Neuronal uptake of pesticides disrupts chemosensory cells of nematodes. *Parasitology* **2002**, *125*, 561–565.
107. Liu, B.; Hibbard, J.K.; Urwin, P.E.; Atkinson, H.J. The production of synthetic chemodisruptive peptides in planta disrupts the establishment of cyst nematodes. *Plant Biotechnol. J.* **2005**, *3*, 487–496. [CrossRef]
108. Costa, J.C.; Lilley, C.J.; Atkinson, H.J.; Urwin, P.E. Functional characterisation of a cyst nematode acetylcholinesterase gene using *Caenorhabditis elegans* as a heterologous system. *Int. J. Parasitol.* **2009**, *39*, 849–858. [CrossRef]
109. Wang, D.; Jones, L.M.; Urwin, P.E.; Atkinson, H.J. A synthetic peptide shows retro- and anterograde neuronal transport before disrupting the chemosensation of plant-pathogenic nematodes. *PLoS ONE* **2011**, *6*, e17475. [CrossRef]
110. Tripathi, L.; Babirye, A.; Roderick, H.; Tripathi, J.N.; Changa, C.; Urwin, P.E.; Tushemereirwe, W.K.; Coyne, D.; Atkinson, H.J. Field resistance of transgenic plantain to nematodes has potential for future African food security. *Sci. Rep.* **2015**, *5*, 8127. [CrossRef]
111. Green, J.; Wang, D.; Lilley, C.J.; Urwin, P.E.; Atkinson, H.J. Transgenic potatoes for potato cyst nematode control can replace pesticide use without impact on soil quality. *PLoS ONE* **2012**, *7*, e30973. [CrossRef]
112. Tripathi, L.; Tripathi, J.N.; Roderick, H.; Atkinson, H.J. Engineering nematode resistant plantains for sub-Saharan Africa. *Acta Hort. (ISHS)* **2013**, *974*, 99–107. [CrossRef]
113. Roderick, H.; Tripathi, L.; Babirye, A.; Wang, D.; Tripathi, J.; Urwin, P.E.; Atkinson, H.J. Generation of transgenic plantain (*Musa* spp.) with resistance to plant pathogenic nematodes. *Mol. Plant Pathol.* **2012**, *13*, 842–851. [CrossRef]
114. Chan, Y.-L.; He, Y.; Hsiao, T.-T.; Wang, C.-J.; Tian, Z.; Yeh, K.-W. Pyramiding taro cystatin and fungal chitinase genes driven by a synthetic promoter enhances resistance in tomato to root-knot nematode *Meloidogyne incognita*. *Plant Sci.* **2015**, *231*, 74–81. [CrossRef]
115. Tripathi, L.; Atkinson, H.; Roderick, H.; Kubiriba, J.; Tripathi, J.N. Genetically engineered bananas resistant to *Xanthomonas* wilt disease and nematodes. *Food Energy Secur.* **2017**, *6*, 37–47. [CrossRef]
116. Waltz, E. Nonbrowning GM apple cleared for market. *Nat. Biotechnol.* **2015**, *33*, 326–327. [CrossRef]
117. Kumar, V.; Jain, M. The CRISPR–Cas system for plant genome editing: Advances and opportunities. *J. Exp. Bot.* **2015**, *66*, 47–57. [CrossRef]
118. Belhaj, K.; Chaparro-Garcia, A.; Kamoun, S.; Nekrasov, V. Plant genome editing made easy: Targeted mutagenesis in model and crop plants using the CRISPR/Cas system. *Plant Methods* **2013**, *9*, 39. [CrossRef]
119. Ali, Z.; Abulfaraj, A.; Idris, A.; Ali, S.; Tashkandi, M.; Mahfouz, M.M. CRISPR/Cas9-mediated viral interference in plants. *Genome Biol.* **2015**, *16*, 238. [CrossRef]
120. Baltes, N.J.; Hummel, A.W.; Konecna, E.; Cegan, R.; Bruns, A.N.; Bisaro, D.M.; Voytas, D.F. Conferring resistance to geminiviruses with the CRISPR–Cas prokaryotic immune system. *Nat. Plants* **2015**, *1*, 15145. [CrossRef]

121. Lozano-Juste, J.; Cutler, S.R. Plant genome engineering in full bloom. *Trends Plant Sci.* **2014**, *19*, 284–287. [CrossRef] [PubMed]
122. Dickinson, D.J.; Goldstein, B. CRISPR-based methods for *Caenorhabditis elegans* genome engineering. *Genetics* **2016**, *202*, 885–901. [CrossRef] [PubMed]
123. Paix, A.; Folkmann, A.; Seydoux, G. Precision genome editing using CRISPR-Cas9 and linear repair templates in *C. elegans*. *Methods* **2017**, *121–122*, 86–93. [CrossRef] [PubMed]
124. Kang, J. *Application of CRISPR/Cas9-Mediated Genome Editing for Studying Soybean Resistance to Soybean Cyst Nematode*; University of Missouri-Columbia, University of Missouri-Columbia: Columbia, MO, USA, 2016.
125. Mayer, K.F.X.; Rogers, J.; Doleel, J.; Pozniak, C.; Eversole, K.; Feuillet, C.; Gill, B.; Friebe, B.; Lukaszewski, A.J.; Sourdille, P.; et al. A chromosome-based draft sequence of the hexaploid bread wheat (*Triticum aestivum*) genome. *Science* **2014**, *345*, 1251788.
126. Consortium, I.B.G.S. A physical, genetic and functional sequence assembly of the barley genome. *Nature* **2012**, *491*, 711–716.
127. International Wheat Genome Sequencing Consortium. Shifting the limits in wheat research and breeding using a fully annotated reference genome. *Science* **2018**, *361*, eaar7191. [CrossRef]
128. Coskun, D.; Deshmukh, R.; Sonah, H.; Menzies, J.G.; Reynolds, O.; Ma, J.F.; Kronzucker, H.J.; Belanger, R.R. The controversies of silicon's role in plant biology. *New Phytol.* **2019**, *221*, 67–85. [CrossRef]
129. Fosu-Nyarko, J.; Jones, M.G.K. Application of biotechnology for nematode control in crop plants. In *Advances in Botanical Research*; Carolina, E., Carmen, F., Eds.; Academic Press: Cambridge, MA, USA, 2015; Volume 73, pp. 339–376.
130. Naz, F.; Fosu-Nyarko, J.; Jones, M. Improving the effectiveness and delivery of gene silencing triggers to control plant nematode pests. In *Proc. 32nd Symp. Eur. Soc. Nematol.*; Murdoch University: Braga, Portugal, 2016; p. 69.
131. Huang, Q.; Li, L.; Zheng, M.; Chen, F.; Long, H.; Deng, G.; Pan, Z.; Liang, J.; Li, Q.; Yu, M.; et al. The Tryptophan decarboxylase 1 gene from *Aegilops variabilis* No.1 regulate the resistance against cereal cyst nematode by altering the downstream secondary metabolite contents rather than auxin synthesis. *Front. Plant Sci.* **2018**, *9*, 1297. [CrossRef]

# 4

# Fine Mapping of the Wheat Leaf Rust Resistance Gene *Lr42*

**Harsimardeep S. Gill [1,†], Chunxin Li [2,†], Jagdeep S. Sidhu [1], Wenxuan Liu [3], Duane Wilson [4], Guihua Bai [5], Bikram S. Gill [4] and Sunish K. Sehgal [1,*]**

1 Department of Agronomy, Horticulture & Plant Science, South Dakota State University, Brookings, SD 57006, USA; harsimardeep.gill@sdstate.edu (H.S.G.); jagdeeproots@gmail.com (J.S.S.)
2 National Engineering Laboratory of Wheat, Wheat Research Institute, Henan Academy of Agricultural Sciences, Zhengzhou 450002, China; licx82@yahoo.com
3 The State Key Laboratory of Wheat and Maize Crop Science, College of Life Sciences, Henan Agricultural University, Zhengzhou 450002, China; wxliu2003@gmail.com
4 Wheat Genetics Resource Center, Department of Plant Pathology, Kansas State University, Manhattan, KS 66506, USA; dlwil@ksu.edu (D.W.); bsgill@ksu.edu (B.S.G.)
5 USDA-ARS, Hard Winter Wheat Genetic Research Unit, Manhattan, KS 66506, USA; guihua.bai@ars.usda.gov
* Correspondence: sunish.sehgal@sdstate.edu
† These authors contributed equally to this work.

**Abstract:** Leaf rust caused by *Puccinia triticina* Eriks is one of the most problematic diseases of wheat throughout the world. The gene *Lr42* confers effective resistance against leaf rust at both seedling and adult plant stages. Previous studies had reported *Lr42* to be both recessive and dominant in hexaploid wheat; however, in diploid *Aegilops tauschii* (TA2450), we found *Lr42* to be dominant by studying segregation in two independent $F_2$ and their $F_{2:3}$ populations. We further fine-mapped *Lr42* in hexaploid wheat using a KS93U50/Morocco $F_5$ recombinant inbred line (RIL) population to a 3.7 cM genetic interval flanked by markers *TC387992* and *WMC432*. The 3.7 cM *Lr42* region physically corresponds to a 3.16 Mb genomic region on chromosome 1DS based on the Chinese Spring reference genome (RefSeq v.1.1) and a 3.5 Mb genomic interval on chromosome 1 in the *Ae. tauschii* reference genome. This region includes nine nucleotide-binding domain leucine-rich repeat (NLR) genes in wheat and seven in *Ae. tauschii*, respectively, and these are the likely candidates for *Lr42*. Furthermore, we developed two kompetitive allele-specific polymorphism (KASP) markers (*SNP113325* and *TC387992*) flanking *Lr42* to facilitate marker-assisted selection for rust resistance in wheat breeding programs.

**Keywords:** wheat; *Aegilops tauschii*; *Lr42*; disease resistance; molecular mapping; KASP markers; marker-assisted selection

## 1. Introduction

Wheat is one of the leading staple foods worldwide, providing one-fifth of the calories and protein to more than 4.5 billion people [1]. Wheat production is constrained not only due to changing climate, but to a great extent by the emergence of new and more virulent races of economically important pathogens. Several diseases and insect pests, including leaf rust (caused by *Puccinia triticina* Eriks), threaten sustainable wheat production in the major wheat-growing areas of the world [2]. In Kansas alone, the leaf rust epidemic of 2007 caused yield losses of 13.9% in winter wheat [3]. Yield losses are attributed to fewer kernels, aggregated by lower kernel weight [2], and losses can be severe if wheat is infected early in development and may reach epidemic proportions in susceptible cultivars under favorable conditions [4]. Diving into history, one can find a reference to leaf rust in the Bible and literature of classical Greece and Rome [5]. Also, the existence of prevalence of this disease from that era until today indicates that this pathogen has evolved along with wheat or other grass species and there has been no permanent solution to control this disease, and likewise for other rusts.

Breeding for rust resistance is considered as one of the most economical approaches to manage rust diseases, and wheat breeding programs throughout the world are deploying rust resistance genes in commercial cultivars. The populations of *P. triticina* are reported to be highly variable in North America [6], with many different virulence pathotypes or races detected annually. Therefore, a more durable approach for long-lasting resistance is the pyramiding of different race-specific and race non-specific genes in single cultivars [7]. However, combining different resistance genes using phenotypic selection is a challenging and cumbersome process. Molecular markers can be very useful for simultaneous stacking of different resistance genes. Thus, the availability of tightly linked molecular markers for different genes is essential for facilitating the stacking of these genes into a single genetic background.

To date, around 80 leaf rust resistance (*Lr*) genes have been formally reported in wheat and its wild relatives [8]. *Aegilops tauschii* Coss, the D-genome donor of wheat, has been a rich source of resistance genes [9,10] and agronomic traits [11]. Several leaf rust resistance genes (*Lr21* (1D), *Lr32* (3D), *Lr39* (2D)) have been transferred into bread wheat from *Ae. tauschii*, including *Lr42* [12]. *Lr42* was introgressed into wheat through a direct cross with *Ae. tauschii* (accession TA 2450) and released as KS91WGRC11 (Century*/TA2450) for further utilization in hexaploid wheat breeding. KS91WGRC11 (carrying *Lr42*) has been successfully used by breeders in several breeding programs [13,14]. Studying near-isogenic lines (NILs) for the *Lr42*, Martin et al. [15] reported that *Lr42* plays an important role in increasing yield, test weight, and kernel weight in wheat. *Lr42* still is one of the highly effective genes, conferring resistance at both seedling and adult plant stages and being used in CIMMYT lines for breeding against leaf rust.

Cox et al. [16] first reported *Lr42* on chromosome 1DS using monosomic analysis in wheat, and it was found to be closely linked to *Lr21*. In their study, Cox et al. [16] reported *Lr42* to be a partially dominant race-specific gene. However, Czembor et al. [17] used Diversity Arrays Technology (DArT) markers to map *Lr42* gene on chromosome 3D and reported that *Lr42* behaved as a dominant gene. By analyzing a set of near-isogenic lines (NILs), Sun et al. [18] mapped *Lr42* on the distal end of chromosome arm 1DS by employing simple sequence repeat (SSR) markers. Using a segregating population of NILs (for *Lr42*) and evaluating for rust infection at both seedling and adult plant stages, they identified three molecular markers—*WMC432*, *CFD15*, and *GDM33*—closely linked to *Lr42*. *WMC432*, about 0.8 cM from *Lr42*, was found to be the closest marker; however, no flanking markers were reported. Further, Liu et al. [19] analyzed an $F_2$ population derived from a cross of KS93U50 (*Lr42*)/Morocco to map *Lr42* on chromosome 1DS using six SSR loci with the quantitative calculation method and reported *Lr42* as a recessive gene. *Lr42* was flanked by *WMC432* and *GDM33* onto a 17 cM region at the distil end of chromosome 1DS with the closest proximal marker (*WMC432*) around 4 cM away. Although the previous studies were able to map *Lr42* to wheat chromosome 1DS, the gene lies in a very gene-rich and recombination-hotspot region at the terminal tip of wheat chromosome 1DS; therefore, SSR markers flanking 17 cM regions are not suitable for marker-assisted selection.

The objectives of this study were to (1) determine the genetic and physical location of *Lr42* on the chromosome 1D and (2) develop kompetitive allele-specific polymorphism (KASP) markers to facilitate marker-assisted selection of *Lr42* in wheat breeding programs. This work will lay the foundation for further cloning of the gene.

## 2. Results

### 2.1. Genetic Analysis of Lr42 in Ae. tauschii

The parental lines TA2433 and TA10132 (AL8/78) showed a highly susceptible response to the leaf rust isolate PNMRJ, with an infection type (IT) score of 3, whereas the *Lr42*-carrying accession

TA2450 showed highly resistant response with an IT score. Of the 66 $F_2$ plants screened from the TA2450/TA2433 population, 50 were resistant and 16 were susceptible, fitting a 3:1 ratio ($\chi^2 = 0.20$, $p = 0.89$) for a single dominant gene (*Lr42*) segregation in this population (Table 1). Further, the 100 $F_{2:3}$ families also exhibited a good fit for the expected 1:2:1 (resistant:segregating:susceptible) ratio ($\chi^2 = 1.92$, $p = 0.38$).

**Table 1.** Segregation of *Lr42* in *Ae. tauschii* and hexaploid wheat populations against leaf rust race PNMRJ. The observed and expected ratios correspond to the resistance:susceptible in $F_2$ generations and homozygous resistant:segregating:homozygous susceptible in $F_{2:3}$ generations.

| Sr No. | Species | Population | Generation | Lines Evaluated | Observed Ratio | Expected Ratio | $\chi^2$ | *p*-Value* |
|---|---|---|---|---|---|---|---|---|
| 1 | *Ae. tauschii* | TA2450/TA2433 | $F_2$ | 66 | 50:16 | 3:1 | 0.20 | 0.89 |
| | | | $F_{2:3}$ | 100 | 27:54:19 | 1:2:1 | 1.92 | 0.38 |
| 2 | *Ae. tauschii* | TA2450/TA10132 (AL8/78) | $F_2$ | 67 | 53:14 | 3:1 | 0.60 | 0.44 |
| | | | $F_{2:3}$ | 100 | 33:53:14 | 1:2:1 | 7.58 | 0.02 |
| 3 | *T. aestivum* | KS93U50/Morocco | $F_5$ RIL | 234 | 99:135 | 1:1 | 5.54 | 0.02 |

* $\alpha = 0.01$.

The second $F_2$ population derived from the cross TA2450/TA10132 showed 53 resistant and 14 susceptible individuals, fitting a 3:1 ratio ($\chi^2 = 0.60$, $p = 0.44$) and confirming that *Lr42* behaves as a dominant gene (Table 1). Similarly, 100 $F_{2:3}$ families (TA2450/TA10132) evaluated also fit the 1:2:1 (resistant:segregating:susceptible) ratio ($\chi^2 = 7.58$, $p = 0.02$), with skewing toward resistant families. Our results from segregation of resistance and susceptibility in these two populations suggest that *Lr42* shows dominant inheritance in *Ae. tauschii* backgrounds.

### 2.2. *Phenotypic Evaluation of KS93U50/Morocco RIL Population*

The hexaploid wheat line KS93U50 carrying *Lr42* showed resistance reaction against isolate PNMRJ producing an infection type (IT) score of 2, whereas Morocco, the susceptible parent of the RIL population, exhibited an IT score of 3+ as expected. The individual plants of 234 $F_5$ RILs from the KS93U50/Morocco population were evaluated for responses to PNMRJ and 99 RILs were found to be rust-susceptible, while the other 135 RILs showed a resistant response. The 1S:1R segregation ratio ($\chi^2 = 5.54$) suggests the presence of a single resistance gene *Lr42* in the RIL population (Table 1).

### 2.3. *Marker Discovery and Molecular Mapping*

Numerous genomic resources were employed to develop new markers to saturate the target *Lr42* region. The flanking markers for *Lr42*, namely *GDM33*, *WMC432*, and *CFD15* [19], were amplified from Chinese Spring (CS) chromosomes 1D, 4D, and 6D bacterial artificial chromosome (BAC) pools. These pools were developed from a BAC library contructed from fraction-I chromosomes (1D, 4D, and 6D) obtained through flow cytometry separation of CS DNA. *CFD15* was physically mapped to two BAC clones (146DhC878D17 and 146DhC799I02), whereas *WMC432* was mapped to four BAC clones (146DhB488K16, 146DhC808O03, 146DhB488K16, and 146DhB458C07). Both these markers were mapped to the single BAC contig ctg1768; however, the distal marker *GWM33* could not be mapped to a unique BAC contig. Our BAC-based physical map of CS chromosome 1D is anchored to *Ae. tauschii* 10K Infinium SNP-based genetic maps, and we identified BAC contigs proximal and distal to ctg1768 in a 1D physical map (https://urgi.versailles.inra.fr/gb2/gbrowse/wheat_phys_1D_v1/). Six BAC contigs were identified spanning this very terminal region of chromosome 1DS. The 24 BACs in these six contigs were end-sequenced to identify five new SSR markers.

We further employed comparative genomic analysis with chromosome 1H of barley [20] for the development of new molecular markers. The collinear interval on chromosome 1H was determined, and 24 genes were predicted to carry plant defense-related domains. Wheat expressed sequence tag (EST) sequences collinear to barley genes were used to develop 19 new EST markers for saturation of the *Lr42* region. In addition, we identified 44 SNPs from *Ae. tauschii* 10K Infinium SNPs mapped on the terminal end of chromosome 1D in a *Prelude* (TA2988)/synthetic wheat (TA8051) RIL population (Figure 1) and also in an *Ae. tauschii* AL8/78 (TA10132)/AS75 $F_2$ population [21]. These SNPs were anchored on a CS 1D physical map. Thus, a total of five SSR markers, 19 EST markers, and 44 KASP markers were designed to enrich the candidate region. Of the 68 markers, 11 were polymorphic between KS93U50 and Morocco and used for mapping of *Lr42* (Figure 1, Table 2). We were able to narrow *Lr42* to a 3.7 cM interval flanked by markers *TC387992* and *WMC432* as against the previously reported 17 cM interval on the terminal end of the short arm of chromosome 1D in wheat (Figure 1).

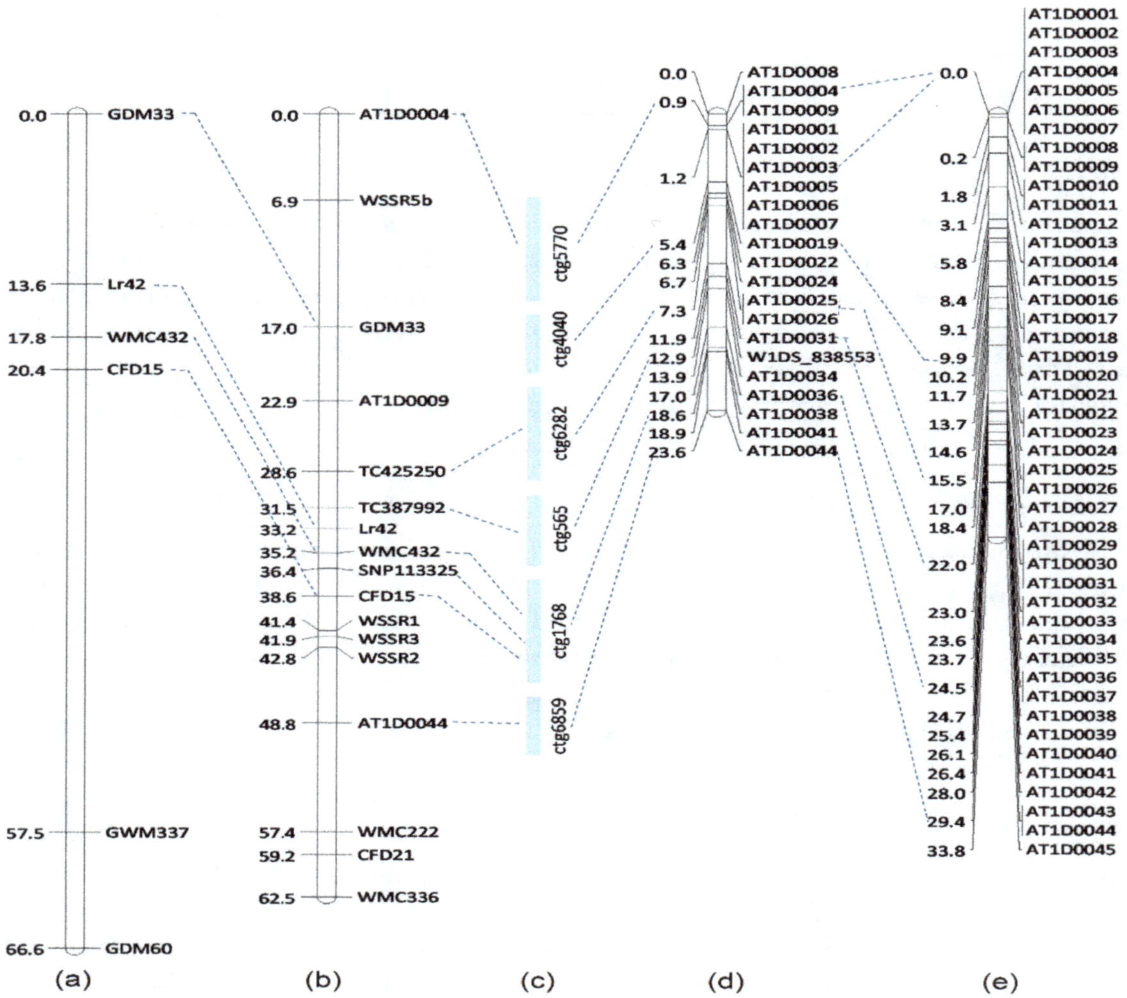


**Figure 1.** Comparative genetic and physical map of the leaf rust resistance gene *Lr42*: (**a**) genetic map in a KS93U50/Morocco $F_2$ population [21]; (**b**) genetic map in a KS93U50/Morocco $F_5$ recombinant inbred line (RIL) population (current study); (**c**) physical mapping of the *Lr42* region on Chinese Spring 1D bacterial artificial chromosome (BAC) contigs (current study), and (**d**) *Prelude* (TA2988)/synthetic wheat (TA8051) RIL population (current study), and (**e**) an *Ae. tauschii* AL8/78 (TA10132)/AS75 $F_2$ population [19].

**Table 2.** Simple sequence repeat (SSR) and KASP markers developed and mapped on the KS93U50/Morocco RIL population.

| Sr. No. | Primer Name | Assay Type | Sequence(s) |
|---|---|---|---|
| 1 | WSSR1 | SSR | ACGACGTTGTAAAACGACTGGAGACAGACGAACGCATA<br>TGCATGCATACACACACCAG |
| 2 | WSSR2 | SSR | ACGACGTTGTAAAACGACAGCAATGCAGTTGCAAAGAG<br>GCAAAGATGGACAGATGGCT |
| 3 | WSSR3 | SSR | ACGACGTTGTAAAACGACAAGATCAGCTCCGACAGCTC<br>CGAAGTCAGCACAAACCAAA |
| 4 | WSSR5 | SSR | ACGACGTTGTAAAACGACTGGTGAATCTTGCACCACAT<br>CTGGACACCGTTCGTTAGGT |
| 5 | AT1D004 | KASP | GGTACCATGTTGTTTCGCATGTCTAT<br>GTACCATGTTGTTTCGCATGTCTAC<br>GGAGGCAGAGACAATAAGTTTATGTTACAA |
| 6 | AT1D0009 | KASP | GGAGATCTTTATATTTGTGGTTTGCCA<br>GAGATCTTTATATTTGTGGTTTGCCG<br>CCAGGTCACAGGCTGTGATGTTTAA |
| 7 | TC425250 | KASP | GCACTACTTTTATTGATGTTGTGTAACC<br>AAGCACTACTTTTATTGATGTTGTGTAACT<br>CAGAGGGAAGAAAACAACACTGAACAAAA |
| 8 | TC387992 | KASP | TTGGATCTGCATTCCTTCTCCCA<br>GGATCTGCATTCCTTCTCCCG<br>CTTTGGGATGTTGCTGCTGGAGAT |
| 9 | SNP113325 | KASP | GGTGTTTGGCAGCATCATCACG<br>GGTGTTTGGCAGCATCATCACC<br>GACAACTTGAGACACTAGATATCAGAGAT |

Further, an additional KASP marker, SNP113325, developed from comparative analysis with barley, was mapped 3.2 cM proximal to *Lr42*, but physically mapped to the same BAC contig as WMC432 and CFD15. The two KASP markers, SNP113325 and TC387992, could be very useful in marker-assisted selection for *Lr42* (Figure 2).

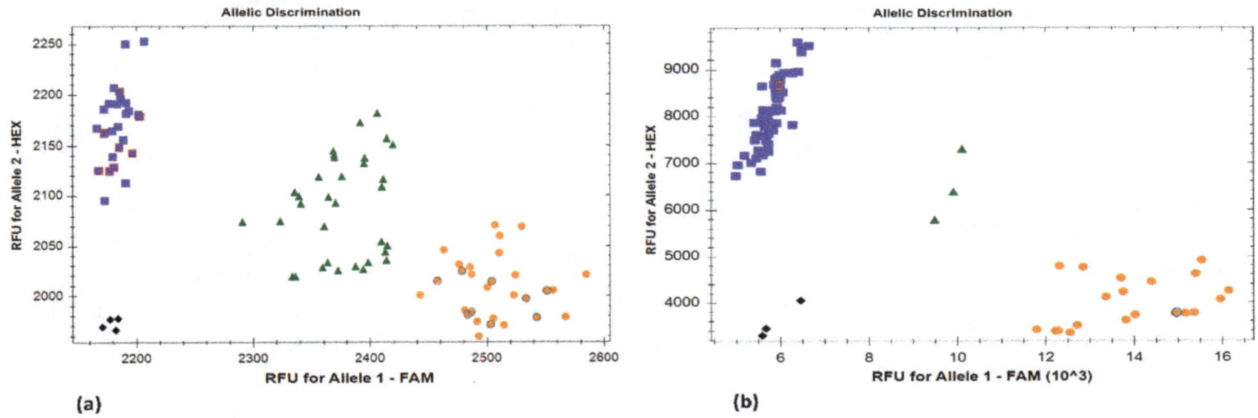


**Figure 2.** Kompetitive allele-specific polymorphism (KASP) marker *SNP113325* identifying the resistant and susceptible parents and selected progenies of (**a**) an *Ae. tauschii* (TA2450/TA2433) $F_2$ population; (**b**) a hexaploid wheat (KS93U50/Morocco) $F_5$ RIL population. Blue: susceptible homozygotes; green: heterozygotes; orange: resistant homozygotes; black: non-template controls. The parental resistant (orange with a blue border) and susceptible (blue with a red border) lines have been highlighted in the respective figures.

## *2.4. Candidate Genes in the Lr42 Region in Wheat and Ae. tauschii*

The flanking markers (*TC42520* and *CFD15*) and BAC end sequences from the region flanking *Lr42* were BLASTN searched against CS Wheat RefSeq v1.1 [22]. We identified a corresponding physical segment of 3.16 Mb (6,327,249 bp to 9,490,443 bp) on the tip of the short arm of chromosome 1D of CS

Wheat. On the other hand, we identified a 3.5 Mb syntenic *Lr42* region in *Ae. tauschii* chromosome 1. There are 109 high-confidence genes in the 3.16 Mb *Lr42* region based on CS Wheat RefSeq v1.1, of which 23 genes were associated with disease resistance function and three were annotated as serine/threonine protein kinase genes (Table S1). Among the 23 disease resistance genes, 19 genes had NLR domains (associated with most of the rust resistance genes cloned in wheat to date) and another four genes had wall-associated kinase (WAK) domains. Further analysis of these 19 NLR genes showed that 10 genes carry pseudo-NLRs (Table S2); therefore, only nine carry functional NLR domains in the *Lr42* region (Figure 3).

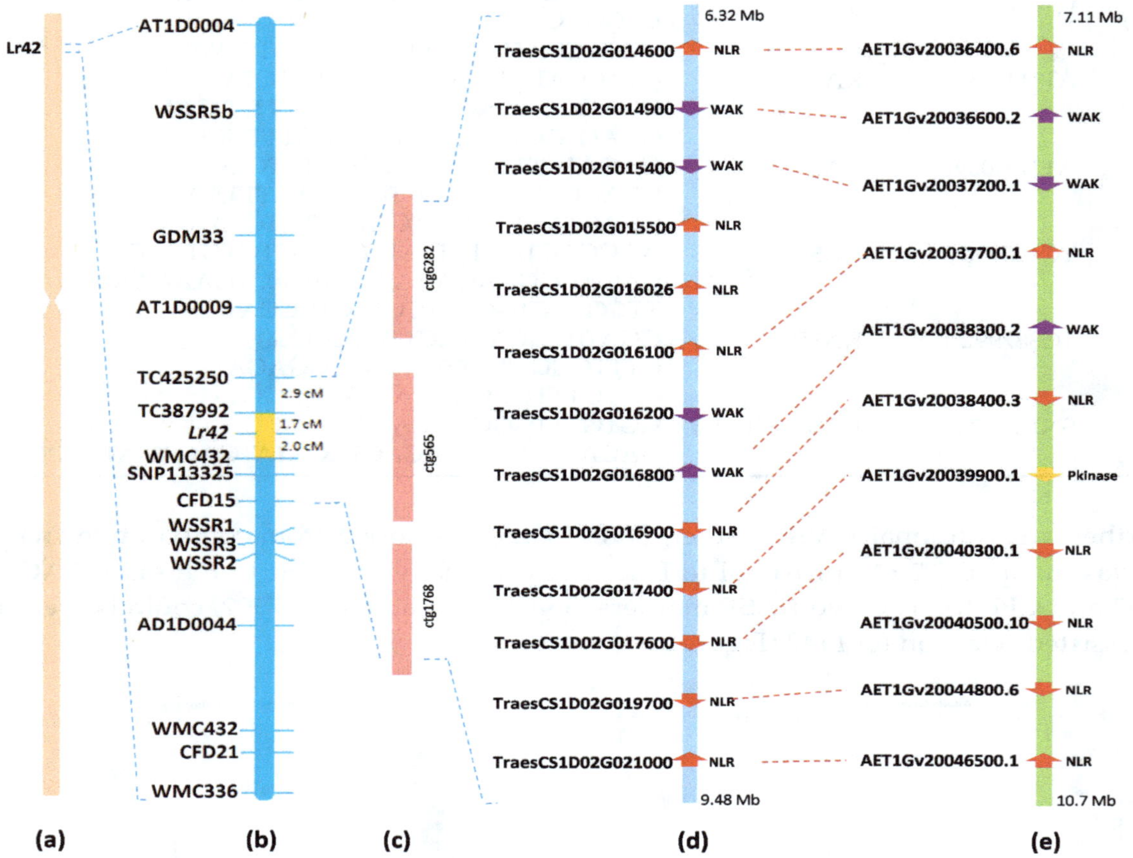


**Figure 3.** Candidate genes in the *Lr42* region in wheat and *Ae. tauschii*. (**a**) Physical location of *Lr42* on chromosome 1DS of wheat; (**b**) genetic location of *Lr42* in KS93U50/Morocco F5 RIL; (**c**) BAC contigs spanning the *Lr42* physical region; (**d**) annotated genes in the *Lr42* region in CS RefSeq v1.1 [22]; (**e**) annotated genes in the *Lr42* region in the *Ae. tauschii* chromosome 1 sequence [21]. NLR: nucleotide-binding leucine-rich repeats; WAK: wall-associated kinase; Pkinase: protein kinase.

In *Ae. tauschii*, we identified 98 genes in the 3.5 Mb *Lr42* region, with 21 genes encoding plant disease defense-related proteins (Table S3), using the PFAM database [23]. Among the 21 genes, three have putative wall-associated kinase (WAK) domains, seven genes had serine/threonine protein kinase (Pkinase) domains, and 11 genes had NB-ARC domains (Figure 3). Comparative analysis of wheat and *Ae. tauschii* genes in the *Lr42* region showed that four of the 11 NLR genes in *Ae. tauschii* are orthologues of pseudo-NLRs in wheat. Furthermore, the comparison of genes from two species showed that six NLRs and two WAKs from wheat have high sequence similarity (>96%) and are orthologues in *Ae. tauschii*. One NLR gene from wheat could be an orthologue of a protein kinase (Pkinase) gene from *Ae. tauschii* as it shared a sequence similarity of 90%. Apart from these genes, no orthologues were found in *Ae. tauschii* for one WAK and two NLR genes that were present in wheat. By contrast, one NLR gene was present in *Ae. tauschii* but absent in wheat (Figure 3).

## 3. Discussion

Leaf rust can cause severe losses in wheat yield and grain quality. Host resistance is a key component in managing leaf rust, and thus, molecular genetic characterization of resistance and identification of tightly linked molecular markers can help achieve better understanding of the mechanism of leaf rust resistance and facilitate marker-assisted breeding and gene pyramiding. Liu et al. [19] mapped *Lr42* on the short arm of wheat chromosome 1D after analyzing $F_2$ and $F_3$ generations of KS93U50/Morocco population. In the current study, we advanced the population to $F_5$ RILs. Segregation observed in $F_2$ and $F_3$ generations of the KS93U50/Morocco population as studied by Liu et al. [19] suggested that *Lr42* was recessive. In the current study, we evaluated two independent *Ae. tauschii* $F_2$ populations (TA2450/TA2433 and TA2450/TA10132) for a response to leaf rust using the same isolate, PNMRJ. In both the populations, the majority of plants were resistant, indicating a dominant monogenic control of the resistance reaction. Further segregation pattern in the $F_{2:3}$ generation in the two *Ae. tauschii* populations was consistent with $F_2$, suggesting that *Lr42* behaves as a dominant gene in *Ae. tauschii*. *Ae. tauschii* (the D-genome donor of wheat) is the source of *Lr42*, however, the segregation behavior of *Lr42* had not been studied earlier in *Ae. tauschii*. There are several conflicting reports regarding the segregation behavior of the *Lr42* gene in hexaploid wheat. Cox et al. [16] reported that *Lr42* was partially dominant, whereas Czembor et al. [17] reported it as being a dominant gene in wheat. However, these studies used several different genetic backgrounds or different leaf rust isolates, which could have resulted in the differences, as demonstrated by Kolmer and Dyck [24].

Applying BAC-based physical mapping, BAC-end sequencing, and comparative genomic analysis with barley [20], we identified 68 new markers and further mapped *Lr42* to a 3.7 cM interval. Liu et al. [19] reported *Lr42* as being located on the distal tip of chromosome 1DS flanked by a 17 cM interval. In the current study, we significantly reduced the size of the *Lr42* flanking segment by mapping *TC387792* and *WMC432*, as being 1.7 cM and 2.0 cM away from *Lr42* on the distal and proximal regions, respectively. Additionally, *SNP113325* lies 3.2 cM distal to *Lr42* and is also tightly linked to *Lr42* in the RIL population. The two KASP markers *SNP113325* and *TC387792* have been used for screening of hard winter wheat in regional nurseries (see 2018 SRPN, Table 10 from https://www.ars.usda.gov/ARSUserFiles/30421000/HardWinterWheatRegionalNurseryProgram/2018%20SRPN%20021519.xlsx) in marker-assisted selection for *Lr42* in wheat breeding programs.

We further physically delimited the *Lr42* region to a 3.16 Mb (6,327,249 bp to 9,490,443 bp) region in chromosome 1D using the Chinese Spring reference genome RefSeq v1.1 [22]. As expected, *Lr42* is present in a high-recombination region with a higher genetic-to-physical map ratio, making the search for the candidate gene relatively easy compared to the centromeric region. Though we identified 109 genes in this region, 23 of these genes have kinase domains mostly associated with disease-related genes in wheat. Besides the 23 genes, three genes were annotated as protein kinases (Pkinases); however, Pkinase genes have not been associated with wheat rust resistance genes to date. Out of the 23 genes, 10 genes were found to be pseudo-NLRs, leaving only 13 candidate genes. Of the final 13 candidate genes, nine genes have NLR domains and four genes have WAK domains. In *Ae. tauschii*, we identified seven genes with NLR domains that could be candidates for *Lr42*. Currently, several rust resistance genes have been cloned in wheat, including *Lr10* [25], *Lr21* [26], *Lr1* [27], *Lr22a* [28], *Sr33* [29], *Sr35* [30], *Sr45* [31], and *Yr10* [32], and all encode an NLR-type protein. Though wall-associated kinase domains (WAKs) have been reported to confer resistance to fungi in wheat, such as *Snn1* against Septoria nodorum blotch [33] and *Stb6* against *Septoria tritici* blotch [34], none of them showed resistance to wheat rusts. Therefore, the nine NLR-type genes in the *Lr42* region are the most probable candidates for the *Lr42* gene. Nonetheless, it is possible that *Lr42* has a regulatory mechanism different from NBS-LRR type genes. Molecular cloning of *Lr42* is thus required to reveal the complete regulation mechanism, which can be facilitated using novel cloning techniques such as MutRenSeq [31] or TACCA [28]. These techniques have been recently used to clone several disease-related genes in wheat, such as

*Lr22a* [28] and *Pm2* [35]. Thus, we have identified an *Lr42* susceptible mutant by EMS mutagenesis of *Ae. tauschii* accession TA2450, and future efforts will be made to further characterize *Lr42*.

*Lr42* is physically located on the distal region of wheat chromosome 1DS, where a number of disease-related genes have been mapped to, such as *Lr21* [26], *Sr33* [29], *Sr45* [31], *Pm24* [36], and several other genes of agronomic importance. In wheat, the terminal gene-rich regions have been found to be positively correlated with recombination frequency [37]. Identification of tightly linked flanking markers could facilitate marker-assisted selection of *Lr42* in wheat breeding programs. The KASP markers developed in our study closely flank *Lr42* and could be efficiently used for marker-assisted selection of *Lr42* and help in stacking of the *Lr42* gene with other race non-specific disease resistance genes to develop wheat cultivars with more durable resistance against leaf rust.

## 4. Materials and Methods

### 4.1. Plant Materials and Rust Resistance Evaluation

*Lr42* was identified from *Ae. tauschii* accession TA2450 [16]. We screened several *Ae. tauschii* accessions with leaf rust race PNMRJ and identified two *Ae. tauschii* accessions, TA2433, and TA10132 (AL8/78), that were susceptible to PNMRJ (PNMRJ is avirulent to *Lr42*). TA2450 was then crossed to TA2433 and TA10132 to develop two $F_2$ populations (TA2450/TA2433 and TA2450/TA10132). A total of 100 $F_2$ plants were derived from each cross (TA2450/TA2433 and TA2450/TA10132), of which a sample of around 70 plants were artificially inoculated with leaf rust race PNMRJ in a greenhouse following the method described by Liu et al. [19]. Briefly, the $F_2$ plants from each of the two crosses, along with the parental lines, were planted in plastic trays. The plants were artificially inoculated with PNMRJ at the two-leaf stage. For artificial inoculation, the PNMRJ spores were suspended in Soltrol 170 mineral oil (CPChemicals LLC, Garland, TX, USA). The suspension was sprayed on the seedlings using a pressure sprayer, followed by incubation at 20 °C for 24 h in a humid chamber. Following incubation, the $F_2$ plants were grown in a greenhouse at 20–24 °C for the establishment of infection. At ten days after inoculation, the seedlings were scored for rust infection type (IT) on a 1 to 4 scale (; = hypersensitive flecks, 1 = small uredinia with necrosis, 2 = moderate size pustules with chlorosis, 3 = moderate-large size uredinia without necrosis or chlorosis, and 4 = large uredinia lacking necrosis or chlorosis) [38,39]. The scoring was repeated after two days for confirmation. For further study, $F_{2:3}$ progenies for each cross were grown and evaluated for leaf rust to identify homozygous non-segregating families. About 20 seeds for each family were grown and evaluated using the same procedure used for the $F_2$ populations. Based on the reactions of parental lines to PNMRJ, the plants with IT $\leq 2$ were considered resistant and those with IT $> 2$ susceptible.

In addition to the two *Ae. tauschii* populations, we used the $F_5$ RIL population of 234 progenies developed from the wheat cross KS93U50/Morocco. The RIL population, along with parents KS93U50 and Morocco, were grown and evaluated for leaf rust resistance under controlled greenhouse conditions with three replications. The $F_5$ RILs were advanced lines from the $F_2$ population analyzed by Liu et al. [19] to map *Lr42*. The inoculation and scoring method was the same as described for *Ae. tauschii* populations. The data from the three populations were used to deduce the inheritance pattern of *Lr42*.

### 4.2. Marker Discovery and Saturation in the Candidate Region

Previously reported *Lr42* flanking markers [19] were mapped on to the physical map of chromosome 1D of Chinese Spring wheat (https://urgi.versailles.inra.fr/gb2/gbrowse/wheat_phys_1D_v1/) by identifying the BACs associated with the candidate region using three-dimensional BAC pools. Identified BACs spanning this region were end-sequenced to generate five new PCR-based SSR markers using the DesignPrimer tool (https://kofler.or.at/bioinformatics/SciRoKo/DesignPrimer.html). Further, the genomic information from the BACs and comparative analysis with chromosome 1H of barley [20] was conducted to develop an additional 19 EST markers. In addition, KASP markers were developed from *Ae. tauschii* 10K Infinium single nucleotide polymorphisms (SNPs) [19] that were mapped onto the BACs of this region. A total of 68 new molecular markers, along with previously reported markers, were used to genotype the $F_5$ RILs along with the parents. The polymorphic markers were used for the construction of a genetic map for the *Lr42* region. The EST markers that were polymorphic between parents were later converted into KASP-based markers for easy and reliable genotyping and to select *Lr42* in breeding programs.

### 4.3. DNA Extraction and Genotyping of the RIL Population

Ten-day-old uninfected seedlings of RILs, along with parental lines, were used for DNA extraction using the cetyl trimethyl ammonium bromide (CTAB) method [40]. Extracted DNA samples were quantified using a Nanodrop ND1000 spectrophotometer (Nanodrop Technologies, US) for normalization of DNA concentration. For SSR and EST genotyping, polymerase chain reaction (PCR) was carried out in a GeneAmp PCR System 9700 (Applied Biosystems, CA, USA) using 20 μL of reaction mixture containing 100 ng genomic DNA, 25 ng each of forward and reverse primers, and 10 μL of 2× PCR Mastermix. Thermocycling profile was set as follows: initial denaturation at 95 °C for 5 min; 30 cycles each at 95 °C for 30 sec, 50–61 °C (depending upon the annealing temperature of particular primers) for 30 sec, 72 °C for 1 min; and a final extension step at 72 °C for 7 min. The EST and SSR PCR products were visualized using 3% high-resolution agarose gel (GeneMate, ISC Bioexpress, Inc, Kaysville, UT, USA). The kompetitive allele-specific polymorphism (KASP) genotyping was carried out using 8 μL of total reaction mixture containing 3 μL of 20–25 ng/μL genomic DNA, 5 μL of KASP Mastermix (LGC Genomics, Teddington, UK) consisting of a FAM and HEX specific FRET cassette, Taq polymerase and optimized buffer; and 0.07 μL of KASP assay mix consisting of two allele-specific primers and one common primer. For accurate distribution of the small volume of assay mix, KASP Mastermix and assay mix were combined before dispensing. PCR was carried out using a Bio-Rad CFX96 Real-time PCR system (Bio-Rad Laboratories, Hercules, CA, USA). The PCR profile was designed as follows: 94 °C for 20 min (hot start activation); 10 touchdown cycles each at 94 °C for 20 sec, 61–55 °C for 60 sec (dropping by 0.6 °C per cycle); followed by 35 cycles at 94 °C for 20 sec and 55 °C for 60 sec. Bio-Rad CFX Manager (Bio-Rad Laboratories, USA) was used for reading the plates after PCR.

### 4.4. Statistical Analysis and Genetic Mapping

Pearson's chi-squared analysis was used to test the goodness of fit for observed frequencies to the expected genetic frequencies. The chi-squared analysis was performed in R ver 3.4 [41] using the function 'chisq.test'. CarthaGene v1.3 was employed for the construction of genetic maps and automatical marker ordering to obtain a multipoint maximum likelihood map [42]. Firstly, a logarithm of odds (LOD) score of 3.0 was used as a threshold value for identifying the linkage groups, followed by ordering of the markers using the Kosambi mapping function [43]. The genetic linkage map was improved using the 'simulated annealing' algorithm and 'verification algorithms' in Carthagene [42]. MapChart version 2.2 [44] was used to draw and combine different genetic maps.

### 4.5. Physical Mapping of the Lr42 Region on Chromosome 1DS

Using BAC-end sequences and the sequences of EST markers flanking *Lr42*, we identified the *Lr42* region in the hexaploid wheat Chinese Spring (CS) chromosome 1D RefSeq v1.1 [22] and the *Ae. tauschii* reference genome sequence [45] using BLASTN [46]. Gene annotation of the *Lr42* genomic interval on CS chromosome 1D of wheat [22] and *Ae. tauschii* chromosome 1 [45] was obtained to identify candidate disease-resistance genes in the region.

## 5. Conclusions

In the current study, we fine-mapped the leaf rust resistance gene *Lr42* to a 3.7 cM region from the previously reported 17 cM region on chromosome 1DS in wheat. We further physically mapped the *Lr42* region on Chinese Spring RefSeq v1.1 and *Ae. tauschii* reference genomes and identified genes with defense-related functions as possible candidates for *Lr42*. The KASP markers flanking *Lr42* developed in the current study will facilitate marker-assisted selection of *Lr42* and pyramiding of the gene with other adult plant resistance genes for effective management of leaf rust in wheat.

**Author Contributions:** S.K.S. and B.S.G. conceptualized the experiment and designed the methodology; S.K.S., H.S.G., and J.S.S. performed data curation and formal analysis; C.L., D.W., H.S.G., J.S.S., W.L., and S.K.S. performed the investigation; G.B., D.W., and S.K.S. developed the mapping populations; H.S.G. and S.K.S. performed the software analysis; S.K.S. supervised the experiment; H.S.G. and S.K.S. wrote the original manuscript; B.S.G., G.B., D.W., W.L., and J.S.S. contributed to the interpretation of results and revision of the manuscript. All authors approved the manuscript.

**Acknowledgments:** The authors would like to thank the South Dakota Agriculture Experimental Station (Brookings, SD, USA) and Kansas State University Agriculture Experimental Station (Manhattan KS) for providing the resources to conduct the experiments.

## Abbreviations

| | |
|---|---|
| MAS | Marker-assisted selection |
| KASP | Kompetitive allele-specific PCR |
| BAC | Bacterial artificial chromosome |
| EST | Expressed sequence tag |
| SSR | Simple sequence repeat |

## References


1. Shiferaw, B.; Smale, M.; Braun, H.J.; Duveiller, E.; Reynolds, M.; Muricho, G. Crops that feed the world 10. Past successes and future challenges to the role played by wheat in global food security. *Food Sec.* **2013**, *5*, 291–317. [CrossRef]
2. Bolton, M.D.; Kolmer, J.A.; Garvin, D.F. Wheat Leaf Rust Caused by *Puccinia Triticina*. *Mol. Plant Pathol.* **2008**, *9*, 563–575. [CrossRef] [PubMed]
3. Appel, J.A.; Dewolf, E.; Todd, T.; Bockus, W.W. *Preliminary 2015 Kansas Wheat Disease Loss Estimates*; Kansas Cooperative Plant Disease Survey Report; Kansas Department of Agriculture: Manhattan, KS, USA, 2015. Available online: http://agriculture.ks.gov/docs/default-source/PP-Disease-Reports-2014/2014-ks-wheat-disease-loss-estimates.pdf (accessed on 18 March 2019).
4. Marasas, C.N.; Melinda, S.; Singh, R.P. *The Economic Impact in Developing Countries of Leaf Rust Resistance Breeding in CIMMYT-Related Spring Bread Wheat*; CIMMYT (International Maize and Wheat Improvement Center): Texcoco, Mexico, 2004.
5. Chester, K.S. *The Nature and Prevention of the Cereal Rusts as Exemplified in the Leaf Rust of Wheat*; Chronica Botanica Company: Walthan, MA, USA, 1946; p. 269.
6. Kolmer, J.A.; Hughes, M.E. Physiologic Specialization of Puccinia Triticina on Wheat in the United States in 2012. *Plant Dis.* **2014**, *98*, 1145–1150. [CrossRef]
7. Kolmer, J.A.; Chen, X.; Jin, Y. Diseases Which Challenge Global Wheat Production the Wheat Rusts. In *Wheat: Science and Trade*; Carver, B.F., Ed.; Wiley-Blackwell: Ames, IA, USA, 2008.
8. Qureshi, N.; Bariana, H.; Kumran, V.V.; Muruga, S.; Forrest, K.L.; Hayden, M.J.; Bansal, U. A New Leaf Rust Resistance Gene Lr79 Mapped in Chromosome 3BL from the Durum Wheat Landrace Aus26582. *Theor. Appl. Genet.* **2018**, *131*, 1091–1098. [CrossRef] [PubMed]
9. Gill, B.S.; Raupp, W.J.; Sharma, H.C.; Browder, L.E.; Hatchett, J.H.; Harvey, T.L.; Moseman, J.G.; Waines, J.G. Resistance in Aegilops Squarrosa to Wheat Leaf Rust, Wheat Powdery Mildew, Greenbug, and Hessian Fly. *Plant Dis.* **1986**, *70*, 553–556. [CrossRef]
10. Dhaliwal, H.S.; Singh, H.; Gupta, S.; Bagga, P.S.; Gill, K.S. Evaluation of Aegilops and wild Triticum species for resistance to leaf rust (Puccinia recondita f. Sp. Tritici) of Wheat. *Int. J. Trop. Agric.* **1991**, *9*, 118–122.
11. Okamoto, Y.; Nguyen, A.T.; Yoshioka, M.; Iehisa, J.C.M.; Takumi, S. Identification of Quantitative Trait Loci Controlling Grain Size and Shape in the D Genome of Synthetic Hexaploid Wheat Lines. *Breed. Sci.* **2013**, *63*, 423–429. [CrossRef]

12. Gill, B.S.; Friebe, B.; Raupp, W.J.; Wilson, D.L.; Cox, T.S.; Sears, R.G.; Brown-Guedira, G.L.; Fritz, A.K. Wheat genetics resource center: The first 25 years. *Adv. Agron.* **2006**, *1*, 73–136. [CrossRef]
13. Bacon, R.K.; Kelly, J.T.; Milus, E.A.; Parsons, C.E. Registration of Soft Wheat Germplasm AR93005 Resistant to Leaf Rust. *Crop Sci.* **2006**, *46*, 1398. [CrossRef]
14. Singh, R.P.; Huerta-Espino, J.; Sharma, R.; Joshi, A.K.; Trethowan, R. High Yielding Spring Bread Wheat Germplasm for Global Irrigated and Rainfed Production Systems. *Euphytica* **2007**, *157*, 351–363. [CrossRef]
15. Martin, J.N.; Carver, B.F.; Hunger, R.M.; Cox, T.S. Contributions of Leaf Rust Resistance and Awns to Agronomic and Grain Quality Performance in Winter Wheat. *Crop Sci.* **2003**, *43*, 1712. [CrossRef]
16. Cox, T.S.; Raupp, W.J.; Gill, B.S. Leaf Rust-Resistance Genes *Lr41*, *Lr42*, and *Lr43* Transferred from Triticum Tauschii to Common Wheat. *Crop Sci.* **1994**, *34*, 339. [CrossRef]
17. Czembor, P.C.; Radecka-Janusik, M.; Pietrusińska, A.; Czembor, H.J. *Proceedings of the 11th International Wheat Genetics Symposium, Brisbane, QLD, Australia, 24–29 August 2008*; Appel, R., Eastwood, R., Lagudah, E., Langridge, P., Mackay, M., McIntyre, L., Sharp, P., Eds.; Sydney University Press: Sidney, Australia, 2008; pp. 739–740.
18. Sun, X.; Bai, G.; Carver, B.F.; Bowden, R. Molecular mapping of wheat leaf rust resistance gene *Lr42*. *Crop Sci.* **2010**, *50*, 59–66. [CrossRef]
19. Liu, Z.; Bowden, R.L.; Bai, G. Molecular markers for leaf rust resistance gene *Lr42* in Wheat. *Crop Sci.* **2013**, *53*, 1566–1570. [CrossRef]
20. International Barley Genome Sequencing Consortium; Mayer, K.F.; Waugh, R.; Brown, J.W.; Schulman, A.; Langridge, P.; Platzer, M.; Fincher, G.B.; Muehlbauer, G.J.; Sato, K.; et al. A physical, genetic and functional sequence assembly of the barley genome. *Nature* **2012**, *491*, 711–716. [CrossRef] [PubMed]
21. Luo, M.-C.; Gu, Y.Q.; You, F.M.; Deal, K.R.; Ma, Y.; Hu, Y.; Huo, N.; Wang, Y.; Wang, J.; Chen, S.; et al. A 4-gigabase physical map unlocks the structure and evolution of the complex genome of *Aegilops tauschii*, the wheat D-genome progenitor. *Proc. Natl. Acad. Sci. USA* **2013**, *110*, 7940–7945. [CrossRef]
22. International Wheat Genome Sequencing Consortium (IWGSC); IWGSC RefSeq principal investigators; Appels, R.; Eversole, K.; Feuillet, C.; Keller, B.; Rogers, J.; Stein, N.; IWGSC whole-genome assembly principal investigators; Pozniak, C.J.; et al. Shifting the Limits in Wheat Research and Breeding Using a Fully Annotated Reference Genome. *Science* **2018**, *361*, eaar7191. [CrossRef]
23. El-Gebali, S.; Mistry, J.; Bateman, A.; Eddy, S.R.; Luciani, A.; Potter, S.C.; Qureshi, M.; Richardson, L.J.; Salazar, G.A.; Smart, A.; et al. The Pfam Protein Families Database in 2019. *Nucleic Acids Res.* **2019**, *47*, D427–D432. [CrossRef] [PubMed]
24. Kolmer, J.; Phytopathology, P.D. Gene Expression in the Triticum Aestivum-Puccinia Recondita f. Sp. tritici gene-for-gene system. *Phytopathology* **1994**, *84*, 437–440. [CrossRef]
25. Feuillet, C.; Travella, S.; Stein, N.; Albar, L.; Nublat, A.; Keller, B. Map-Based Isolation of the Leaf Rust Disease Resistance Gene Lr10 from the Hexaploid Wheat (*Triticum aestivum*, L.) Genome. *Proc. Natl. Acad. Sci. USA* **2003**, *100*, 15253–15258. [CrossRef]
26. Huang, L.; Brooks, S.A.; Li, W.; Fellers, J.P.; Trick, H.N.; Gill, B.S. Map-Based cloning of leaf rust resistance gene lr21 from the large and polyploid genome of bread wheat. *Genetics* **2003**, *164*, 655–664. [PubMed]
27. Cloutier, S.; McCallum, B.D.; Loutre, C.; Banks, T.W.; Wicker, T.; Feuillet, C.; Keller, B.; Jordan, M.C. Leaf rust resistance gene Lr1, isolated from bread wheat (*Triticum aestivum*, L.) is a member of the large psr567 gene family. *Plant Mol. Biol.* **2007**, *65*, 93–106. [CrossRef] [PubMed]
28. Thind, A.K.; Wicker, T.; Šimková, H.; Fossati, D.; Moullet, O.; Brabant, C.; Vrána, J.; Doležel, J.; Krattinger, S.G. Rapid Cloning of Genes in Hexaploid Wheat Using Cultivar-Specific Long-Range Chromosome Assembly. *Nat. Biotechnol.* **2017**, *35*, 793–796. [CrossRef] [PubMed]
29. Periyannan, S.; Moore, J.; Ayliffe, M.; Bansal, U.; Wang, X.; Huang, L.; Deal, K.; Luo, M.; Kong, X.; Bariana, H.; et al. the gene Sr33, an ortholog of barley mla genes, encodes resistance to wheat stem rust race Ug99. *Science* **2013**, *341*, 786–788. [CrossRef]
30. Saintenac, C.; Zhang, W.; Salcedo, A.; Rouse, M.N.; Trick, H.N.; Akhunov, E.; Dubcovsky, J. Identification of wheat gene Sr35 that confers resistance to Ug99 stem rust race group. *Science* **2013**, *341*, 783–786. [CrossRef]
31. Steuernagel, B.; Periyannan, S.K.; Hernández-Pinzón, I.; Witek, K.; Rouse, M.N.; Yu, G.; Hatta, A.; Ayliffe, M.; Bariana, H.; Jones, J.D.G.; et al. Rapid cloning of disease-resistance genes in plants using mutagenesis and sequence capture. *Nat. Biotechnol.* **2016**, *34*, 652–655. [CrossRef]

32. Liu, W.; Frick, M.; Huel, R.; Nykiforuk, C.L.; Wang, X.; Gaudet, D.A.; Eudes, F.; Conner, R.L.; Kuzyk, A.; Chen, Q.; et al. The stripe rust resistance gene Yr10 encodes an evolutionary-conserved and unique CC–NBS–LRR sequence in wheat. *Mol. Plant* **2014**, *7*, 1740–1755. [CrossRef]
33. Shi, G.; Zhang, Z.; Friesen, T.L.; Raats, D.; Fahima, T.; Brueggeman, R.S.; Lu, S.; Trick, H.N.; Liu, Z.; Chao, W.; et al. The hijacking of a receptor kinase–driven pathway by a wheat fungal pathogen leads to disease. *Sci. Adv.* **2016**, *2*, e1600822. [CrossRef] [PubMed]
34. Saintenac, C.; Lee, W.-S.; Cambon, F.; Rudd, J.J.; King, R.C.; Marande, W.; Powers, S.J.; Bergès, H.; Phillips, A.L.; Uauy, C.; et al. Wheat receptor-kinase-like protein stb6 controls gene-for-gene resistance to fungal pathogen zymoseptoria tritici. *Nat. Genet.* **2018**, *50*, 368–374. [CrossRef] [PubMed]
35. Sánchez-Martín, J.; Steuernagel, B.; Ghosh, S.; Herren, G.; Hurni, S.; Adamski, N.; Vrána, J.; Kubaláková, M.; Krattinger, S.G.; Wicker, T.; et al. Rapid gene isolation in barley and wheat by mutant chromosome sequencing. *Genome Biol.* **2016**, *17*, 221. [CrossRef]
36. Huang, X.-Q.; Röder, M.S. High-density genetic and physical bin mapping of wheat chromosome 1D reveals that the powdery mildew resistance gene Pm24 is located in a highly recombinogenic region. *Genetica* **2011**, *139*, 1179–1187. [CrossRef]
37. Gill, K.S.; Gill, B.S.; Endo, T.R.; Taylor, T. Identification and high-density mapping of gene-rich regions in chromosome group 1 of wheat. *Genetics* **1996**, *144*, 1001–1012.
38. Stakman, E.C.; Stewart, D.M.; Loegering, W.Q. Identification of physiologic races of Puccinia graminis var. tritici. *Can. J. Plant Pathol.* **1981**, *3*, 33–39.
39. Roelfs, A.P.; Bushnell, W.R.; San, O.; New, D.; London, Y.; Montreal, T.; Tokyo, S. *The Cereal Rusts: Diseases, Distribution, Epidemiology, and Control*; Elsevier: Amsterdam, The Netherlands, 1985.
40. Saghai-Maroof, M.A.; Soliman, K.M.; Jorgensen, R.A.; Allard, R.W. Ribosomal DNA Spacer-Length Polymorphisms in Barley: Mendelian Inheritance, Chromosomal Location, and Population Dynamics. *Proc. Natl. Acad. Sci. USA* **1984**, *81*, 8014–8018. [CrossRef]
41. The R Development Core Team. *R: A Language and Environment for Statistical Computing*; R Foundation for Statistical Computing: Vienna, Austria, 2017. Available online: https://www.R-project.org (accessed on 18 March 2019).
42. de Givry, S.; Bouchez, M.; Chabrier, P.; Milan, D.; Schiex, T. CARHTA GENE: Multipopulation integrated genetic and radiation hybrid mapping. *Bioinformatics* **2005**, *21*, 1703–1704. [CrossRef] [PubMed]
43. Kosambi, D.D. The estimation of map distances from recombination values. *Ann. Eugen.* **1943**, *12*, 172–175. [CrossRef]
44. Voorrips, R.E. MapChart: Software for the graphical presentation of linkage maps and QTLs. *J. Hered.* **2002**, *93*, 77–78. [CrossRef]
45. Luo, M.-C.; Gu, Y.Q.; Puiu, D.; Wang, H.; Twardziok, S.O.; Deal, K.R.; Huo, N.; Zhu, T.; Wang, L.; Wang, Y.; et al. Genome sequence of the progenitor of the wheat d genome aegilops tauschii. *Nature* **2017**, *551*, 498. [CrossRef]
46. Altschul, S.F.; Gish, W.; Miller, W.; Myers, E.W.; Lipman, D.J. Basic local alignment search tool. *J. Mol. Biol.* **1990**, *215*, 403–410. [CrossRef]

# Comparative Proteomic Analysis of Wheat Carrying *Pm40* Response to *Blumeria graminis* f. sp. *tritici* using Two-Dimensional Electrophoresis

**Yinping Liang [1,2], Ye Xia [2], Xiaoli Chang [1,3], Guoshu Gong [1], Jizhi Yang [1], Yuting Hu [1], Madison Cahill [2], Liya Luo [1], Tao Li [1], Lu He [1] and Min Zhang [1,*]**

1 College of Agronomy & Key Laboratory for Major Crop Diseases, Sichuan Agricultural University, Chengdu 611130, China; liangyinping3@163.com (Y.L.); xl_changkit@126.com (X.C.); guoshugong@126.com (G.G.); yjzgxy@126.com (J.Y.); yutinghu116sicau@163.com (Y.H.); scmlsjc@163.com (L.L.); 13187874590@126.com (T.L.); hl415676745@163.com (L.H.)

2 Department of Plant Pathology, The Ohio State University, Columbus, OH 43202, USA; xia.374@osu.edu (Y.X.); madisontaylor313@hotmail.com (M.C.)

3 Institute of Ecological Agriculture, Sichuan Agricultural University, Chengdu 611130, China

* Correspondence: yalanmin@126.com.

**Abstract:** Wheat powdery mildew caused by *Blumeria graminis* f. sp. *tritici* (*Bgt*) is considered a major wheat leaf disease in the main wheat producing regions of the world. Although many resistant wheat cultivars to this disease have been developed, little is known about their resistance mechanisms. *Pm40* is a broad, effective resistance gene against powdery mildew in wheat line *L699*. The aim of this study was to investigate the resistance proteins after *Bgt* inoculation in wheat lines *L699*, *Neimai836*, and *Chuannong26*. *Neimai836* with *Pm21* was used as the resistant control, and *Chuannong26* without any effective *Pm* genes was the susceptible control. Proteins were extracted from wheat leaves sampled 2, 4, 8, 12, and 24 h after *Bgt* inoculation, separated by two-dimensional electrophoresis, and stained with Coomassie brilliant blue G-250. The results showed that different proteins were upregulated and downregulated in three wheat cultivars at different time points. For the wheat cultivar *L699*, a total of 62 proteins were upregulated and 71 proteins were downregulated after *Bgt* inoculation. Among these, 46 upregulated proteins were identified by mass spectrometry analysis using the NCBI nr database of *Triticum*. The identified proteins were predicted to be associated with the defense response, photosynthesis, signal transduction, carbohydrate metabolism, energy pathway, protein turnover, and cell structure functions. It is inferred that the proteins are not only involved in defense response, but also other physiological and cellular processes to confer wheat resistance against *Bgt*. Therefore, the resistance products potentially mediate the immune response and coordinate other physiological and cellular processes during the resistance response to *Bgt*. The lipoxygenase, glucan exohydrolase, glucose adenylyltransferasesmall, phosphoribulokinase, and phosphoglucomutase are first reported to be involved in the interactions of wheat-*Bgt* at early stage. The further study of these proteins will deepen our understanding of their detailed functions and potentially develop more efficient disease control strategies.

**Keywords:** *Blumeria graminis* f. sp. *tritici*; protein two-dimensional electrophoresis; mass spectrometry; *Pm40*

---

## 1. Introduction

Wheat powdery mildew caused by the obligate fungus *Blumeria graminis* f. sp. *tritici (Bgt)* is a major wheat leaf disease in the main wheat producing regions of world, leading to significant yield loss each year [1]. Agricultural and chemical methods are widely used to combat the disease. Wheat powdery mildew is an airborne disease and the chemical control methods for *Bgt* seriously pollute environments. Therefore, planting resistant cultivars is the most economical, most effective, and safest method to control wheat powdery mildew [2]. To date, approximately 90 formally designated powdery mildew resistance genes (*Pm* genes) are catalogued at 58 loci (*Pm1–Pm62*, *Pm18* = *Pm1c*, *Pm22* = *Pm1e*, *Pm23* = *Pm4c*, *Pm31* = *Pm21*) with the loci of *Pm1*, *Pm3*, *Pm4*, *Pm5*, and *Pm24* having 5, 17, 4, 5, and 2 alleles, respectively [3–13]. However, resistance genes often become ineffective due to the enrichment and variation of virulent races, particularly when a single resistance gene is used in large areas for long periods of time. Therefore, it is very important to identify effective resistance genes and develop multiple resistance cultivars in wheat breeding [14].

The resistant mechanisms of wheat cultivars against *Bgt* are not well-known. Bread wheat (*Triticum aestivum* L.) is a hexaploid ($2n = 42$; AABBDD) with a 17-gigabase genome that contains 124 201 genes [15]. Due to this complexity, cloning wheat genes by the standard map-based cloning strategy remains challenging. Although many powdery mildew resistance genes were identified and mapped in wheat, to date, only five genes, *Pm2*, *Pm3*, *Pm8*, *Pm21*, and *Pm60* have been cloned [9,16–20]. The resistance gene *Pm40* was transferred from *Elytrigia intermedium* into wheat line *GRY19* and mapped on chromosome arm 7BS [21]. *Pm40* is flanked by *Xwmc335* and *BF291338* at genetic distances of 0.58 cM and 0.26 cM, respectively, in deletion bin C-7BS-1-0.27 [22]. Wheat line *L699*, which is the high generation of wheat line *GRY19*, carries the resistance gene *Pm40* and confers resistance to all available isolates of *Bgt* in China [23].

Proteins are not only the final executant of life functions but also the key to understanding physiological, pathological, and pharmacological functions of plants [24]. Therefore, it is difficult to thoroughly explain the powdery mildew resistance mechanism using genomic and transcriptomic methods. Proteomic approaches have been extensively applied in plant pathology research [25,26]. However, only a few studies examined the changes of plant proteome in response to *Bgt*. Wheat cultivars Bainong/W2132 (*Pm21*), JD8/JD8-*Pm30*, N8038 (*PmG25*), N9134 (*PmAS846*), and Xinong979 (without effective *Pm* genes) were used to analyze the effect of *Bgt* on wheat protein expression. These studies showed that most of the upregulated proteins were involved in stress responses and primary metabolic pathways [24,27–30]. However, there is no such study investigating the differences of protein expressions in the period before *Bgt* haustoria formation, which is very critical for us to better understand the interactions of this pathogen with different wheat cultivars at early stage. To understand the molecular recognition of wheat-*Bgt* during the contact period and penetration period, we identified a set of proteins in wheat inoculated with *Bgt* using two-dimensional electrophoresis (2-DE). The possible roles of the identified proteins in the defense response at early interaction stage were discussed according to their functional implications. This study deepens and extends our knowledge on the interactions of wheat with *Bgt* and allows us to further understand the wheat immune systems against *Bgt*. All these will facilitate the development of more efficient strategies to control this devastating pathogen for enhancing wheat production, which can also potentially provide insights for the control of different plant diseases caused by diverse powdery mildew pathogens.

## 2. Results

### 2.1. Phenotypic Differences of Leaves Affected by Bgt

The bioassay revealed differences in resistance to *Bgt* among *L699*, *Chuannong26* and *Neimai836* (Figure 1). The susceptible cultivar *Chuannong26* was covered by a high number of sori and had the white powdery appearance due to the abundant conidia and conidiophores production on the leaf surface after 6 days of *Bgt* infection, with the infection type (IT) = 9 (Figure 1a). Meanwhile, the resistant wheat lines *L699* and *Neimai836* were observed to be healthy without any epidermal cell necrosis, chlorotic patches, and powdery appearance on the leaf surface, with the IT = 0 (*L699*: Figure 1b, *Neimai836*: Figure 1c) [23].

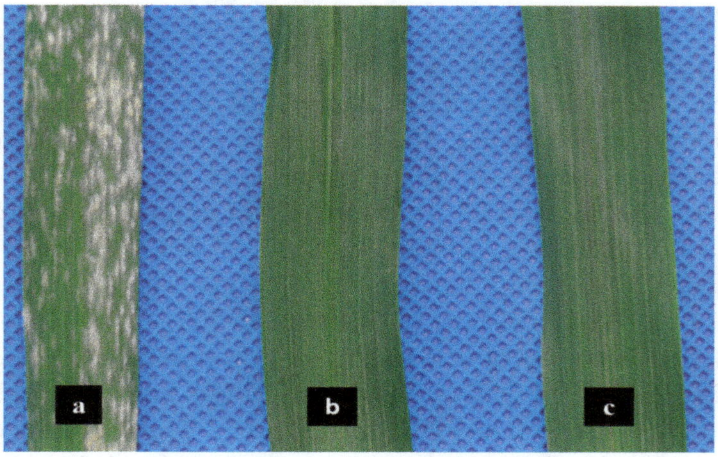


**Figure 1.** Different responses of wheat leaves to *Blumeria graminis* f. sp. *tritici (Bgt)* infection after six days. (**a**) Susceptible wheat cultivar *Chuannong26*. (**b**) Resistant wheat cultivar *L699*. (**c**) Resistant wheat cultivar *Neimai836*.

## 2.2. Estimation of Wheat-Bgt Interactions

To examine the development of *Bgt* and immune responses of wheat at 2, 4, 8, 12, and 24 h post-inoculation (hpi), the cytological observations of wheat samples were carried out. *Bgt* conidia successively formed primary germ tubes, appressorium germ tubes, appressoria, penetration pegs, and haustoria at 2, 4, 8, 12, and 24 hpi in susceptible wheat cultivar *Chuannong26*. However, in resistant wheat cultivars *L699* and *Neimai836*, only a small number of conidia successfully penetrated the epidermal cells at 24 hpi, and the hypersensitive reaction (HR) and formation of papilla (PA) effectively suppressed the development of haustoria and hyphae [31]. In addition, the appressoria of some conidia sprouted another lobe and stopped growing because of the lack of nutrition at 24 hpi (Figure 2).

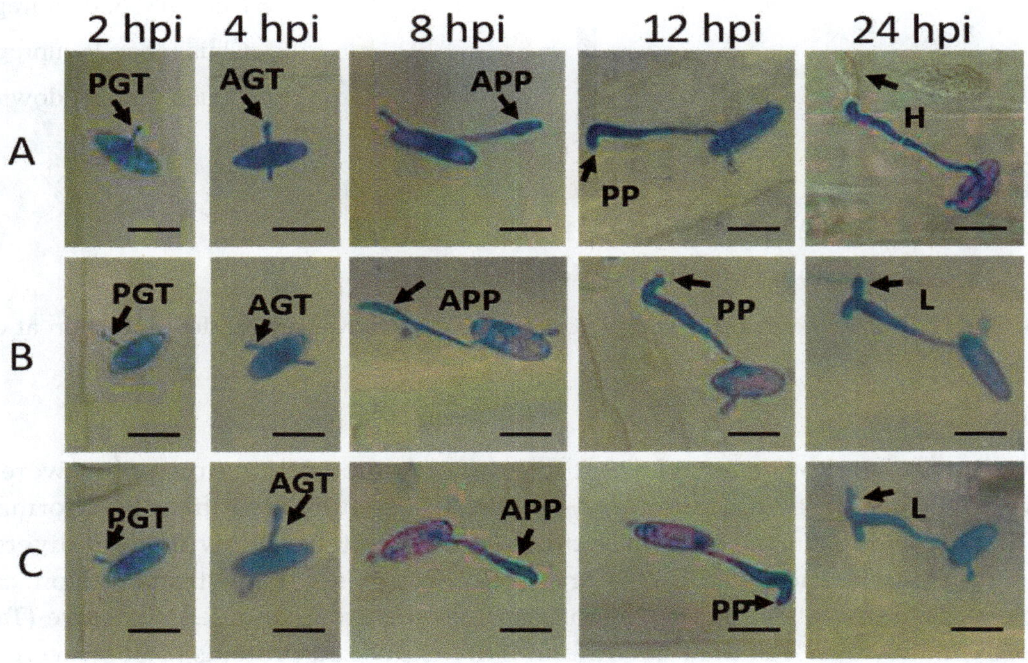


**Figure 2.** Microscopic observations of wheat-*Bgt* interactions on the leaf surface. The development of *Bgt* at 2, 4, 8, 12, and 24 hpi in wheat cultivars *Chuannong26* (**A**), *L699* (**B**), and *Neimai836* (**C**). PGT: primary germ tube, AGT: appressorium germ tube, APP: appressorium, PP: penetration peg, H: haustorium and L: lobe. Scale bar: 20 μm.

### *2.3. Detection of Differential Proteins by 2-DE*

Approximately 500 protein spots were detected in all gels in this study. Using a twofold change cutoff, we found wheat cultivars *L699*, *Neimai836*, and *Chuannong26* all had upregulated and downregulated proteins affected by *Bgt*. The numbers of upregulated proteins were seven (spot 01, 02, 24, 27, 45, 46, 50), five (spot 12, 17, 36, 37, 44), 18 (spot 11, 13, 29, 30, 32–35, 38–46, 55), 26 (spot 02–06, 08–12, 14–16, 18–28, 48, 56), and 18 (spot 07, 31, 47–62). The numbers of downregulated proteins were four (spot 68–71), 12 (spot 45, 48, 58–67), 12 (spot 09, 12, 14, 49–57), 10 (spot 39–48), and 38 (spot 01–38) in wheat cultivar *L699* at 2, 4, 8, 12, 24 hpi, respectively (Figures 3 and 4). The numbers of upregulated proteins were 10 (spot 14, 19, 20, 22, 58–63), 13 (spot 32, 46–57), 15 (spot 31–45), 23 (spot 1–23), and nine (spot 19, 21, 24–30). The numbers of downregulated proteins were 11 (spot 01, 36, 43, 52, 69–75), 11 (spot 23, 28, 31, 50, 62–68), 11 (spot 50, 52–61), 35 (spot 1–35), and 16 (spot 36–51) in wheat cultivar *Neimai836* at 2, 4, 8, 12, 24 hpi, respectively (Figures 3 and 5). The numbers of upregulated proteins were three (spot 71, 109–110), nine (spot 33, 101–108), seven (spot 33, 34, 66, 97–100), 25 (spot 23, 34, 59, 75–96), and 74 (spot 1–74). The numbers of downregulated proteins were four (spot 42–45), 13 (spot 04, 25, 31–41), nine (spot 16, 19, 24–30), 7 (spot 01, 04, 19–23), and 18 (spot 1–18) in wheat cultivar *Chuannong26* at 2, 4, 8, 12, 24 hpi, respectively (Figures 3 and 6).

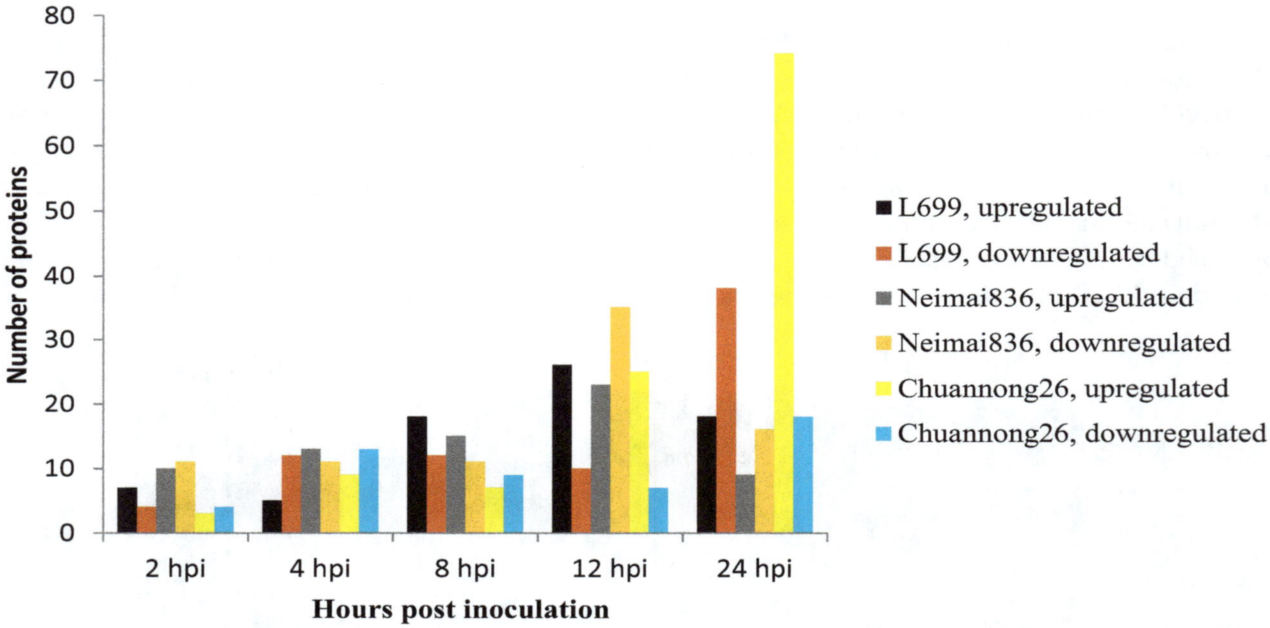


**Figure 3.** Number of proteins differentially expressed after powdery mildew infection at different time points.

### *2.4. Protein Identification*

Sixty-two upregulated proteins in *L699* at five different inoculation time points were eluted from representative 2-DE gels for identification, and 46 were successfully identified. Bioinformatics analysis of the identified proteins revealed that these proteins were putatively involved in diverse biological processes including stress and disease resistance, photosynthesis, signal transduction, carbohydrate metabolism, energy pathway, gene expression, protein turnover, and cell structure (Table 1). Five proteins (approximately 16.5% of the total differentially expressed proteins (DEPs)) were unnamed or hypothetical proteins. The largest category of these upregulated proteins was protein turnover (28%, thirteen), followed by carbohydrate metabolism (22%, ten), stress and disease resistance (13%, six), photosynthesis (13%, six), energy pathway (6.5%, three), signal transduction (2%, one), gene expression (2%, one), and cell structure (2%, one).

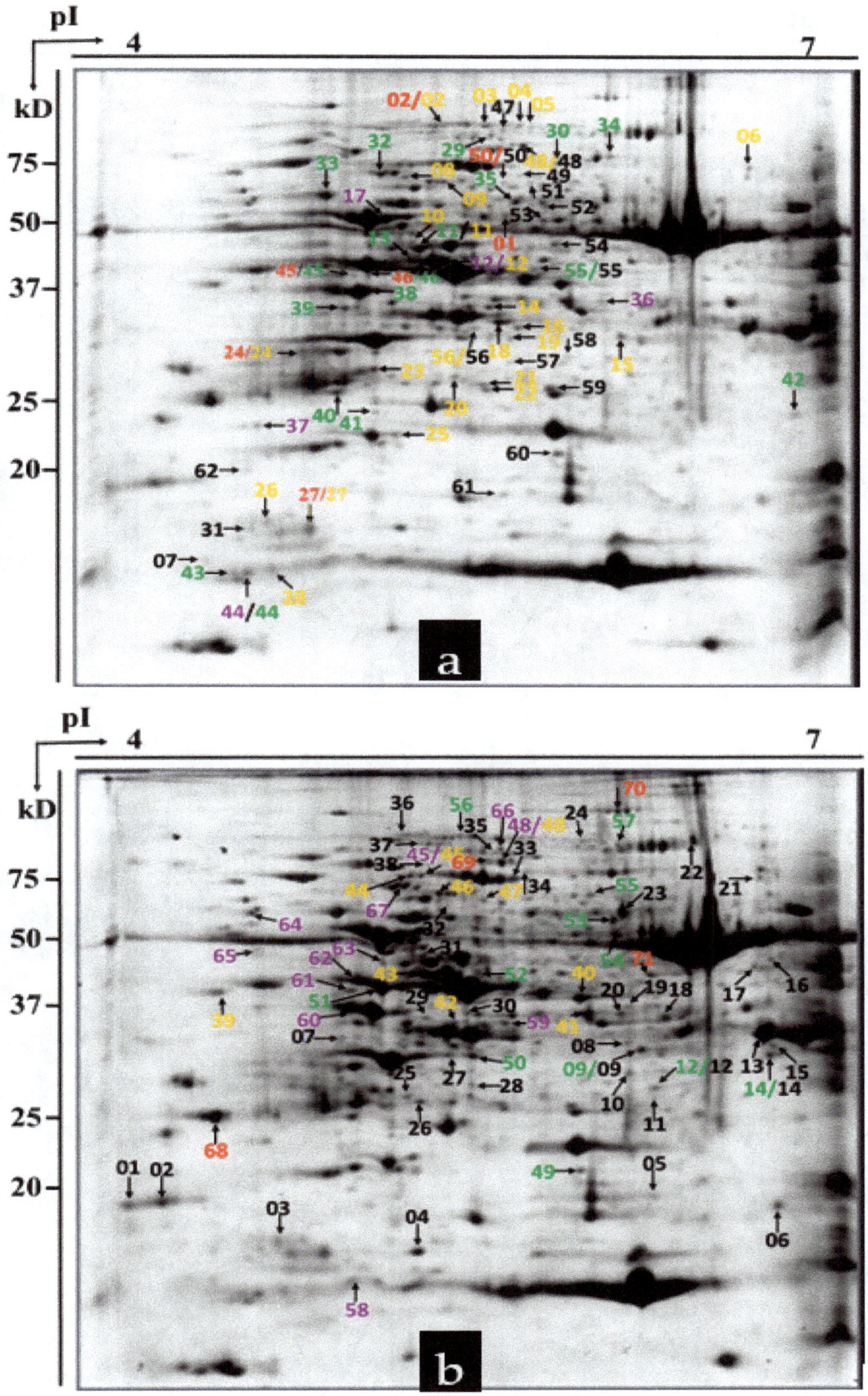


**Figure 4.** Differences in protein expression between resistant wheat cultivar *L699* wheat with and without powdery mildew infection. (**a**) Upregulated proteins are labeled in the representative 2-DE gel of *Bgt*-inoculated *L699* wheat at 24 hpi. (**b**) Downregulated proteins are labeled in the representative 2-DE gel of mock-inoculated *L699* wheat at 24 hpi. Red, 2 hpi; purple, 4 hpi; green, 8 hpi; yellow, 12 hpi; black, 24 hpi.

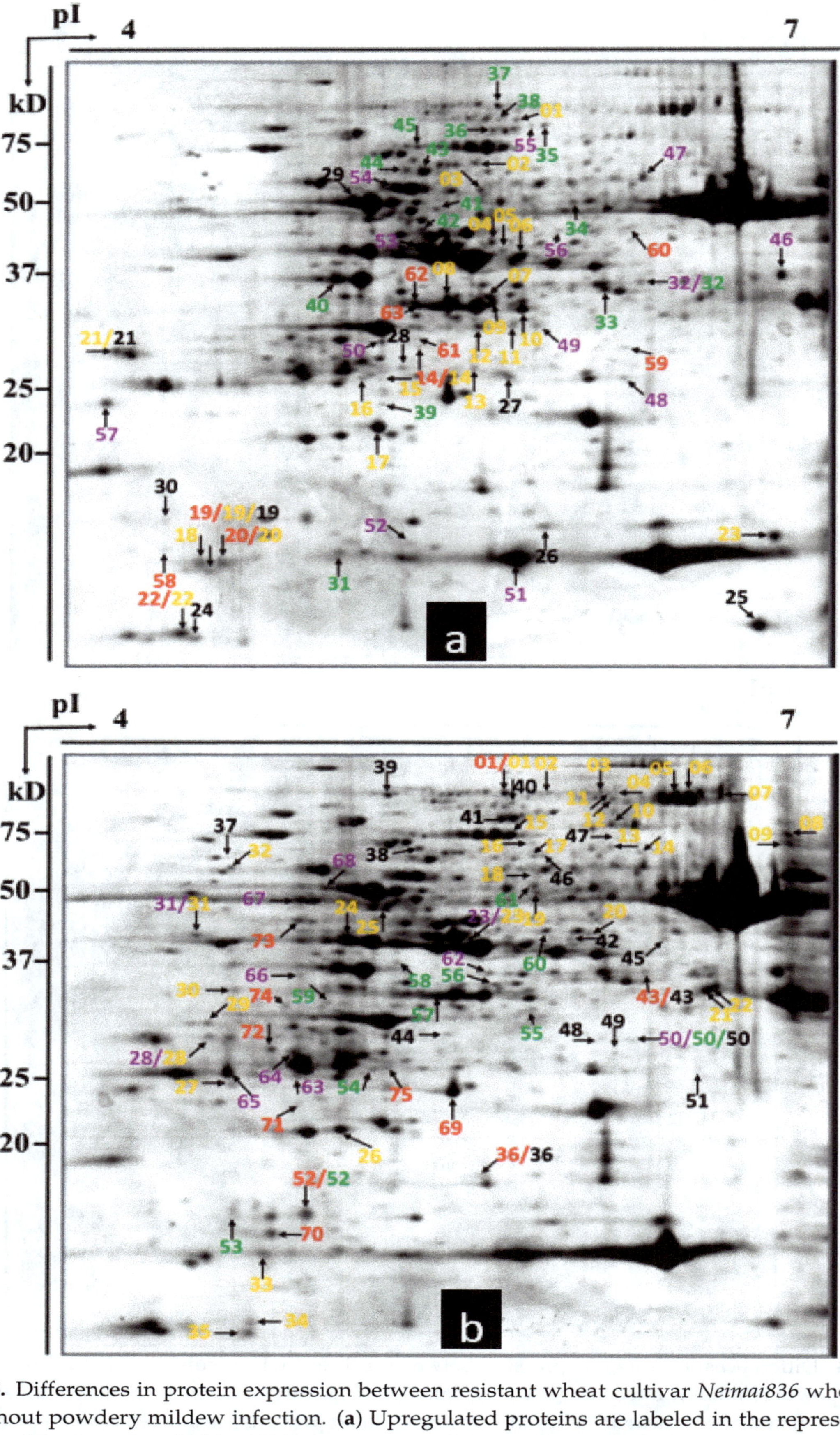


**Figure 5.** Differences in protein expression between resistant wheat cultivar *Neimai836* wheat with and without powdery mildew infection. (**a**) Upregulated proteins are labeled in the representative 2-DE gel of *Bgt*-inoculated *Neimai836* wheat at 12 hpi. (**b**) Downregulated proteins are labeled in the representative 2-DE gel of mock-inoculated *Neimai836* wheat at 12 hpi. Red, 2 hpi; purple, 4 hpi; green, 8 hpi; yellow, 12 hpi; black, 24 hpi.

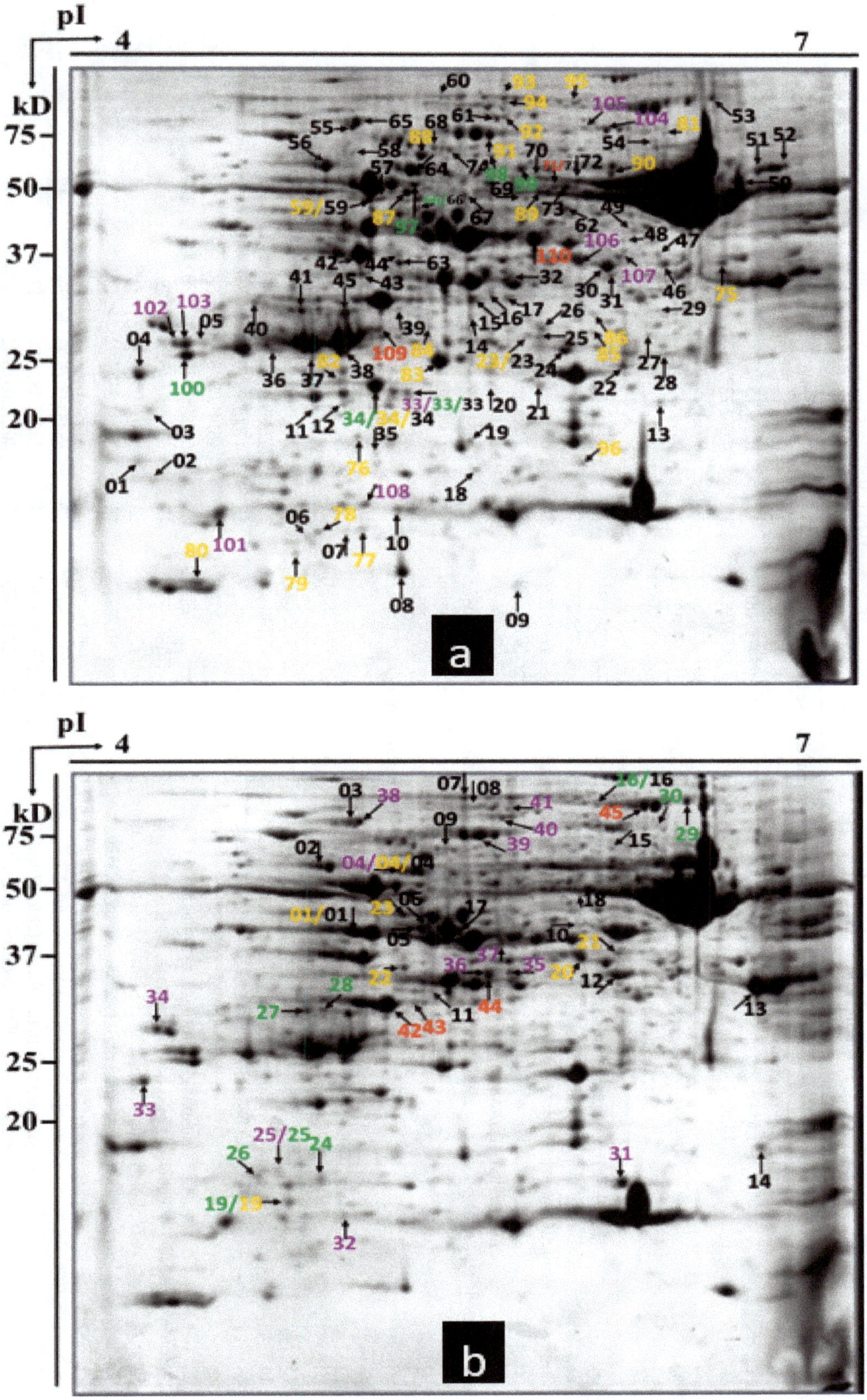


**Figure 6.** Differences in protein expression between susceptible wheat cultivar *Chuannong26* wheat with and without powdery mildew infection. (**a**) Upregulated proteins are labeled in the representative 2-DE gel of *Bgt*-inoculated *Chuannong26* wheat at 24 hpi. (**b**) Downregulated proteins are labeled in the representative 2-DE gel of mock-inoculated *Chuangnong26* wheat at 24 hpi. Red, 2 hpi; purple, 4 hpi; green, 8 hpi; yellow, 12 hpi; black, 24 hpi.

**Table 1.** Identification of differentially upregulated proteins in wheat resistance cultivar *L699* by MALDI-TOF-MS.

| Spot | Protein Name | Accession | Matched Peptides | MW/PI | Score | Time (h) |
|---|---|---|---|---|---|---|
| | Proteins involved in disease defense response | | | | | |
| 3 | Lipoxygenase 2.1, chloroplastic | gi∣473948122 | 21 | 105625.33/5.70 | 185 | 12 |
| 4 | Lipoxygenase 2.1, chloroplastic | gi∣473948122 | 41 | 105625.33/5.70 | 591 | 12 |
| 5 | Lipoxygenase 2.1, chloroplastic | gi∣473948122 | 17 | 105625.33/5.70 | 196 | 12 |
| 29 | Lipoxygenase 1 | gi∣474399175 | 16 | 96333.65/5.91 | 299 | 8 |
| 32 | Heat shock cognate 70 kDa protein 1 | gi∣474012573 | 37 | 71123.58/5.06 | 641 | 8 |
| 61 | Germin-like protein 8-14 | gi∣473963025 | 4 | 21939.25/5.36 | 174 | 24 |
| | Photosynthesis-related proteins | | | | | |
| 11 | Ribulose-1,5-bisphosphate carboxylase activase, partial | gi∣37783283 | 10 | 22336.08/4.98 | 309 | 8, 12 |
| 28 | Ribulose bisphosphate carboxylase small chain, chloroplastic | gi∣473882355 | 14 | 18526.35/8.65 | 208 | 12 |
| 33 | RuBisCO large subunit-binding protein subunit alpha, chloroplastic | gi∣474113969 | 34 | 65380.60/5.17 | 864 | 8 |
| 36 | Ribulose-1,5-bisphosphate carboxylase activase, partial | gi∣37783283 | 10 | 22336.08/4.98 | 330 | 4 |
| 37 | Photosystem II cytochrome b559 alpha subunit (chloroplast) | gi∣699976019 | 6 | 9444.60/4.64 | 196 | 4 |
| 44 | Ribulose bisphosphate carboxylase small chain PWS4.3, chloroplastic | gi∣132087 | 2 | 19417.36/8.99 | 92 | 4, 8 |
| | Proteins involved in Signal transduction | | | | | |
| 24 | 14-3-3 protein | gi∣431822520 | 16 | 29264.88/4.83 | 434 | 12 |
| | Carbohydrate metabolism-related proteins | | | | | |
| 6 | Beta-D-glucan exohydrolase | gi∣20259685 | 14 | 67301.15/6.87 | 74 | 12 |
| 10 | Glucose-1-phosphate adenylyltransferasesmall subunit, chloroplastic/amyloplastic | gi∣474108293 | 23 | 64723.14/7.91 | 266 | 12 |
| 13 | Phosphoglycerate kinase | gi∣3293043 | 16 | 49839.53/6.57 | 580 | 8, 12 |
| 14 | Glycerophosphodiester phosphodiesterase GDE1 | gi∣473847956 | 13 | 52899.68/5.69 | 43 | 12 |
| 18 | Fructose-bisphosphate aldolase, chloroplastic | gi∣473848356 | 15 | 42002.99/5.94 | 358 | 12 |
| 38 | Phosphoribulokinase | gi∣5924030 | 22 | 45141.39/5.72 | 587 | 8 |
| 50 | Phosphoglucomutase, cytoplasmic | gi∣473763033 | 18 | 63499.68/5.14 | 302 | 2, 24 |
| 51 | Phosphoglucomutase, partial | gi∣18076790 | 15 | 62789.15/5.66 | 218 | 24 |
| 54 | 6-phosphogluconate dehydrogenase, decarboxylating | gi∣474379872 | 23 | 81169.95/8.56 | 608 | 24 |
| 55 | Cytosolic 3-phosphoglycerate kinase, partial | gi∣28172911 | 16 | 31334.35/4.98 | 291 | 8, 24 |
| | Proteins involved in energy pathway | | | | | |
| 9 | Vacuolar proton-ATPase subunit A | gi∣90025017 | 37 | 68454.90/5.23 | 583 | 12 |
| 17 | ATP synthase CF1 beta subunit (chloroplast) | gi∣667669997 | 33 | 53857.48/5.06 | 1200 | 4 |
| 39 | Ferredoxin-NADP(H) oxidoreductase | gi∣20302473 | 10 | 40232.03/6.92 | 120 | 8 |
| | Proteins involved in gene expression and DNA remodeling | | | | | |
| 15 | Guanine nucleotide-binding protein subunit beta-like | gi∣473957859 | 6 | 27150.69/6.29 | 211 | 12 |

**Table 1.** *Cont.*

| Spot | Protein Name | Accession | Matched Peptides | MW/PI | Score | Time (h) |
|---|---|---|---|---|---|---|
| | Proteins involved in protein turnover | | | | | |
| 8 | ATP-dependent zinc metalloprotease FTSH 1, chloroplastic | gi\|474350516 | 29 | 54477.49/5.58 | 673 | 12 |
| 27 | 50S Ribosomal protein L12-2, chloroplastic | gi\|475532245 | 10 | 21837.90/5.35 | 452 | 12 |
| 30 | Tyrosine phosphorylation protein A | gi\|548319365 | 25 | 74252.07/6.61 | 434 | 8 |
| 34 | 5-methyltetrahydropteroyltriglutamate-Homocysteine methyltransferase | gi\|473993302 | 14 | 84552.49/5.74 | 423 | 8 |
| 35 | 5, 10-methylene-tetrahydrofolate reductase | gi\|115589742 | 12 | 64875.07/5.86 | 83 | 8 |
| 40 | 20 kDa chaperonin, chloroplastic | gi\|474407512 | 10 | 29710.03/6.76 | 154 | 8 |
| 47 | Putative alanyl-tRNA synthetase, chloroplastic | gi\|474142555 | 12 | 111648.20/5.62 | 198 | 24 |
| 48 | ATP-dependent Clp protease ATP-binding subunit ClpA-like protein CD4B, chloroplastic | gi\|474241774 | 33 | 82735.21/5.16 | 513 | 12, 24 |
| 49 | Lysyl-tRNA synthetase | gi\|474147702 | 8 | 132545.46/6.28 | 92 | 24 |
| 52 | Putative mitochondrial-processing peptidase subunit beta | gi\|474142281 | 30 | 43290.34/5.41 | 486 | 24 |
| 53 | Adenosylhomocysteinase | gi\|474154141 | 8 | 45700.84/6.48 | 56 | 24 |
| 56 | Cysteine synthase | gi\|474315986 | 13 | 35583.27/5.82 | 216 | 12, 24 |
| 60 | Ribosome-recycling factor, chloroplastic | gi\|474043078 | 15 | 24770.60/8.92 | 504 | 24 |
| | Cell structure-related proteins | | | | | |
| 12 | Actin-3 | gi\|474259583 | 18 | 44367.62/5.26 | 376 | 8, 12 |
| | Proteins of unknow function | | | | | |
| 2 | Hypothetical protein TRIUR3_05354 | gi\|473755342 | 27 | 104676.25/5.87 | 368 | 2, 12 |
| 26 | Unnamed protein product | gi\|669029445 | 4 | 18152.74/5.60 | 244 | 12 |
| 58 | Hypothetical protein TRIUR3_21449 | gi\|474384687 | 14 | 32942.26/9.31 | 129 | 24 |
| 59 | Unnamed protein product | gi\|669027704 | 13 | 26818.76/5.57 | 335 | 24 |
| 63 | Unnamed protein product | gi\|669029445 | 5 | 18152.74/5.60 | 269 | 24 |

### 2.5. Validation of Upregulated Proteins by qRT-PCR

To confirm the changes in protein abundance, qRT-PCR was used to analyze the mRNA expression levels of protein-coding genes after inoculation with *Bgt* in wheat cultivars *L699*, *Neimai836*, and *Chuannong26*. Six upregulated proteins were randomly selected and primers for the mRNAs of these proteins were specifically designed (shown in Table 2). The mRNA levels for all the six proteins were significantly increased in at least one sampling time in wheat *L699*. These changes reflected the increases in proteins and the differentially expressed proteins identified by two-dimensional electrophoresis were validated (Figure 7). However, the mRNAs of four proteins, i.e. fructose-bisphosphate aldolase, phosphoglycerate kinase, 5-methyltetrahydropteroyltriglutamate-homocysteine methyltransferase, and the Germin-like protein (GLP) 8-14, were upregulated earlier than proteins in wheat *L699*. The protein level of fructose-bisphosphate aldolase increased at 12 hpi, but the mRNA level did not exhibit a significant increase at any sampling time point in wheat *Neimai836*.

**Table 2.** List of primers used for qRT-PCR amplification.

| Spot | Protein Name | Accession No. | Primer Sequence 5′-3′ |
|---|---|---|---|
| Reference gene | *18S rRNA* | AY049040 | Sense: 5′-GTGACGGGTGACGGAGAATT-3′<br>Antisense: 5′-GACACTAATGCGCCCGGTAT-3′ |
| 9 | Vacuolar proton-ATPase | ABD85016 | Sense: 5′-TATGAACGTGCTGGGAAGGT-3′<br>Antisense: 5′-GGGTTGCAGAGGTAACAGGA-3′ |
| 18 | Fructose-bisphosphate aldolase | EMS47455 | Sense: 5′-TCTTGTCTGGTGGTCAGTCG-3′<br>Antisense: 5′-CGTCTTGAGGCAGGTGTTCT-3′ |
| 13 | Phosphoglycerate kinase | CAA51931 | Sense: 5′-AATGGTGCTGTTTTGCTCCT-3′<br>Antisense: 5′-TGTTCCGAATGCATCGTTTA-3′ |
| 36 | Ribulose-1,5-bisphosphate carboxylase activase | AAP72270 | Sense: 5′-ACGGACCAGTGACCTTTGAG-3′<br>Antisense: 5′-ACCAGTCTTCATCGCATCCT-3′ |
| 34 | 5-methyltetrahydropteroyltriglutamate-homocysteine methyltransferase | EMS51950 | Sense: 5′-TGTGTTCTGGTCCAAGATGG-3′<br>Antisense: 5′-CTCAAACCTCGGTTGGTCAT-3′ |
| 61 | Germin-like protein 8-14 | EMS51159 | Sense: 5′-TGCAGATCACCGACTACGC-3′<br>Antisense: 5′-CACGGACTTGAGCTTCTTGAC-3′ |

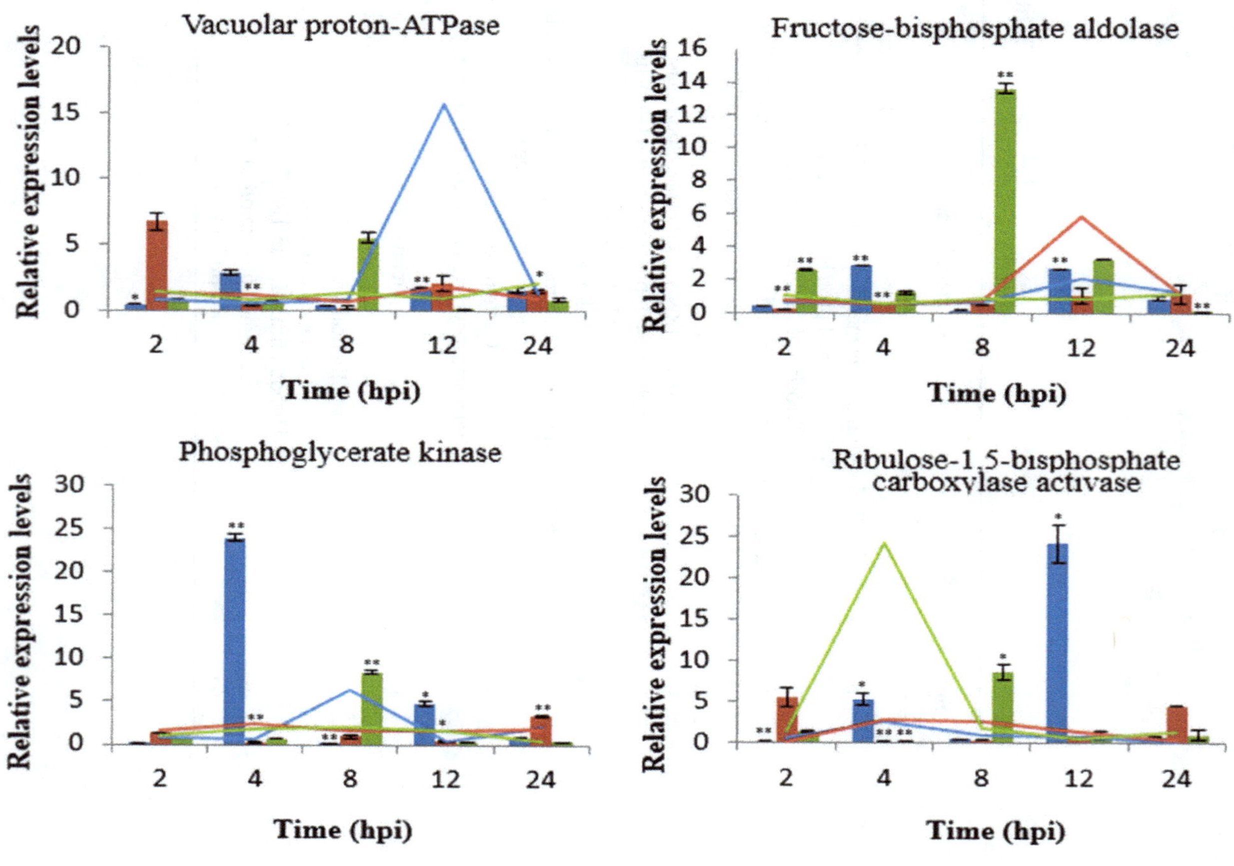


**Figure 7.** *Cont.*

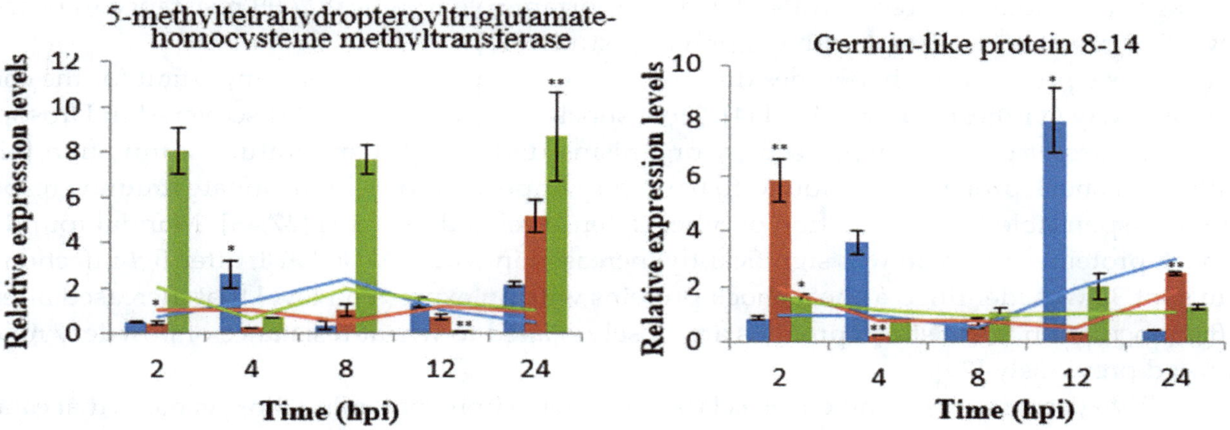


**Figure 7.** Quantification of six gene transcripts and protein levels at different time points post inoculation with *Bgt* in wheat. The bar graph shows the fold changes of the mRNA expression levels in inoculation vs control samples at five time points. The blue, red, and green columns are representatives of wheat *L699*, *Neimai836*, and *Chuannong26*, respectively. The lines show the fold changes of the protein expression levels in inoculation vs control samples at five time points. The blue, red, and green lines are representatives of wheat *L699*, *Neimai836*, and *Chuannong26*, respectively. The mRNA expression levels were quantified by qRT-PCR normalized against *18S rRNA*. Asterisks indicate statistically significant differences (*, $p < 0.05$; **, $p < 0.01$) of mRNA expression levels between the inoculation and control samples.

## 3. Discussion

Plants employ two levels of immunity to encounter pathogen invasion: Pathogen-associated molecular pattern (PAMP)-triggered immunity (PTI) and effector-triggered immunity (ETI). In the early phase of defense, PAPMs are recognized as 'non-self' molecules by the host plants. This induces downstream defense signaling, such as the generation of reactive oxygen species (ROS) and the transcription of genes encoding pathogenesis-related proteins (PRs). The pathogens release effector proteins to oppose PTI, and then the plant resistance proteins recognize the effector, which stimulates the plant's ETI, leading to the hypersensitive response (HR) and activating other plant defense pathways [32–34].

However, not only the specific signaling mediated by resistance genes, but also the other basic cellular processes, are involved in the effective defense to support the plant innate immune system [35]. In our results, the differentially expressed proteins, including both resistance proteins against *Bgt* and other proteins related to the direct and indirect defensive processes. The potential roles of these proteins in the defense response are discussed below.

### *3.1. Stress- and Defense-Related Proteins*

Plants experience a variety of biotic and abiotic stresses during the growth and development periods. Studies on the plant stress response found many stress response related proteins. For example, the Germin-like proteins are important stress-related proteins.

Protein spot 61, with an increasing expression level 24 h after *Bgt* infection, was identified as GLPs. GLPs as extracellular glycoproteins are important components of the plant PRs [36]. Recently, GLPs were reported to be involved in the stress responses of Arabidopsis, pepper, barely, and rice [37–40]. GLPs can remove excess ROS generated by plants in the form of enzymes, receptors, or structural proteins in various physiological and biochemical processes. The expression of GLPs increased significantly and potentially catalyzing the production of $H_2O_2$, in plants infected by fungi, bacteria, viruses, or other pathogens [41–43]. $H_2O_2$ can selectively participate in the signaling cascade pathway, which can stimulate plant self-defense reactions. In addition, $H_2O_2$ is able to use the cellulose crosslinking action to strengthen the structure of plant cell walls, which is very important in plant defense against oxidative stress. GLPs play an important role in wheat *L699* resistance to powdery mildew. This result is consistent with the previous study [43].

Heat shock proteins, as chaperones during the stress response, are very important for the correct folding of newly synthesized proteins [44]. Heat shock proteins were first discovered in Drosophila and were a class of proteins expressed by organisms under high temperature stimulation [45,46]. Recently, heat shock proteins were found to have very important roles in the innate immune response and are indispensable for the function of other defense-related proteins [47,48]. Mandal found that heat shock protein expression was significantly increased in wheat *N0308* 72 h after *Bgt* infection [29]. Protein spot 32 was identified as heat shock proteins with an expression level that increased after 8 h with *Bgt* inoculation. Heat shock proteins are closely related to wheat resistance of powdery mildew, as reported previously [29].

Plant lipoxygenases are members of a class of nonheme iron-containing dioxygenases that catalyze the addition of molecular oxygen to fatty acids containing a cis, cis-1,4-pentadiene system, which produces an unsaturated fatty acid hydroperoxide [49]. Currently, an increasing number of studies show that there are many similarities between the plant defense mechanisms and the animal defense mechanisms under adverse conditions [50]. Lipoxygenases in animals and plants play an important role in withstanding adverse environments. Protein spots 3, 4, 5, and 29, which were identified as lipoxygenase, were upregulated 8 h and 12 h after *Bgt* inoculation compared to non-inoculated wheat. In the lipoxygenase pathway, the polyunsaturated fatty acids are catalyzed by lipoxygenases to generate hydrogen peroxide and subsequently form compounds with a specific mass of physiological functions by the catalytic reaction of other enzymes, such as jasmonic acid and guaiac acid, which induces the synthesis of resistance proteins against stresses [51,52].

### 3.2. Proteins Related to Photosynthesis

Plant defense reactions are closely related to photosynthesis. It is generally believed that plant photosynthesis-related protein biosynthesis is reduced and resources are allocated to the defense response when plants are infected by a pathogen. The plant defense responses to pathogens is known as the "hidden costs" defense [53]. Plants affected by pathogens active the HR response, which is considered as another reason for weakening the plant photosynthesis after the original infection. However, protein spots 11, 28, 33, 36, 44 comprise ribulose carboxylase, which is an indicator of photosynthesis. These spots were upregulated 4 h, 8 h, and 12 h after powdery mildew infection, which could indicate that photosynthesis is increased. It was reported that photosynthesis is enhanced in early plant pathogen infections and weakened on later stage during the infection [54].

### 3.3. Proteins Involved in Carbohydrate Metabolism and Energy Pathways

The expression of several proteins involved in glucose metabolism, including β-D-glucose hydrolase (spot 6), phosphoglycerate kinase (spot 13, 55), glycerol phosphodiester enzyme (spot 14), diphosphate aldolase (spot 18), ribulose kinase (spot 38), glucose phosphate mutase (50, 51), and six-glucose phosphate decarboxylase (spot 54), were increased in response to wheat powdery mildew in wheat *L699*. Previous studies have shown that hexose can provide extra energy and serve as a signal for activating resistant response. For instance, in response to barley powdery mildew infection, the expression of hexose metabolizing enzymes significantly increased [55].

### 3.4. Proteins Involved in Gene Expression and Protein Turnover

Each step in the flow of genetic information is very strict, so the error rate of protein synthesis in this process is very low. However, protein synthesis has a certain error rate that is the net result of several processes. Aminoacyl-tRNA synthetases and ribosomes play important roles in protein synthesis [56,57]. Studies have shown that aminoacy-tRNA synthetase is not only involved in protein synthesis, but also participates in other activities, including the regulation of transcription and translation, RNA splicing, signal transduction, and immune response [58]. Current research is focused on the relationship between the function and structure of new amino acid-tRNA synthetases, especially the aminoacyl-tRNA synthetase, which is related to diseases. After powdery mildew infection in wheat *L699*, the expression levels of alanyl-tRNA synthetase (spot47), lysyl-tRNA synthetase (spot49), and ribosomal protein (spot27) increased, suggesting that these proteins may be important in the wheat anti-powdery mildew responses.

### *3.5. Proteins Associated with Cell Organization*

In powdery mildew-infected wheat *L699*, actin (spot 12) was upregulated. The actin cytoskeleton is an essential dynamic component for cells and is highly conserved in eukaryotic cells. The cytoskeleton is closely linked with the membrane and is involved in various cellular processes, including defense signaling based on actin cytoskeletal structures after pathogen infection [59].

### *3.6. The Correlation of mRNA and Protein Expression*

The analysis of six proteins and the expression of the protein-coding regions of their genes showed that the proteins and mRNA levels had a certain uniformity in our study. However, there exists post-transcriptional regulation after translational regulation in wheat, which may lead to differences between protein expression levels and mRNA levels.

### *3.7. The Novel Proteins Potentially Involed in the Response of Wheat Against Bgt*

Some of the identified proteins, such as the lipoxygenase, glucan exohydrolase, glucose adenylyltransferasesmall, phosphoribulokinase, and phosphoglucomutase, are first reported during the interaction of wheat-*Bgt* in this study. These proteins are potentially very critical for the wheat-*Bgt* interaction at early stage. For future research, the defense functions of these novel proteins deserve further investigation by using the integrative approaches, such as the comparative metabolomics, gene overexpression, and silencing methods. The related study will lead to deeper understanding of the detailed functions of these important proteins and more efficient disease control strategies.

## **4. Materials and Methods**

### *4.1. Plant Materials and Inoculation*

*L699*, *Neimai836*, and *Chuannong26* were used in this study. *L699* carries the resistance gene *Pm40* and shows resistance to most powdery mildew isolates in China. *Neimai836* carries the resistance gene *Pm21*, but not *Pm40*, and also shows resistance to most powdery mildew isolates in China. Nevertheless, *Chuannong26* is highly susceptible to powdery mildew without any effective resistance gene. Plants were cultivated in 30 cm pots in a growth chamber at 18 °C under a 12 h/12 h dark photoperiod. These pots were divided into the *Bgt*-inoculated group and the mock-inoculated group with nonopaque and breathable hoods. Seedlings of the *Bgt*-inoculated group were artificially inoculated by dusting with *Bgt* conidia from sporulating seedlings of *Chuannong26* at two to three leaf stages. Leaf samples were harvested at 2, 4, 8, 12, and 24 hpi with liquid nitrogen and immediately stored at −80 °C. Samples were collected from three biological replicates at each time point and every sample protein was run on three gels. For the analysis, one best gel was selected from three gels.

### *4.2. Cytological Observation of the Interaction between Wheat and Bgt*

Leaves of the *Bgt*-inoculated and mock-inoculated wheat were sampled at 2, 4, 8, 12, and 24 hpi and cut into 2–3 cm leaf fragments. The leaf fragments collected at 2, 4, 8, and 12 hpi were destained using isopropanol fumigation. The leaf fragments collected at 24 hpi were destained with AA solution (ethanol:glacial acetic acid = 1:1, *v*/*v*). Then the sample were stained with Coomassie blue staining solution (0.15% trichloroacetic acid aqueous solution:0.6% Coomassie brilliant blue R-250 methanol solution = 1:1, *v*/*v*) for 4 h. After rinsing with distilled water, the leaves were saved in a mix solution (glacial acetic acid:glycerol:distilled water = 1:4:15, *v*/*v*/*v*). The infection structures including germ tubes, appressoria and haustoria of *Bgt* were observed under electron microscope (40×, Nikon Eclipse 80i, Nikon Corporation, Tokyo, Japan).

### 4.3. Protein Extraction

Leaf tissue (1 g) was ground in a prechilled mortar with liquid nitrogen. Then, the powder was transferred to a 1.5 mL centrifuge tube with the addition of 1 mL acetone containing 10% trichloroacetic acid (TCA) and 0.07% β-mercaptoethanol. The samples were vortexed and chilled for 1 h at −20 °C. Then the homogenate was centrifuged at 13,000× *g* for 30 min at 4 °C. After the supernatant was gently decanted, the pellet was washed four times with chilled acetone containing 0.07% β-mercaptoethanol, then dried until all the acetone was removed by a vacuum drying instrument. The resulting powder was dissolved in 1 mL of IEF buffer (7 M urea, 2 M thiourea, 4% CHAPS, 20 mM DTT, 0.001% bromophenol blue, and 0.5% ampholyte (pH 3–10)). After centrifugation at 13,000× *g* for 20 min twice, the leaf proteins were obtained from the supernatant, and their concentration was determined using a Bradford dye binding assay [60].

### 4.4. Two-Dimensional Electrophoresis, Protein Visualization, and Image Analysis

The protein mixture was loaded onto an IPG strip (17 cm, pH 4-7, linear gradient (Bio-Rad, California, CA, USA)) by active rehydration at 50 V for 14 h (20 °C) on a Protein Isoelectric Focusing (IEF) Cell (Bio-Rad). The following conditions were used for the IEF: 20 °C, 50 μA/strip, 250 V for 1 h, 1000 V for 1 h, 10,000 V for 5 h, and 10,000 V, with a total of 60,000 vhs. The focused IPG strips were equilibrated in buffer containing 5 mL 6 M urea, 2% SDS, 20% glycerol, 375 mM Tris-HCl (pH 8.8), and 200 mM DTT for 15 min, then re-equilibrated in a similar buffer whose 200 mM DTT was replaced by 250 mM iodoacetamide for 15 min. Proteins were separated on the second dimension on vertical 12% sodium dodecyl sulfate-polyacrylamide (SDS-PAGE) gel in a Protean II XI Cell (Bio-Rad) at 25 mA/gel. Proteins in the gels were stained by the "blue silver" protocol as described by Candiano et al. [61]. Gels were scanned by a GS-800 scanner (Bio-Rad) and the proteins in the images were analyzed using PDQuest software with version 8.0 (Bio-Rad). There were variations due to sample loading, the 2-DE techniques and staining. To minimize these variations, each spot intensity was normalized according to its percent volume of all protein spots on the gel. The proteins showing at least a twofold change in abundance were considered as differentially expressed proteins (DEPs).

### 4.5. MS and Database Searches

Protein slices in fresh blue silver-stained gel were excised and plated into a 96-well microtiter plate. Excised slices were first distained twice with 60 μL 50 mM $NH_4HCO_3$ and 50% acetonitrile, then dried twice with 60 μL acetonitrile. Afterwards, the dried pieces of gels were incubated in ice-cold digestion solution (12.5 ng/μL trypsin and 20 mM $NH_4HCO_3$) for 20 min, then transferred into a 37 °C incubator for digestion overnight. Finally, peptides in the supernatant were collected after extraction twice with 60 μL extraction solution (5% formic acid in 50% acetonitrile).

The peptide solution described above was dried under the protection of $N_2$. A 0.8 μL matrix solution (5 mg/mL α-cyano-4-hydroxy-cinnamic acid diluted in 0.1% TFA, 50% ACN) was pipetted to dissolve the peptides. Then, the mixture was spotted on a MALDI target plate (AB SCIEX, Framingham, Massachusetts, MA, USA). MS analysis of the peptides was performed on an AB SCIEX 5800 TOF/TOF. The UV laser was operated at a 400 Hz repetition rate with a wavelength of 355 nm. The accelerated voltage was operated at 20 kV, and the mass resolution was maximized at 1600 Da. The mass instrument with internal calibration mode was calibrated by myoglobin digested with trypsin. All acquired spectra of samples were processed using TOF/TOF Explorer$^{TM}$ Software (AB SCIEX) in default mode. The data were searched by GPS Explorer (V3.6) with the search engine MASCOT (V2.3, Matrix Science, London,

UK). The search parameters were as follows: dates were compared against the NCBI nr database, trypsin was digested with one missing cleavage, MS tolerance was set at 100 ppm, and MS/MS tolerance was set to 0.6 Da. Functional annotation of identified proteins based on gene ontology was performed using the Protein Information Resource (https://proteininformationresource.org).

### *4.6. RNA Isolation and qRT-PCR Assays*

Total RNA from *Bgt*-inoculated or mock-inoculated wheat leaves was sampled at 2, 4, 8, 12, and 24 hpi and extracted using Trizol reagent (Tiangen Biotech, Beijing, China). First strand cDNA was synthesized with Transcript One-Step gDNA removal and cDNA Synthesis Supermix (Transgen Biotech, Beijing, China). Primers were specifically designed to anneal to each of the selected genes and the endogenous reference gene *18S rRNA* (GenBank accession No. AY049040) [62]. The expression patterns of selected genes were analyzed with a Bio-Rad iQ5 system. Relative gene quantification was calculated by the comparative $2^{-\Delta\Delta Ct}$ method [63] and normalized to the corresponding expression level of the *18S rRNA*. All reactions were performed in triplicate, including three no-template controls.

## 5. Conclusions

In summary, we identified 46 differentially expressed proteins in wheat in response to *Bgt* inoculation using 2-DE and mass spectrometry. Among these identified proteins, the lipoxygenase, glucan exohydrolase, glucose adenylyltransferasesmall, phosphoribulokinase, and phosphoglucomutase are first reported during the interaction of wheat-*Bgt*. We inferred that these proteins are not only involved in defense response but also physiology and cellular process for wheat to confer resistance against *Bgt*. The wheat resistance gene products potentially mediate the immune response and coordinate other physiological and cellular processes during the resistance response to *Bgt*.

**Author Contributions:** Y.L., M.Z., G.G. and X.C. designed the research. Y.L., J.Y., Y.H., L.L., T.L. and L.H. performed the research. Y.L., X.C. and Y.H. analyzed the data. Y.L., Y.X., X.C., M.C. and M.Z. wrote the paper. All authors read and approved the final version of the manuscript.

**Acknowledgments:** We thank Peigao Luo for providing the wheat cultivar *L699*.

## References

1. Wang, Z.; Li, L. Seedling and adult plant resistance to powdery mildew in Chinese bread wheat cultivars and *lines*. *Plant Dis.* **2005**, *89*, 457–463. [CrossRef]
2. Duveiller, E.; Singh, R. The challenges of maintaining wheat productivity: Pests, diseases, and potential epidemics. *Euphytica* **2007**, *157*, 417–430. [CrossRef]
3. Hao, Y.; Parks, R. Molecular characterization of a new powdery mildew resistance gene *Pm54* in soft red winter wheat. *Theor. Appl. Genet.* **2015**, *128*, 465–476. [CrossRef] [PubMed]
4. Zhang, R.; Sun, B. *Pm55*, a developmental-stage and tissue-specific powdery mildew resistance gene introgressed from Dasypyrum villosum into common wheat. *Theor. Appl. Genet.* **2016**, *129*, 1975–1984. [CrossRef] [PubMed]
5. Hsam, S.; Huang, X. Chromosomal location of genes for resistance to powdery mildew in common wheat (*Triticum aestivum* L. em Thell.). 5. Alleles at the *Pm1* locus. *Theor. Appl. Genet.* **1998**, *96*, 1129–1134. [CrossRef]
6. Singrün, C.; Hsam, S. Powdery mildew resistance gene *Pm22* in cultivar *Virest* is a member of the complex *Pm1* locus in common wheat (*Triticum aestivum* L. em Thell.). *Theor. Appl. Genet.* **2003**, *106*, 1420–1424. [CrossRef] [PubMed]
7. Hao, Y.; Liu, A. *Pm23*: A new allele of *Pm4* located on chromosome 2AL in wheat. *Theor. Appl. Genet.* **2008**, *117*, 1205–1212. [CrossRef] [PubMed]
8. Xie, W.; Ben-David, R. Suppressed recombination rate in 6VS/6AL translocation region carrying the *Pm21* locus introgressed from *Haynaldia villosa* into hexaploid wheat. *Mol. Breed.* **2012**, *29*, 399–412. [CrossRef]

9. Zou, S.; Wang, H. The NB-LRR gene *Pm60* confers powdery mildew resistance in wheat. *New Phytol.* **2018**, *218*, 298–309. [CrossRef] [PubMed]
10. Wiersma, A.T.; Pulman, J.A. Identification of *Pm58* from *Aegilops tauschii*. *Theor. Appl. Genet.* **2017**, *130*, 1123–1133. [CrossRef]
11. Tan, C.; Li, G. Characterization of *Pm59*, a novel powdery mildew resistance gene in Afghanistan wheat landrace PI 181356. *Theor. Appl. Genet.* **2018**, *131*, 1145–1152. [CrossRef] [PubMed]
12. Sun, H.; Hu, J. *Pm61*: A recessive gene for resistance to powdery mildew in wheat landrace Xuxusanyuehuang identified by comparative genomics analysis. *Theor. Appl. Genet.* **2018**, *131*, 2085–2097. [CrossRef] [PubMed]
13. Zhang, R.; Fan, Y. *Pm62*, an adult-plant powdery mildew resistance gene introgressed from *Dasypyrum villosum* chromosome arm 2VL into wheat. *Theor. Appl. Genet.* **2018**, *131*, 2613–2620. [CrossRef] [PubMed]
14. Huang, X.Q.; Hsam, S.L.K. Molecular mapping of the wheat powdery mildew resistance gene *Pm24* and marker validation for molecular breeding. *Theor. Appl. Genet.* **2000**, *101*, 407–414. [CrossRef]
15. Consortium, I.W.G.S. A chromosome-based draft sequence of the hexaploid bread wheat (*Triticum aestivum*) genome. *Science* **2014**, *345*, 1251788.
16. Yahiaoui, N.; Srichumpa, P. Genome analysis at different ploidy levels allows cloning of the powdery mildew resistance gene *Pm3b* from hexaploid wheat. *Plant J.* **2004**, *37*, 528–538. [CrossRef] [PubMed]
17. Cao, A.; Xing, L. Serine/threonine kinase gene *Stpk-V*, a key member of powdery mildew resistance gene *Pm21*, confers powdery mildew resistance in wheat. *Proc. Natl. Acad. Sci. USA* **2011**, *108*, 7727–7732. [CrossRef]
18. Hurni, S.; Brunner, S. Rye *Pm8* and wheat *Pm3* are orthologous genes and show evolutionary conservation of resistance function against powdery mildew. *Plant J.* **2013**, *76*, 957–969. [CrossRef]
19. Sánchez-Martín, J.; Steuernagel, B. Rapid gene isolation in barley and wheat by mutant chromosome sequencing. *Genome Biol.* **2016**, *17*, 221. [CrossRef]
20. Xing, L.; Hu, P. NLR1-V, a CC-NBS-LRR encoding gene, is a potential candidate gene of the wheat powdery mildew resistance gene *Pm21*. *bioRxiv* **2017**, 114058. [CrossRef]
21. Luo, P.G.; Luo, H. Characterization and chromosomal location of *Pm40* in common wheat: A new gene for resistance to powdery mildew derived from *Elytrigia intermedium*. *Theor. Appl. Genet.* **2009**, *118*, 1059–1064. [CrossRef]
22. Zhong, S.; Ma, L. Collinearity analysis and high-density genetic mapping of the wheat powdery mildew resistance gene *Pm40* in PI 672538. *PLoS ONE* **2016**, *11*, e0164815. [CrossRef] [PubMed]
23. Liu, Z.H.; Xu, M. Registration of the novel wheat lines *L658*, *L693*, *L696*, and *L699*, with resistance to Fusarium Head blight, stripe rust, and powdery mildew. *J. Plant Registrat.* **2015**, *9*, 121–124. [CrossRef]
24. Li, J.; Yang, X. Proteomic analysis of the compatible interaction of wheat and powdery mildew (*Blumeria graminis* f. sp. *tritici*). *Plant Physiol. Biochem.* **2017**, *111*, 234–243. [CrossRef] [PubMed]
25. Lim, M.S.; Elenitoba-Johnson, K.S.J. Proteomics in pathology research. *Lab. Investig.* **2004**, *84*, 1227–1244. [CrossRef] [PubMed]
26. Marra, R.; Ambrosino, P. Study of the three-way interaction between *Trichoderma atroviride*, plant and fungal pathogens by using a proteomic approach. *Curr. Genet.* **2006**, *50*, 307–321. [CrossRef] [PubMed]
27. Li, Q.; Chen, X. Differences in protein expression and ultrastructure between two wheat near-isogenic lines affected by powdery mildew. *Russ. J. Plant Physiol.* **2011**, *58*, 686. [CrossRef]
28. Wang, B.; Xie, C. Comparative proteomic analysis of wheat response to powdery mildew infection in wheat *Pm30* near-lsogenic lines. *J. Phytopathol.* **2012**, *160*, 229–236. [CrossRef]
29. Mandal, M.S.N.; Fu, Y. Proteomic analysis of the defense response of wheat to the powdery mildew fungus, *Blumeria graminis* f. sp. *tritici*. *Protein J.* **2014**, *33*, 513–524. [CrossRef]
30. Fu, Y.; Zhang, H. Quantitative proteomics reveals the central changes of wheat in response to powdery mildew. *J. Proteom.* **2016**, *130*, 108–119. [CrossRef]
31. Luo, L.Y. *Defense Responses Induced by Bgt in The Novel Wheat Line L699 Carrying a Pm40 Gene*; Sichuan Agricultural University: Ya'an, China, 2016.
32. Spoel, S.H.; Dong, X. How do plants achieve immunity? Defence without specialized immune cells. *Nat. Rev. Immunol.* **2012**, *12*, 89–100. [CrossRef] [PubMed]
33. Chisholm, S.T.; Coaker, G. Host-microbe interactions: Shaping the evolution of the plant immune response. *Cell* **2006**, *124*, 803–814. [CrossRef] [PubMed]

34. Hammond-Kosack, K.E.; Jones, J.D. Resistance gene-dependent plant defense responses. *Plant Cell* **1996**, *8*, 1773–1791. [CrossRef] [PubMed]
35. Rojas, C.M.; Senthil-Kumar, M. Regulation of primary plant metabolism during plant-pathogen interactions and its contribution to plant defense. *Front. Plant Sci.* **2014**, *5*, 1–12. [CrossRef] [PubMed]
36. Bernier, F.; Berna, A. Germins and germin-like proteins: Plant do-all proteins. But what do they do exactly? *Plant Physiol. Biochem.* **2001**, *39*, 545–554. [CrossRef]
37. Membré, N.; Bernier, F. Arabidopsis thaliana germin-like proteins: Common and specific features point to a variety of functions. *Planta* **2000**, *211*, 345–354. [CrossRef] [PubMed]
38. Park, C.-J.; An, J.-M. Molecular characterization of pepper germin-like protein as the novel PR-16 family of pathogenesis-related proteins isolated during the resistance response to viral and bacterial infection. *Planta* **2004**, *219*, 797–806. [CrossRef]
39. Zimmermann, G.; Bäumlein, H. The multigene family encoding germin-like proteins of barley. Regulation and function in basal host resistance. *Plant Physiol.* **2006**, *142*, 181–192. [CrossRef]
40. Deeba, F.; Sultana, T. Involvement of WRKY, MYB and DOF DNA-binding proteins in interaction with a rice germin-like protein gene promoter. *Acta Physiol. Plant.* **2017**, *39*, 189. [CrossRef]
41. Camejo, D.; Guzmán-Cedeño, Á. Reactive oxygen species, essential molecules, during plant-pathogen interactions. *Plant Physiol. Biochem.* **2016**, *103*, 10–23.
42. Dunwell, J.M.; Gibbings, J.G. Germin and germin-like proteins: Evolution, structure, and function. *Crit. Rev. Plant Sci.* **2008**, *27*, 342–375. [CrossRef]
43. Schweizer, P.; Christoffel, A. Transient expression of members of the germin-like gene family in epidermal cells of wheat confers disease resistance. *Plant J.* **1999**, *20*, 541–552. [CrossRef] [PubMed]
44. Feder, M.E.; Hofmann, G.E. Heat-shock proteins, molecular chaperones, and the stress response: Evolutionary and ecological physiology. *Annu. Rev. Physiol.* **1999**, *61*, 243–282. [CrossRef] [PubMed]
45. Ritossa, F. A new puffing pattern induced by temperature shock and DNP in Drosophila. *Experientia* **1962**, *18*, 571–573. [CrossRef]
46. Tissiéres, A.; Mitchell, H.K. Protein synthesis in salivary glands of Drosophila melanogaster: Relation to chromosome puffs. *J. Mol. Biol.* **1974**, *84*, 389–398. [CrossRef]
47. Neckers, L.; Tatu, U. Molecular chaperones in pathogen virulence: Emerging new targets for therapy. *Cell Host Microbe* **2008**, *4*, 519–527. [CrossRef] [PubMed]
48. Bakthisaran, R.; Tangirala, R. Small heat shock proteins: Role in cellular functions and pathology. *Biochim. Biophys. Acta (BBA) Proteins Proteom.* **2015**, *1854*, 291–319. [CrossRef]
49. Siedow, J.N. Plant lipoxygenase: Structure and function. *Annu. Rev. Plant Biol.* **1991**, *42*, 145–188. [CrossRef]
50. Apel, K.; Hirt, H. Reactive oxygen species: Metabolism, oxidative stress, and signal transduction. *Annu. Rev. Plant Biol.* **2004**, *55*, 373–399. [CrossRef]
51. Vick, B.A.; Zimmerman, D. Oxidative systems for modification of fatty acids: The lipoxygenase pathway. *Biochem. Plants* **1987**, *9*, 53–90.
52. Feussner, L.; Wasternack, C. The lipoxygenase pathway. *Annu. Rev. Plant Biol.* **2002**, *53*, 275–297. [CrossRef] [PubMed]
53. Bilgin, D.D.; Zavala, J.A. Biotic stress globally downregulates photosynthesis genes. *Plant Cell Environ.* **2010**, *33*, 1597–1613. [CrossRef] [PubMed]
54. Scholes, J.; Lee, P. Photosynthetic metabolism in leaves infected with powdery mildew. *Curr. Res. Photosynth.* **1990**, *4*, 219–222.
55. Swarbrick, P.J.; Schulze-lefert, P. Metabolic consequences of susceptibility and resistance (race-specific and broad-spectrum) in barley leaves challenged with powdery mildew. *Plant Cell Environ.* **2006**, *29*, 1061–1076. [CrossRef] [PubMed]
56. Ling, J.; So, B.R. Resampling and editing of mischarged tRNA prior to translation elongation. *Mol. Cell* **2009**, *33*, 654–660. [CrossRef] [PubMed]
57. Zaher, H.S.; Green, R. Quality control by the ribosome following peptide bond formation. *Nature* **2009**, *457*, 161–166. [CrossRef] [PubMed]
58. Park, S.G.; Ewalt, K.L. Functional expansion of aminoacyl-tRNA synthetases and their interacting factors: New perspectives on housekeepers. *Trends Biochem. Sci.* **2005**, *30*, 569–574. [CrossRef] [PubMed]
59. Henty-Ridilla, J.L.; Shimono, M. The plant actin cytoskeleton responds to signals from microbe-associated molecular patterns. *PLoS Pathog.* **2013**, *9*, e1003290. [CrossRef] [PubMed]

60. Bradford, M.M. A rapid and sensitive method for the quantitation of microgram quantities of protein utilizing the principle of protein-dye binding. *Analy. Biochem.* **1976**, *72*, 248–254. [CrossRef]
61. Candiano, G.; Bruschi, M. Blue silver: A very sensitive colloidal Coomassie G-250 staining for proteome analysis. *Electrophoresis* **2004**, *25*, 1327–1333. [CrossRef]
62. Balaji, B.; Bucholtz, D.B. Barley yellow dwarf virus and Cereal yellow dwarf virus quantification by real-time polymerase chain reaction in resistant and susceptible plants. *Phytopathology* **2003**, *93*, 1386–1392. [CrossRef] [PubMed]
63. Livaka, K.J.; Schmittgen, T.D. Analysis of relative gene expression data using real-time quantitative PCR and the $2^{-\Delta\Delta CT}$ method. *Methods* **2001**, *25*, 402–408. [CrossRef] [PubMed]

# 6

# The Dark Matter of Large Cereal Genomes: Long Tandem Repeats

**Veronika Kapustová [1], Zuzana Tulpová [1], Helena Toegelová [1], Petr Novák [2], Jiří Macas [2], Miroslava Karafiátová [1], Eva Hřibová [1], Jaroslav Doležel [1] and Hana Šimková [1,*]**

[1] Institute of Experimental Botany, Centre of the Region Haná for Biotechnological and Agricultural Research, Šlechtitelů 31, CZ-78371 Olomouc, Czech Republic; kapustova@ueb.cas.cz (V.K.); tulpova@ueb.cas.cz (Z.T.); toegelova@ueb.cas.cz (H.T.); karafiatova@ueb.cas.cz (M.K.); hribova@ueb.cas.cz (E.H.); dolezel@ueb.cas.cz (J.D.)

[2] Biology Centre, Czech Academy of Sciences, Institute of Plant Molecular Biology, Branišovská 31, CZ-37005 České Budějovice, Czech Republic; petr@umbr.cas.cz (P.N.); macas@umbr.cas.cz (J.M.)

* Correspondence: simkovah@ueb.cas.cz.

**Abstract:** Reference genomes of important cereals, including barley, emmer wheat and bread wheat, were released recently. Their comparison with genome size estimates obtained by flow cytometry indicated that the assemblies represent not more than 88–98% of the complete genome. This work is aimed at identifying the missing parts in two cereal genomes and proposing techniques to make the assemblies more complete. We focused on tandemly organised repetitive sequences, known to be underrepresented in genome assemblies generated from short-read sequence data. Our study found arrays of three tandem repeats with unit sizes of 1242 to 2726 bp present in the bread wheat reference genome generated from short reads. However, this and another wheat genome assembly employing long PacBio reads failed in integrating correctly the 2726-bp repeat in the pseudomolecule context. This suggests that tandem repeats of this size, frequently incorporated in unassigned scaffolds, may contribute to shrinking of pseudomolecules without reducing size of the entire assembly. We demonstrate how this missing information may be added to the pseudomolecules with the aid of nanopore sequencing of individual BAC clones and optical mapping. Using the latter technique, we identified and localised a 470-kb long array of 45S ribosomal DNA absent from the reference genome of barley.

**Keywords:** genome assembly; bread wheat; barley; optical mapping; BAC; ribosomal DNA

---

## 1. Introduction

Small grain cereals, such as bread wheat, durum wheat, barley and rye are crucial crops for the European population and most of them are grown worldwide. Despite their socio-economic importance, reference genomes of these cereals were only recently generated [1–5], which can be attributed to their high genome complexities, ranging from ~5 Gb for barley to ~16 Gb for bread wheat [6], and enormous proportion of repetitive DNA (85–90%). These assemblies are characterized by high contiguity and low proportion of internal gaps. However, a comparison of assembly lengths with genome-size estimates obtained by flow cytometry indicated that the reference genomes of barley, wild emmer wheat, bread wheat and rye represented no more than 98%, 88%, 90% and 90% of the estimated genome sizes, respectively [6]. This gives rise to an obvious question as to what the missing part of the reference genomes is, and stimulates efforts towards complementing it.

Low-copy genome regions are known to be a relatively easy target for genome assemblers and were found well represented even in early cereal genome sequences that were based on low-coverage Roche/454 data [7–9] or Illumina pair-end reads only [10]. On the contrary, large regions of repeats are known to pose a challenge and result in gaps, mis-assemblies and collapsed tandem repeats in a majority of genome sequences [11]. Dispersed repeats, represented by various types of transposable elements, have been largely resolved in the recent assemblies thanks to the combination of pair-end and mate-pair Illumina reads and sophisticated assembling algorithms [1,3] or implementation of long-read PacBio data [12].

On the other hand, tandem repeats organized as arrays of multiple units (microsatellites, macrosatellites and centromeric satellite repeats) tend to collapse in assemblies into fewer copies. Consequently, they are under-represented in reference genomes and pose a significant source of gaps and assembling errors in de novo assemblies, including that of humans [11]. The repeat-associated gaps are abundant in heterochromatic regions, making it impossible to completely assemble these genome parts. This usually results in genome assemblies missing a majority of (peri)centromeric regions and secondary constrictions [11,13]. To resolve the copy numbers, it is essential to use reads longer than the total array length. To some extent, this can be sorted out by using long-read DNA sequencing technologies, such as PacBio or nanopore sequencing, which produce reads of tens and up to hundreds kilobases, respectively. Nevertheless, only short arrays of simpler repeats, such as 5S rRNA multigene loci spanning over several kilobases, can be tackled by these approaches [14]. Arrays spanning over hundreds to thousands kilobases and consisting of units that are several kilobases long, such as loci coding for 45S rRNA, cannot be resolved by any of the current sequencing technologies.

This shortage can be compensated for by other approaches that facilitate identification, positioning and characterization of long arrays of tandem repeats, such as in situ hybridisation, application of dedicated bioinformatics tools, and optical mapping. The initial methods to investigating distribution of tandem repeats in cereal genomes included in situ hybridisation (ISH) [15–17] and fluorescence in situ hybridisation (FISH) [18,19]. Although the cytogenetic techniques provided first insights into the genome evolution [20] and facilitated chromosome identification and construction of molecular karyotypes [16,19,21], they did not provide information at the DNA sequence level. This could be obtained by application of dedicated bioinformatics tools, such as RepeatExplorer [22,23]. This computational pipeline utilizes a graph-based sequence clustering algorithm to de-novo assemble tandem repeats from raw next-generation sequencing data, without the need of a reference database of known elements. It has been used to identify and characterize repetitive elements in several complex plant genomes, including that of rye [24].

Optical mapping in nanochannel arrays, also known as Bionano genome (BNG) mapping, is a high-throughput long-read technology that generates genome maps of a short sequence motif—the recognition site of an enzyme used for labelling [25]. It has been used to support and validate physical-map and genome assemblies of several complex cereal genomes [2,3,26–28]. The ability of optical mapping to size gaps, cover some types of tandem DNA repeats [26] and identify misassemblies due to collapsed duplicated sequences [29] makes this technology a promising tool for identifying and characterizing missing parts of genome assemblies.

In this study, we identified three new tandem repeats specific for the short arm of wheat chromosome 7D (7DS) and interrogated their representation in recently published bread wheat assemblies, including (i) Triticum 3.1 [12], which combines short Illumina and long PacBio reads, (ii) IWGSC RefSeq v1.0 [3], which is based on short reads only, and (iii) Illumina assemblies of physical map-ordered 7DS-specific BAC clones [30]. While these assemblies comprised of arrays of all three repeats, they failed in unravelling organization of a repeat with unit size of 2726 bp, whose genome arrangement could only be resolved after adding information from nanopore sequencing of two BAC clones bearing arrays of this repeat and by using an optical map (OM) of the wheat 7DS arm [26] as a reference. Besides, we employed an OM of barley cv. Morex [31] to investigate a minor 45S rDNA locus in barley chromosome 1H, which was identified by in situ hybridisation in various barley cultivars [32,33] and underrepresented in the 'Morex' BAC-by-BAC genome assembly [2]. Thus, we demonstrated that optical mapping is a suitable tool to identify the missing parts of the assemblies, and, in some cases, can reveal overall organization of the repeat array. Targeted long-read nanopore sequencing was confirmed as a promising approach to complementing the missing sequences to the genome assemblies.

## 2. Results and Discussion

### 2.1. Chromosome-Specific Tandem Repeats in Wheat

Using RepeatExplorer pipeline to cluster Illumina raw data from flow-sorted wheat chromosome arm 7DS, we identified four tandem repeats with monomer lengths ranging from 1167 bp to 2726 bp (Table 1, File S1).

**Table 1.** Monomer sizes and distribution of identified tandem repeats.

| Tandem Repeat | Monomer Size | Distribution |
|---|---|---|
| TaeCsTr163 | 1390 bp | 7D subtelomere |
| TaeCsTr230 | 1242 bp | 7D subtelomere |
| TaeCsTr99 | 2726 bp | 7D subtelomere |
| TaeCsTr111 | 1167 bp | All chromosomes Dispersed |

Cytogenetic mapping revealed that three of the repeats, TaeCsTr163, TaeCsTr230 and TaeCsTr99, provided unique FISH signals specific for subtelomeric region of wheat 7DS chromosome arm (Figure 1; Figure S1). A probe derived from the TaeCsTr111 sequence provided dispersed hybridization signals on multiple chromosomes, predominantly in pericentromeric and subtelomeric regions (Figure S1). The clustered hybridization signals obtained with three probes are supportive of the tandem organisation of these repeats. Application of RepeatExplorer on raw data obtained from flow-sorted chromosomes thus proved a suitable approach to identify chromosome-specific tandem repeats. The tool worked efficiently, despite using DNA amplified by multiple displacement amplification, which is known to introduce a quantitative amplification bias [34].

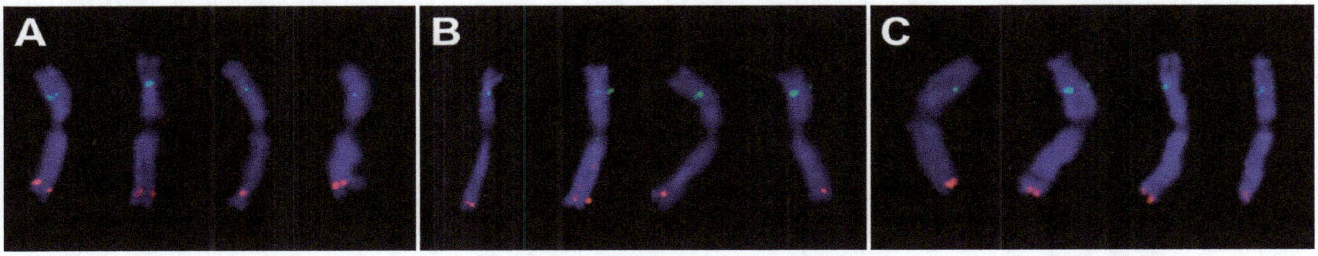


**Figure 1.** FISH on metaphase 7D chromosomes of bread wheat cv. Chinese Spring with probes for GAA microsatellite (green) and three tandem repeats (red). (**A**) TaeCsTr163, (**B**) TaeCsTr230 and (**C**) TaeCsTr99 repeats localized in subtelomeric region of the 7DS chromosome arm. 7D chromosomes were identified based on the GAA hybridization signal on the 7DL arm. The chromosomes were counterstained by DAPI (blue).

To assess the representation of the tandem repeats in the recently published wheat reference genome, we performed blastn search on IWGSC RefSeq v1.0 assembly [3]. Out of the three repeats assigned specifically to the 7DS arm, only two, TaeCsTr163 and TaeCsTr230, could be reliably identified in the IWGSC RefSeqv1.0 assembly of the 7D chromosome. Using blastn search, we found a cluster of 39 complete and several incomplete units of the TaeCsTr163 repeat, partially tandemly organised, that spanned over 260 kb in the interval of 49.08–49.34 Mb of the 7D pseudomolecule. The TaeCsTr230 repeat was identified as an array of ten complete and three incomplete units located in the interval of 33.254–33.27 Mb of the 7D pseudomolecule (Figure S2). Considering the entire 7DS arm length of 338 Mb [3], the positions of both repeats in the assembly are in agreement with their cytogenetic locations. On the contrary, we failed to find a significant blastn hit in the 7D pseudomolecule of RefSeq v1.0 for the repeat TaeCsTr99 that provided the strongest FISH signal on mitotic chromosomes.

Additional search in the unassigned scaffolds (ChrUn) revealed 12 of them containing TaeCsTr99 units (Figure 2A, Figure S3). Scaffold lengths varied from 3 kb to 163 kb, and they comprised from one to eleven complete TaeCsTr99 units. Altogether, we identified in unassigned scaffolds 36 complete tandemly organised units totalling 98 kb of length, accompanied by several incomplete units and unit fragments. Based on a high sequence homology (Figure S3), some of the scaffolds could be overlapping. Out of the twelve scaffolds, only ChrUn8536 carried an 82.3-kb non-repetitive segment (Figure S4A), which enabled its positioning in the context of the 7D pseudomolecule utilizing the OM of the 7DS chromosome arm. OM contig 77 placed the ChrUn8536 to position 14.4 Mb in the 7D pseudomolecule and revealed an additional 272 kb gap distal of the ChrUn8536 (Figure S5A, Figure 3). The identified position is consistent with the (sub)telomeric location of the repeat indicated by FISH (Figure 1C).

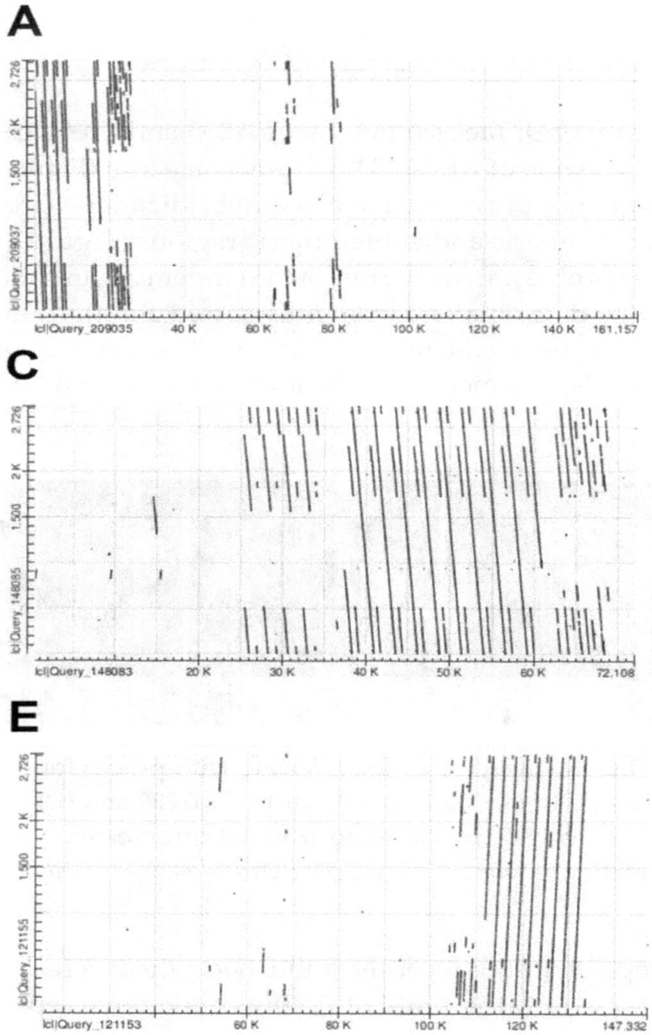

**Figure 2.** Dot plots showing position and arrangement of TaeCsTr99 repeat in (**A**) ChrUn8536, (**B**) interval 9.06 Mb to 9.16 Mb of 7D pseudomolecule of Triticum 3.1 assembly [12] (**C**) 72-kb Illumina contig of BAC clone 28N04, (**D**) 99.8-kb nanopore read 51ef9015 of BAC clone 28N04, (**E**) 148-kb nanopore read f24cdcf5 of BAC clone 104G18.

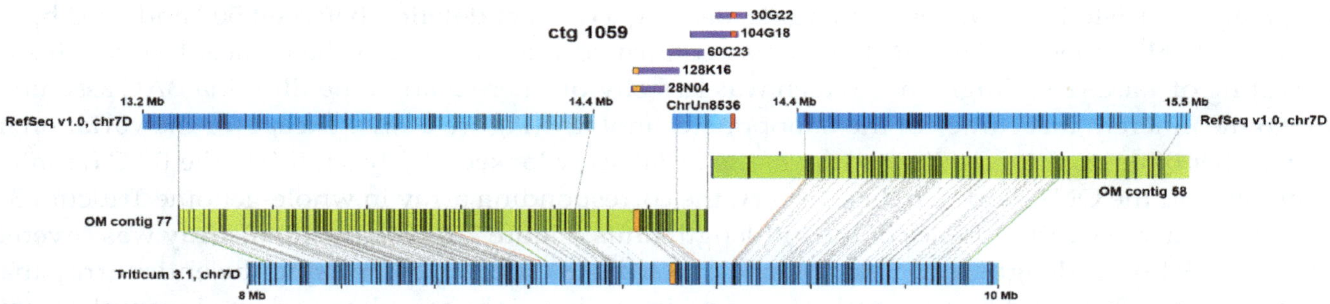


**Figure 3.** Organization of the TaeCsTr99 region. Positions of two TaeCsTr99 arrays (highlighted in orange and red) and an overall arrangement of the region were obtained by aligning ChrUn8536, IWGSC RefSeq v1.0 [3] and Triticum 3.1 [12] 7D pseudomolecules (blue bars) and BAC clones of 7DS physical-map contig 1059 [30] (violet bars) to 7DS optical map [26] (green bars). Numbers at the 7D pseudomolecules indicate assembly coordinates.

In order to span the entire TaeCsTr99 array, we employed short-read sequence assemblies of physical-map ordered 7DS BAC clones [30]. TaeCsTr99 was identified in four BAC clones belonging to 7DS physical-map contig 1059. Clones 104G18 and 30G22 overlap and show sequence homology with ChrUn8536 (Figure 3). Illumina assemblies of these BAC clones were rather fragmented and did not allow reconstructing the entire TaeCsTr99 array. Surprisingly, an array of the TaeCsTr99 repeat was also found in overlapping BAC clones 28N04 and 128K16. The latter could be aligned to the OM 77 (Figure S5B), but it was separated from the TaeCsTr99 array identified in clones 104G18 and 30G22 by a 225-kb non-repetitive segment (Figure 3, Figures S4 and S5). This suggested the presence of two separate TaeCsTr99 arrays in a close proximity, which we termed distal (covered by clones 128K16 and 28N04) and proximal (covered by ChrUn8536, 104G18 and 30G22), respectively. The size of the distal array was deduced from the assembly of BAC clone 28N04, which appeared to comprise the entire array in one scaffold, and was estimated to be ~44 kb. This array was composed of nine complete units, four units comprising a 1056-bp deletion and an additional cluster of repeat fragments spanning over ~6 kb (Figure 2C). The size of the proximal array in ChrUn8536 was 26 kb and this, likely incomplete array sequence comprised of three complete and three partial units and a ~6-kb cluster of repeat fragments (Figure 2A). Apparently, the organization of the repeat array in BAC clones 28N04 and 128K16 differed from that in ChrUn8536 (Figure 2, Figure S4A), which supported our hypothesis of two spatially separated arrays of TaeCsTr99 repeat in 7DS.

We hypothesized that the difficulties in assembling and incorporating of the TaeCsTr99 arrays into the pseudomolecule could be overcome by employing longer reads, such as those generated by SMRT sequencing (PacBio technology). To verify this, we explored bread wheat Triticum 3.1 assembly [12] that combines short-read Illumina and long-read PacBio data to search the 7D pseudomolecule for the TaeCsTr99 sequence. The blastn search revealed in the position 9.11 - 9.14 Mb a ~30-kb array composed of 14 incomplete TaeCsTr99 units, 13 of which carried the 1056-bp deletion observed in BAC clone 28N04 (Figure 2). We also observed the ~6-kb cluster of repeat fragments located distal of the array. Alignment of this region to 7DS OM placed the array to position ~1.2 Mb in OM contig 77 (Figure S5C), which corresponded to the position of the distal array identified in the short-read assemblies (Figure 3). On the contrary, we did not find any evidence of the proximal array in the Triticum 3.1 assembly.

To resolve the discrepancies in the location and organization of the TaeCsTr99 arrays identified in various assemblies, we made use of the long-read platform of Oxford Nanopore Technologies (ONT) and generated nanopore reads from BAC clones 28N04 and 104G18, which cover the distal and proximal array, respectively. For each of the clones, we obtained two reads that spanned over the entire insert and showed a consistent array structure. ONT read 51ef9015 (File S2) of 99,802 bp covering clone 28N04 confirmed the complex structure of the distal array composed of two sub-arrays with differently organised units (Figure 2D, Figure S4B). The distally located sub-array, approximately 27 kb in size, comprised of 10 complete units of the TaeCsTr99. The proximal sub-array was of similar length and consisted of 12 incomplete units, bearing a distinct deletion between 507 and 1563 bp of the TaeCsTr99 sequence. The ONT read also confirmed the presence of the adjacent ~6-kb cluster consisting of TaeCsTr99 fragments, which was less obvious here than in the Illumina BAC assembly due to the inherent inaccuracy of the nanopore technology (Figure 2C,D). Except for the variation in the number of units, the overall structure of the distal array looked highly similar in the BAC Illumina assembly and the ONT read. On the contrary, the corresponding array in whole-genome Triticum 3.1 assembly differed by the absence of the full-length units (Figure 2B). The proximal array was covered by a 148,009 bp full-length read f24cdcf5 of clone 104G18 (File S2), comprising the entire array that spanned over ~30kb and had a simple structure, including eight complete and one incomplete unit and the cluster of repeat fragments (Figure 2E). The total number of TaeCsTr99 units in the ONT reads covering the distal and the proximal array (18) was smaller than that identified in unassigned scaffolds (36). This could be due to non-recognized overlaps between the scaffolds, which may have resulted in overestimating the number of the repeats. Alternatively, we cannot exclude the presence of additional TaeCsTr99 array(s), missing both in the pseudomolecules and in the 7DS BAC assemblies that might be located in proximity of the confirmed ones.

The data obtained in our study suggest that tandemly organised repeats with unit size of 1–3 kb are not the major contributor to the missing part of the wheat IWGSC RefSeq v1.0 assembly as three of such repeats were well represented in the wheat reference genome obtained from short read data. Nevertheless, a more detailed analysis of a repeat with unit size of 2726 bp revealed that it was completely missing from the 7D pseudomolecule and was found in unassigned scaffolds (ChrUn) of the RefSeq v1.0 only. Thus, we concluded that this type of repeats may cause shrinking of pseudomolecules without impacting size of the entire assembly. Our results are in line with a finding that 27% centromeric sequences, identified by association with a centromere-specific histone H3 variant and highly enriched in centromere-specific repeats, were found in ChrUn of the RefSeq v1.0 [3]. This indicates that advanced assemblers can to some extent assemble shorter arrays of tandemly organised repeats but integration of these arrays in the pseudomolecule context may still pose a substantial challenge.

The TaeCsTr99 repeat was also found underrepresented and likely misassembled in the 7D pseudomolecule of Triticum v3.1, generated from both Illumina and PacBio reads. Moreover, both tested wheat whole-genome assemblies failed in discriminating two similar arrays located 225 kb apart, which could only be resolved after nanopore sequencing of BAC clones. This approach was successful not only because it employed a technology that provides reads exceeding the length of the whole array, but also because it leveraged the separation of the two arrays into the individually sequenced BAC clones. Interestingly, the identification of relatively long arrays of tandem repeats in BAC clones contradicts the finding of [11] that the tandem repeats are underrepresented in BAC libraries because of their toxicity for bacteria. The organisation of the whole tandem arranged region was resolved thanks to the application of the OM of the 7DS arm, which provided a reference for alignment of various sequences and revealed existing gaps and misassemblies. Nevertheless, the full potential of this genomic resource could not be exploited because none of the repeats analysed comprised a *Bsp*QI site (GCTCTTC) labelled in the 7DS OM. Consequently, the repeat arrays could not be recognised in the map through a specific labelling pattern, but appeared as longer regions devoid of labels. This shortage of the method might be overcome by the application of a new approach based on CRISPR-mediated labelling of specific sequences in the context of the optical map [35], which may facilitate straightforward mapping and quantifying of any repeat of interest.

### 2.2. Minor 45S rDNA Locus in Barley Chromosome 1H

Our second target was a minor 45S ribosomal DNA locus in barley chromosome 1H, identified by in situ hybridisation in various barley cultivars [32,33]. To access this locus, we first reconstructed the 1H-specific rDNA unit from 1H-specific paired-end Illumina reads (File S3). The unit sequence with the length of 8407 bp was then used to search 1H pseudomolecule of the barley 'Morex' reference genome [2]. Fragments of the unit were found between 139.05 Mb and 139.33 Mb of the 1H pseudomolecule, which fits well with the cytogenetic location of the rDNA locus at ~60% of the short arm of 1H [32], but we did not identify a regular rDNA array at this locus. To investigate completeness of the sequence in this region, we aligned it to available OM of barley cv. Morex [2], which identified OM contig 310 spanning over the region (Figure 4). Central part of the contig 310 did not align to the pseudomolecule and showed a regular labelling pattern with label spacing of approximately 5 kb. We compared it with the label pattern predicted for tandemly organised rDNA units. The reconstructed rDNA unit sequence comprised of three *Bsp*QI sites (File S3), but two of them were located just 1133 bp apart, which is too close for them to be discriminated in optical maps generated on the Bionano Genomics Irys platform. Thus, the labels associated with the *Bsp*QI sites were predicted to generate a composed pattern alternating ~3.5- and ~4.9-kb units (Figure S6A). This roughly corresponded to the pattern seen in the OM, with a discrepancy relating to the predicted ~3.5 kb "restriction fragment", which was not apparent in the optical map. This fragment covers the intergenic spacer (IGS) that comprises of two types of shorter tandem repeats with 78- and 135-bp unit length, respectively. It is likely that these repeats are collapsed in our consensus sequence and their real number is larger, extending the proposed IGS size by as much as 1.5 kb. This hypothesis was supported by the analysis of several partial rDNA units found in the 1H pseudomolecule, which were showing for both spacer repeats a higher number than included in our rDNA consensus sequence. Thus, we suggest that the complete size of the 1H rDNA unit is ~9.9 kb, which is supported by findings of [36] who identified in barley ribosomal DNA units of two sizes, 9.9 kb and 9 kb. Co-localisation of the blastn hits for 45S rDNA with the array in the optical map and the reported cytogenetic position lead us to the conclusion that the ~470-kb long array comprising ~47 putative rDNA units represents the minor 1H rDNA locus detected by in situ hybridisation. Our copy number estimate is close to that of [32] who quantified rRNA genes in the 1H chromosomes by in situ hybridization and proposed 50-100 copies in this locus. The slight discrepancy could be due to using a different barley cultivar (Morex vs. Sultan).

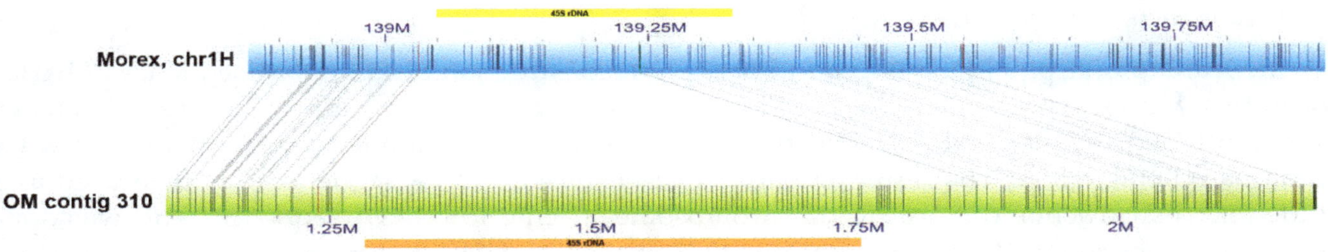


**Figure 4.** Positioning of the 45S rDNA locus in barley chromosome1H. Alignment of barley 'Morex' 1H pseudomolecule [2] (blue bar) to 'Morex' OM contig 310 (green bar) revealed a tandemly organized repeat with ~5 kb label spacing (highlighted by orange bar) missing in the pseudomolecule. Co-localisation with a cluster of 45S rDNA fragments in the sequence assembly (highlighted by yellow bar) indicates that the array represents the 1H rDNA locus.

Using the optical map, we identified a 470-kb segment that is absent from the Morex 1H pseudomolecule. We also performed blastn search for the consensus 45S rDNA sequence in chromosomes 5H and 6H, which are known to harbour major barley rDNA loci comprising thousands of genes [32,36], and in unassigned scaffolds of the 'Morex' assembly [2]. The search failed in revealing rDNA arrays in any of the datasets and identified fragments of the rDNA units only. This indicates that the missing rDNA loci contribute significantly to the dark matter of the cereal genomes.

## 3. Materials and Methods

### 3.1. *De Novo Identification of Wheat Tandem Repeats*

In order to identify new tandem repeats specific for the short arm of wheat chromosome 7D (7DS), we randomly selected 3.6 million reads obtained by Illumina sequencing multiple-displacement-amplified (MDA) DNA of flow-sorted 7DS [37]. Raw reads were examined and filtered by quality using FastQC and Trimmomatic tool. Repeat identification was performed employing similarity-based clustering of paired-end (2 × 100 nt) Illumina reads using local installation of the RepeatExplorer pipeline [23]. The pipeline employs graph representation of read similarities to find clusters of frequently overlapping reads corresponding to various repetitive elements or their parts. Putative tandem repeats were identified based on circular topology of their graphs [22] and tandem structure of contigs assembled from the reads within individual clusters. Sequences of the assembled contigs were then used to design PCR primers to verify the presence of corresponding sequences in the wheat genome (Table S1). In addition, the amplified fragments were cloned using the TOPO-TA Cloning Kit for Sequencing (Invitrogen, Carlsbad, CA, USA), selected clones were verified by sequencing and used as probes for in situ hybridisation experiments.

### 3.2. In Situ Hybridisation

For the in situ hybridisation experiment, we employed seeds of *Triticum aestivum* L., cv. Chinese Spring, kindly provided by Dr. Pierre Sourdille (INRA, Clermont-Ferrand, France). Seed germination, cell cycle synchronisation, metaphase accumulation and squash preparations were performed from wheat root tip meristems according to [38] with minor modifications. Metaphase accumulation was done by incubating root tips in 2.5 μM amiprophos-methyl for 2h in the dark at 25 °C. Inserts of clones bearing particular repeats were amplified using M13 primers and the PCR products were labeled by biotin using BioNick™ Labeling System (Invitrogen, Carlsbad, CA, USA). GAA microsatellite, used for identification of wheat chromosomes, was labelled by digoxigenin. Biotin- and digoxigenin-labeled probes were detected using streptavidin-Cy3 (Invitrogen, Carlsbad, CA, USA) and anti-digoxigenin-fluorescein (Roche, Basel, Switzerland), respectively. Chromosomes were counterstained with 4′,6′-diamidino-2-phenylindole (DAPI) and the preparations were imaged using Axio Imager Z.2 Zeiss microscope (Zeiss, Oberkochen, Germany) equipped with a CCD camera.

### 3.3. Reconstruction of Barley 1H rDNA Unit

RepeatExplorer pipeline was used to perform reconstruction of 45S rDNA sequence of barley. To do this, whole-genome paired-end (2 × 100 nt) Illumina reads of barley (*Hordeum vulgare*) cv. Morex (SRR490932) were downloaded from the SRA database, trimmed to quality and used for graph-based clustering. The resulting barley consensus 45S rDNA sequence was then used as a guide for reconstruction of a barley 1H chromosome-specific 45S rDNA sequence. This was done using online version of RepeatExplorer pipeline on the Galaxy platform and applying paired-end (2 × 100 nt) Illumina reads from flow-sorted 1H chromosome [39] of *H. vulgare* cv. Morex (SRR490144). The graph-based clustering resulted in five clusters homologous to the barley consensus 45S rDNA. The 1H-specific rDNA unit was then assembled manually utilizing the barley consensus 45S rDNA as a reference.

### 3.4. Application of Optical Maps

To validate sequences and analyse repeats in wheat 7DS and barley 1H chromosome, we employed available optical (BNG) maps constructed from 7DS chromosome arm of wheat cv. Chinese Spring [26] and the whole genome of barley cv. Morex [31], respectively. Both maps were assembled from single molecule data obtained after labelling molecules at Nt.*Bsp*QI nicking sites (motif GCTCTTC). Comparison of the optical maps with sequences was carried out using the IrysView 2.5.1 software package (Bionano Genomics, San Diego, CA, USA). For the alignment, cmap files were generated from fasta files of particular sequences. Query-to-anchor comparison was done with default parameters and *P*-value threshold of $1e^{-10}$.

### 3.5. Nanopore Sequencing

To resolve organization of the TaeCsTr99 repeats in the wheat genome, nanopore sequencing was conducted on 7DS BAC clones TaaCsp7DS028N04 (28N04) and TaaCsp7DS104G18 (104G18) from the 'Chinese Spring' 7DS arm-specific BAC library [40]. BAC DNA was extracted using alkaline lysis method followed by phenol-chloroform extraction and ethanol precipitation. Finally, the DNA was purified by incubating with 1:1 AMPure XP beads (Beckman Coulter, Miami, FL, USA) for 5 min and eluted into 30 μl 10 mM Tris, pH 8.5. Barcoded sequencing libraries were prepared from 700 ng DNA per BAC clone using Rapid Barcoding Sequencing Kit (SQK-RBK004; Oxford Nanopore Technologies, Oxford, UK) and sequenced together with additional ten clones on the MinION platform (Oxford Nanopore Technologies, Oxford, UK). Raw data were basecalled using Poretools 0.6.0 (https://github.com/arq5x/poretools, accessed on: 30 May 2019), demultiplexed using Porechop 0.2.3 (https://github.com/rrwick/Porechop, accessed on: 30 May 2019) and size-filtered >10 kb, which yielded 315 reads ranging from 10,003 to 101,160 bp, and 62 reads ranging from 10,082 to 149, 812 bp for the clone 28N04 and 104G18, respectively. Selected reads of 99,802 bp and 148,009 bp for 28N04 and 104G18, respectively, spanned the entire lengths of the respective clones.

## 4. Conclusions

Our study on tandem organised DNA repeats with unit sizes of 1.2–2.7 kb suggested that such repeats might be present in genome assemblies of large cereal genomes even if generated from short-read data. Nevertheless, they are typically comprised in short sequence contigs or scaffolds and thus may be difficult to incorporate into the pseudomolecules. We demonstrated that tandem repeats could be identified by a dedicated bioinformatics tool—RepeatExplorer—on a chromosome-specific basis and that nanopore sequencing of BAC clones provided a reliable approach to analysing organization of particular repeat arrays. We showed that an optical map might be useful for anchoring unassigned repeat-bearing scaffolds and for validating sequence assemblies in the problematic regions. The potential of the method was confirmed in our attempt to localise and characterise a minor 45S ribosomal DNA locus, which is missing in the reference genome of barley. Since BAC resources and optical maps are available for many plant species including major crops, the approaches presented in our study are widely applicable.

**Author Contributions:** Conceptualization, H.Š. and J.M.; FISH experiment, M.K., Repeat assembling P.N., J.M., V.K., E.H.; Repeat analyses V.K., Z.T., H.T.; Nanopore sequencing, Z.T.; Writing—Original Draft Preparation, V.K.; Writing—Review & Editing, H.Š., J.D.; Supervision, H.Š.; Funding Acquisition, H.Š., J.D. All authors have read and approved the manuscript.

**Acknowledgments:** We acknowledge the excellent assistance of Andrea Koblížková in validating the wheat tandem repeats. Computational resources were provided by the ELIXIR-CZ project (LM2015047).

## Abbreviations

| | |
|---|---|
| 7DS | Short arm of wheat chromosome 7D |
| BAC | Bacterial Artificial Chromosome |
| BNG mapping | Bionano genome mapping |
| CRISPR | Clustered Regularly Interspaced Short Palindromic Repeats |
| DAPI | 4,6-diamidino-2-phenylindole |
| FISH | Fluorescence in situ hybridisation |
| IGS | Intergenic spacer |
| ISH | In situ hybridisation |
| IWGSC | International Wheat Genome Sequencing Consortium |
| MDA | Multiple Displacement Amplification |
| OM | Optical map |
| ONT | Oxford Nanopore Technologies |

| | |
|---|---|
| PacBio | Pacific Biosciences |
| rDNA | Ribosomal DNA |
| SMRT | Single-molecule real-time |
| SRA | Sequence read archive |

## References


1. Avni, R.; Nave, M.; Barad, O.; Baruch, K.; Twardziok, S.O.; Gundlach, H.; Hale, I.; Mascher, M.; Spannagl, M.; Wiebe, K.; et al. Wild emmer genome architecture and diversity elucidate wheat evolution and domestication. *Science* **2017**, *357*, 93–97. [CrossRef] [PubMed]
2. Mascher, M.; Gundlach, H.; Himmelbach, A.; Beier, S.; Twardziok, S.O.; Wicker, T.; Radchuk, V.; Dockter, C.; Hedley, P.E.; Russell, J.; et al. A chromosome conformation capture ordered sequence of the barley genome. *Nature* **2017**, *544*, 427–433. [CrossRef] [PubMed]
3. The International Wheat Genome Sequencing Consortium (IWGSC). Shifting the limits in wheat research and breeding using a fully annotated reference genome. *Science* **2018**, *361*. [CrossRef]
4. Maccaferri, M.; Harris, N.S.; Twardziok, S.O.; Pasam, R.K.; Gundlach, H.; Spannagl, M.; Ormanbekova, D.; Lux, T.; Prade, V.M.; Milner, S.G.; et al. Durum wheat genome highlights past domestication signatures and future improvement targets. *Nat Genet.* **2019**, *51*, 885–895. [CrossRef] [PubMed]
5. Stein, N. (IPK, Gatersleben, Germany). Personal communication, 2018.
6. Doležel, J.; Čížková, J.; Šimková, H.; Bartoš, J. One major challenge of sequencing large plant genomes is to know how big they really are. *Int. J. Mol. Sci.* **2018**, *19*, 3554. [CrossRef]
7. Brenchley, R.; Spannagl, M.; Pfeifer, M.; Barker, G.L.; D'Amore, R.; Allen, A.M.; McKenzie, N.; Kramer, M.; Kerhornou, A.; Bolser, D.; et al. Analysis of the bread wheat genome using whole-genome shotgun sequencing. *Nature* **2012**, *491*, 705–710. [CrossRef]
8. Martis, M.M.; Zhou, R.; Haseneyer, G.; Schmutzer, T.; Vrána, J.; Kubaláková, M.; König, S.; Kugler, K.G.; Scholz, U.; Hackauf, B.; et al. Reticulate evolution of the rye genome. *Plant Cell* **2013**, *25*, 3685–3698. [CrossRef]
9. Mayer, K.F.X.; Martis, M.; Hedley, P.E.; Šimková, H.; Liu, H.; Morris, J.A.; Steuernagel, B.; Taudien, S.; Roessner, S.; Gundlach, H.; et al. Unlocking the barley genome by chromosomal and comparative genomics. *Plant Cell* **2011**, *23*, 1249–1263. [CrossRef]
10. Mayer, K.F.X.; Rogers, J.; Doležel, J.; Pozniak, C.; Eversole, K.; Feuillet, C.; Gill, B.; Friebe, B.; Lukaszewski, A.J.; Sourdille, P.; et al. A chromosome-based draft sequence of the hexaploid bread wheat (*Triticum aestivum*) genome. *Science* **2014**, *345*, 1251788. [CrossRef]
11. Chaisson, M.J.; Wilson, R.K.; Eichler, E.E. Genetic variation and the *de novo* assembly of human genomes. *Nat. Rev. Genet.* **2015**, *16*, 627–640. [CrossRef]
12. Zimin, A.V.; Puiu, D.; Hall, R.; Kingan, S.; Clavijo, B.J.; Salzberg, S.L. The first near-complete assembly of the hexaploid bread wheat genome, *Triticum aestivum*. *Gigascience* **2017**, *6*, 1–7. [CrossRef]
13. Handa, H.; Kanamori, H.; Tanaka, T.; Murata, K.; Kobayashi, F.; Robinson, S.J.; Koh, C.S.; Pozniak, C.J.; Sharpe, A.G.; Paux, E.; et al. Structural features of two major nucleolar organizer regions (NORs), Nor-B1 and Nor-B2, and chromosome-specific rRNA gene expression in wheat. *Plant J.* **2018**, *96*, 1148–1159. [CrossRef]
14. Symonová, R.; Ocalewicz, K.; Kirtiklis, L.; Delmastro, G.B.; Pelikánová, Š.; Garcia, S.; Kovařík, A. Higher-order organisation of extremely amplified, potentially functional and massively methylated 5S rDNA in European pikes (*Esox* sp.). *BMC Genom.* **2017**, *18*, 391. [CrossRef] [PubMed]
15. Appels, R.; Gerlach, W.L.; Dennis, E.S.; Swift, H.; Peacock, W.J. Molecular and Chromosomal Organization of DNA Sequences Coding for the Ribosomal RNAs in Cereals. *Chromosoma* **1980**, *78*, 293–311. [CrossRef]
16. Rayburn, A.L.; Gill, B.S. Use of biotin-labeled probes to map specific DNA sequences on wheat chromosomes. *Heredity* **1985**, *76*, 78–81. [CrossRef]
17. Mukai, Y.; Endo, T.R.; Gill, B.S. Physical mapping of the 18S.26S rRNA multigene family in common wheat: Identification of a new locus. *Chromosoma* **1991**, *100*, 71–78. [CrossRef]
18. Leitch, I.J.; Leitch, A.R.; Heslop-Harrison, J.S. Physical mapping of plant DNA sequences by simultaneous in situ hybridization of two differently labelled fluorescent probes. *Genome* **1991**, *34*, 329–333. [CrossRef]

19. Mukai, Y.; Nakahara, Y.; Yamamoto, M. Simultaneous discrimination of the three genomes in hexaploid wheat by multicolor fluorescence in situ hybridization using total genomic and highly repeated DNA probes. *Genome* **1993**, *36*, 489–494. [CrossRef] [PubMed]
20. Jiang, J.; Gill, B.S. New 18S. 26S ribosomal RNA gene loci: Chromosomal landmarks for the evolution of polyploid wheats. *Chromosoma* **1994**, *103*, 179–185. [CrossRef]
21. Brandes, A.; Röder, M.S.; Ganal, M.W. Barley telomeres are associated with two different types of satellite DNA sequences. *Chromosome Res.* **1995**, *3*, 315–320. [CrossRef]
22. Novák, P.; Neumann, P.; Macas, J. Graph-based clustering and characterization of repetitive sequences in next-generation sequencing data. *BMC Bioinform.* **2010**, *11*, 378. [CrossRef]
23. Novák, P.; Neumann, P.; Pech, J.; Steinhaisl, J.; Macas, J. RepeatExplorer: A Galaxy-based web server for genome-wide characterization of eukaryotic repetitive elements from next-generation sequence reads. *Bioinformatics* **2013**, *29*, 792–793. [CrossRef]
24. Martis, M.M.; Klemme, S.; Banaei-Moghaddam, A.M.; Blattner, F.R.; Macas, J.; Schmutzer, T.; Scholz, U.; Gundlach, H.; Wicker, T.; Šimková, H.; et al. Selfish supernumerary chromosome reveals its origin as a mosaic of host genome and organellar sequences. *Proc. Natl. Acad. Sci. USA* **2012**, *109*, 13343–13346. [CrossRef]
25. Lam, E.T.; Hastie, A.; Lin, C.; Ehrlich, D.; Das, A.K.; Austin, M.D.; Deshpande, P.; Cao, H.; Nagarajan, N.; Xiao, M.; et al. Genome mapping on nanochannel arrays for structural variation analysis and sequence assembly. *Nat. Biotechnol.* **2012**, *30*, 771–777. [CrossRef]
26. Staňková, H.; Hastie, A.R.; Chan, S.; Vrána, J.; Tulpová, Z.; Kubaláková, M.; Visendi, P.; Hayashi, S.; Luo, M.C.; Batley, J.; et al. BioNano genome mapping of individual chromosomes supports physical mapping and sequence assembly in complex plant genomes. *Plant Biotechnol. J.* **2016**, *14*, 1523–1531. [CrossRef]
27. Luo, MC.; Gu, Y.Q.; Puiu, D.; Wang, H.; Twardziok, S.O.; Deal, K.R.; Huo, N.; Zhu, T.; Wang, L.; Wang, Y.; et al. Genome sequence of the progenitor of the wheat D genome *Aegilops tauschii*. *Nature* **2017**, *551*, 498–502. [CrossRef]
28. Zhu, T.; Wang, L.; Rodriguez, J.C.; Deal, K.R.; Avni, R.; Distelfeld, A.; McGuire, P.E.; Dvorak, J.; Luo, MC. Improved genome sequence of wild emmer wheat Zavitan with the aid of optical maps. *G3 (Bethesda)* **2019**, *9*, 619–624. [CrossRef]
29. Tulpová, Z.; Toegelová, H.; Lapitan, N.L.V.; Peairs, F.B.; Macas, J.; Novák, P.; Lukaszewski, A.J.; Kopecký, D.; Mazáčová, M.; Vrána, J.; et al. Accessing a Russian wheat aphid resistance gene in bread wheat by long-read technologies. *Plant Genome* **2019**, *12*, 1–11. [CrossRef]
30. Tulpová, Z.; Luo, M.C.; Toegelová, H.; Visendi, P.; Hayashi, S.; Vojta, P.; Paux, E.; Kilian, A.; Abrouk, M.; Bartoš, J.; et al. Integrated physical map of bread wheat chromosome arm 7DS to facilitate gene cloning and comparative studies. *N. Biotechnol.* **2019**, *48*, 12–19. [CrossRef]
31. Beier, S.; Himmelbach, A.; Colmsee, C.; Zhang, X.Q.; Barrero, R.A.; Zhang, Q.; Li, L.; Bayer, M.; Bolser, D.; Taudien, S.; et al. Construction of a map-based reference genome sequence for barley, *Hordeum vulgare* L. *Sci. Data* **2017**, *4*, 170044. [CrossRef]
32. Leitch, I.J.; Heslop-Harrison, J.S. Physical mapping of the 18S-5.8S-26S rRNA genes in barley by in situ hybridization. *Genome* **1992**, *35*, 1013–1018. [CrossRef]
33. Szakács, É.; Kruppa, K.; Molnár-Láng, M. Analysis of chromosomal polymorphism in barley (*Hordeum vulgare* L. ssp. *vulgare*) and between *H. vulgare* and *H. chilense* using three-color fluorescence in situ hybridization (FISH). *J. Appl. Genet.* **2013**, *54*, 427–433. [CrossRef]
34. Shoaib, M.; Baconnais, S.; Mechold, U.; Le Cam, E.; Lipinski, M.; Ogryzko, V. Multiple displacement amplification for complex mixtures of DNA fragments. *BMC Genom.* **2008**, *9*, 415. [CrossRef]
35. Zhang, D.; Chan, S.; Sugerman, K.; Lee, J.; Lam, E.T.; Bocklandt, S.; Cao, H.; Hastie, A.R. CRISPR-bind: A simple, custom CRISPR/dCas9-mediated labeling of genomic DNA for mapping in nanochannel arrays. *bioRxiv* **2018**, preprint. [CrossRef]
36. Gerlach, W.L.; Bedbrook, J.R. Cloning and characterization of ribosomal RNA genes from wheat and barley. *Nucleic Acid Res.* **1979**, *7*, 1869–1886. [CrossRef]
37. Berkman, P.J.; Skarshewski, A.; Lorenc, M.T.; Lai, K.; Duran, C.; Ling, E.Y.; Stiller, J.; Smits, L.; Imelfort, M.; Manoli, S.; et al. Sequencing and assembly of low copy and genic regions of isolated *Triticum aestivum* chromosome arm 7DS. *Plant Biotechnol. J.* **2011**, *9*, 768–775. [CrossRef]

38. Karafiátová, M.; Bartoš, J.; Doležel, J. Localization of low-copy DNA sequences on mitotic chromosomes by FISH. In *Plant cytogenetics. Methods and Protocols*; Kianian, S.F., Kianian, P.M.A., Eds.; Humana Press: New York, NY, USA, 2016; Volume 1429, pp. 49–64.
39. Muñoz-Amatriaín, M.; Lonardi, S.; Luo, MC.; Madishetty, K.; Svensson, J.T.; Moscou, M.J.; Wanamaker, S.; Jiang, T.; Kleinhofs, A.; Muehlbauer, G.J.; et al. Sequencing of 15 622 gene-bearing BACs clarifies the gene-dense regions of the barley genome. *Plant J.* **2015**, *84*, 216–227. [CrossRef]
40. Šimková, H.; Šafář, J.; Kubaláková, M.; Suchánková, P.; Čihalíková, J.; Robert-Quatre, H.; Azhaguvel, P.; Weng, Y.; Peng, J.; Lapitan, N.L.V.; et al. BAC Libraries from wheat chromosome 7D: Efficient tool for positional cloning of aphid resistance genes. *J. Biomed. Biotechnol.* **2011**, *2011*, 302543. [CrossRef]

7

# Molecular Advances on Phytases in Barley and Wheat

**Claus Krogh Madsen and Henrik Brinch-Pedersen ***

Department of Molecular Biology and Genetics, Research Center Flakkebjerg, Aarhus University, 4200-Slagelse, Denmark; ClausKrogh.Madsen@mbg.au.dk
* Correspondence: hbp@mbg.au.dk

**Abstract:** Phytases are pro-nutritional enzymes that hydrolyze phytate and make associated nutrients, such as phosphorous, iron, and zinc, bioavailable. Single-stomached animals and humans depend on phytase supplied through the diet or the action of phytase on the food before ingestion. As a result, phytases—or lack thereof—have a profound impact on agricultural ecosystems, resource management, animal health, and public health. Wheat, barley and their Triticeae relatives make exceptionally good natural sources of phytase. This review highlights advances in the understanding of the molecular basis of the phytase activity in wheat and barley, which has taken place over the past decade. It is shown how the phytase activity in the mature grains of wheat and barley can be ascribed to the *PAPhy_a* gene, which exists as a single gene in barley and in two or three homeologous copies in tetra- and hexaploid wheat, respectively. It is discussed how understanding the function and regulation of *PAPhy_a* may support the development of improved wheat and barley with even higher phytase activity.

**Keywords:** phytase; wheat; barley; purple acid phosphatase phytase; PAPhy; mature grain phytase activity (MGPA)

---

## 1. Introduction

Phytases (myo-inositol hexakisphosphate 3-,6- and 5-phosphohydrolase, EC 3.1.3.8, EC 3.1.3.26 and EC 3.1.3.72) are phosphatases that can initiate the stepwise hydrolysis of phytate (IP6, myoinositol-(1,2,3,4,5,6)-hexakisphosphate) and thereby provide phosphate (P), inositol phosphates, and inositol for a range of cellular activities [1]. In addition to purely scientific inquiries, phytase research has for many years been driven by the urgent need for improving utilization of phytate-phosphorus in diets for single-stomached animals, such as pigs and poultry, and to reduce the anti-nutritional effect of non-digested IP6 chelating micronutrients in the digestive tracts of humans and animals. As such, phytases can be regarded as tools for managing global phosphate resources and for alleviating human micro-nutrient deficiencies mainly in the developing world.

IP6 is the main storage form of phosphate in plants, typically amounting 2/3 of the total P content in the seed (Table 1). IP6 is a strong chelator and exists in the plant seeds as an insoluble mixed salt with cations called phytin. In cereals and many other plant seeds, phytin forms spherical crystalloid inclusions called globoids inside protein storage vacuoles. The globoids are the principal site of phosphorous (P), potassium (K) and magnesium (Mg) in the mature cereal grain but they also contain calcium (Ca), iron (Fe), zinc (Zn), copper (Cu), manganese (Mn), sodium (Na), sulfur (S), and protein [2,3].

**Table 1.** Total P, IP6 bound P, proportion of IP6 bound P, and phytase activity in wheat, barley and other cereals seeds. The number of samples is n and ± denotes the standard deviation. The range is given if $n \geq 2$. Data from [4] [1], [5] [2], [6] [3].

| Cereal | $n$ | Total P (% of Dry Matter) | Total IP6 P (% of Dry Matter) | Percent IP6 out of Total P | Phytase Activity (FTU/kg) * |
|---|---|---|---|---|---|
| Wheat [1] | 13 | 0.33 ± 0.02 | 0.22 ± 0.02 | 67 ± 4.8 | 1193 ± 223 |
| Wheat [2] | 18 | 0.40 ± 0.04 | 0.29 ± 0.04 | 73 ± 8.1 | 2886 ± 645 |
| Wheat [3] | 30 | 0.29 ± 0.03 | 0.23 ± 0.03 | 79 ± 0.07 | 1637 ± 275 |
| Barley [1] | 9 | 0.37 ± 0.02 | 0.22 ± 0.01 | 60 ± 2.4 | 582 ± 178 |
| Barley [2] | 15 | 0.42 ± 0.4 | 0.26 ± 0.03 | 63 ± 3.5 | 2323 ± 648 |
| Barley [3] | 21 | 0.31 ± 0.03 | 0.19 ± 0.02 | 61 ± 0.04 | 1016 ± 330 |
| Rye [1] | 2 | 0.36 (0.35-0.36) | 0.22 (0.20-0.23) | 61 ± (56 -66) | 5130 (4132-6127) |
| Rye [2] | 13 | 0.36 ± 0.02 | 0.24 ± 0.2 | 67 ± 5.0 | 6016 ± 1578 |
| Rye [3] | 6 | 0.34 ± 0.03 | 0.20 ± 0.01 | 59 ± 0.02 | 5147 ± 649 |
| Triticale [1] | 6 | 0.37 ± 0.02 | 0.25 ± 0.02 | 67 ± 3.7 | 1688 ± 227 |
| Triticale [2] | 12 | 0.40 ± 0.03 | 0.28 ± 0.03 | 70 ± 5.4 | 2799 ± 501 |
| Oats [1] | 6 | 0.36 ± 0.03 | 0.21 ± 0.04 | 59 ± 11 | 42 ± 50 |
| Oats [2] | 6 | 0.37 ± 0.01 | 0.25 ± 0.02 | 67 ± 5.4 | 496 ± 35 |
| Oats [3] | 9 | 0.29 ± 0.02 | 0.17 ± 0.03 | 59 ± 0.07 | 84 ± 39 |
| Maize [1] | 11 | 0.28 ± 0.03 | 0.19 ± 0.03 | 68 ± 5.9 | 15 ± 18 |
| Maize [3] | 7 | 0.32 ± 0.01 | 0.18 ± 0.01 | 78 ± 0.01 | 70 ± 7 |
| Rice [4] | 1 | | | | 72 |

* One FTU is the amount of enzyme that liberates 1 μmol of inorganic phosphorus per minute from sodium phytate at pH 5.5 and 37 °C.

Micronutrients are also chelated by phytate in food and feed, and hydrolysis is most wanted for improving micronutrient bioavailability. The anti-nutritional effect of phytate is in particular regarded as critical for Fe and Zn, where phytate is considered the single most important anti-nutritional compound for the bioavailability of these two micronutrients [7]. Iron deficiency is the primary cause of anemia and ranks among the most widespread nutrient deficiencies, estimated to affect 1.6 billion people worldwide [8]. Iron deficiency anemia has been linked to maternal and prenatal mortality, and to impairment of cognitive skills and physical activity [9]. For zinc, around 800,000 child deaths worldwide per year are attributable to Zn deficiency [10] because it significantly increases the risk of diarrhea, pneumonia, and malaria. Moreover, Zn deficiency has been linked to the morbidity and mortality of children younger than five [11].

Humans and single-stomached animals have insufficient phytase activity in their digestive tract unless it is provided by the diet. Unfortunately, major food and feed components like rice, maize and soybeans contribute with phytate but negligible phytase [4]. Because of the missing phytase activity, the phytate passes largely un-digested through the single-stomached animals' digestive tract and enters the environment when their manure is spread on agricultural fields. Moreover, to ensure that farm animals get the phosphate needed, bio-available mined P is added to the feed. However, this strategy has become critical in many regions of the world where intense livestock production and spreading of manure with high levels of undigested phytate P on oversupplied agricultural soil leads to run-off of phosphorus to aquatic ecosystems. The resulting eutrophication is a severe environmental risk [12]. However, also from a resource perspective, inefficient utilization of plant phytate P is inappropriate. P is a non-renewable resource, essential for efficient agricultural production, and complete depletion of mined P will have unmanageable consequences for global food production [13].

## 2. Plant and Microbial Phytases

IP6 is resistant to most phosphatases whereas the lower inositol phosphates can be degraded by a wider range of phosphatases. Phosphatases that can initiate the dephosphorylation of IP6 are classified as phytases. So far, four classes of phytases have been identified: (1) Histidine acid phosphatase (HAP), (2) purple acid phosphatase (PAP), (3) cysteine phosphatase (CP) and (4) β-propeller phytase (BPP). Each phytase type has unique structural features due to their distinct catalytic apparatus that allows them to utilize phytate as a substrate in various environments [14].

For decades, applied phytase research was focusing mainly on microbial phytases and to our knowledge, all commercial phytases currently used for feed supplementation are microbial enzymes belonging to the HAP class [14]. Similarly, until recently, microbial HAP phytases were used exclusively for increasing plant seeds phytase activity through transgenesis. However, scientific achievements in recent years have led to a substantially increased knowledge based on the complements of phytases, in particular barley and wheat, and have demonstrated significant potentials of their phytases as highly stable and potent enzymes with potentials both in feed and food (see later).

## 3. Mature Grain Phytase Activity

When hydrated, the mature seed tissues activate a battery of preformed hydrolytic enzymes that degrades the large internal pool of IP6 but also storage compounds like lipids, carbohydrates, and proteins. When ungerminated seeds are used as feedstuffs, this battery of enzymes constitutes all plant-derived hydrolytic activities. We refer to this as the first wave of activity and for phytase, it constitutes what is called the mature grain phytase activity (MGPA). In parallel with imbibition, the embryo synthesizes and secretes the plant hormone gibberellic acid. The aleurone and the scutellum layer of the embryo are thereby turned into secretory tissues where a wide range of hydrolytic enzymes are synthesized and secreted into the endosperm for degradation of cell walls, starch grains and storage proteins—the second wave of hydrolysis [2].

Cereals generally express phytases to assist in their IP6 metabolism but the MGPA varies several orders of magnitude. This is in strong contrast to the modest variation in total and proportional content of seed IP6 (Table1).

## 4. Classes of Phytases in Barley and Wheat

In wheat and barley, two types of phytases have been described, phytases belonging to the HAP class and phytases belonging to the PAP class of phosphatases [1]. The HAP phytases belong to the multiple inositol polyphosphate phosphatase (MINPP) group [15]. MINPP phytases have been reported to be expressed both during grain development, and thereby, potentially contribute to the MGPA, and during germination contributing to the second wave of phytate hydrolysis. The PAP phytases are represented by the TaPAPhy_a/bs from wheat and the HvPAPhy_a/bs from barley, respectively. Expression analysis showed that *PAPhy_a* genes are preferentially expressed during grain filling whereas the *PAPhy_b* genes are preferentially expressed during germination [16]. The Km value with phytate as a substrate for recombinant wheat MINPP rTaPhyIIa2 phytase and barley rHvPhyIIb phytase is around ten-fold higher than for the rTaPAPhy_a/b and rHvPAPhy_a/b PAP phytases, indicating PAPhys to be more potent phytases than the HAPhys (Table 2).

**Table 2.** Kinetic parameters from HAP and PAPhy phytases from wheat, barley, and *Aspergillus ficuum*. Data from [1] [15], [2] [16], [3] [17].

| Class | Enzyme | *Km* (μM) | *Vmax* (μmol/(min × mg)) | *Kcat* ($s^{-1}$) | *Kcat/Km* ($s^{-1}M^{-1}$) | pH Optimum |
|---|---|---|---|---|---|---|
| PAP Phytases | rTaPAPhy_a1 [2] | 35 | 223 | 279 | $796 \times 10^4$ | 5.5 |
| | rTaPAPhy_b1 [2] | 45 | 216 | 270 | $600 \times 10^4$ | 5 |
| | rHvPAPhy_a [2] | 36 | 208 | 260 | $722 \times 10^4$ | |
| | rHvPAPhy_b [2] | 46 | 202 | 253 | $550 \times 10^4$ | |
| HAP Phytases | rTaPhyIIa2 [1] | 246 | | | | 4.5 |
| | rHvPhyIIb [1] | 334 | | | | 4.5 |
| | *A. ficuum* phytase [3] | 27 | - | 348 | $129 \times 10^5$ | 5.5 |

The contribution of HvPAPhy_a and HvPAPhy_b to the total MGPA in barley was recently evaluated using CRISPR/Cas and TALEN [18]. TALEN- and CRISPR/Cas9 were used for introducing targeted mutations in the promoter of the barley phytase gene *HvPAPhy_a*. Barley lines with substantial deletions in the *HvPAPhy_a* promoter and 5′CDS retained <5% normal MGPA. This confirms that the barley PAPhy_a enzyme is the main contributor to the MGPA and can be regarded as the main target for modulating MGPA.

## 5. Biochemical Properties and Storage of the PAPhys

Wheat and barley mainly store phytate in the protein storage vacuoles (PSVs) of the aleurone layer. PAPhy accumulated during grain filling is localized in the same organelles [16]. This suggests that some mechanism protects the phytate from hydrolysis during grain filling. The PSV's are rapidly acidified in response to gibberellic acid as germination commences. A decrease from pH 6.6 to 5.9 was reported in the PSV's of barley protoplasts incubated with 5 μM $GA_3$ and the authors speculated that pH might play a crucial role in regulating vacuolar hydrolases [19]. Recombinant wheat phytases rTaPAPhy_a1 and rTaPAPhy_b1 showed pH optima of 5.5 and 5, respectively [16]. Optima of pH 5 and 6 were measured for seed purified barley phytases P1 (=HvPAPhy_b) and P2 (=HvPAPhy_a) respectively [20]. This shows that the PAPhy's are most active when the PSV is in the acidified lytic, state and the higher pH optimum of preformed PAPhy_a may even be an adaptation, which enhances activity in the earliest stages of germination. Nevertheless, rTaPAPhy_a retained some activity up to pH 7.5 so pH regulation of the enzyme alone does not offer a satisfying explanation for the protection of phytate against premature hydrolysis. The second layer of pH-dependent protection is provided by the substrate's organization into globoids, which provides some degree of water exclusion and steric hindrance [3]. A membrane surrounds the globoids and immuno-gold localization suggests that PAPhy is located outside this membrane [16,21]. The membrane, therefore, seems to provide an additional layer of protection, by physical separation. The temperature optimum of the recombinant wheat enzymes was 50 and 55 °C, respectively. The temperature curves showed a broad peak with 50% activity already at 30–35 °C but decreasing sharply around 60 °C. Optima of 55 and 45 °C were reported for the seed purified HvPAPhy_a and HvPAPhy_b respectively [20]. The kinetic parameters of recombinant wheat and barley phytases at 36 °C and pH 5 are summarized in Table 2. The corresponding values for *Aspergillus ficuum* phytase are given for comparison [17]. The values for the PAPhys are very similar.

## 6. PAPhy Genetics

The *PAPhy_a* and *PAPhy_b* genes are paralogs, which originate from gene duplication in a common ancestor of wheat, barley, and rye (i.e., the Triticeae tribe) [22]. Rice, maize, and sorghum diverged from the Triticeae earlier and carry only one PAPhy gene whereas *Brachypodum distachyon* has the duplication but lack the conserved *PAPhy_a* promoter of the Triticeae (Figure 1) [22]. Allopolyploidzation has united the A and B genomes to form tetraploid wheat (e.g., durum wheat). *Triticum urartu* and

*Aegilops speltoides* are the closest living relatives of the A and B genomes, respectively. Additional hybridizations have added the *Ae. tauschii* derived D genome to produce hexaploid wheat (bread wheat, spelt) and the rye derived R genome to produce triticale (Figure 1) [23]. Allopolyploidization results in large-scale gene duplication because most genes in one parent will have a homolog in the other parent species. In allopolyploids, such sets of genes are termed homeologs. Some homeologs may be lost or translocated as the polyploid species continues to evolve but the Triticeae *PAPhy* gene copy number and chromosomal localization are highly conserved. A single locus of *PAPhy_a* and *PAPhy_b* resides on chromosome 5 and 3, respectively in barley and on the homologous chromosomes on the three subgenomes of wheat [22]. The sequenced diploid members of the *Triticum* and *Aegilops*, i.e., *Ae. tauschii*, *Ae. speltoides*, *Ae. sharonensis*, *T. Urartu*, and *T. monococcum* also have one *PAPhy_a* and one *PAPhy_b* [24]. The chromosomal localization in these relatives has not been determined but it is reasonable to expect a conserved synteny between wheat and its ancestors *T. urartu*, *Ae. Speltoides*, and *Ae. tauschii*. *Secale* provides an exception since some members of this tribe have two *PAPhy_a* loci. In the case of domesticated rye, one and two *PAPhy_a* variants were isolated from the cultivars Imperial and Picasso, respectively [25]. Rye is an outbreeding species, unlike wheat and barley, and tends to have higher allele heterogeneity. Therefore, it cannot be excluded that the two variants from Picasso are alleles of the same gene even though phylogeny suggests that one allele may have been introgressed from *S. strictum* [25]. Thus, it is not known with certainty if domesticated rye has one or two *PAPhy_a* loci or whether it is cultivar-dependent.

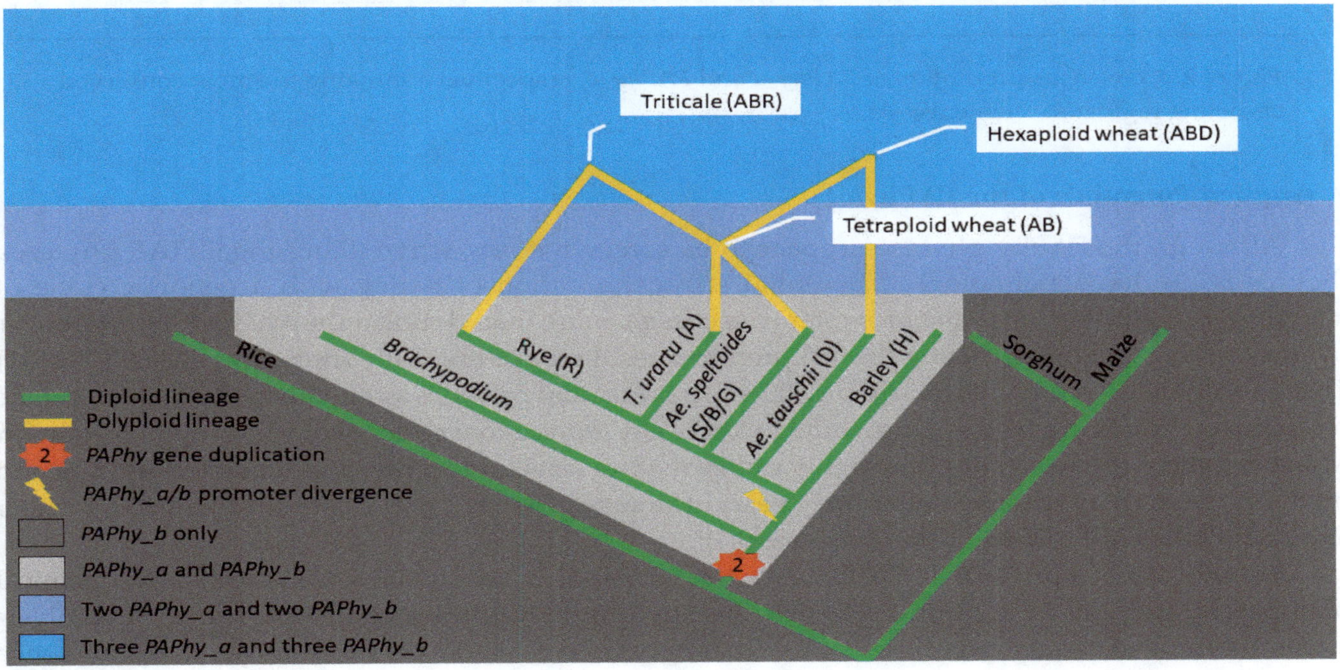


**Figure 1.** Key events in *PAPhy* evolution. Gene duplication and divergence of the PAPhy_a/b promoters and further duplications through polyploidization. For simplicity, rye is assumed to have just one *PAPhy_a* locus, and phylogenetic distance is not drawn to scale.

The intron/exon structure and the respective promoters of *PAPhy_a* and *PAPhy_b* are highly conserved. *PAPhy_a* has four introns and *PAPhy_b* has five. The position and, to a large extent, the size of the introns is conserved between Triticeae species and between the two paralogs [22]. Both genes have a core promoter of 3–400 base pairs which is conserved in all studied Triticeae. Both promoters have two TATA-boxes and upstream of those reside cis-acting regulatory elements consistent with the differential expression pattern of the two genes (Figure 2). For *PAPhy_b*, they are ABRE (abscisic-acid-responsive), TGACG (methyl-jasmonate-responsive) and—conserved in all examined *PAPhy_b* promoters—GARE (gibberellic acid responsive) [22]. The *PAPhy_a* promoter lacks these hormone responsive elements, except TGACG which is found at the very beginning of the conserved sequence. Instead, the most notable feature is a composite element with the consensus 5′ **GAACATG**AGT*CATGCATG* 3′ which is made of the GAMYB binding motif (bold) [26], the odd base palindrome/GCN4 (underlined) [27,28] and the RY element (italic) [29]. These are all elements associated with seed development and storage proteins. Between the composite element and the first TATA box is a G-box motif [22]. Deletion of the odd base palindrome and the RY-element reduce barley MGPA by approximately 40% whereas deletion of the whole element and an additional 10 base pairs 3′ reduce MGPA by 75%. Deletions immediately 3′ of the composite element have even more severe effects [14]. It is not clear if this is caused by an unknown cis-acting regulatory element at that position or by the change in distance between the composite element and elements further downstream.

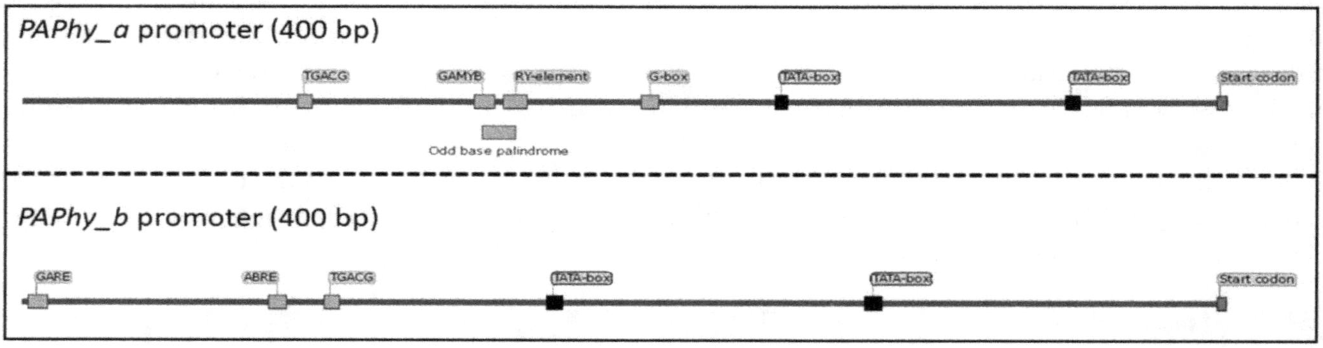


**Figure 2.** Consensus promoters of *PAPhy_a* and *PAPhy_b*, respectively, showing the most conserved cis-acting regulatory elements.

## 7. Applied Potentials of the PAPhys

Although the PAPhy phytases appear to be somewhat less active than fungal HAP phytases, they certainly have technological potential. Transformation of barley with a genomic clone of *HvPAPhy_a*, including its native promoter (cisgenesis), more than doubled the MGPA [30]. Moreover, overexpressing *HvPAPhy_a* using the 35S promoter resulted in a line with impressive 40.000 FTU [31]. This level of MGPA represents more than a 20 fold increase and provides so much activity that <2% of the recombinant grains in a feed mixture could theoretically replace conventional phytase supplementation. Moreover, the enzyme activity was highly stable, mature leaves and straw accumulated PAPhy_a and high phytase activity remained after three years of storage. The PAPhy_a was easily extracted in water and could be added to feed or other processes.

The efficacy of a phytase in the digestive tract depends in part on how well intrinsic biochemical parameters, such as pH optimum, match the environment but also on resistance to proteolysis and inhibition. Feeding experiments are, therefore, necessary to prove efficacy. Early evidence for the efficacy of PAPhy_a was provided by comparing phosphorous and calcium utilization in pigs fed maize and triticale, respectively [32]. In this case, triticale-based feed performed better on both parameters, indicating an effect of the higher MGPA associated with triticale. In humans, an intervention study compared wheat bran with or without phytase (native vs. heat treated) in ileostomy patients. This resulted in recoveries of 40% vs. 95% phytate in the ileostomy content [33]. It is also possible to utilize the MGPA to achieve dephytination during food or feed preparation. Up to 96% of the phytate in whole grain wheat bread could be removed by simply adjusting the pH during proofing to five [34]. Similarly, fonio (*Digitaria exilis*) porridge could be dephytinized in one hour at 50 °C and pH 5 when 25% of wheat was included. This was demonstrated to significantly increase iron absorption in West African women [35]. Taken together, these studies demonstrate a lot of potential for the utilization of endogenous Triticeae PAPhy_a in food and feed.

## 8. Achieving Higher MGPA with the PAPhys

Given the high conservation of the *PAPhy_a* coding sequence and highly similar enzymatic properties of all examined PAPhys, including PAPhy_b as well as rice and maize PAPhy, it seems unlikely that alternative *PAPhy_a* alleles, encoding decisively better enzymes, exist. On the other hand, increasing *PAPhy_a* expression has proven a viable strategy, as discussed. Efforts to increase MGPA by conventional breeding should, therefore, focus on the discovery of more highly expressed *PAPhy_a* alleles and the elimination of defective alleles.

Polyploids (e.g., wheat) are prone to acquire defective alleles because homeologs on the other subgenomes provide functional redundancy. Nevertheless, defective *PAPhy_a* alleles may reduce MGPA because the gene copies appear to function in an additive manner as demonstrated by cisgenic gene duplication [30]. Further support for this hypothesis is provided by the discovery of a defective *TaPAPhy_a2* gene in the wheat cultivars Skagen and Bob White that correlated with a lower MGPA compared to Chinese Spring, which has three normal loci [22,25]. A similar reduction of phytase activity has been observed in wheat lines with induced deleterious mutations in the *PAPhy_a* genes (author's unpublished results). It is not known how commonly and how many different defective alleles exist in the elite wheat gene pool. A systematic investigation of this combined with the development of markers for the defective alleles would be helpful in ensuring a minimum MGPA in new cultivars.

More active alleles might be found in crop relatives. Wild emmer (*T. dicocoides*) can be crossed directly with other tetraploid wheat (e.g., durum wheat) and with hexaploid wheat trough synthetic hexaploids. Alleles from *Ae. tauscii* can also be introduced by this route or through the "octo-amphiploid bridge" [36]. A higher allele diversity is expected for these species because they have not gone through the bottlenecks of domestication and hexaploidization [37]. A small sampling of wheat and the closest relatives support this assumption [25]. Furthermore, it was found that at least one exotic allele had already been introduced in the elite wheat gene pool in the pursuit of other breeding goals—from *T. timopheevii*. In addition to the abovementioned, einkorn (wild and domesticated), *T. urartu*, rye, and *Ae. speltoides* could be used but here the introgression is more complicated.

Another strategy is mutagenesis. A wheat mutant with two to three times increased MGPA has been patented by the authors and co-workers [38]. This "HighPhy" wheat has an SNP in the composite element discussed above. Incremental replacement (1/3, 2/3, and 3/3) of conventional wheat with HighPhy all improved calcium and phosphorous digestibility compared to the control diet in a comprehensive feeding experiment with broiler chicken. Complete replacement with HighPhy wheat even significantly outperformed a control to which the standard dose of microbial phytase was added [39].

With the exception of the Triticeae cereals, none of the major food and feed grains accumulate nutritionally sufficient amounts of phytase during grain development [4,5]. Rather, phytase is synthesized de novo during germination in response to gibberellic acid. Introducing a transgene driven by a seed development specific promoter has, therefore, been considered the only realistic approach to introduce preformed phytase in major grains, such as maize, rice, and soybean. However, the recent advances in genome editing may change this because they enable targeted modification of endogenous genes [40]. In theory, this can introduce preformed phytase activity in one of two ways (a) by modifying the promoter of endogenous phytase genes to become active during grain development or (b) by modifying the active site of phosphatases already expressed during grain development so they become phytases. It seems possible to pursue option "a" by combining the lessons of the natural evolution of the *PAPhy_a* promoter from the *PAPhy_b* counterpart with target species-specific information on cis-acting regulatory elements involved in seed development-specific expression. The pursuit of option "b" would have to be guided by structural biology but has the advantage that mutations could be tested in a heterologous system before the more laborious plant gene editing. It is also likely that this approach would require smaller changes to the target gene sequence because a single base mutation is enough to change one critical amino acid.

Proteinaceous inhibitors are known for many hydrolases including amylases, proteases, and xylanases. To our knowledge, such an inhibitor has never been reported for a phytase of microbial or plant origin. However, it was recently discovered that barley grain extracts do inhibit *Aspergillus* phytase. The inhibition could be reduced by the addition of pepstatin A. This suggests that the apparent inhibition is caused by an aspartic acid proteolytic activity [41]. It would represent a new and important variable, if plants use specialized proteases rather than conventional inhibitors to counteract phytases, e.g., from pathogens.

In conclusion, preformed wheat and barley phytase has the potential to counter the negative effects of phytate in the nutrition of single-stomached farm animals and humans alike. Realizing this potential depends on awareness because minor adjustments to processing may be needed. Generally, these adjustments should ensure that the phytase is not inactivated before ingestion or, alternatively, that the phytase has optimal conditions to perform the dephytination before inactivation. The higher the MGPA, the better the chances of a successful dephytination. Characterization of the *PAPhy_a* gene over the past decade should be most helpful for plant breeders who make MGPA a priority.

## References

1. Brinch-Pedersen, H.; Madsen, C.K.; Holme, I.B.; Dionisio, G. Increased understanding of the cereal phytase complement for better mineral bio-availability and resource management. *J. Cereal Sci.* **2014**, *59*, 373–381. [CrossRef]
2. Bethke, P. From Storage Compartment to Lytic Organelle: The Metamorphosis of the Aleurone Protein Storage Vacuole. *Ann. Bot.* **1998**, *82*, 399–412. [CrossRef]
3. Bohn, L.; Josefsen, L.; Meyer, A.S.; Rasmussen, S.K. Quantitative Analysis of Phytate Globoids Isolated from Wheat Bran and Characterization of Their Sequential Dephosphorylation by Wheat Phytase. *J. Agric. Food Chem.* **2007**, *55*, 7547–7552. [CrossRef] [PubMed]
4. Eeckhout, W.; De Paepe, M. Total phosphorus, phytate-phosphorus and phytase activity in plant feedstuffs. *Anim. Feed Sci. Technol.* **1994**, *47*, 19–29. [CrossRef]
5. Steiner, T.; Mosenthin, R.; Zimmermann, B.; Greiner, R.; Roth, S. Distribution of phytase activity, total phosphorus and phytate phosphorus in legume seeds, cereals and cereal by-products as influenced by harvest year and cultivar. *Anim. Feed Sci. Technol.* **2007**, *133*, 320–334. [CrossRef]
6. Viveros, A.; Centeno, C.; Brenes, A.; Canales, R.; Lozano, A. Phytase and Acid Phosphatase Activities in Plant Feedstuffs. *J. Agric. Food Chem.* **2000**, *48*, 4009–4013. [CrossRef]
7. Bouis, H.E. Special Issue on Improving Human Nutrition Through Agriculture. *Food Nutr. Bull.* **2000**, *21*, 351–578.
8. De Benoist, B.; Cogswell, M.; Egli, I.; McLean, E. Worldwide prevalence of anaemia 1993–2005. In *WHO Global Database of Anaemia*; WHO: Geneva, Switzerland, 2008.
9. Stoltzfus, R.J.; Mullany, L.; Black, R.E. Iron deficiency anemia. In *Comparative Quantification of Health Risks: Global and Regional Burden of Disease Attribution to Selected Major Risk Factors*; Ezzati, M., Lopez, A.D., Rodgers, A., Murray, C.J., Eds.; World Health Organization: Geneva, Switzerland, 2004; ISBN 9241580313.
10. Brown, K.H.; Rivera, J.A.; Bhutta, Z.A.; Gibson, R.S.; King, J.C.; Lonnerdal, B. Chapter 2: Assessment of the Risk of Zinc Deficiency in Populations. *Food Nutr. Bull.* **2004**, *25*, S130.
11. Yakoob, M.Y.; Theodoratou, E.; Jabeen, A.; Imdad, A.; Eisele, T.P.; Ferguson, J.; Jhass, A.; Rudan, I.; Campbell, H.; Black, R.E.; et al. Preventive zinc supplementation in developing countries: impact on mortality and morbidity due to diarrhea, pneumonia and malaria. *BMC Public Health* **2011**, *11*, S23. [CrossRef] [PubMed]
12. Schindler, D.W.; Carpenter, S.R.; Chapra, S.C.; Hecky, R.E.; Orihel, D.M. Reducing Phosphorus to Curb Lake Eutrophication is a Success. *Environ. Sci. Technol.* **2016**, *50*, 8923–8929. [CrossRef] [PubMed]
13. Cordell, D.; Drangert, J.-O.; White, S. The story of phosphorus: Global food security and food for thought. *Glob. Environ. Chang.* **2009**, *19*, 292–305. [CrossRef]

14. Lei, X.G.; Porres, J.M.; Mullaney, E.J.; Brinch-Pedersen, H. Phytase: Source, Structure and Application. In *Industrial Enzymes: Structure, Function and Applications*; Polaina, J., MacCabe, A., Eds.; Springer: Dordrecht, The Netherlands, 2007; pp. 505–530. ISBN 978-1-4020-5376-4.
15. Dionisio, G.; Holm, P.B.; Brinch-Pedersen, H. Wheat (*Triticum aestivum* L.) and barley (*Hordeum vulgare* L.) multiple inositol polyphosphate phosphatases (MINPPs) are phytases expressed during grain filling and germination. *Plant Biotechnol. J.* **2007**, *5*, 325–338. [CrossRef]
16. Dionisio, G.; Madsen, C.K.; Holm, P.B.; Welinder, K.G.; Jørgensen, M.; Stoger, E.; Arcalis, E.; Brinch-Pedersen, H. Cloning and characterization of purple acid phosphatase phytases from wheat, barley, maize, and rice. *Plant Physiol.* **2011**, *156*. [CrossRef] [PubMed]
17. Ullah, A.H.J.; Phillippy, B.Q. Substrate selectivity in Aspergillus ficuum phytase and acid phosphatases using myo-inositol phosphates. *J. Agric. Food Chem.* **1994**, *42*, 423–425. [CrossRef]
18. Holme, I.B.; Wendt, T.; Gil-Humanes, J.; Deleuran, L.C.; Starker, C.G.; Voytas, D.F.; Brinch-Pedersen, H. Evaluation of the mature grain phytase candidate HvPAPhy_a gene in barley (Hordeum vulgare L.) using CRISPR/Cas9 and TALENs. *Plant Mol. Biol.* **2017**, *95*, 111–121. [CrossRef]
19. Swanson, S.J.; Jones, R.L. Gibberellic Acid Induces Vacuolar Acidification in Barley Aleurone. *Plant Cell* **1996**, *8*, 2211, LP-2221. [CrossRef] [PubMed]
20. Greiner, R.; Jany, K.-D.; Larsson Alminger, M. Identification and Properties of myo -Inositol Hexakisphosphate Phosphohydrolases (Phytases) from Barley (Hordeum vulgare). *J. Cereal Sci.* **2000**, *31*, 127–139. [CrossRef]
21. Jiang, L.; Phillips, T.E.; Hamm, C.A.; Drozdowicz, Y.M.; Rea, P.A.; Maeshima, M.; Rogers, S.W.; Rogers, J.C. The protein storage vacuole. *J. Cell Biol.* **2001**, *155*, 991, LP-1002. [CrossRef]
22. Madsen, C.K.; Dionisio, G.; Holme, I.B.; Holm, P.B.; Brinch-Pedersen, H. High mature grain phytase activity in the Triticeae has evolved by duplication followed by neofunctionalization of the purple acid phosphatase phytase (PAPhy) gene. *J. Exp. Bot.* **2013**, *64*. [CrossRef]
23. IWGSC A chromosome-based draft sequence of the hexaploid bread wheat (Triticum aestivum) genome. *Science* **2014**, *345*, 1251788. [CrossRef]
24. IWGSC IWGSC-Blast. Available online: https://urgi.versailles.inra.fr/blast_iwgsc/blast.php (accessed on 17 May 2019).
25. Madsen, C.K.; Petersen, G.; Seberg, O.; Brinch-Pedersen, H. Evolution and diversity of PAPhy_a phytase in the genepool of wheat (Triticum aestivum L., Poaceae). *Genet. Resour. Crop Evol.* **2017**, *64*. [CrossRef]
26. Guo, W.; Yang, H.; Liu, Y.; Gao, Y.; Ni, Z.; Peng, H.; Xin, M.; Hu, Z.; Sun, Q.; Yao, Y. The wheat transcription factor TaGAMyb recruits histone acetyltransferase and activates the expression of a high-molecular-weight glutenin subunit gene. *Plant J.* **2015**, *84*, 347–359. [CrossRef]
27. De Pater, S.; Katagiri, F.; Kijne, J.; Chua, N.-H. bZIP proteins bind to a palindromic sequence without an ACGT core located in a seed-specific element of the pea lectin promoter. *Plant J.* **1994**, *6*, 133–140. [CrossRef] [PubMed]
28. Müller, M.; Knudsen, S. The nitrogen response of a barley C-hordein promoter is controlled by positive and negative regulation of the GCN4 and endosperm box. *Plant J.* **1993**, *4*, 343–355. [CrossRef]
29. Bäumlein, H.; Nagy†, I.; Villarroel, R.; Inzé, D.; Wobus, U. Cis-analysis of a seed protein gene promoter: the conservative RY repeat CATGCATG within the legumin box is essential for tissue-specific expression of a legumin gene. *Plant J.* **1992**, *2*, 233–239.
30. Holme, I.B.; Dionisio, G.; Brinch-Pedersen, H.; Wendt, T.; Madsen, C.K.; Vincze, E.; Holm, P.B. Cisgenic barley with improved phytase activity. *Plant Biotechnol. J.* **2012**, *10*. [CrossRef]
31. Holme, I.B.; Dionisio, G.; Madsen, C.K.; Brinch-Pedersen, H. Barley HvPAPhy_a as transgene provides high and stable phytase activities in mature barley straw and in grains. *Plant Biotechnol. J.* **2017**, *15*. [CrossRef]
32. Pointillart, A.; Fourdin, A.; Fontaine, N. Importance of Cereal Phytase Activity for Phytate Phosphorus Utilization by Growing Pigs Fed Diets Containing Triticale or Corn. *J. Nutr.* **1987**, *117*, 907–913. [CrossRef]
33. Sandberg, A.-S.; Andersson, H. Effect of Dietary Phytase on the Digestion of Phytate in the Stomach and Small Intestine of Humans. *J. Nutr.* **1988**, *118*, 469–473. [CrossRef] [PubMed]
34. Türk, M.; Carlsson, N.-G.; Sandberg, A.-S. Reduction in the Levels of Phytate During Wholemeal Bread Making; Effect of Yeast and Wheat Phytases. *J. Cereal Sci.* **1996**, *23*, 257–264. [CrossRef]

35. Koréissi-Dembélé, Y.; Fanou-Fogny, N.; Moretti, D.; Schuth, S.; Dossa, R.A.M.; Egli, I.; Zimmermann, M.B.; Brouwer, I.D. Dephytinisation with intrinsic wheat phytase and iron fortification significantly increase iron absorption from fonio (Digitaria exilis) meals in West African women. *PLoS ONE* **2013**, *8*, e70613. [CrossRef] [PubMed]
36. Singh, N.; Wu, S.; Tiwari, V.; Sehgal, S.; Raupp, J.; Wilson, D.; Abbasov, M.; Gill, B.; Poland, J. Genomic Analysis Confirms Population Structure and Identifies Inter-Lineage Hybrids in Aegilops tauschii. *Front. Plant Sci.* **2019**, *10*, 9. [CrossRef]
37. Feuillet, C.; Langridge, P.; Waugh, R. Cereal breeding takes a walk on the wild side. *Trends Genet.* **2008**, *24*, 24–32. [CrossRef] [PubMed]
38. Brinch-Pedersen, H.; Madsen, C.K.; Dionisio, G.; Holm, P.B. High expression cereal phytase gene. WO 2012/146597 Al. 2012. Available online: https://patents.google.com/patent/WO2012146597A1/en (accessed on 17 May 2019).
39. Scholey, D.; Burton, E.; Morgan, N.; Sanni, C.; Madsen, C.K.; Dionisio, G.; Brinch-Pedersen, H. P and Ca digestibility is increased in broiler diets supplemented with the high-phytase HIGHPHY wheat. *Animal* **2017**, *11*. [CrossRef]
40. Yin, K.; Gao, C.; Qiu, J.-L. Progress and prospects in plant genome editing. *Nat. Plants* **2017**, *3*, 17107. [CrossRef] [PubMed]
41. Bekalu, Z.E.; Madsen, C.K.; Dionisio, G.; Brinch-Pedersen, H. Aspergillus ficuum phytase activity is inhibited by cereal grain components. *PLoS ONE* **2017**, *12*. [CrossRef]

# 8

# RNA Sequencing-Based Bulked Segregant Analysis Facilitates Efficient D-genome Marker Development for a Specific Chromosomal Region of Synthetic Hexaploid Wheat

**Ryo Nishijima [1], Kentaro Yoshida [1,*], Kohei Sakaguchi [1], Shin-ichi Yoshimura [1], Kazuhiro Sato [2] and Shigeo Takumi [1,*]**

1 Graduate School of Agricultural Science, Kobe University, Rokkodai 1-1, Nada, Kobe 657-8501, Japan; ryosnishijima@gmail.com (R.N.); k.sakaguchi@brown.plala.or.jp (K.S.); shin.82353@gmail.com (S.-i.Y.)

2 Institute of Plant Science and Resources, Okayama University, Kurashiki, Okayama 710-0046, Japan; kazsato@rib.okayama-u.ac.jp

* Correspondence: kentaro.yoshida@port.kobe-u.ac.jp (K.Y.); takumi@kobe-u.ac.jp (S.T.).

**Abstract:** Common wheat originated from interspecific hybridization between cultivated tetraploid wheat and its wild diploid relative *Aegilops tauschii* followed by amphidiploidization. This evolutionary process can be reproduced artificially, resulting in synthetic hexaploid wheat lines. Here we performed RNA sequencing (RNA-seq)-based bulked segregant analysis (BSA) using a bi-parental mapping population of two synthetic hexaploid wheat lines that shared identical A and B genomes but included with D-genomes of distinct origins. This analysis permitted identification of D-genome-specific polymorphisms around the *Net2* gene, a causative locus to hybrid necrosis. The resulting single nucleotide polymorphisms (SNPs) were classified into homoeologous polymorphisms and D-genome allelic variations, based on the RNA-seq results of a parental tetraploid and two *Ae. tauschii* accessions. The difference in allele frequency at the D-genome-specific SNP sites between the contrasting bulks (ΔSNP-index) was higher on the target chromosome than on the other chromosomes. Several SNPs with the highest ΔSNP-indices were converted into molecular markers and assigned to the *Net2* chromosomal region. These results indicated that RNA-seq-based BSA can be applied efficiently to a synthetic hexaploid wheat population to permit molecular marker development in a specific chromosomal region of the D genome.

**Keywords:** allohexaploid; homoeolog; hybrid necrosis; molecular marker; wheat

---

## 1. Introduction

Wild relatives of common wheat (*Triticum aestivum* L.), including *Aegilops* species, constitute important genetic resources for common wheat breeding. Notably, *Aegilops tauschii* Coss. is the D-genome progenitor of common wheat, and a limited number of *Ae. tauschii* strains appear to have contributed to the speciation of common wheat through interspecific crossing to the cultivated tetraploid wheat (*Triticum turgidum* L. ssp. *durum*) strains carrying the A and B genomes and subsequent amphidiploidization about 8000 years ago in the Fertile Crescent [1,2]. Artificial replication of this evolutionary process can be achieved by generation of synthetic hexaploid wheat derived from the interspecific crosses between cultivated tetraploid wheat and *Ae. tauschii* [3]. Thus, agriculturally useful alleles of *Ae. tauschii* are directly available for common wheat breeding via the transmission from synthetic hexaploid wheat to common wheat [4–10]. In common wheat, the D genome is known to harbor much lower genetic diversity than do the A and B genomes, whereas introgression of the

*Ae. tauschii* natural variation via synthetic hexaploid wheat has been used successfully to enlarge D-genome diversity [11–14].

*Ae. tauschii* is widely distributed, and is present from northern Syria and southeastern Turkey to western China [15,16]. Recent studies on the population structure of *Ae. tauschii* showed that only one (TauL1) of three major lineages has contributed to the wide species range [2,17,18]. The other two major lineages, TauL2 and TauL3, are restricted to the Transcaucasus/Middle East region. TauL2 includes both subspecies *tauschii* and subspecies *strangulata*, and the TauL3 accessions are limited in Georgia [18].

The *Ae. tauschii* strains associated with the origin of common wheat are assumed to be the TauL2 lineage [2,17], and only limited reproductive barriers are thought to exist between tetraploid wheat and many of the TauL2 accessions [19]. TauL1 accessions frequently carry a hybrid incompatibility gene, designated *Net2*, that triggers low-temperature-induced necrotic cell death upon interspecific hybridization with tetraploid wheat, impeding the generation of synthetic hexaploid wheat [20,21]. Thus, *Net2* is a major reproductive barrier for breeding use of the TauL1 accessions, and the development of markers closely linked to *Net2* is needed for efficient use of the TauL1 gene pool. A fine map for the *Net2* chromosomal region already has been constructed on the short arm of chromosome 2D [22].

Genomic approaches using next-generation sequencing (NGS) techniques have been applied to analysis of the genomes of the wild relatives of domesticated crops, expanding the genetic resources available for crop improvement [23]. In common wheat, the D-genome markers remain much less developed than those of the A and B genomes, whereas recent progress using the NGS technique has facilitated an increase in the number of D-genome markers [24]. RNA sequencing (RNA-seq) of the *Ae. tauschii* accessions has generated a huge number of genome-wide polymorphisms, including single nucleotide polymorphisms (SNPs) and insertions/deletions (indels); the genome-wide SNPs and indels can be efficiently anchored to the chromosomes of barley and *Ae. tauschii* [25–27]. The SNP- and indel-based markers are available for construction of linkage maps in the target chromosomal regions of not only *Ae. tauschii* but also the D genome of hexaploid wheat including synthetic wheat [26,27].

Bulked segregant analysis (BSA) combined with NGS allows efficient development of molecular markers linked to a genomic region associated with the target phenotype in cereals [28–31]. In maize, for example, an RNA-seq-based BSA approach has been used to construct high-density linkage maps and to screen among candidate genes for a target locus [32–34].

Hexaploid wheat has three closely related genomes (designated A, B, and D), each of which carries a set of highly similar genes (homoeology). Due to the genome complexity via allopolyploidy and the large proportion of repetitive DNA in polyploid wheat, whole-genome resequencing is still unviable and reduced-representation methods of NGS data have been employed in this species [35]. Recently, RNA-seq-based BSA was employed successfully for the development of molecular markers closely linked to target chromosomal genes such as a grain protein content gene (*GPC-B1*), a yellow rust resistance gene (*Yr15*), and a powdery mildew resistance gene (*Pm4b*) in tetraploid and common wheat [36–38]. However, a limited amount of information has been reported for the RNA-seq-based BSA approach in polyploid wheat. Here, we employed the RNA-seq-based BSA method to develop a molecular marker closely linked to *Net2*. This process used a mapping population generated from a cross of two synthetic hexaploid wheat lines that shared identical A and B genomes but contained diverse D genomes originating from two distinct pollen parents.

## 2. Results and Discussion

Two synthetic hexaploid wheat lines were derived from interspecies crosses of tetraploid wheat cultivar Langdon (Ldn) and two *Ae. tauschii* accessions (KU-2075 and KU-2025). Ldn/KU-2075 and Ldn/KU-2025, respectively, showed normal phenotype (wild-type) and type-II necrosis phenotype [20]. To obtain novel molecular markers tightly linked to *Net2*, RNA sequencing of four bulks of synthetic hexaploid wheat was performed (Figure 1). Each bulk was composed of ten *Net2*-homozygous individuals or non-carriers (*net2*-homozygous) selected from two $F_5$ populations between Ldn/KU-2075 and Ldn/KU-2025 (hereafter referred to as Segregating Populations SP1 and SP2). The two bulks of each SP (*Net2*-SP1, non-carrier-SP1, *Net2*-SP2, and non-carrier-SP2) were sequenced twice; the bulks were designated as follows: *Net2*-SP1-1st, *Net2*-SP1-2nd, non-carrier-SP1-1st, non-carrier-SP1-2nd, *Net2*-SP2-1st, *Net2*-SP2-2nd, non-carrier-SP2-1st, and non-carrier-SP2-2nd (Table 1). In each experiment, 4.06 million to 5.22 million paired-end reads were generated. After quality filtering, 2.80 million to 3.71 million high-quality reads were acquired. These reads were aligned with transcripts of the two parental synthetic hexaploid *Ae. tauschii* accessions, KU-2075 and KU-2025, transcriptomes that had been de novo assembled in our previous study [23]. In each experiment, 2.02 million to 2.76 million and 1.93 million to 2.68 million reads were aligned to the KU-2075 and KU-2025 transcriptomes, respectively. The alignment output files of the first and second runs were merged for each alignment pair of bulk and transcript. On average, 292,678 and 278,690 SNPs were detected on KU-2075 and KU-2025 transcripts, respectively (Table 2). Of these, 290,712 (99.33%) and 276,700 (99.29%) SNPs on the KU-2075 and KU-2025 transcripts (respectively) were anchored to the *Ae. tauschii* genome. SNP sites in the eight alignment pairs of bulk and transcript (e.g., *Net2*-SP1 vs. KU-2075), as shown in Table 3, were integrated with the genome sequence, yielding a total of 319,808 non-redundant (NR) SNP sites on the seven chromosomes (Table 3). This section may be divided by subheadings. It should provide a concise and precise description of the experimental results, their interpretation as well as the experimental conclusions that can be drawn.

To distinguish D-genome-specific variations from homoeologous polymorphisms between the tetraploid AB and diploid D genomes, the RNA-seq raw read data of Ldn, the tetraploid parental accession of the synthetic wheat lines, were processed as described above. Out of the total of 6.32 million read pairs obtained, 4.37 million read pairs passed the quality filtering (Table 1). To KU-2075 and KU-2025 transcripts, 2.97 million and 2.66 million reads were aligned, respectively.

**Table 1.** Summary of RNA sequencing results for four pairs of bulks of synthetic hexaploid wheat and for tetraploid wheat cv. Langdon.

| Samples | Total Read Pairs | Filtered Read Pairs (%) [a] | Aligned to the *Ae. tauschii* Transcripts [b] (%) [c] | |
|---|---|---|---|---|
| | | | KU-2075 | KU-2025 |
| Synthetic hexaploids | | | | |
| non-carrier-SP1-1st | 4,202,114 | 2,799,202 (66.61%) | 2,020,405 (72.18%) | 1,930,400.5 (68.96%) |
| non-carrier-SP1-2nd | 4,059,840 | 2,858,956 (70.42%) | 2,062,291 (72.13%) | 1,971,110.5 (68.95%) |
| non-carrier-SP2-1st | 4,492,358 | 2,953,088 (65.74%) | 2,164,656 (73.3%) | 2,083,120 (70.54%) |
| non-carrier-SP2-2nd | 4,115,352 | 2,864,271 (69.60%) | 2,098,383 (73.26%) | 2,020,752.5 (70.55%) |
| *Net2*-SP1-1st | 4,710,499 | 3,148,652 (66.84%) | 2,208,392 (70.14%) | 2,110,452.5 (67.03%) |
| *Net2*-SP1-2nd | 4,403,630 | 3,108,568 (70.59%) | 2,178,403 (70.08%) | 2,082,004 (66.98%) |
| *Net2*-SP2-1st | 4,828,182 | 3,249,596 (67.30%) | 2,420,056 (74.47%) | 2,348,814 (72.28%) |
| *Net2*-SP2-2nd | 5,216,082 | 3,709,478 (71.12%) | 2,763,471 (74.5%) | 2,684,019 (72.36%) |
| Tetraploid wheat | | | | |
| cv. Langdon | 6,316,174 | 4,372,660 (69.23%) | 2,974,277 (68.02%) | 2,661,487 (60.87%) |

[a] The ratio of the filtered read pairs to the total read pairs. [b] Nishijima et al. [27]. [c] The ratio of the aligned reads to the filtered read pairs.

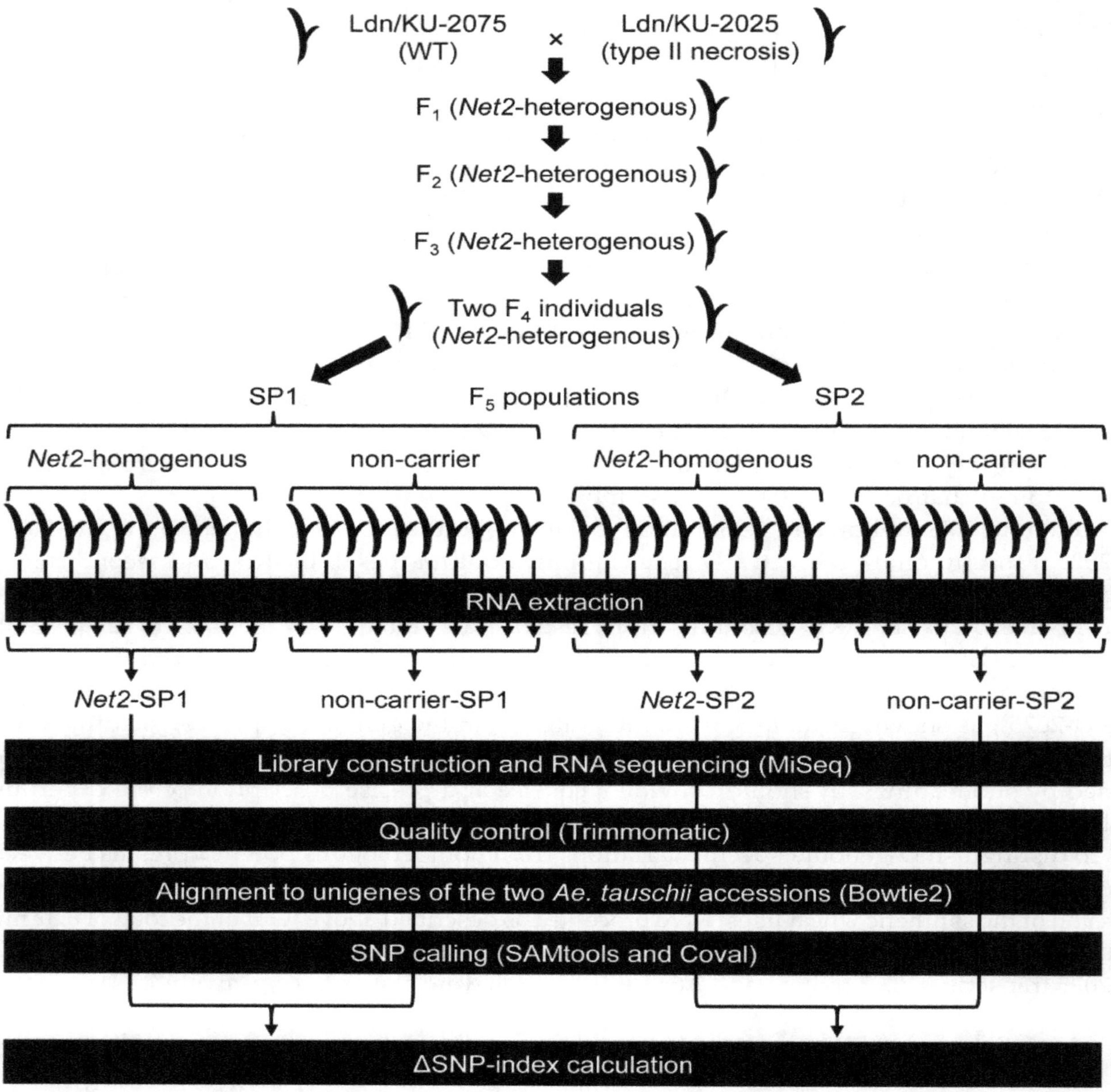


**Figure 1.** Workflow of the RNA sequencing analysis in this study.

**Table 2.** The number of single nucleotide polymorphisms (SNPs) detected in four bulks of synthetic hexaploid wheat and tetraploid wheat cv. Langdon compared to the two parental *Ae. tauschii* transcriptomes.

| Transcripts [a] | KU-2075 | | KU-2025 | |
|---|---|---|---|---|
| **The Number of SNP** | **Total** | **Anchored to the Genome [b] (%)** | **Total** | **Anchored to the Genome [b] (%)** |
| Synthetic hexaploids | | | | |
| non-carrier-SP1 | 277,605 | 275,799 (99.35%) | 262,966 | 261,128 (99.30%) |
| non-carrier-SP2 | 276,564 | 274,772 (99.35%) | 269,175 | 267,249 (99.28%) |
| *Net2*-SP1 | 318,046 | 315,859 (99.31%) | 296,819 | 294,739 (99.30%) |
| *Net2*-SP2 | 298,496 | 296,419 (99.30%) | 285,798 | 283,684 (99.26%) |
| Tetraploid wheat | | | | |
| cv. Langdon | 429,346 | 421,957 (98.28%) | 350,871 | 345,657 (98.51%) |

[a] Nishijima et al. [27]. [b] Luo et al. [39].

**Table 3.** The number of SNPs classified into three categories, including D-genome-specific allelic variations, homoeologous polymorphisms, and unclassified (those falling into neither of the other two classes).

| Chr. | D-genome-Specific | Homoeologous | Unclassified | Total |
|---|---|---|---|---|
| 1D | 1674 | 29,307 | 11,760 | 42,741 |
| 2D | 2295 | 33,822 | 13,975 | 50,092 |
| 3D | 1611 | 31,932 | 15,532 | 49,075 |
| 4D | 1698 | 28,781 | 11,776 | 42,255 |
| 5D | 2936 | 34,961 | 14,336 | 52,233 |
| 6D | 3730 | 24,534 | 10,966 | 39,230 |
| 7D | 3983 | 28,593 | 11,606 | 44,182 |
| Total | 17,927 | 211,930 | 89,951 | 319,808 |

Out of the 429,346 and 350,871 SNPs identified on KU-2075 and KU-2025 transcripts (respectively), 421,957 (98.28%) and 345,657 SNPs (98.51%) (respectively) were anchored to the *Ae. tauschii* genome (Table 2). Based on these SNPs derived from tetraploid wheat reads, and on the SNPs between the two *Ae. tauschii* accessions reported in our previous study [27], NR SNPs identified in synthetic hexaploid bulks were classified into homoeologous polymorphisms and D-genome-specific allelic variation. Out of the 319,808 SNPs, 17,927 were assigned to the allelic variation on the D genome, and 211,930 were grouped into the homoeologous polymorphisms, while the remained 89,951 did not fall into either of these two classes ("unclassified"; Table 3).

SNP-index values were computed for each of the eight alignment pairs, and then ΔSNP-index values were calculated for the comparisons between *Net2*-homogenous bulks and non-carrier bulks, yielding indices for *Net2*-SP1 minus non-carrier-SP1 and *Net2*-SP2 minus non-carrier-SP2. The average of the four ΔSNP-index values at each SNP site was mapped to the *Ae. tauschii* genome (Figure 2). The ΔSNP-index distribution on the chromosomes indicated that SNPs with high ΔSNP-index values tended to be located on the short arm of chromosome 2D, where the *Net2* locus resides (Figure 2). Comparison of ΔSNP-index values among D-genome-specific SNPs showed that the D-genome-specific SNPs on chromosome 2D possessed significantly higher (Steel-Dwass test, $p < 0.001$) ΔSNP-index values than those on the other six chromosomes (Figure 3).

Theoretically, homozygous chromosomal regions account for 87.5% of the total genome in an $F_4$ individual derived from two parents. D-genome-specific SNP sites with low ΔSNP-index values appeared to already have been fixed as homozygous for the KU-2075 or KU-2025 allele by the $F_4$ generation of the synthetic hexaploid populations. The ΔSNP-index values of D-genome-specific SNPs on chromosome 2D also were significantly higher (Steel-Dwass test, $p < 0.001$) than those of homoeologous and unclassified SNPs (Figure 3). Moreover, most of the homoeologous and unclassified polymorphisms showed ΔSNP-index values of approximately zero (Figure 3), indicating that homoeologous polymorphisms had been efficiently removed from candidate SNPs for development of molecular markers, since such candidate SNPs should have possessed high ΔSNP-index values. Low ΔSNP-index values at homoeologous polymorphic sites were likely due to the similar allele frequency of A- and B-genome-derived reads between the *Net2*-homogenous and non-carrier bulks, which thereby offset the SNP-index values of each other.

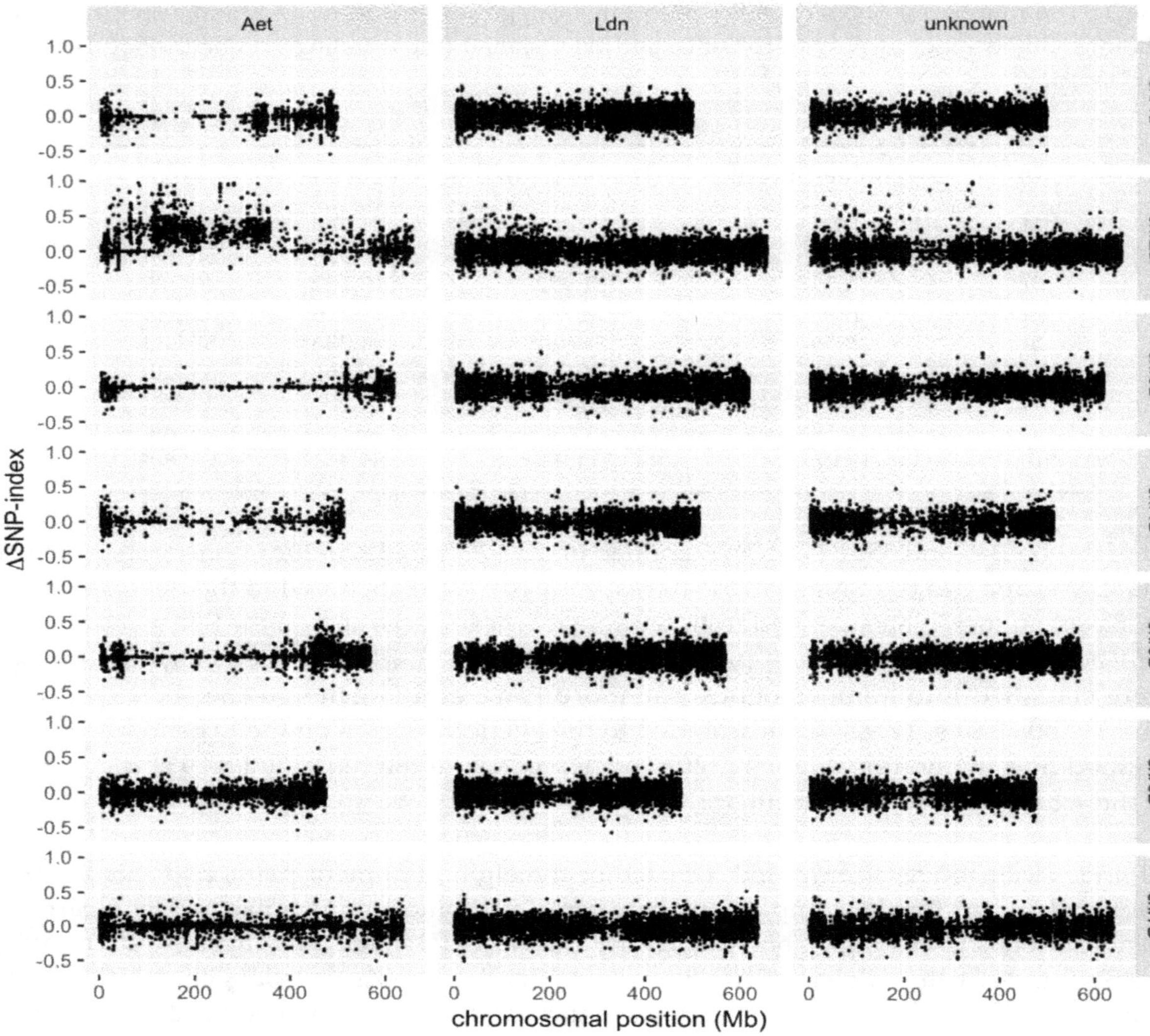


**Figure 2.** Distribution of ΔSNP-index values along the *Ae. tauschii* chromosomes. The three categories of SNPs (D-genome-specific allelic variations, homoeologous polymorphisms, and unclassified SNPs) are designated as "Aet", "Ldn", and "Unknown", respectively.

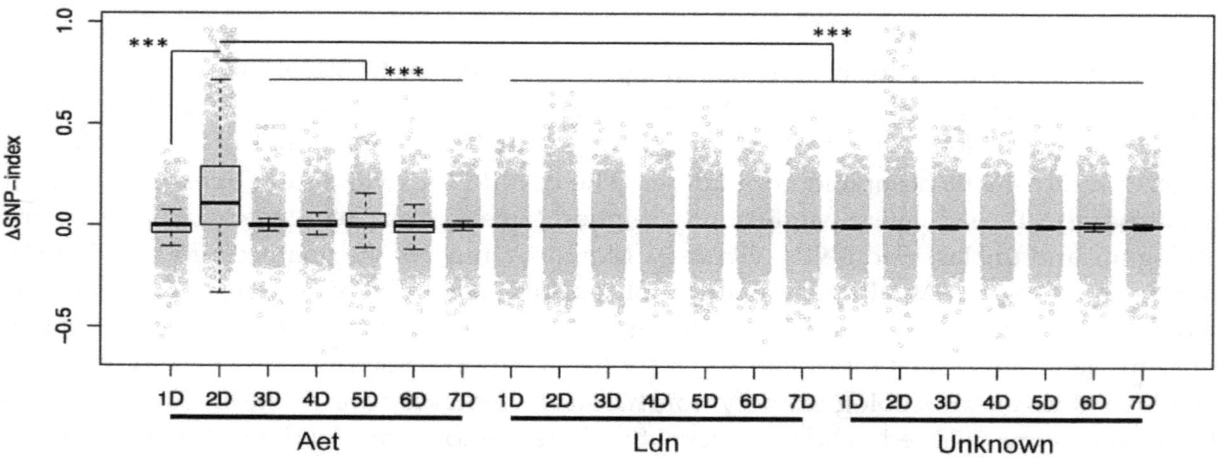


**Figure 3.** Box and dot plots for ΔSNP-index values based on the seven chromosomes and on the SNP classifications in Figure 2. D-genome-specific SNPs on chromosome 2D had significantly higher ΔSNP-index values than SNPs on the other six chromosomes, homoeologous SNPs, and unclassified SNPs. *** $p < 0.001$ by the Steel-Dwass test.

To assess whether SNPs with high ΔSNP-index values can be used as molecular markers for the *Net2* locus, derived cleaved amplified polymorphic sequence (dCAPS) markers were designed from these SNPs. Within the physical region from 81.8 Mb to 83.3 Mb on chromosome 2D that was defined by the two markers noted in our previous study [22], 208 SNPs were detected in the present study, including 10 D-genome specific, 144 homoeologous, and 54 unclassified SNPs. The ΔSNP-index values of this subset of SNPs ranged from −0.4025 to 0.5425. Four of the D-genome-specific SNPs with ΔSNP-index values higher than 0.38 were converted into dCAPS markers (Table 4), and three of these markers (designated *bsa1*, *bsa2*, and *bsa4*) were successfully genotyped and mapped in the Ldn/KU-2025//Ldn/KU-2075 population (Figure 4). All three of these markers were anchored within 2.5 cM distal to the *Net2* locus, and two of these markers were closer to *Net2* than to *scaf52*, the most closely linked markers in our previous study [22]. This result indicates that SNPs derived from RNA-seq-based BSA of synthetic hexaploid wheat can be used for molecular marker development in a specific chromosomal region of the D genome.

**Table 4.** List of the derived cleaved amplified polymorphic sequence (dCAPS) markers developed in this study.

| Marker Name | Primer Sequence (5′ to 3′) | Restriction Enzyme |
|---|---|---|
| *bsa1* | TCATGACCTGCTGGTTTGTT<br>GATTCCAATGTTATTTCTGAACCCT | *Sty*I |
| *bsa2* | TCACAACATTCGCAGGTCAT<br>TGGTTCTGTTGATCTCACTGCC | *Hpa*II |
| *bsa4* | ACAAGTCGGATATCGCCAAA<br>CAGCTAAAAACTGTTTGCTTGAGA | *Hin*fI |

The distribution of ΔSNP-index values was not continuous (Figure 2), even in a single transcript. The discontinuous pattern was due to the homoeologous and nonpolymorphic RNA-seq reads, which could be derived from homoeologous regions in the A and B genomes and overlap with the reads containing the D-genome-specific SNPs. For complete masking of the overlapping reads from the A and B genomes, precise alignment to all three genomes of hexaploid wheat would be needed. In the present study, although transcripts of the parental diploid species were used as the reference sequences, our strategy succeeded in efficiently detecting candidate SNPs on a specific chromosomal region of synthetic hexaploid wheat. Therefore, this strategy of RNA-seq-based BSA is expected to be applicable to other synthetic polyploids derived from crosses among tetraploid and/or diploid wheat relatives.

Taken together, these data demonstrate that RNA-seq-based BSA can be applied to synthetic hexaploid wheat for the development of molecular markers in specific chromosomal regions. The use of a single tetraploid wheat cultivar as a parental line for synthetic hexaploid wheat effectively cancelled the increase in ΔSNP-index values at homoeologous polymorphic sites. To date, a large number of synthetic wheat hexaploids have been produced as part of efforts to enlarge the D-genome genetic diversity [5,6,40]. Epistatic interactions can occur between the parental genomes in the newly synthesized allopolyploids [41], and several phenotypes specific to the synthetic wheat hexaploids have been reported [20,42]. RNA-seq and subsequent de novo transcriptome assembly can be performed even in wild diploid wheat relatives with no reference sequences. This strategy would facilitate molecular marker development in diverse wheat synthetics with various genome constitutions. Thus, RNA-seq-based BSA is expected to serve as a rapid and efficient approach for genetic evaluation of target traits from wild wheat relatives with allopolyploid backgrounds.

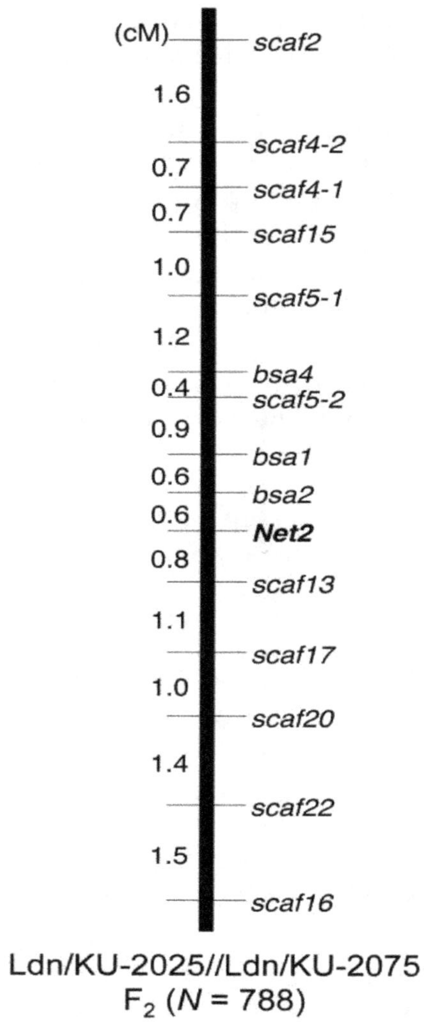


**Figure 4.** Genetic map of the short arm of chromosome 2D in the Ldn/KU-2025//Ldn/KU-2075 population. The three newly developed markers are designated with "*bsa*" prefixes.

## 3. Materials and Methods

### 3.1. Plant Materials

Two synthetic hexaploid wheat lines derived from interspecies crosses of tetraploid wheat cultivar Langdon (Ldn) and two *Ae. tauschii* accessions (KU-2075 and KU-2025) were used in this study. These lines were produced in our previous study [21]. A synthetic hexaploid line (Ldn/KU-2075) showed normal growth, whereas another synthetic hexaploid line (Ldn/KU-2025) exhibited the type-II necrosis phenotype when grown at low temperature [21]. An $F_2$ mapping population was generated through a cross of the two synthetic hexaploid lines [22]. Subsequently, the $F_3$ and $F_4$ generations were developed from selfing $F_2$ and $F_3$ individuals (respectively) heterozygous for the chromosomal region of *Net2*, a type-II necrosis-causing gene [22]. Two $F_5$ populations (SP1 and SP2) were generated through selfing of two independent $F_4$ individuals heterozygous for the *Net2* region (Figure 1).

### 3.2. Library Construction and RNA Sequencing

Ten *Net2*-homozygous individuals and ten non-carriers (*net2*-homozygous) each were selected from the SP1 and SP2 populations. Total RNA was extracted from leaves of each of the forty selected plants at the seedling stage using a Plant Total RNA Extraction Miniprep System (Viogene, Taipei Hsien, Taiwan, ROC). The resulting RNA preparations were pooled independently into four bulk preparations according to the genotype of the *Net2* region and to the SP source population (*Net2*-SP1, non-carrier-SP1,

*Net2*-SP2, and non-carrier-SP2). A total of 8 μg of each bulk preparation of RNA was used to construct paired-end libraries using TruSeq RNA Library Preparation Kit v2 (Illumina, San Diego, CA, USA). The libraries were sequenced twice for 300 cycles × 2 on an Illumina MiSeq sequencer with 300-bp paired-end reads. The read data was deposited to the DDBJ Sequence Read Archive under the accession number DRA007501. RNA sequencing data of Ldn, which was obtained in our previous report [43] and deposited under the accession number DRA007097, also was used for the subsequent analysis.

### *3.3. Alignment of RNA-seq Reads to de Novo Assembled Transcripts of the Parental Ae. tauschii Accessions*

Low-quality bases (average Phred quality score per 4 bp < 30), adapter sequences, and reads < 100 bp were removed from the short reads using the Trimmomatic version 0.32 tool [44]. RNA-seq analysis of KU-2075 and KU-2025 for de novo transcriptome assembly was described in our previous study [27], and the corresponding sequencing data was obtained from the DDBJ Sequence Read Archive DRA004604. The quality-filtered reads of each bulk and Ldn were aligned to the transcripts of KU-2075 and KU-2025 using the Bowtie2 tool [45] with local alignment, obtaining bam outputs.

### *3.4. Identification of D-genome Specific SNPs and Calculation of ΔSNP-Index*

After merging the two bam files of each bulk library using SAMtools software (version 1.9; La Jolla, CA, USA) with the command "samtools merge", SNPs were called from the alignment files using SAMtools and Coval software (version 1.5; Iwate, Japan) with option "-freq 0.1 -m 1000000 -n 10" [46,47]. The bam files of the synthetic hexaploid bulks were supposed to hold homoeologous polymorphisms and allelic variations on the D genome, whereas the alignment outputs of Ldn would contain homoeologous polymorphisms only. We assumed that, of the SNPs derived from the bulks, SNPs also detected in the pairwise comparison of KU-2075 and KU-2025 would represent D-genome specific polymorphisms, and SNPs also found in the alignment files of Ldn would represent homoeologous polymorphisms. The transcript sequences of the two *Ae. tauschii* accessions were mapped to the reference genome of *Ae. tauschii* [39] using GMAP version 2013-03-31 software (South San Francisco, CA, USA) [48,49], and the identified SNPs were anchored to the *Ae. tauschii* genome based on the GMAP outputs. The allele frequency at the SNP sites, a value designated as the SNP-index [28,29], was calculated in each of the eight respective alignment pairs, and the difference in the SNP-index values of the contrasting bulks (*Net2*-homogenous *minus* non-carrier bulk alignment, e.g., *Net2*-SP1-KU-2075 *minus* non-carrier-SP1-KU-2075) was defined as the ΔSNP-index by analogy to the work of Takagi et al. [29]. RStudio ver. 0.99.902 [50] with R software ver. 3.3.1 [51] was used for statistical analyses and ΔSNP-index plotting of the *Ae. tauschii* chromosomes.

### *3.5. Molecular Marker Development and Genotyping*

In our previous study [22], the *Net2* locus was fine-mapped within a 0.6-cM region on the short arm of chromosome 2D. The two most closely linked markers were assigned to the *Ae. tauschii* genome [39] using the BLASTN search function of the BLAST+ software (Bethesda, MD, USA) [52] to define the physical region where the *Net2* locus was assumed to reside. D-genome-specific SNPs of high ΔSNP-index values within this chromosomal region were converted into derived cleaved amplified polymorphic sequence (dCAPS) markers (Table 4), and these novel markers were used to genotype 788 individuals of the Ldn/KU-2025//Ldn/KU-2075 population, as described in our previous report [22]. The genotyped markers were assigned to the genetic map around *Net2* using the MAPMAKER/EXP version 3.0 package (Cambridge, MA, USA) [53].

**Author Contributions:** Conceptualization, S.T. and K.Y.; methodology, R.N. and K.Y.; formal analysis, R.N., K.S., S.-i.Y. and K.S.; resources, S.T.; writing—original draft preparation, R.N. and S.T.; writing—review and editing, K.Y., K.S. and S.T.; project administration, S.T.; funding acquisition, S.T., R.N. and K.S.

## Abbreviations

| | |
|---|---|
| RNA-seq | RNA sequencing |
| NGS | next-generation sequencing |
| BSA | bulked segregant analysis |
| SNP | single nucleotide polymorphism |
| Indel | insertion/deletion |
| NR | non-redundant |
| SP | Segregating Population |
| dCAPS | derived cleaved amplified polymorphic sequence |
| Ldn | Langdon |

## References

1. Matsuoka, Y. Evolution of polyploid *Triticum* wheats under cultivation: The role of domestication, natural hybridization and allopolyploid speciation in their diversification. *Plant Cell Physiol.* **2011**, *52*, 750–764. [CrossRef] [PubMed]
2. Wang, J.; Luo, M.C.; Chen, Z.; You, F.M.; Wei, Y.; Zheng, Y.; Dvorak, J. *Aegilops tauschii* single nucleotide polymorphisms shed light on the origins of wheat D-genome genetic diversity and pinpoint the geographic origin of hexaploid wheat. *New Phytol.* **2013**, *198*, 925–937. [CrossRef] [PubMed]
3. Matsuoka, Y.; Nasuda, S. Durum wheat as a candidate for the unknown female progenitor of bread wheat: An empirical study with a highly fertile $F_1$ hybrid with *Aegilops tauschii* Coss. *Theor. Appl. Genet.* **2004**, *109*, 1710–1717. [CrossRef] [PubMed]
4. Zohary, D.; Harlan, J.R.; Vardi, A. The wild diploid progenitors of wheat and their breeding value. *Euphytica* **1969**, *18*, 58–65. [CrossRef]
5. Mujeeb-Kazi, A.; Rosas, V.; Roldan, S. Conservation of the genetic variation of *Triticum tauschii* (Coss.) Schmalh. (*Aegilops squarrosa* auct. non L.) in synthetic hexaploid wheats (*T. turgidum* L. s.lat. x *T. tauschii*; 2n=6x=42, AABBDD) and its potential utilization for wheat improvement. *Genet. Resour. Crop Evol.* **1996**, *43*, 129–134. [CrossRef]
6. Jones, H.; Gosman, N.; Horsnell, R.; Rose, G.A.; Everst, L.A.; Bentley, A.R.; Tha, S.; Uauy, C.; Kowalski, A.; Novoselovic, D.; et al. Strategy for exploiting exotic germplasm using genetic, morphological, and environmental diversity: The *Aegilops tauschii* Coss. example. *Theor. Appl. Genet.* **2013**, *126*, 1793–1808. [CrossRef] [PubMed]
7. Bhatta, M.; Morgounov, A.; Belamkar, V.; Yorgancilar, A.; Baenziger, P.S. Genome-wide association study reveals favorable alleles associated with common bunt resistance in synthetic hexaploid wheat. *Euphytica* **2018**, *214*, 200. [CrossRef]
8. Bhatta, M.; Morgounov, A.; Belamkar, V.; Baenziger, P.S. Genome-wide association study reveals novel genomic regions for grain yield and yield-related traits in drought-stressed synthetic hexaploid wheat. *Int. J. Mol. Sci.* **2018**, *19*, 3011. [CrossRef] [PubMed]
9. Bhatta, M.; Baenziger, P.S.; Waters, B.M.; Poudel, R.; Belamkar, V.; Poland, J.; Morgounov, A. Genome-wide association study reveals novel genomic regions associated with 10 grain minerals in synthetic hexaploid wheat. *Int. J. Mol. Sci.* **2018**, *19*, 3237. [CrossRef] [PubMed]
10. Gorafi, Y.S.A.; Kim, J.S.; Elbashir, A.A.E.; Tsujimoto, H. A population of wheat multiple synthetic derivatives: An effective platform to explore, harness and utilize genetic diversity of *Aegilops tauschii* for wheat improvement. *Theor. Appl. Genet.* **2018**, *131*, 1615–1626. [CrossRef] [PubMed]
11. Zhang, P.; Dreisigacker, S.; Melchinger, A.E.; Reif, J.C.; Mujeeb-Kazi, A.; Van Ginkel, M.; Hoisington, D.; Warburton, M.L. Quantifying novel sequence variation and selective advantage in synthetic hexaploid wheats and their backcross-derived lines using SSR markers. *Mol. Breed.* **2005**, *15*, 1–10. [CrossRef]
12. Jafarzadeh, J.; Bonnett, D.; Jannink, J.L.; Akdemir, D.; Dreisigacker, S.; Sorrells, M.E. Breeding value of primary synthetic wheat genotypes for grain yield. *PLoS ONE* **2016**, *11*, e0162860. [CrossRef] [PubMed]

13. Bhatta, M.; Morgounov, A.; Belamkar, V.; Poland, J.; Baenziger, P.S. Unlocking the novel genetic diversity and population structure of synthetic hexaploid wheat. *BMC Genom.* **2018**, *19*, 591. [CrossRef] [PubMed]
14. Rasheed, A.; Mujeeb-Kazi, A.; Ogbonnaya, F.C.; He, Z.; Rajaram, S. Wheat genetic resources in the post-genomics era: Promise and challenges. *Ann. Bot.* **2018**, *121*, 603–616. [CrossRef] [PubMed]
15. Van Slageren, M.W. *Wild Wheats: A Monograph of Aegilops L. and Amblyopyrum (Jaub. & Spach) Eig (Poaceae)*; Wageningen Agricultural University: Wageningen, The Netherlands, 1994; pp. 326–344, ISBN 90-6754-377-2.
16. Matsuoka, Y.; Takumi, S.; Kawahara, T. Flowering time diversification and dispersal in central Eurasian wild wheat *Aegilops tauschii* Coss.: Genealogical and ecological framework. *PLoS ONE* **2008**, *3*, e3138. [CrossRef] [PubMed]
17. Mizuno, N.; Yamasaki, M.; Matsuoka, Y.; Kawahara, T.; Takumi, S. Population structure of wild wheat D-genome progenitor *Aegilops tauschii* Coss.: Implications for intraspecific lineage diversification and evolution of common wheat. *Mol. Ecol.* **2010**, *19*, 999–1013. [CrossRef] [PubMed]
18. Matsuoka, Y.; Kawahara, T.; Takumi, S. Intraspecific lineage divergence and its association with reproductive trait change during species range expansion in central Eurasian wild wheat *Aegilops tauschii* Coss. (Poaceae). *BMC Evol. Biol.* **2015**, *15*, 213. [CrossRef] [PubMed]
19. Matsuoka, Y.; Takumi, S. The role of reproductive isolation in allohexaploid speciation pattern: Empirical insights from the progenitors of common wheat. *Sci. Rep.* **2017**, *7*, 16004. [CrossRef] [PubMed]
20. Mizuno, N.; Hosogi, N.; Park, P.; Takumi, S. Hypersensitive response-like reaction is associated with hybrid necrosis in interspecific crosses between tetraploid wheat and *Aegilops tauschii* Coss. *PLoS ONE* **2010**, *5*, e11326. [CrossRef] [PubMed]
21. Mizuno, N.; Shitsukawa, N.; Hosogi, N.; Park, P.; Takumi, S. Autoimmune response and repression of mitotic cell division occur in inter-specific crosses between tetraploid wheat and *Aegilops tauschii* Coss. that show low temperature-induced hybrid necrosis. *Plant J.* **2011**, *68*, 114–128. [CrossRef] [PubMed]
22. Sakaguchi, K.; Nishijima, R.; Iehisa, J.C.M.; Takumi, S. Fine mapping and genetic association analysis of *Net2*, the causative D-genome locus of low temperature-induced hybrid necrosis in interspecific crosses between tetraploid wheat and *Aegilops tauschii*. *Genetica* **2016**, *144*, 523–533. [CrossRef] [PubMed]
23. Brozynska, M.; Furtado, A.; Henry, R.J. Genomics of crop wild relatives: Expanding the gene pool for crop improvement. *Plant Biotechnol. J.* **2016**, *14*, 1070–1085. [CrossRef] [PubMed]
24. Ishikawa, G.; Saito, M.; Tanaka, T.; Katayose, Y.; Kanamori, H.; Kurita, K.; Nakamura, T. An efficient approach for the development of genome-specific markers in allohexaploid wheat (*Triticum aestivum* L.) and its application in the construction of high-density linkage maps of the D genome. *DNA Res.* **2018**, *25*, 317–326. [CrossRef] [PubMed]
25. Iehisa, J.C.M.; Shimizu, A.; Sato, K.; Nasuda, S.; Takumi, S. Discovery of high-confidence single nucleotide polymorphisms from large-scale de novo analysis of leaf transcripts of *Aegilops tauschii*, a wild wheat progenitor. *DNA Res.* **2012**, *19*, 487–497. [CrossRef] [PubMed]
26. Iehisa, J.C.M.; Shimizu, A.; Sato, K.; Nishijima, R.; Sakaguchi, K.; Matsuda, R.; Nasuda, S.; Takumi, S. Genome-wide marker development for the wheat D genome based on single nucleotide polymorphisms identified from transcripts in the wild wheat progenitor *Aegilops tauschii*. *Theor. Appl. Genet.* **2014**, *127*, 261–271. [CrossRef] [PubMed]
27. Nishijima, R.; Yoshida, K.; Motoi, Y.; Sato, K.; Takumi, S. Genome-wide identification of novel genetic markers from RNA sequencing assembly of diverse *Aegilops tauschii* accessions. *Mol. Genet. Genom.* **2016**, *291*, 1681–1694. [CrossRef] [PubMed]
28. Abe, A.; Kosugi, S.; Yoshida, K.; Natsume, S.; Takagi, H.; Kanzaki, H.; Matsumura, H.; Yoshida, K.; Mitsuoka, C.; Tamiru, M.; et al. Genome sequencing reveals agronomically important loci in rice using MutMap. *Nat. Biotechnol.* **2012**, *30*, 174–178. [CrossRef] [PubMed]
29. Takagi, H.; Abe, A.; Yoshida, K.; Kosugi, S.; Natsume, S.; Mitsuoka, C.; Uemura, A.; Utsushi, H.; Tamiru, M.; Takuno, S.; et al. QTL-seq: Rapid mapping of quantitative trait loci in rice by whole genome resequencing of DNA from two bulked populations. *Plant J.* **2013**, *74*, 174–183. [CrossRef] [PubMed]
30. Schneeberge, K. Using next-generation sequencing to isolate mutant genes from forward genetic screens. *Nat. Rev. Genet.* **2014**, *15*, 662–676. [CrossRef] [PubMed]
31. Zou, C.; Wang, P.; Xu, Y. Bulked sample analysis in genetics, genomics and crop improvement. *Plant Biotechnol. J.* **2016**, *14*, 1941–1955. [CrossRef] [PubMed]

32. Liu, S.; Yeh, C.T.; Tang, H.M.; Nettleton, D.; Schnable, P.S. Gene mapping via bulked segregant RNA-Seq (BSR-Seq). *PLoS ONE* **2012**, *7*, e36406. [CrossRef] [PubMed]
33. Su, A.; Song, W.; Xing, J.; Zhao, Y.; Zhang, R.; Li, C.; Duan, M.; Luo, M.; Shi, Z.; Zhao, J. Identification of genes potentially associated with the fertility instability of S-type cytoplasmic male sterility in maize via bulked segregant RNA-Seq. *PLoS ONE* **2016**, *11*, e0163489. [CrossRef] [PubMed]
34. Du, H.; Zhu, J.; Su, H.; Huang, M.; Wang, H.; Ding, S.; Zhang, B.; Luo, A.; Wei, S.; Tian, X.; et al. Bulked segregant RNA-seq revealed differential expression and SNPs of candidate genes associated with waterlogging tolerance in maize. *Front. Plant Sci.* **2017**, *8*, 1022. [CrossRef] [PubMed]
35. Borrill, P.; Adamski, N.; Uauy, C. Genomics as the key to unlocking the polyploid potential of wheat. *New Phytol.* **2015**, *208*, 1008–1022. [CrossRef] [PubMed]
36. Trick, M.; Adamski, N.M.; Mugford, S.G.; Jiang, C.C.; Febrer, M.; Uauy, C. Combining SNP discovery from next-generation sequencing data with bulked segregant analysis (BSA) to fine-map genes in polyploid wheat. *BMC Plant Biol.* **2012**, *12*, 14. [CrossRef] [PubMed]
37. Ramirez-Gonzalez, R.H.; Segovia, V.; Bird, N.; Fenwick, P.; Holdgate, S.; Bery, S.; Jack, P.; Caccamo, M.; Uauy, C. RNA-Seq bulked segregant analysis enables the identification of high-resolution genetic markers for breeding in hexaploid wheat. *Plant Biotechnol. J.* **2015**, *13*, 613–624. [CrossRef] [PubMed]
38. Wu, P.; Xie, J.; Hu, J.; Qiu, D.; Liu, Z.; Li, J.; Li, M.; Zhang, H.; Yang, L.; Liu, H.; et al. Development of molecular markers linked to powdery mildew resistance gene *Pm4b* by combining SNP discovery from transcriptome sequencing data with bulked segregant analysis (BSR-Seq) in wheat. *Front. Plant Sci.* **2018**, *9*, 95. [CrossRef] [PubMed]
39. Luo, M.-C.; Gu, Y.Q.; Puiu, D.; Wang, H.; Ywardziok, S.O.; Deal, K.R.; Huo, N.; Zhu, T.; Wang, L.; Wang, Y.; et al. Genome sequence of the progenitor of the wheat D genome *Aegilops tauschii*. *Nature* **2017**, *551*, 498–502. [CrossRef] [PubMed]
40. Comai, L. The advantages and disadvantages of being polyploid. *Nat. Rev. Genet.* **2005**, *6*, 836–846. [CrossRef] [PubMed]
41. Okada, M.; Yoshida, K.; Takumi, S. Hybrid incompatibilities in interspecific crosses between tetraploid wheat and its wild relative *Aegilops umbellulata*. *Plant Mol. Biol.* **2017**, *95*, 625–645. [CrossRef] [PubMed]
42. Miki, Y.; Yoshida, K.; Mizuno, N.; Nasuda, S.; Sato, K.; Takumi, S. Origin of the wheat B-genome chromosomes conferred by RNA sequencing analysis of leaf transcripts in the section Sitopsis species of *Aegilops*. *DNA Res.* **2018**, under review.
43. Bolger, A.M.; Lohse, M.; Usadel, B. Trimmomatic: A flexible trimmer for Illumina sequence data. *Bioinformatics* **2014**, *30*, 2114–2120. [CrossRef] [PubMed]
44. Langmead, B.; Salzberg, S.L. Fast gapped-read alignment with Bowtie 2. *Nat. Methods* **2012**, *9*, 357–359. [CrossRef] [PubMed]
45. Li, H.; Handsaker, B.; Wysoker, A.; Fennell, T.; Ruan, J.; Homer, N.; Marth, G.; Abecasis, G.; Durbin, R. Subgroup 1000 Genome Project Data Processing. The sequence alignment/map format and SAMtools. *Bioinformatics* **2009**, *25*, 2078–2079. [CrossRef] [PubMed]
46. Kosugi, S.; Natsume, S.; Yoshida, K.; MacLean, D.; Cano, L.; Kamoun, S.; Terauchi, R. Coval: Improving alignment quality and variant calling accuracy for next-generation sequencing data. *PLoS ONE* **2013**, *8*, e75402. [CrossRef] [PubMed]
47. Kajimura, T.; Murai, K.; Takumi, S. Distinct genetic regulation of flowering time and grain-filling period based on empirical study of D genome diversity in synthetic hexaploid wheat lines. *Breed. Sci.* **2011**, *61*, 130–141. [CrossRef]
48. Wu, T.D.; Watanabe, C.K. GMAP: A genomic mapping and alignment program for mRNA and EST sequences. *Bioinformatics* **2005**, *21*, 1859–1875. [CrossRef] [PubMed]
49. Quinlan, A.R.; Hall, I.M. BEDTools: A flexible suite of utilities for comparing genomic features. *Bioinformatics* **2010**, *26*, 841–842. [CrossRef] [PubMed]
50. RStudio Team. *RStudio: Integrated Development for R*; RStudio, Inc.: Boston, MA, USA, 2016; Available online: http://www.rstudio.com/ (accessed on 1 November 2016).
51. R Core Team. *R: A Language and Environment for Statistical Computing*; R Foundation for Statistical Computing: Vienna, Austria; Available online: https://www.R-project.org/ (accessed on 1 November 2016).
52. Altschul, S.F.; Gish, W.; Miller, W.; Myers, E.W.; Lipman, D.J. Basic local alignment search tool. *J. Mol. Biol.* **1990**, *215*, 403–410. [CrossRef]

# 9

# Genome-Wide Genetic Diversity and Population Structure of Tunisian Durum Wheat Landraces Based on DArTseq Technology

**Cyrine Robbana [1,2,*], Zakaria Kehel [3], M'barek Ben Naceur [2], Carolina Sansaloni [4], Filippo Bassi [3] and Ahmed Amri [3]**

1 Faculté des Sciences de Bizerte, Jarzouna, 7021 Bizerte, Tunisia
2 Banque Nationale de Gènes, Boulevard du Leader Yasser Arafat Z. I Charguia 1, 1080 Tunis, Tunisia; nour3alanour@yahoo.com
3 International Center for Agricultural Research in the Dry Areas (ICARDA), 10112 Rabat, Morocco; Z.Kehel@cgiar.org (Z.K.); F.Bassi@cgiar.org (F.B.); A.Amri@cgiar.org (A.A.)
4 Centro Internacional de Mejoramiento de Maíz y Trigo (CIMMYT), El Batán, 56237 Texcoco, Mexico; C.Sansaloni@cgiar.org
* Correspondence: cyrine_rob@yahoo.fr.

**Abstract:** Tunisia, being part of the secondary center of diversity for durum wheat, has rich unexploited landraces that are being continuously lost and replaced by high yielding modern cultivars. This study aimed to investigate the genetic diversity and population structure of 196 durum wheat lines issued from landraces collected from Tunisia using Diversity Array Technology sequencing (DArTseq) and to understand possible ways of introduction in comparing them to landraces from surrounding countries. A total of 16,148 polymorphic DArTseq markers covering equally the A and B genomes were effective to assess the genetic diversity and to classify the accessions. Cluster analysis and discriminant analysis of principal components (DAPC) allowed us to distinguish five distinct groups that matched well with the farmer's variety nomenclature. Interestingly, Mahmoudi and Biskri landraces constitute the same gene pool while Jenah Zarzoura constitutes a completely different group. Analysis of molecular variance (AMOVA) showed that the genetic variation was among rather than within the landraces. DAPC analysis of the Tunisian, Mediterranean and West Asian landraces confirmed our previous population structure and showed a genetic similarity between the Tunisian and the North African landraces with the exception of Jenah Zarzoura being the most distant. The genomic characterization of the Tunisian collection will enhance their conservation and sustainable use.

**Keywords:** durum wheat; Tunisian landraces; center of diversity; genetic diversity; population structure; DArTseq technology

## 1. Introduction

Durum wheat is the tenth most important crop worldwide, grown on 13 M ha with a production of 39.1 Mt in 2017 (International Grain Council, Grain Market reports, 2017) and is mainly used for pasta production. It is mainly grown in the countries around the Mediterranean basin and its cultivation has extended to India, Mexico, North America, and Russia [1,2]. North Africa grows around 2.5 M ha of durum wheat with around 1 M ha in Tunisia. Landraces are still grown under traditional farming systems and are highly appreciated for local dishes such as frikeh, burghul, couscous, but their acreage is reduced significantly with the large adoption of new cultivars released since the 1970s [3,4].

Durum wheat (*Triticum turgidum* ssp. *durum*) is a self-pollinated allotetraploid cereal (2n = 4x = 28, AABB) originating from a cross between *Aegilops speltoides* (SS) and *Triticum urartu* (AA) and domesticated from primitive wheat (*Triticum turgidum* ssp. *dicoccum*) in the Fertile Crescent around 8000 years ago [5–9]. North Africa and Abyssinian regions are considered as secondary centers of diversity for durum wheat [10]. Tunisia, being part of the secondary center of diversity for durum wheat, has a rich diversity in terms of landraces and wild relatives [11,12].

Durum wheat landraces are valuable parental germplasm for wheat improvement programs around the world. Genesys database (www.genesys-pgr.com) contains 60,488 durum wheat accessions of which 22,600 are landraces. The International Center for Agricultural Research in the Dry Areas (ICARDA) genebank holds one of the largest collections of durum wheat accounting more than 20,531 accessions of which more than 65% are landraces. The Tunisian genebank (NGBT) holds a total of 4000 accessions of durum wheat most of which are pure lines collected from the predominant remaining landraces.

The Tunisian landraces are well adapted to a broad range of environments, genetically diverse, and are considered as an important reservoir of useful genes that could be exploited in wheat breeding programs [3,13,14]. A high genetic diversity of local Tunisian durum wheat has been reported using morphological, agronomic, physiological, and biochemical traits [4–15]. Sourour et al. [16] has shown important phenotypic diversity of the Tunisian germplasm estimated to 0.77 using Shannon–Weaver indices. Several molecular markers techniques were used to assess the genetic diversity in wheat including RFLP, RAPD, ISSR, AFLP, EST–SSR [17], DArT, and SNP [18,19]. A significant genetic difference using SSR markers was detected between landraces of durum wheat originated from Morocco and Syria using the Bayesian method and the Eigen analysis [20]. Medini et al. [21] reported high genetic diversity among 33 old Tunisian cultivars using 15 SSR markers. More recently, progress has been made for wheat in genomic research and genetic diversity analysis methods through the construction of a high-density SNP-based consensus map for durum wheat and the development of several next-generation sequencing (NGS) platforms [22,23]. An Axiom 35K array used to genotype 370 entries of durum wheat panel allowed us to differentiate among improved varieties and landraces and to show that Middle East and Ethiopia had the lowest level of allelic diversity compared to other regions [24]. Many studies have shown that genotyping-by-sequencing (GBS) has been increasingly adopted as a rapid and low-cost molecular technique for whole-genome SNP coverage [25], SNP discovery, genotyping, and genetic variability analysis for various crop species including durum wheat landraces [26–28]. The development of DArTseq™ technology has been applied successfully for large genomes such as barley [29] and polyploid or/and complex genomes such as tetraploid and hexaploid wheat [18,29,30]. The power of the DArTseq™ approach based on Illumina short read sequencing have been proven to be effective when used to study the genetic diversities of Syrian and Turkish durum wheat landraces [31,32], bread wheat [33], watermelon, and common bean landraces [34,35].

The present study aimed at (1) characterizing for the first time the population structure and the genetic diversity within and among the Tunisian durum wheat collection, (2) looking at potential mis-classification by linking off types of landraces to local modern durum wheat varieties and/or other landraces, and (3) comparing the Tunisian durum wheat landraces to landraces from the countries around the Mediterranean basin and West Asia region using high throughput DArTseq™ technology.

## 2. Results

### 2.1. DArTseq Marker Characteristics

In this study, a total of 110,856 DArTseq markers were identified in a set of 196 lines issued from six Tunisian durum wheat landraces, of which 16,148 markers were found to be of high quality and polymorphic, after removing markers with more than 20% missing data and with less than 5% minor allele frequency (MAF). Around 10% of the markers had 0 heterozygote alleles. The average of the polymorphism information content (PIC) value of the DArTseq markers was 0.165, and the median was 0.105. The distribution of the majority of MAFs was between 0.05 and 0.15 (Figure 1).

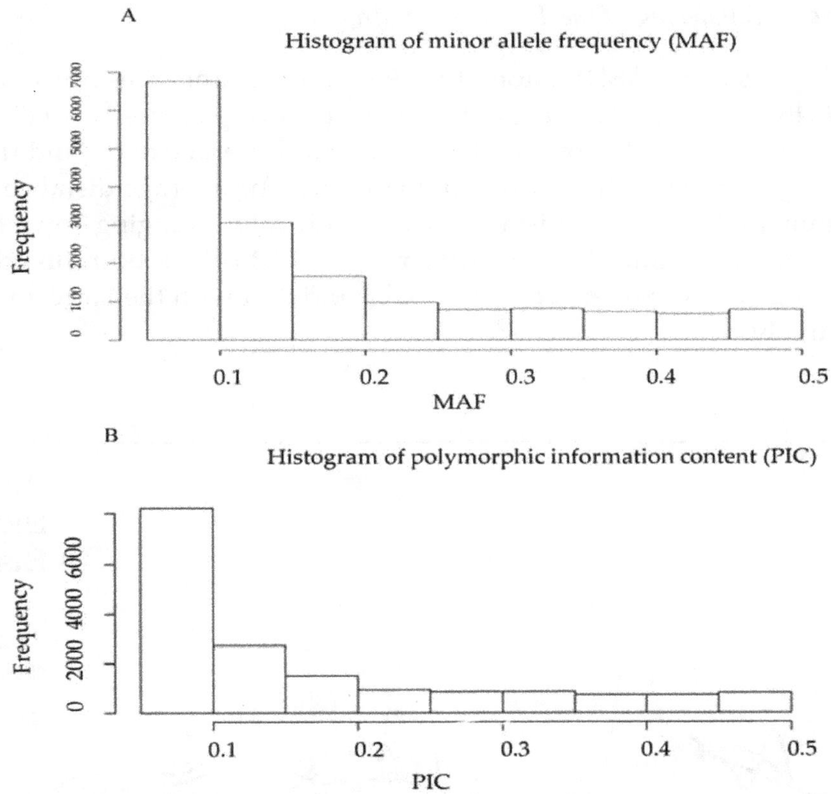


**Figure 1.** Frequency distribution of (**A**) minor allele frequency (MAF) and (**B**) polymorphic information content (PIC) of 16,148 DArtseq markers.

The DArTseq markers are well distributed across all the 14 chromosomes of durum wheat genomes based on the consensus bread wheat genetic map obtained from the International Wheat Genome Sequencing Consortium database (https://urgi.versailles.inra.fr/download/iwgsc/). The distribution is almost equal between A (4201 markers) and B (4659 markers) genomes with the highest number of markers observed on the chromosomes 7A and 2B. However, 7324 markers are still unassigned to any of the chromosomes (Table 1). A similar range of genetic diversity values ($H_e$) is observed for both A and B genomes with an average of around 0.25. The maximum value of the expected heterozygosity ($H_e$) was observed at chromosome 6B (0.28) and the minimum value was seen for chromosomes 3A, 4A, 5A, and 5B (0.24). For all chromosomes, the expected heterozygosity values ($H_e$) were higher than the observed heterozygosity values ($H_O$).

**Table 1.** Number of the selected DArTseq markers ($n$), the observed heterozygosity ($H_O$), and the expected heterozygosity ($H_e$) across the 14 chromosomes in 196 Tunisian durum wheat landraces lines based on the consensus bread wheat genetic map.

| | A Genome | | | B Genome | | |
|---|---|---|---|---|---|---|
| **Chromosome** | $n$ | $H_e$ | $H_o$ | $n$ | $H_e$ | $H_o$ |
| **1** | 562 | 0.27 | 0.15 | 658 | 0.26 | 0.14 |
| **2** | 569 | 0.25 | 0.14 | 864 | 0.27 | 0.13 |
| **3** | 633 | 0.24 | 0.14 | 777 | 0.26 | 0.14 |
| **4** | 511 | 0.24 | 0.16 | 429 | 0.25 | 0.16 |
| **5** | 587 | 0.24 | 0.18 | 639 | 0.24 | 0.16 |
| **6** | 452 | 0.25 | 0.15 | 573 | 0.28 | 0.15 |
| **7** | 797 | 0.25 | 0.14 | 719 | 0.26 | 0.15 |
| **Total** | 4201 | | | 4659 | | |
| **Unassigned** | 7324 | 0.25 | 0.18 | | | |

### *2.2. Genetic Distance and Clustering of the Tunisian Landraces*

The allele sharing distance (ASD) among the 196 Tunisian landraces lines ranged from 0 to 0.79 with an average of 0.46 (Figure 2). However, different distance patterns were observed for different landraces. Jenah Zarzoura shows the lowest distance with an average of 0.1 and the lowest variability among its lines ranging from 0 to 0.16. For the Bidi landrace, the average distance is 0.17 with a range of 0.02–0.48. For Rommani, the average distance is 0.25 with values ranging between 0.02 and 0.63 and for Mahmoudi the average distance is 0.30 with a range of 0.35–0.58. Both Biskri and Jenah Khotifa show the highest average distances respectively of 0.43 and 0.44 with the largest variabilities reaching 0.78 and 0.61 respectively.

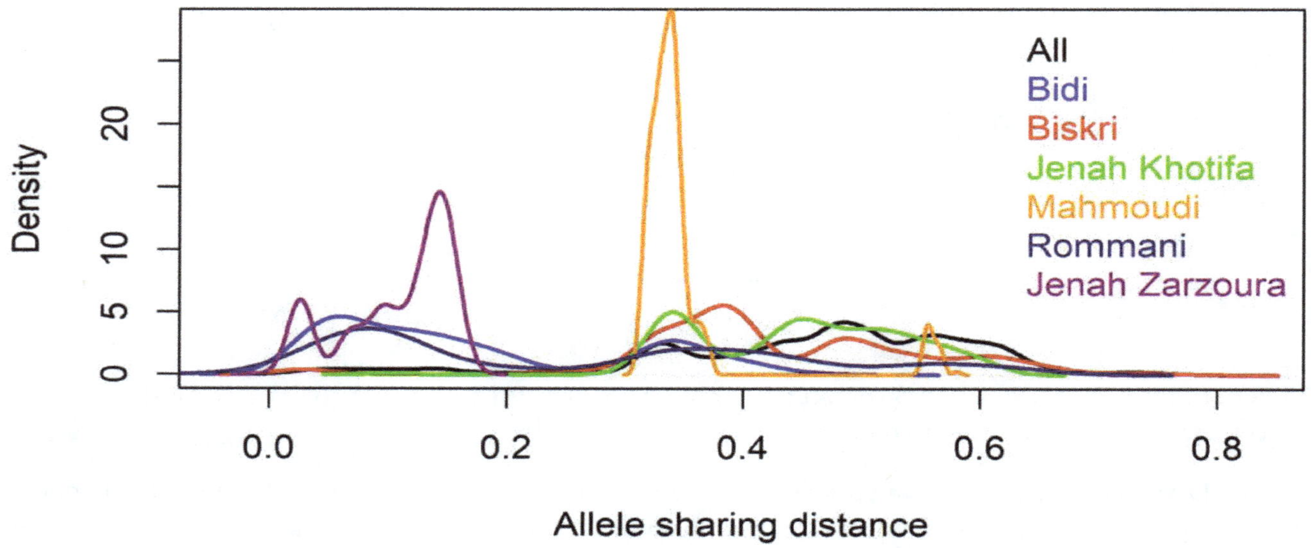


**Figure 2.** Distribution of allele sharing distance of 196 pure lines derived from six Tunisian durum wheat landraces using DArTseq markers.

Cluster analysis of the 196 lines derived from the six Tunisian landraces was performed using the allele sharing distance (ASD) method and the results allowed us to group them into five clusters and to identify lines wrongly assigned to some landraces. The first cluster comprised all the lines of Jenah Zarzoura (Zar). The second cluster contained mainly Bidi lines (Bid). The third cluster is subdivided into two sub-groups, one containing Rommani lines (Rom) and the other having a mixture of lines from other landraces (seven from Biskri (Bis), three from Jenah Khotifa (jkf), and two from Bidi (Bid)). Cluster 4 had the majority of Jenah Khotifa lines (jkf) but contained also four lines of Rommani, two of Biskri, and one of Mahmoudi. The last cluster constitutes the largest one and gathered almost all lines of Biskri (Bis) and Mahmoudi (Mah) (Figure 3).

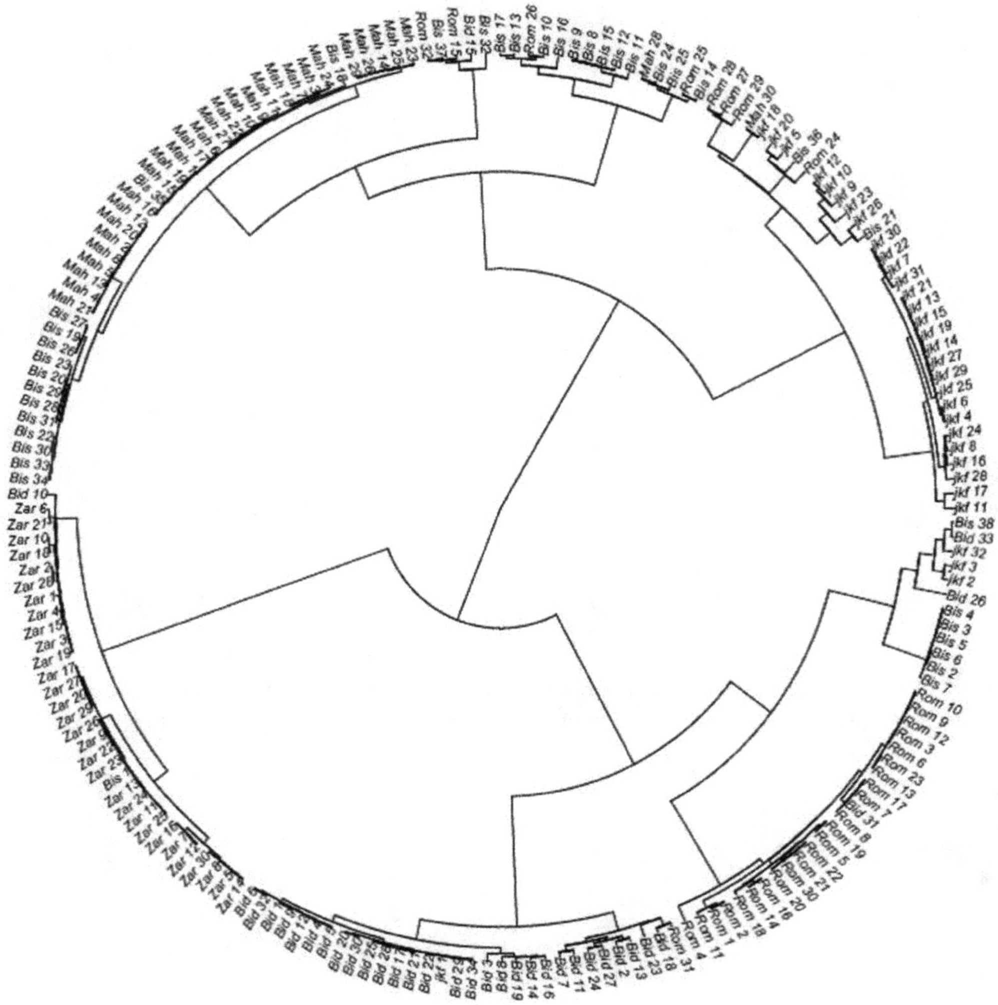

**Figure 3.** Cluster tree of 196 lines derived from six Tunisian durum wheat landraces based on allele sharing genetic distance. Bid: Bidi; Mah: Mahmoudi; Rom: Rommani; Bis: Biskri; jkf: Jenah Khotifa; Zar: Jenah Zarzoura.

### 2.3. *Population Structure of the Tunisian Landraces*

The DAPC (discriminant analysis of principal components) results showed that 88 Principal Components (PCs) explain 82% of the total molecular variance. The optimum number of clusters was obtained with K = 6 using the Bayesian Information Criterion (BIC), which divided the lines into six sub-populations (Figure 4). The first subdivision level using the hierarchical population structure K = 2 separated Jenah Zarzoura lines (Zar) clearly from all the lines of the other landraces (Figure 5A). When K = 3, in addition to the separate sub-population of Jenah Zarzoura, two other sub-populations were identified, one combining most Mahmoudi (Mah) and Biskri (Bis) lines and the other included lines of landraces Bidi (Bid), Jenah Khotifa (jkf), and Rommani (Rom) (Figure 5B). With K = 4, a separate sub-population containing Jenah Khotifa lines appeared (Figure 5C); when K = 5, all the landraces were assigned to different sub-populations, except for Biskri and Mahmoudi, which remained grouped in the same sub-population (Figure 5D). The number of sub-population K = 5 was then chosen to differentiate between different landraces and the sub-populations were given the following names: BID for Bidi, BIS+MAH for Biskri and Mahmoudi, JKF for Jenah Khotifa, ROM for Rommani, and ZAR for Jenah Zarzoura.

Increasing K to 6 resulted into the formation of an additional small group composed by 13 lines mainly of Biskri (7) and Jenah Khotifa (4) showing that these lines could be off-types within their respective landraces, which need further characterization (Figure 5E). The global population structure image was maintained for the values of K equal to 7 and 8 with all different landraces assigned to different sub-populations except the landraces Biskri and Mahmoudi that are still grouped in the same sub-population (Figure 5F).

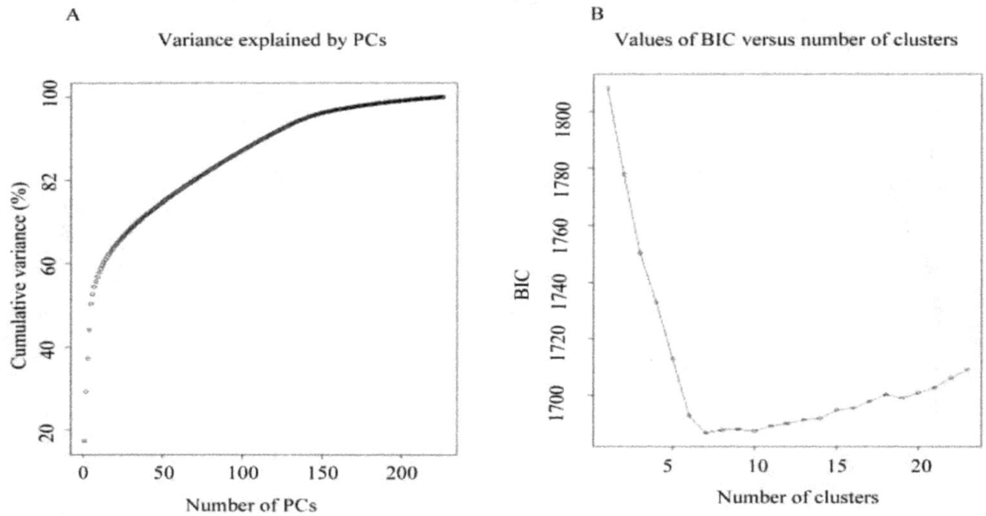


**Figure 4.** Optimal number of sub-populations using discriminant analysis of principal components (DAPC). (**A**) Cumulative variance explained by PCs; (**B**) BIC versus number of clusters.

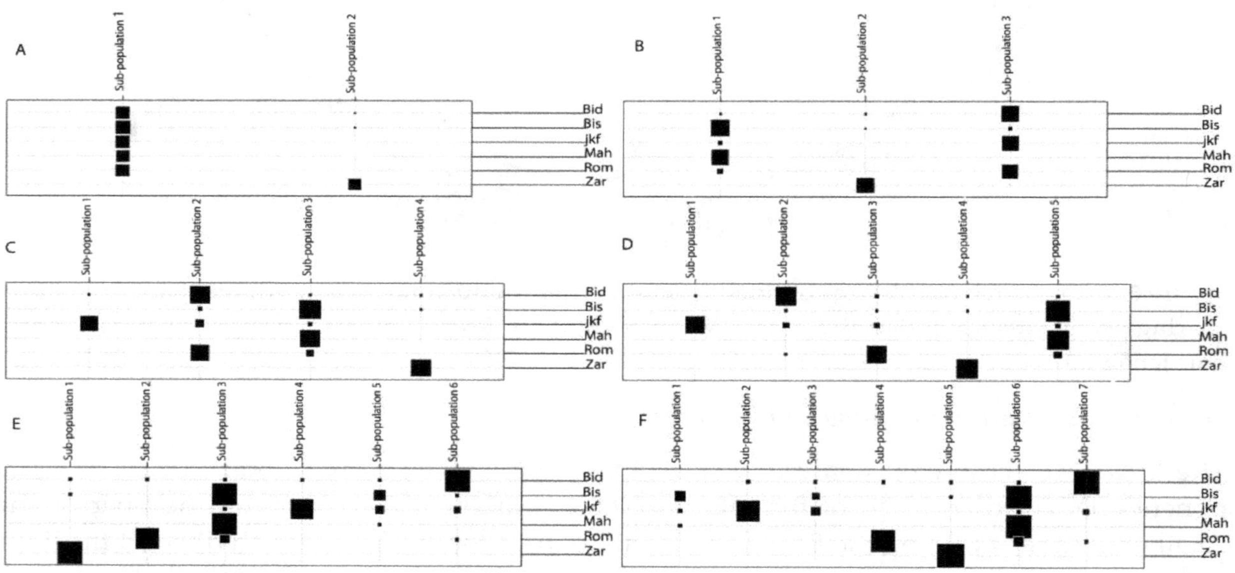


**Figure 5.** Table graphs comparing distribution of the original landrace classification to the sub-populations using DAPC with (**A**) K = 2; (**B**) K = 3; (**C**) K = 4; (**D**) K = 5; (**E**) K = 6; (**F**) K = 7. K: Number of sub-populations; Bid: Bidi; Mah: Mahmoudi; Rom: Rommani; Bis: Biskri; jkf: Jenah Khotifa; Zar: Jenah Zarzoura.

The results of analysis of molecular variance (AMOVA) for K = 2 to K = 7 showed that the variance between populations was high for K = 2 (31.92%), and it decreased for K = 3 and K = 4 (28.75% and 31.29%, respectively) to increase and reach a plateau for K = 5 with a slightly higher value (31.77%). The variance between lines within populations decreased from 11.32% for K = 2 to reach values close to 0% for K = 5. Comparing AMOVA results between populations with K = 5 and the real groups using the landrace's name clearly support the hypothesis that K = 5 better differentiates genetically the lines than the characterization of landraces by name, resulting in higher variance between populations (31.77% for K = 5 and 28.79% for real groups), lower variance between lines 68.23% for K = 5 and 70.39%

for real groups, and lower variance between lines within populations (Table 2). Thus, these findings indicated a higher genetic variation among rather than within Tunisian landraces.

**Table 2.** AMOVA summary for comparison between different numbers of sub-populations K and the real groups of lines issued from six Tunisian durum wheat landraces.

| | **Variance** | | |
|---|---|---|---|
| **Number of Sub-Populations** | **Among Landraces (%)** | **Among Lines within Landrace (%)** | **Among Lines (%)** |
| K = 2 | 31.92 | 11.32 | 56.76 |
| K = 3 | 28.75 | 5.60 | 65.65 |
| K = 4 | 31.29 | 1.59 | 67.12 |
| K = 5 | 31.77 | 0.00 | 68.23 |
| K = 6 | 31.71 | 0.00 | 68.29 |
| K = 7 | 31.51 | 0.00 | 68.49 |
| Real groups | 28.79 | 0.82 | 70.39 |

Results from hierarchical AMOVA using hierarchical subdivision strata from K = 2 up to K = 5 indicated that most of the molecular variance was explained with the first level K = 2 separating Jenah Zarzoura from the other landraces (17.18%) and the fifth level K = 5 within the four groups (26.62%) separating Bidi, Rommani, and Jenah Khotifa and the large group constituted by Biskri and Mahmoudi. The remaining molecular variance (variance between lines) was also low compared to non-hierarchical AMOVA results (60.55%) using K = 5 (Table 3).

**Table 3.** Hierarchical AMOVA results from K = 2 to K = 5 using hierarchical subdivision strata of 196 lines derived from the six durum wheat Tunisian landraces.

| **Subdivision Strata** | **Variance Components** | **Percentage** |
|---|---|---|
| Variations between K = 2 | 271.63 | 17.18 |
| Variations between K = 3 and K = 2 | 0 | 0 |
| Variations between K = 4 and K = 3 | 13.46 | 0.85 |
| Variations between K = 5 and K = 4 | 420.82 | 26.62 |
| Variations between lines and K = 5 | 957.20 | 60.56 |
| Variations between lines | 0 | 0 |
| Total variations | 1580.63 | 100 |

Finally, the resulted variance components from AMOVA analysis with K = 5 were significant when they were tested using permutations (Figure 6). These results showed that the structure found by number of groups with K = 5 was valid and not a result of a random effect.

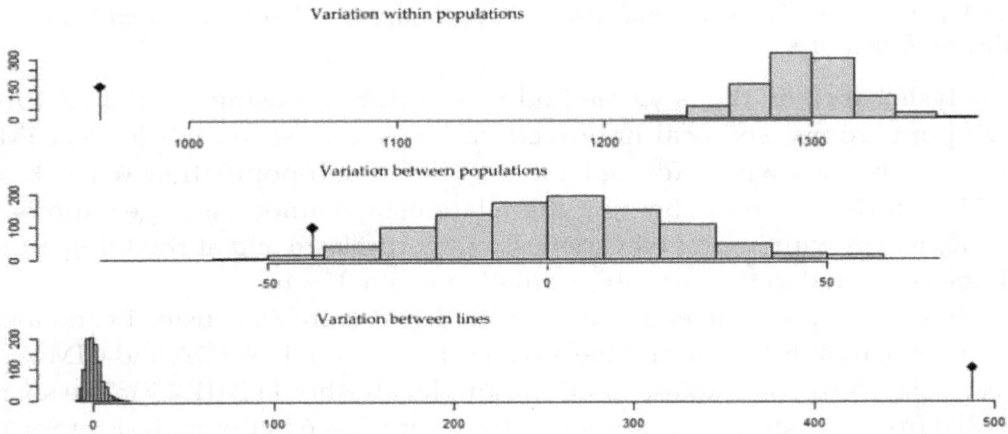


**Figure 6.** AMOVA permutations test using 1000 permutations for number of sub-population K = 5.

### 2.4. Genetic Diversity and Genetic Distance between Tunisian Landraces

The genetic distance between landrace populations showed that Jenah Zarzoura sub-population is the farthest from all the other landraces, from Rommani (0.820), from Biskri and Mahmoudi (0.730), from Jenah Khotifa (0.682), and from Bidi (0.816) (Table 4). Furthermore, the Jenah Khotifa sub-population was equally distant to the landraces Bidi and Rommani (0.557). These results confirmed the population subdivision found by DAPC, which separated the Jenah Zarzoura sub-population from the other landraces when K = 2 and grouped the sub-populations Jenah Khotifa, Bidi, and Rommani when K = 3.

**Table 4.** Reynolds genetic distance between Tunisian durum wheat landrace populations based on sub-populations number K = 5.

| Landraces | BID | ZAR | BIS + MAH | JFK |
|---|---|---|---|---|
| **ZAR** | 0.816 | | | |
| **BIS+MAH** | 0.624 | 0.730 | | |
| **JFK** | 0.557 | 0.682 | 0.466 | |
| **ROM** | 0.729 | 0.820 | 0.645 | 0.577 |

BID: Bidi; JKF: Jenah Khotifa; ZAR: Jenah Zarzoura; BIS: Biskri; ROM: Rommani; MAH: Mahmoudi.

Results of genetic diversity estimates in each sub-population obtained based on DAPC with K = 5 show that the highest genetic diversity was observed within the Jenah Zarzoura and Rommani populations ($H_e$ = 0.27). The lowest genetic diversity was observed within the Bidi population ($H_e$ = 0.12). Biskri and Mahmoudi forming the same group with 73 lines showed a moderate genetic diversity ($H_e$ = 0.23). Based on $F_{ST}$ values, Jenah Zarzoura showed a low $F_{ST}$ value of 0.05. Thus, this population presents a genetic drift compared to the others, having a fixation of alternate alleles with $F_{ST}$ values superior to 0.25 (Table 5). The results from genetic diversity confirmed the results obtained using the allele sharing distance (ASD) between lines within the same population as shown in Figure 1.

**Table 5.** Genetic diversity among Tunisian durum wheat landrace populations based on sub-populations number K = 5.

| Landraces | $n$ | $H_e$ | $F_{ST}$ |
|---|---|---|---|
| **ZAR** | 32 | 0.27 | 0.05 |
| **JFK** | 25 | 0.20 | 0.47 |
| **BID** | 34 | 0.12 | 0.57 |
| **ROM** | 32 | 0.27 | 0.30 |
| **BIS+MAH** | 73 | 0.24 | 0.26 |

$n$: Number of lines; $H_e$: Expected heterozygosity (genetic diversity); $F_{ST}$: Measure of genetic differentiation. BID: Bidi; JKF: Jenah Khotifa; ZAR: Jenah Zarzoura; BIS: Biskri; ROM: Rommani; MAH: Mahmoudi.

### 2.5. Linking the Mis-Classified Lines to Local Landraces and Improved Varieties or/and to the ICARDA/CIMMYT Elite Lines

Cluster analysis based on the ASD method was performed using a set of 27 Tunisian durum wheat landrace populations, six local improved varieties, and seven ICARDA/CIMMYT inbred elite lines, along with lines which formed the additional sub-population when K = 6 from the previous set. This study aimed at showing the relationships among a larger number of Tunisian landraces, the differences with improved varieties, and germplasm and at shedding more light on the mis-classified lines included in the last sub-population where K = 6.

Results of cluster analysis showed three main clusters (Figure 7). Cluster 1 contained most of the improved varieties released in Tunisia and the advanced lines from ICARDA and CIMMYT, along with three landraces; Azizi (Aziz P), Chetla 1(Chet1 P), and Jenah Khotifa 2 (JK2 P). The six mis-classified lines of Biskri (Bis) from the additional sub-population when K = 6 in the analysis of Set 1 are included in this cluster and are grouped with the improved variety Khiar (Khia T). Cluster 2 contained the majority of the Tunisian landraces along with two improved varieties: INRAT69 (INRA T) and Karim (Kari T). This cluster could be subdivided into two groups: a small group composed of six local landraces (Arbi (Arbi P), Rommani 3 (Rom3 P), Biskri 1(Bis1 P), Chetla 2 (Chet2 P), Richi (Richi P), and Agili (Agil P)) and the two improved varieties, Karim (Kar T) and INRAT 69 (INRA T) and a large group that can be further divided into two sub-groups, one of which constitutes a separate group with only Jenah Zarzoura lines (Zar1 L, Zar2 L, and Zar3 L) and the other of which is composed of the majority of local landraces: Jenah Khotifa 1(jkf1 P), Sbei (Sbei P), Mahmoudi (Mah P), Souri (Sour P), Bayadha (Baya P), Swabei Algia (SA P), Aouadi (Aoua P), Hamira (Hmir P), Rommani (Rom1 P), Bidi 1(Bid1 P), Bidi 3 ( Bid3 P), Rommani 2 (Rom2 P), Wardbled (WB P), Derbessi (Derb P), Chili (Chil P), Biskri 2 (Bis2 P), and Bidi 2 (Bid2 P). This last sub-group showed that the two lines of Jenah Khotifa (Jk2 L and Jk4 L) from Set 1 are clustered with Jenah Khotifa population 1(jkf1 P). The Rommani line (Rom1 L) is grouped with the Rommani population (Rom1 P) and the line Bidi 25 (Bid25 L) is grouped with Bidi populations (Bid1 P and Bid3 P). Cluster 3 was divided into two groups: the first group contained a mixture of lines including Jenah Khotifa 11 line (Jk11 L), Biskri 12 line (Bis12 L), and Mahmoudi 8 line (Mah8 L) and the second group showed that the mis-classified lines of Jenah Khotifa 19 (Jk19 L) and Mahmoudi 30 (Mah30 L) included in the additional sub-population when K = 6 of Set 1 analysis are clustered with the ICARDA/CIMMYT inbred line (MCHCB-102). These results showed that most of the lines included in the mixed sub-population of the DAPC analysis with K = 6 are grouped with improved varieties and germplasm and showed that some landraces having the same local name are classified in different clusters.

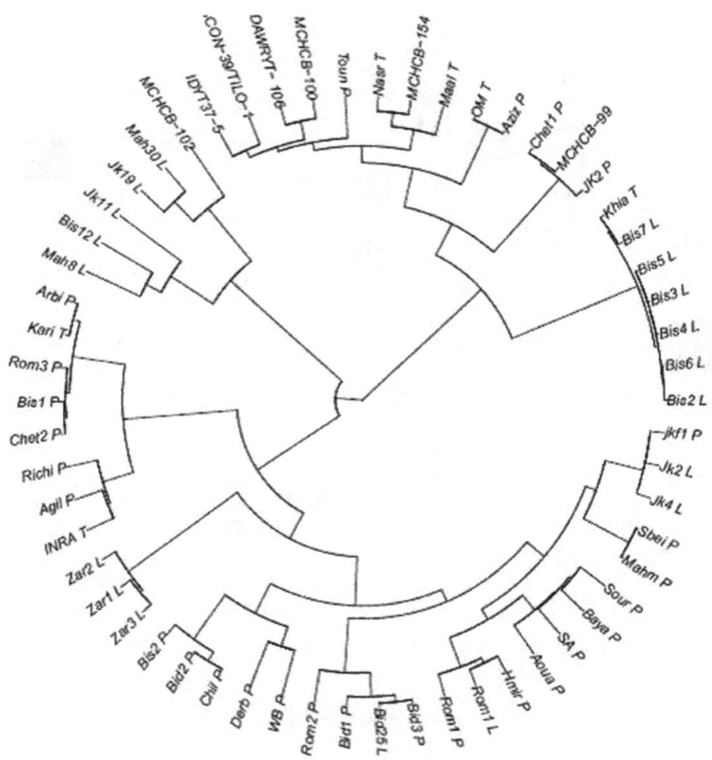


**Figure 7.** Dendrogram of Tunisian durum wheat landraces, local improved varieties, and ICARDA/CIMMYT elite lines based on allele sharing genetic distance. L: Line; P: Population. Local landraces: Bid: Bidi; Mah: Mahmoudi; Rom: Rommani; Bis: Biskri; jkf: Jenah Khotifa; Zar: Jenah Zarzoura; Aziz: Azizi; Chet: Chetla; Sbei; Sour: Souri; Baya: Bayadha; SA: Swabei Algia; Aoua: Aouadhi; Hmir: Hamira; WB: WardBled; Derb: Derbessi; Chil: Chili; Agil: Agili; Richi; Arbi; Toun: Tounsia. Local improved varieties: OM: OmRabii; Nasr; Maal: Maali; Khia: Khiar; INRAT: INRAT69; Kari: Karim. ICARDA/CIMMYT elites lines: MCHCB-102: ICARDA inbred line; MCHCB-100: IcaJoudy1; DAWRYT-106: Nachit; MCHCB-154: Zeina4; Con39/Tilo-1: Louiza; IDYT37-5: Ammar 6; MCHCB-99: Ammar 10.

*2.6. Comparison of Tunisian Landraces to Landraces from Mediterranean and West Asia Regions*

The population structure of the Mediterranean and West Asian landraces along with four lines representing each of six Tunisian landraces included in Set 1 was assessed using the DAPC method. The optimum number of groupings is determined with K = 12 based on the BIC criterion using 66 PCs, which explained 82% of the total molecular variation. Sub-population 6 showed that the Tunisian landraces Rommani (Rom), Jenah Khotifa (jkf), and Bidi (Bid) were grouped with the majority of the Algerian landraces (8); Sub-population 12 included all Ethiopian landraces (11), along with one accession from Afghanistan and two accessions from Yemen; Sub-population 9 included the Tunisian landraces Biskri (Bis) and Mahmoudi (Mah) that were grouped with the remaining five landraces from Tunisia, with the majority of Algerian and Lebanese landraces (6), and with some Moroccan landraces (3). Sub-population 11 included exclusively the Tunisian landrace Jenah Zarzoura (Zar) with two accessions from Jordan (Figure 8). The population structure results confirmed that the newly collected landrace Jenah Zarzoura constitutes a new gene pool and that the Tunisian landraces are genetically closer to North African landraces.

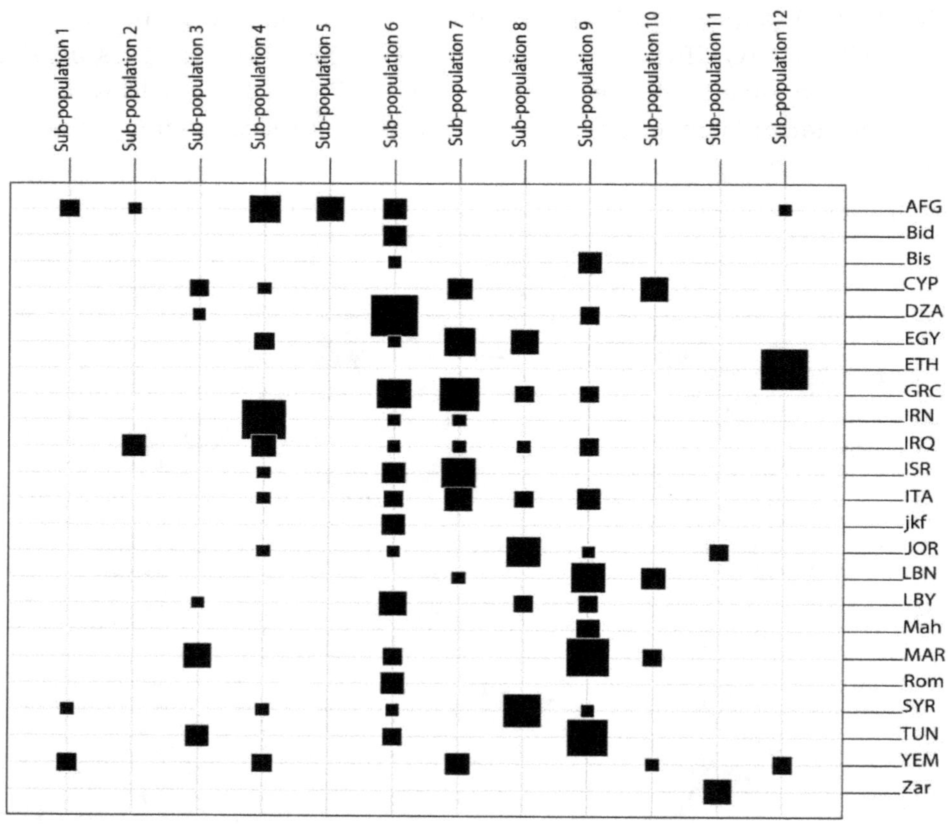


**Figure 8.** Table graphs comparing Mediterranean, West Asian, and Tunisian landraces classification to the sub-populations using DAPC with K = 12. AFG: Afghanistan; CYP: Cyprus; DZA: Algeria; EGY: Egypt; ETH: Ethiopia; GRC: Greece; IRN: Iran, IRQ: Iraq; ISR: Palestine/Israel; ITA: Italy; JOR: Jordan; LBN: Lebanon; LBY: Libya; MAR: Morocco; SYR: Syria; TUN: Tunisia; YEM: Yemen.

## 3. Discussion

The molecular markers techniques are important tools for better understanding genetic diversity, undertaking association mapping and ensuring efficient conservation and management of genetic resources. This study demonstrated the relevance of DArTseq technology as a reliable and cost-effective tool for assessing the diversity within and between landraces and for comparing them with other germplasm of durum wheat. This technique yielded a large number of polymorphic and informative markers equally spread in the A and B genomes allowing high coverage of the genomes of the Tunisian durum wheat germplasm compared to other molecular techniques used previously. Similar genomes coverage was found on a panel of 170 durum wheat entries [27]. The large coverage of the genomes can serve to undertake association mapping in the studied germplasm, including finding new allelic variations for major breeders sought traits such as QTLs found by other studies in chromosomes 7A and 2B linked to protein content [36], gluten strength and yellow pigment [37,38], and salinity tolerance and yield components [39,40]. DArTseq technology along with other high throughput and genotyping by sequencing molecular techniques are increasingly used to study the genetic diversity of different crops as they allow to study the genetic diversity of large number entries and complex genomes [41,42].

### 3.1. *DArTseq Polymorphism of Tunisian Durum Wheat Landraces*

Polymorphic information content (PIC) values revealed by DArTseq markers averaged 0.165 with an asymmetric distribution skewed towards low values. The same distribution tendency was found using the same approach for 91 durum wheat landraces from the Fertile Crescent and for 138 wheat germplasm from Southwest China [32,41]. Ren et al. [43] have shown the same PIC value for North African durum accessions (0.168) using 946 SNP markers as part of a genetic diversity study in a worldwide durum wheat germplasm collection. Recent studies using DArTseq markers reported higher PIC values for worldwide durum wheat accessions (0.35) and for accessions originating from central Fertile Crescent (0.26) [27,32]. A previous study on 34 Tunisian durum wheat old varieties reported a PIC value of 0.68 [25]. This difference is explained by the bi-allelic nature of DArT markers for which the maximum value for PIC is 0.5 compared to multi-allelic SSR markers with the maximum PIC value of 1 [44].

The lines of the six Tunisian durum wheat landraces showed a moderate level of genetic diversity ($H_e$ = 0.25), which is higher than the one exhibited by a set of world-wide durum wheat collection ($H_e$ = 0.224) using SNP markers [43]. Our results confirmed previous findings showing that the Fertile Crescent and the eastern Mediterranean durum landraces are more diverse than those from the Western Mediterranean regions [45,46]. Although several studies revealed high genetic diversity in the Tunisian landraces based on phenotypic characterization [15,16], the moderate level of genetic diversity could be explained by the fact that the Tunisian durum wheat lines included in this study are derived from six landraces collected from a limited geographic area in the center and the south parts of Tunisia.

### 3.2. *Genetic Diversity and Population Structure of the Tunisian Durum Wheat Landraces*

Allele sharing distance using DArTseq markers allowed us to differentiate among the six landraces Jenah Zarzoura, Biskri, Jenah Khotifa, Mahmoudi, Bidi, and Rommani by showing different distance patterns. Two methods, clustering analysis based on ASD (allele sharing distance) and DAPC (discriminant analysis of principal components) were used for depicting the genetic relationship and structure of the Tunisian collection. The first method classified the panel into five groups matching mostly with the farmer's landraces names, with Jenah Zarzoura being the most distant and Mahmoudi and Biskri included in the same cluster. The closeness of Mahmoudi and Biskri is for a long time reported by Boeuf [11] based on glume and spike color, which could be due to the exchange of these landraces among farmers from Algeria and Tunisia [47]. The clear distinction of Jenah Zarzoura could be due to its confinement to a geographically limited area of the Mareth oasis or to a different pattern of introduction.

The use of the multivariate method DAPC to evaluate the population structure showed better performance and allowed for a population subdivision similar to other studies [24,48]. Several molecular approaches were used to assess the population structure in durum wheat landraces from SSRs to DArTs [49,50]. More recently, GBS-SNPs and DArTseq approaches were mainly used for hexaploid wheat population structure studies [28,51], and some reports have described a durum wheat population structure based on the DArTseq technique that allowed to classify the Turkish and Syrian durum wheat landraces in the same gene pool [32]. The DAPC analysis results were in concordance with those of clustering analysis, despite minor differences. Both showed a good fit between the grouping and the names of the varieties, which reflects the ability of farmers to differentiate among the landraces. However, a small group composed of a mixture of landraces appeared, and some lines of the differentiated landraces are included in different clusters, which could be due mainly to mis-naming of the landraces during the collecting missions and to possible mixtures in the landraces. These findings confirm the multiline nature of landraces that is also found through morphological and agronomic characterization [52]. This heterogeneity offers an important buffering capacity of landraces in drought-prone and fluctuating environments.

Hierarchical AMOVA analysis based on hierarchical subdivision strata from K = 2 up to K = 5 agreed with DAPC analysis results and supported the high level of molecular variance to K = 2 (17.18%) and to K = 5 (26.62%). AMOVA analysis results on the basis of the landraces' names indicated a higher genetic variation among (28.79%) rather than within (0.82%) Tunisian landraces. Taking in consideration the structure of the population based on the optimal number of grouping K = 5, AMOVA results showed an increase in percent of the explained genetic variance among landraces (~31.77%) and a decrease of the genetic variance within them. Soriano et al. [49] revealed much variability within sub-populations (83%) than between them (17%) and Mangini et al. [53] found higher genetic diversity within the two Italian durum wheat landraces "Bianchetta" and "Grano Ricco" (9.5 and 9.4%, respectively) and low genetic diversity within "Dauno III.".

This study allowed us to assess for the first time the genetic diversity within and among the Tunisian durum wheat landraces using the DArTseq technique, which allowed us to show different levels of genetic diversity within landraces. Jenah Zarzoura and Rommani populations showed the highest genetic diversity ($H_e$ = 0.27), and the Bidi landrace showed the lowest genetic diversity ($H_e$ = 0.12). The low genetic variation within the landraces could be explained by the selection from farmers for desirable traits and/or from environmental conditions pressure. Compared to the other landraces, the Jenah Zarzoura landrace showed a high expected heterozygosity ($H_e$ = 0.268) and a low fixation index ($F_{ST}$ = 0.05). Thus, this small differentiation could be explained by the confinement of this landrace to a specific environment in the oasis of Mareth, which reflects the geographic isolation of the oasis of Mareth, limited seed exchange, and selection pressure by farmers. Most often, farmers are selecting the best representative spikes from a landrace to form the seed lots.

### *3.3. Origin of Tunisian Durum Wheat Landraces*

The cluster analysis extended to 27 other Tunisian landraces, improved varieties, and germplasm along with the lines included in the mixed cluster when K = 6 from Set 1 showed large genetic diversity among the germplasm studied. The six lines of landrace Biskri mis-classified in the additional sub-population when K = 6 in the Set 1 analysis were grouped with the modern cultivar Khiar, and the two mis-classified lines of Mahmoudi and Jenah Khotifa were grouped with the elite germplasm (MCHCB-102). These results confirm the possibility of mixture in some landraces, which could be due to seed exchange and threshing practices as suggested by other studies [52,54]. This extended genetic study confirms the uniqueness of the Jenah Zarzoura landrace and the classification of Jenah Khotifa, Rommani, and Bidi lines with their respective landraces, but not with the other populations with the same name. This could be due to the mis-naming of landraces during seed exchange among farmers or during the collecting missions undertaken by the genebank teams.

Tunisia is considered as a secondary center of diversity for durum wheat, and the introduction patterns of durum wheat landraces into Tunisia and North Africa are still under discussion. The DAPC analysis including landraces from Tunisia and landraces from the countries around the Mediterranean and West Asia regions allowed us to define 12 distinct groups, which can be used to highlight the relationships between Tunisian landraces and other landraces. Most Tunisian landraces held at the ICARDA genebank as well as the lines derived from landraces collected recently in Tunisia were grouped with landraces from North Africa neighboring countries and with landraces from Greece, Italy, and Lebanon. Jenah Zarzoura remained distinct from all landraces studied except for the two landraces from Jordan. These results suggest that most Tunisian landraces could be obtained through Lebanon via Greece and Italy, while Jenah Zarzoura was obtained through another introduction pathway. Previous reports have demonstrated two dispersal patterns of durum wheat in the Mediterranean Basin from the Fertile Crescent: over the north side via Turkey, Greece, and Italy and the south side via North Africa [55]. Moragues et al. [56] supported this hypothesis and classified a collection of durum wheat landraces originating from different Mediterranean countries in two groups: (i) landraces from the North and East Mediterranean basin and (ii) landraces from North Africa and the Iberian Peninsula. More recently, Soriano et al. [49] showed an eastern–western dispersal of the Mediterranean durum landraces, which have been classified into four sub-populations: (i) Eastern Mediterranean, (ii) the Eastern Balkans and Turkey, (iii) the Western Balkans, and (iv) Egypt and the Western Mediterranean. The grouping of landraces from North Africa with Italy and Greece was also confirmed by Olivera et al. [57] and Nazco et al. [58], and could be explained by the Roman influence on durum wheat cultivation in North Africa. The early development of Carthage trade maritime activity in the Mediterranean sea enhanced seed exchanges between Tunisia and the Mediterranean countries [59], which might explain the grouping of the Bidi, Jenah Khoutifa, and Rommani lines with the majority of Algerian landraces (Sub-population 6) and that of the Biskri and Mahmoudi lines with Lebanese and Moroccan landraces (Sub-population 9). Moreover, our work confirmed that Biskri and Mahmoudi lines constitute the same gene pool and that Jenah Zarzoura lines constitute a new gene pool distant from the other Tunisian and foreign landraces, which were grouped with only two accessions from Jordan (Sub-population 11). A possible explanation is that the Jenah Zarzoura population, which was collected from the oasis of Mareth, located in the south of Tunisia, near the Mediterranean Sea, might have been introduced from the east through different paths, probably from Egypt to neighboring countries and possibly received from Palestine and Jordan [60,61], or through the introduction by the Phoenicians coming from Lebanon to Carthage between the 9th and 2nd centuries B.C [59]. During the Roman period (7th to 3rd centuries B.C.), Tunisia became the breadbasket of the Italian peninsula and the source of the excellent semolina quality from durum wheat grown in North-African countries [62]. The landraces of Tunisia and North Africa have also be traced to the introductions by Romans, who contributed to the modernization of the irrigation systems and extended wheat cultivation to Southern Tunisia [59].

### *3.4. Implications on Conservation and Use of Genetic Resources*

The results of this study can be used to direct future activities of collection, conservation, and the use of durum wheat genetic resources in Tunisia. In terms of adding new diversity to the existing collections, more collection is needed in Tunisia mainly in the oasis areas to collect different landraces like Jenah Zarzoura. Future studies of this kind of germplasm will shed more light on the specific nature of this germplasm as in the case of durum wheat germplasm from Ethiopia [24]. Additionally, more landraces from other regions, even if they have the same local names, should be collected and given different identifiers within the genebank database and considered as different accessions in the ex situ collection. When collecting, the team should avoid plants with characteristics of improved varieties to avoid mixtures. For ex situ conservation, the genebank in Tunisia should conserve the bulk seeds of each landrace instead of seeds of many individual lines constituting each landrace. This will reduce the cost of conservation and avoid conserving several copies of the same line. DArTseq markers have also allowed us to identify outlier lines within the landraces and can therefore be eliminated during multiplication and characterization. For landraces still prevailing under traditional farming systems and under harsh conditions, on-farm conservation could be promoted to conserve a larger genetic base and the associated local knowledge. The need for ensuring long-term conservation of Tunisian durum wheat landraces is dictated by the on-going genetic erosion due to the spread of newly released cultivars and by their richness in genes for tolerance to drought, heat, and salinity and their quality attributes for different end uses [63].

## 4. Materials and Methods

### 4.1. Plant Material

This study used three sets of germplasm:

- Set 1: A total of 196 pure lines issued from six Tunisian durum wheat landraces known as Mahmoudi, Rommani, Jenah Zarzoura, Bidi, Jenah Khotifa, and Biskri, collected from 5 regions between 2009 and 2010 and conserved by the National Genebank of Tunisia (NGBT), were used for assessing the intra and inter genetic diversity (Table 6). The landraces Mahmoudi and Jenah Zarzoura are constituted by 30 lines each, and Rommani and Bidi are constituted by 33 lines each. The landraces Biskri and Jenah Khotifa are represented by two populations each: Biskri1 (31), Biskri2 (7), Jenah Khotifa1 (29), and Jenah Khotifa2 (3).

- Set 2: A total of 40 accessions composed of six improved varieties released in Tunisia, seven ICARDA/CIMMYT elite inbred lines and 27 Tunisian durum wheat landraces (15 landraces are represented by one accession; Jenah Khotifa, Biskri, and Chetla are represented by two accessions each; Bidi and Rommani are represented by three accessions each) were used for identification of potential mis-classified lines from Set 1 and for comparison between Tunisian durum wheat landraces and improved germplasm (Table 7).

- Set 3: A total of 207 durum wheat landraces collected from Mediterranean and West Asia countries—Morocco (17), Algeria (14), Tunisia (13), Libya (9), Egypt (12), Lebanon (10), Syria (11), Jordan (11), Israel/Palestine (10), Iraq (11), Iran (12), Afghanistan (16), Yemen (10), Greece (18), Cyprus (10), Italy (12), and Ethiopia (11)—randomly chosen from the ICARDA genebank collection along with four lines representing each Tunisian landraces from Set 1 were used for a genetic relationship study ( Table 8).

The seeds for Sets 1 and 2 were taken from single plant spikes of each line, landrace, and germplasm grown at the Mornag INRA-Tunisia experiment station during the 2014–2015 growing season, while the seeds for Set 3 were send to CIMMYT from the ICARDA genebank within the joint effort to genotype wheat genetic resources.

**Table 6.** List and sample size of the Tunisian durum wheat landraces collected from the center, the south and the Oasis of Tunisia (Set 1).

| Landraces | Number of Lines | Province |
|---|---|---|
| Bidi | 33 | Kairouan |
| Biskri1 | 31 | Gafsa (Djebel ouled ouhiba) |
| Biskri2 | 7 | Medenine (Zarzis) |
| Jenah Khotifa1 | 29 | Kairouan |
| Jenah Khotifa2 | 3 | Tozeur (El Frid) |
| Jenah Zarzoura | 30 | Matmata (Oasis of Mareth) |
| Mahmoudi | 30 | Gafsa (Snad) |
| Rommani | 33 | Gafsa (Djebel Ouled Ouhiba) |
| Total | 196 | |

**Table 7.** List of the Tunisian durum wheat landraces, Tunisia released varieties, and ICARDA/CIMMYT elite lines (Set 2).

| Tunisian Landraces | N | Origin | Accession Identifier | Pedigree |
|---|---|---|---|---|
| Agili | 1 | NGBT | IG 23903 | - |
| Aouadi | 1 | NGBT | IG 23908 | - |
| Arbi | 1 | NGBT | IG 23903 | - |
| Azizi | 1 | NGBT | IG 23904 | - |
| Bayadha | 1 | NGBT | IG 23905 | - |
| Bidi | 3 | NGBT | IG 19553; IG 23906; IG 23929 | - |
| Biskri | 2 | NGBT | IG 19551; IG 23907 | - |
| Chetla | 2 | NGBT | IG 19555; IG 19557 | - |
| Chili | 1 | NGBT | IG 23908 | - |
| Derbessi | 1 | NGBT | IG 23909 | - |
| Hmira | 1 | NGBT | IG 23910 | - |
| Jneh khotifa | 2 | NGBT | IG 23915; IG 999 | - |
| Mahmoudi | 1 | NGBT | IG 23911 | - |
| Richi | 1 | NGBT | IG 23912 | - |
| Rommani | 3 | NGBT | IG 19552; IG 19554; IG 19558 | - |
| Sbei | 1 | NGBT | IG 23913 | - |
| Souri | 1 | NGBT | IG 23914 | - |
| Swabei algia | 1 | NGBT | IG 23916 | - |
| Tounsia | 1 | NGBT | IG 19559 | - |
| Ward bled | 1 | NGBT | IG 23917 | - |
| **Tunisian Improved Varieties** | **N** | **Origin** | **Accession Identifier** | **Pedigree** |
| Inrat69 * | 1 | NGBT | IG 23919 | Mahamoudi/Kyperounda |
| Karim * | 1 | NGBT | IG 23924 | Jori"S"/Anhinga"S"//Flamingo"S" |
| Khiar * | 1 | NGBT | IG 23922 | Chen/Altar 84 |
| Om Rabii * | 1 | NGBT | IG 23921 | Jori C69/ Hau |
| Nasr * | 1 | NGBT | IG 23923 | GoVZ512/Cit//Ruff/Fg/3/Pin/Gre//Trob |
| Maali * | 1 | NGBT | IG 23920 | CMH80A.1060/4/T.TURA/CMH74A.370// CMH77.774/3/YAV79/5/RAZZAK/6/DACK /YEL//KHIAR |
| **ICARDA/CIMMYT Elite Lines** | **N** | **Origin** | **Accession Identifier** | **Pedigree** |
|  | 1 | ICARDA | MCHCB-102 | OmRabi3/*T.urartu*500651/ch5//980947/3/ Otb4//Ossl1/Rfm6 |
| IcaJoudy1 * | 1 | ICARDA | MCHCB-100 | Atlast1/961081//Icasyr1 |
| Nachit * | 1 | ICARDA | DAWRYT-106 | Ameddkul1/*T. dicoccoides* Syrian collection//Loukos |
| Zeina4 * | 1 | ICARDA | MCHCB-154 | GdoVZ512/Cit//Ruff/Fg/3/Src3 |
| Louiza * | 1 | ICARDA | – | Rscn39/Til1 |
| Ammar 6 * | 1 | ICARDA | IDYT37-5 | ICAMORTA0472/Ammar7 |
| Ammar 10 * | 1 | ICARDA | MCHCB-99 | Lgt3/Bicrecham1 |

N: Number of accessions; *: Accessions marked with * are advanced lines; –: Inbred line without identifier; -: Accession without pedigree (Landrace), NGBT: National Gene Bank of Tunisia; ICARDA: International Center for Agricultural Research in Dry Areas.

**Table 8.** Number of durum wheat landraces from the Mediterranean and West Asian countries provided by the ICARDA genebank (Set 3).

| Geographical Origin | Number of Landraces |
|---|---|
| Afghanistan | 16 |
| Cyprus | 10 |
| Algeria | 14 |
| Egypt | 12 |
| Ethiopia | 11 |
| Greece | 18 |
| Iran | 12 |
| Iraq | 11 |
| Israel | 10 |
| Italy | 12 |
| Jordan | 11 |
| Lebanon | 10 |
| Libya | 9 |
| Morocco | 17 |
| Syria | 11 |
| Tunisia | 13 |
| Yemen | 10 |

*4.2. Genotypic Characterization Using the DArtseq™ Method*

Fresh young leaves from a single individual plant per accession have been used for genomic DNA extraction performed through a modified CTAB (cetyltrimethylammonium bromide) method [64]. The DNA quality was determined by electrophoresis using 1% agarose gel and quantified with NanoDrop 8000 spectrophotometer V 2.1.0.

A high-throughput genotyping method using DArTseq™ technology was employed to generate a genomic profile of the germplasm at the Genetic Analysis Service for Agriculture (SAGA) facility at CIMMYT, Mexico. This method used a combination of two restriction enzymes (PstI and HpaII) to reduce the genome complexity and to generate a genomic representation of the samples [31]. The genomic DNA was submitted to a process of digestion and ligation with a specific PstI-RE site-adapter tagged with 96 different barcodes, which allow to multiplex 96 DNA samples in a single lane of Illumina HiSeq2500 instrument (Illumina Inc., San Diego, CA, USA). The amplified fragments were sequenced up to 77 bases, generating around 500,000 unique reads per sample. A FASTQ files (full reads of 77 bp) were quality filtered using a Phred quality score of 30, which represents a 90% base call accuracy for at least 50% of the bases. An additional filter has been applied on barcode sequences. DArTsoft 14 was used to generate Silico-DArT score tables as data (1/0), indicating the presence/absence variation (PAV) markers and SNP markers.

*4.3. DArTseq Markers and Cluster Analysis*

DArTseq markers were mapped using the consensus map version 4.0 (www.diversityarrays.com) developed by DArT Pty. Ltd., Australia, and the reference wheat genome issued from the International Wheat Genome Sequencing Consortium database (IWGSC WGAv0.4), available online at https://urgi.versailles.inra.fr/download/iwgsc/.

DArTseq raw data was filtered according to markers criterion; minor allele frequency >5% and missing data $\leq$20%.

The summary statistics of the filtered DArTseq markers such as the expected heterozygosity (He) or genetic diversity (GD), minor allele frequency (MAF), and the polymorphic information content (PIC) [44], were calculated using R-project (http://www.r-project.org/) [65].

For cluster analysis of the collection, allele sharing distance matrix was computed as described by Goa et al. [66]. The distance between individuals i and j was defined as

$$D_{ij} = \frac{1}{L}\sum_{l=1}^{L} d_{ij}(l)$$

where L is the total number of markers, and dij(l) is 0, 1, or 2 if individuals i and j have zero, one, or two allele(s) in common at Locus l. Classification of the individuals into groups was performed using the allele sharing matrix and Ward's minimum variance algorithm [67]. The clustering algorithm was run using the hclust function within the R package [68].

*4.4. Population Structure and Genetic Differentiation*

Discriminant analysis of principal components (DAPC) was used to infer the number of clusters of genetically related individuals [48], using the *adegenet* package and *popr* in R-project [68]. DAPC is a multivariate analysis requiring three steps; first the data is transformed using principal component analysis, sub-groups are then identified using k-mean clustering, and discrimination between the sub-groups is then optimized using discriminant analysis. For the k-mean clustering, the optimal number of groups was identified using the Bayesian information criterion (BIC) as a measure of goodness of fit. The number of sub-groups (K) was set from 2 to 8 and the *K*-value with the lowest BIC was retained as the optimal number of clusters. A discriminant analysis was then implemented using the groups found by k-mean clustering [69].

For detecting the genetic variation among and within population(s) and supporting the hypothesis of the population structure, analysis of molecular variance (AMOVA) was performed for different hierarchical subdivision levels as well as for the full population structure strata from K = 2 to K = 8. Significance levels for variance component were estimated based on 10,000 permutations using the *randtest* function in the R-project as described by Excoffier et al. [70].

For genetic differentiation and relationships among the sub-populations issued from the population structure analysis, the genetic distance between the sub-populations using Reynolds genetic distance was computed [71], and the genetic indices, such as the observed heterozygosity ($H_O$, the proportion of loci that are heterozygote for a population), the expected heterozygosity or genetic diversity (*He*, the fraction of all landraces which would be heterozygote for any randomly chosen locus), and the F-statistics ($F_{ST}$) as developed by Wright [72], were calculated.

**Author Contributions:** Conceptualization, C.R., Z.K. and A.A.; data curation, C.R., Z.K. and A.A.; formal analysis, C.R., Z.K. and A.A.; funding acquisition, A.A.; investigation, C.R. and A.A.; methodology, C.R., Z.K. and A.A.; project administration, A.A.; resources, M'B.B.N.; Software, Z.K.; supervision, A.A.; Validation, Z.K. and A.A.; visualization, Z.K., C.S., F.B. and A.A.; writing—original draft, C.R., Z.K. and A.A.; writing—review & editing, C.R., Z.K. and A.A.

**Acknowledgments:** We would like to thank the General Director of the National Gene bank of Tunisia, M. Ben Naceur, for providing the germplasm used in this study. We would also like to thank Athanasios Tsivelikas for field assistance and the technical staff of Mornag INRAT station, Tunisia, for handling field and laboratory activities. The genotyping characterization work was implemented by CIMMYT as part of the MasAgro Biodiversidad project in collaboration with ICARDA, made possible by the generous support of the Mexican Secretariat of Agriculture, Livestock, Rural Development, Fisheries and Food (SAGARPA) and CRP Wheat.

## Abbreviations

| | |
|---|---|
| DarTseq | diversity array technology sequencing |
| DAPC | discriminant analysis of principal components |
| AMOVA | analysis of molecular variance |
| RE | restriction enzyme |

## References

1. Gioia, T.; Nagel, K.A.; Beleggia, R.; Fragasso, M.; Ficco, D.B.M.; Pieruschka, R.; De Vita, P.; Fiorani, F.; Papa, R. Impact of Domestication on the Phenotypic Architecture of Durum Wheat under Contrasting Nitrogen Fertilization. *J. Exp. Bot.* **2015**, *66*, 5519–5530. [CrossRef] [PubMed]
2. *The World Wheat Book: A History of Wheat Breeding*; Bonjean, A.P.; Angus, W.J.; van Ginkel, M. (Eds.) Lavoisier: Paris, France, 2016; Volume 3, ISBN 9782743020910.
3. Daaloul, A.; Harrabi, M.; Amara, H. Evaluation de la collection nationale de blé dur. In *Revure de l'INAT*; numéro spécial; INAT: Tunis, Tunisia, 1998; pp. 337–358. (In French)
4. Deghaïs, M.; Kouki M, M.; Gharbi M, M.; El Felah, M. *Les Variétés de Céréales Cultivées en Tunisie (blé dur, blé Tendre, orge et Triticale)*; INRAT: Tunis, Tunisia, 2007. (In French)
5. Dvorak, J.; Akhunov, E.D. Tempos of Gene Locus Deletions and Duplications and Their Relationship to Recombination Rate During Diploid and Polyploid Evolution in the *Aegilops-Triticum* Alliance. *Genetics* **2005**, *171*, 323–332. [CrossRef] [PubMed]
6. Feldman, M.; Levy, A.A. Allopolyploidy—A Shaping Force in the Evolution of Wheat Genomes. *Cytogenet. Genome Res.* **2005**, *109*, 250–258. [CrossRef] [PubMed]
7. Salamini, F.; Özkan, H.; Brandolini, A.; Schäfer-Pregl, R.; Martin, W. Genetics and Geography of Wild Cereal Domestication in the near East. *Nat. Rev. Genet.* **2002**, *3*, 429–441. [CrossRef] [PubMed]
8. Matsuoka, Y. Evolution of Polyploid Triticum Wheats under Cultivation: The Role of Domestication, Natural Hybridization and Allopolyploid Speciation in Their Diversification. *Plant Cell Physiol.* **2011**, *52*, 750–764. [CrossRef] [PubMed]

9. Zohary, D.; Hopf, M. *Domestication of Plants in the Old World: The Origin and Spread of Cultivated Plants in West Asia, Europe, and the Nile Valley*, 3rd ed.; Oxford University Press: Oxford, UK, 2000; ISBN 9780198503576 9780198503569.
10. Vavilov, N.I. The Origin, Variation, Immunity and Breeding of Cultivated Plants. *Soil Sci.* **1951**, *72*, 482. [CrossRef]
11. Boeuf, F. *Le blé en Tunisie: La Plante. Le Milieu Physico-Chimique*; Société Anonyme de L'Imprimerie Rapide de Tunis: Tunis, Tunisia, 1932; Volume 1.
12. Lala, S.; Amri, A.; Maxted, N. Towards the Conservation of Crop Wild Relative Diversity in North Africa: Checklist, Prioritisation and Inventory. *Genet. Resour. Crop Evol.* **2018**, *65*, 113–124. [CrossRef]
13. Khoufi, S. Morphological and Molecular Characterization of Six of the Most Frequently Cultivated Hard Wheat Varieties in Tunisia. *J. Plant Breed. Crop Sci.* **2012**, *4*. [CrossRef]
14. Lopes, M.S.; El-Basyoni, I.; Baenziger, P.S.; Singh, S.; Royo, C.; Ozbek, K.; Aktas, H.; Ozer, E.; Ozdemir, F.; Manickavelu, A.; et al. Exploiting Genetic Diversity from Landraces in Wheat Breeding for Adaptation to Climate Change. *J. Exp. Bot.* **2015**, *66*, 3477–3486. [CrossRef]
15. Sourour, A.; Chahine, K.; Youssef, T.; Olfa, S.; Hajer, S. Phenotypic Diversity of Tunisian Durum Wheat Landraces. *Afr. Crop Sci. J.* **2010**, *18*. [CrossRef]
16. Sourour, A.; Hajer, S. Distribution and Phenotypic Variability Aspects of Some Quantitiative Traits among Durum Wheat Accessiions. *Afr. Crop Sci. J.* **2010**, *16*. [CrossRef]
17. Ya, N.; Raveendar, S.; Bayarsukh, N.; Ya, M.; Lee, J.-R.; Lee, K.-J.; Shin, M.-J.; Cho, G.-T.; Ma, K.-H.; Lee, G.-A. Genetic Diversity and Population Structure of Mongolian Wheat Based on SSR Markers: Implications for Conservation and Management. *Plant Breed. Biotechnol.* **2017**, *5*, 213–220. [CrossRef]
18. Sohail, Q.; Manickavelu, A.; Ban, T. Genetic Diversity Analysis of Afghan Wheat Landraces (*Triticum aestivum*) Using DArT Markers. *Genet. Resour. Crop Evol.* **2015**, *62*, 1147–1157. [CrossRef]
19. Khan, M.K.; Pandey, A.; Choudhary, S.; Hakki, E.E.; Akkaya, M.S.; Thomas, G. From RFLP to DArT: Molecular Tools for Wheat (*Triticum* spp.) Diversity Analysis. *Genet. Resour. Crop Evol.* **2014**, *61*, 1001–1032. [CrossRef]
20. Kehel, Z.; Garcia-Ferrer, A.; Nachit, M.M. Using Bayesian and Eigen Approaches to Study Spatial Genetic Structure of Moroccan and Syrian Durum Wheat Landraces. *AJMB* **2013**, *03*, 17–31. [CrossRef]
21. Medini, M.; Hamza, S.; Rebai, A.; Baum, M. Analysis of Genetic Diversity in Tunisian Durum Wheat Cultivars and Related Wild Species by SSR and AFLP Markers. *Genet. Resour. Crop Evol.* **2005**, *52*, 21–31. [CrossRef]
22. Maccaferri, M.; Ricci, A.; Salvi, S.; Milner, S.G.; Noli, E.; Martelli, P.L.; Casadio, R.; Akhunov, E.; Scalabrin, S.; Vendramin, V.; et al. A High-Density, SNP-Based Consensus Map of Tetraploid Wheat as a Bridge to Integrate Durum and Bread Wheat Genomics and Breeding. *Plant Biotechnol. J.* **2015**, *13*, 648–663. [CrossRef]
23. Borrill, P.; Adamski, N.; Uauy, C. Genomics as the Key to Unlocking the Polyploid Potential of Wheat. *New Phytol.* **2015**, *208*, 1008–1022. [CrossRef]
24. Kabbaj, H.; Sall, A.T.; Al-Abdallat, A.; Geleta, M.; Amri, A.; Filali-Maltouf, A.; Belkadi, B.; Ortiz, R.; Bassi, F.M. Genetic Diversity within a Global Panel of Durum Wheat (*Triticum durum*) Landraces and Modern Germplasm Reveals the History of Alleles Exchange. *Front. Plant Sci.* **2017**, *8*, 1277. [CrossRef]
25. Holtz, Y.; Ardisson, M.; Ranwez, V.; Besnard, A.; Leroy, P.; Poux, G.; Roumet, P.; Viader, V.; Santoni, S.; David, J. Genotyping by Sequencing Using Specific Allelic Capture to Build a High-Density Genetic Map of Durum Wheat. *PLoS ONE* **2016**, *11*, e0154609. [CrossRef]
26. He, J.; Zhao, X.; Laroche, A.; Lu, Z.-X.; Liu, H.; Li, Z. Genotyping-by-Sequencing (GBS), an Ultimate Marker-Assisted Selection (MAS) Tool to Accelerate Plant Breeding. *Front. Plant Sci.* **2014**, *5*, 484. [CrossRef]
27. Sieber, A.-N.; Longin, C.F.H.; Würschum, T. Molecular Characterization of Winter Durum Wheat (*Triticum durum*) Based on a Genotyping-by-Sequencing Approach. *Plant Genet. Resour.* **2017**, *15*, 36–44. [CrossRef]
28. Alipour, H.; Bihamta, M.R.; Mohammadi, V.; Peyghambari, S.A.; Bai, G.; Zhang, G. Genotyping-by-Sequencing (GBS) Revealed Molecular Genetic Diversity of Iranian Wheat Landraces and Cultivars. *Front. Plant Sci.* **2017**, *8*, 1293. [CrossRef]
29. Wenzl, P.; Carling, J.; Kudrna, D.; Jaccoud, D.; Huttner, E.; Kleinhofs, A.; Kilian, A. Diversity Arrays Technology (DArT) for Whole-Genome Profiling of Barley. *Proc. Natl. Acad. Sci. USA* **2004**, *101*, 9915–9920. [CrossRef] [PubMed]

30. Akbari, M.; Wenzl, P.; Caig, V.; Carling, J.; Xia, L.; Yang, S.; Uszynski, G.; Mohler, V.; Lehmensiek, A.; Kuchel, H.; et al. Diversity Arrays Technology (DArT) for High-Throughput Profiling of the Hexaploid Wheat Genome. *Theor. Appl. Genet.* **2006**, *113*, 1409–1420. [CrossRef]
31. Sansaloni, C.; Petroli, C.; Jaccoud, D.; Carling, J.; Detering, F.; Grattapaglia, D.; Kilian, A. Diversity Arrays Technology (DArT) and next-Generation Sequencing Combined: Genome-Wide, High Throughput, Highly Informative Genotyping for Molecular Breeding of Eucalyptus. *BMC Proc.* **2011**, *5* (Suppl. 7), P54. [CrossRef]
32. Baloch, F.S.; Alsaleh, A.; Shahid, M.Q.; Çiftçi, V.E.; Sáenz de Miera, L.; Aasim, M.; Nadeem, M.A.; Aktaş, H.; Özkan, H.; Hatipoğlu, R. A Whole Genome DArTseq and SNP Analysis for Genetic Diversity Assessment in Durum Wheat from Central Fertile Crescent. *PLoS ONE* **2017**, *12*, e0167821. [CrossRef] [PubMed]
33. Li, H.; Vikram, P.; Singh, R.P.; Kilian, A.; Carling, J.; Song, J.; Burgueno-Ferreira, J.A.; Bhavani, S.; Huerta-Espino, J.; Payne, T.; et al. A High Density GBS Map of Bread Wheat and Its Application for Dissecting Complex Disease Resistance Traits. *BMC Genom.* **2015**, *16*. [CrossRef]
34. Yang, X.; Ren, R.; Ray, R.; Xu, J.; Li, P.; Zhang, M.; Liu, G.; Yao, X.; Kilian, A. Genetic Diversity and Population Structure of Core Watermelon (*Citrullus lanatus*) Genotypes Using DArTseq-Based SNPs. *Plant Genet. Resour.* **2016**, *14*, 226–233. [CrossRef]
35. Valdisser, P.A.M.R.; Pereira, W.J.; Almeida Filho, J.E.; Müller, B.S.F.; Coelho, G.R.C.; de Menezes, I.P.P.; Vianna, J.P.G.; Zucchi, M.I.; Lanna, A.C.; Coelho, A.S.G.; et al. In-Depth Genome Characterization of a Brazilian Common Bean Core Collection Using DArTseq High-Density SNP Genotyping. *BMC Genom.* **2017**, *18*. [CrossRef]
36. Marcotuli, I.; Gadaleta, A.; Mangini, G.; Signorile, A.; Zacheo, S.; Blanco, A.; Simeone, R.; Colasuonno, P. Development of a High-Density SNP-Based Linkage Map and Detection of QTL for β-Glucans, Protein Content, Grain Yield per Spike and Heading Time in Durum Wheat. *Int. J. Mol. Sci.* **2017**, *18*, 1329. [CrossRef] [PubMed]
37. Patil, R.M.; Oak, M.D.; Tamhankar, S.A.; Sourdille, P.; Rao, V.S. Mapping and Validation of a Major QTL for Yellow Pigment Content on 7AL in Durum Wheat (*Triticum turgidum* L. ssp. *durum*). *Mol. Breed* **2008**, *21*, 485–496. [CrossRef]
38. Colasuonno, P.; Lozito, M.L.; Marcotuli, I.; Nigro, D.; Giancaspro, A.; Mangini, G.; De Vita, P.; Mastrangelo, A.M.; Pecchioni, N.; Houston, K.; et al. The Carotenoid Biosynthetic and Catabolic Genes in Wheat and Their Association with Yellow Pigments. *BMC Genom.* **2017**, *18*. [CrossRef]
39. Turki, N.; Shehzad, T.; Harrabi, M.; Okuno, K. Detection of QTLs Associated with Salinity Tolerance in Durum Wheat Based on Association Analysis. *Euphytica* **2015**, *201*, 29–41. [CrossRef]
40. Mangini, G.; Gadaleta, A.; Colasuonno, P.; Marcotuli, I.; Signorile, A.M.; Simeone, R.; De Vita, P.; Mastrangelo, A.M.; Laidò, G.; Pecchioni, N.; et al. Genetic Dissection of the Relationships between Grain Yield Components by Genome-Wide Association Mapping in a Collection of Tetraploid Wheats. *PLoS ONE* **2018**, *13*, e0190162. [CrossRef]
41. Chen, T.; Tantasawat, P.A.; Wang, W.; Gao, X.; Zhang, L. Population Structure of Chinese Southwest Wheat Germplasms Resistant to Stripe Rust and Powdery Mildew Using the DArT-Seq Technique. *Cienc. Rural* **2018**, *48*. [CrossRef]
42. Nadeem, M.A.; Nawaz, M.A.; Shahid, M.Q.; Doğan, Y.; Comertpay, G.; Yıldız, M.; Hatipoğlu, R.; Ahmad, F.; Alsaleh, A.; Labhane, N.; et al. DNA Molecular Markers in Plant Breeding: Current Status and Recent Advancements in Genomic Selection and Genome Editing. *Biotechnol. Biotechnol. Equip.* **2018**, *32*, 261–285. [CrossRef]
43. Ren, J.; Sun, D.; Chen, L.; You, F.; Wang, J.; Peng, Y.; Nevo, E.; Sun, D.; Luo, M.-C.; Peng, J. Genetic Diversity Revealed by Single Nucleotide Polymorphism Markers in a Worldwide Germplasm Collection of Durum Wheat. *Int. J. Mol. Sci.* **2013**, *14*, 7061–7088. [CrossRef]
44. Botstein, D.; White, R.L.; Skolnick, M.; Davis, R.W. Construction of a Genetic Linkage Map in Man Using Restriction Fragment Length Polymorphisms. *Am. J. Hum. Genet.* **1980**, *32*, 314–331.
45. Soriano, J.M.; Villegas, D.; Sorrells, M.E.; Royo, C. Durum Wheat Landraces from East and West Regions of the Mediterranean Basin Are Genetically Distinct for Yield Components and Phenology. *Front. Plant Sci.* **2018**, *9*, 80. [CrossRef]
46. Oliveira, H.R.; Hagenblad, J.; Leino, M.W.; Leigh, F.J.; Lister, D.L.; Penã-Chocarro, L.; Jones, M.K. Wheat in the Mediterranean Revisited – Tetraploid Wheat Landraces Assessed with Elite Bread Wheat Single Nucleotide Polymorphism Markers. *BMC Genet.* **2014**, *15*, 54. [CrossRef] [PubMed]

47. Miège, E. Les principales espèces et variétés de Blé cultivées en Afrique du Nord (Suite et Fin). *Revue Internationale de Botanique Appliquée et d'Agriculture Tropicale* **1950**, *30*, 203–215. (In French) [CrossRef]
48. Jombart, T.; Devillard, S.; Balloux, F. Discriminant Analysis of Principal Components: A New Method for the Analysis of Genetically Structured Populations. *BMC Genet.* **2010**, *11*, 94. [CrossRef] [PubMed]
49. Soriano, J.M.; Villegas, D.; Aranzana, M.J.; García del Moral, L.F.; Royo, C. Genetic Structure of Modern Durum Wheat Cultivars and Mediterranean Landraces Matches with Their Agronomic Performance. *PLoS ONE* **2016**, *11*, e0160983. [CrossRef]
50. Ruiz, M.; Giraldo, P.; Royo, C.; Villegas, D.; Aranzana, M.J.; Carrillo, J.M. Diversity and Genetic Structure of a Collection of Spanish Durum Wheat Landraces. *Crop Sci.* **2012**, *52*, 2262. [CrossRef]
51. Eltaher, S.; Sallam, A.; Belamkar, V.; Emara, H.A.; Nower, A.A.; Salem, K.F.M.; Poland, J.; Baenziger, P.S. Genetic Diversity and Population Structure of F3:6 Nebraska Winter Wheat Genotypes Using Genotyping-By-Sequencing. *Front. Genet.* **2018**, *9*, 76. [CrossRef]
52. Jaradat, A. Wheat Landraces: A Mini Review. *Emir. J. Food Agric.* **2013**, *25*, 20. [CrossRef]
53. Mangini, G.; Margiotta, B.; Marcotuli, I.; Signorile, M.A.; Gadaleta, A.; Blanco, A. Genetic Diversity and Phenetic Analysis in Wheat (*Triticum turgidum* subsp. *durum* and *Triticum aestivum* subsp. *aestivum*) Landraces Based on SNP Markers. *Genet. Resour. Crop Evol.* **2017**, *64*, 1269–1280. [CrossRef]
54. Sahri, A.; Chentoufi, L.; Arbaoui, M.; Ardisson, M.; Belqadi, L.; Birouk, A.; Roumet, P.; Muller, M.-H. Towards a Comprehensive Characterization of Durum Wheat Landraces in Moroccan Traditional Agrosystems: Analysing Genetic Diversity in the Light of Geography, Farmers' Taxonomy and Tetraploid Wheat Domestication History. *BMC Evol. Biol.* **2014**, *14*. [CrossRef]
55. Mac Key, J. Wheat: Its concept, evolution and taxonomy. In *Durum Wheat Breeding: Current Approaches and Future Strategies*; Royo, C., Nachit, M., Di Fonzo, N., Araus, J.L., Pfeiffer, W.H., Slafer, G.A., Eds.; Food Products Press: Binghamton, NY, USA, 2005; pp. 3–61. ISBN 9781482277883.
56. Moragues, M.; Moralejo, M.; Sorrells, M.E.; Royo, C. Dispersal of Durum Wheat [*Triticum turgidum* L. ssp. *turgidum* Convar. *durum* (Desf.) MacKey] Landraces across the Mediterranean Basin Assessed by AFLPs and Microsatellites. *Genet. Resour. Crop Evol.* **2007**, *54*, 1133–1144. [CrossRef]
57. Oliveira, H.R.; Campana, M.G.; Jones, H.; Hunt, H.V.; Leigh, F.; Redhouse, D.I.; Lister, D.L.; Jones, M.K. Tetraploid Wheat Landraces in the Mediterranean Basin: Taxonomy, Evolution and Genetic Diversity. *PLoS ONE* **2012**, *7*, e37063. [CrossRef] [PubMed]
58. Nazco, R.; Villegas, D.; Ammar, K.; Peña, R.J.; Moragues, M.; Royo, C. Can Mediterranean Durum Wheat Landraces Contribute to Improved Grain Quality Attributes in Modern Cultivars? *Euphytica* **2012**, *185*, 1–17. [CrossRef]
59. Essid, M.Y. Chapter 2. History of Mediterranean food, CIHEAM ed. In *MediTERRA 2012 (english): The Mediterranean Diet for Sustainable Regional Development*; Presses de Sciences Po (P.F.N.S.P.): France, Paris, 2012; pp. 51–69.
60. Perrino, P.; Porceddu, E. Wheat Genetic Resources in Ethiopia and the Mediterranean Region. In *Wheat Genetic Resources: Meeting Diverse Needs*; John Wiley & Sons: Chichester, UK, 1990; Volume 27, pp. 161–178, 364–365. ISBN 0471928801.
61. Feldman, M. Origin of Cultivated Wheat. In *The World Wheat Book: A History of Wheat Breeding*; Bonjean, A.P., Angus, W.J., Eds.; Lavoisier: Paris, France, 2001; pp. 3–56.
62. Thompson, D.A.W. Wheat in Antiquity - Naum Jasny: The Wheats of Classical Antiquity. (Johns Hopkins University Studies in Historical and Political Science, Series LXII, No. 3.) Pp. 176; 2 Plates. Baltimore: Johns Hopkins Press, 1944. Paper, $1.75. *Class. Rev.* **1946**, *60*, 120–122. [CrossRef]
63. Zaharieva, M.; Bonjean, A.; Monneveux, P. Saharan Wheats: Before They Disappear. *Genet. Resour. Crop Evol.* **2014**, *61*, 1065–1084. [CrossRef]
64. Hoisington, D.; Khairallah, M.; Gonzalez-de-Leon, D. *Laboratory Protocols, CIMMYT Applied Molecular Genetics Laboratory*, 2nd ed.; CIMMYT: Texcoco, Mexico, 1994; Available online: http://repository.cimmyt.org/xmlui/bitstream/handle/10883/1333/91195.pdf (accessed on 20 April 2015).
65. R core team. *R: A Language and Environment for Statistical Computing*; R Foundation for Statistical Computing: Vienna, Austria, 2015; Available online: http://www.R-project.org/ (accessed on 3 November 2018).
66. Gao, X.; Martin, E.R. Using Allele Sharing Distance for Detecting Human Population Stratification. *Hum. Hered* **2009**, *68*, 182–191. [CrossRef] [PubMed]

67. Ward, J.H. Hierarchical Grouping to Optimize an Objective Function. *J. Am. Stat. Assoc.* **1963**, *58*, 236–244. [CrossRef]
68. Jombart, T. Multivariate Analysis of Genetic Data: Exploring Group Diversity. Available online: http://adegenet.r-forge.r-project.org/files/PRstats/practical-MVAgroups.1.0.pdf (accessed on 18 August 2016).
69. Schwarz, G. Estimating the Dimension of a Model. *Ann. Stat.* **1978**, *6*, 461–464. [CrossRef]
70. Excoffier, L.; Smouse, P.E.; Quattro, J.M. Analysis of Molecular Variance Inferred from Metric Distances among DNA Haplotypes: Application to Human Mitochondrial DNA Restriction Data. *Genetics* **1992**, *131*, 479–491.
71. Reynolds, J.; Weir, B.S.; Cockerham, C.C. Estimation of The Coancestry Coefficient: Basis For A Short-Term Genetic Distance. *Genetics* **1983**, *105*, 767–779.
72. Wright, S. Coefficients of Inbreeding and Relationship. *Am. Nat.* **1922**, *56*, 330–338. [CrossRef]

# 10

# Cas Endonuclease Technology—A Quantum Leap in the Advancement of Barley and Wheat Genetic Engineering

**Iris Koeppel †, Christian Hertig †, Robert Hoffie † and Jochen Kumlehn ***

Plant Reproductive Biology, Leibniz Institute of Plant Genetics and Crop Plant Research (IPK) Gatersleben, 06466 Seeland, Germany; koeppel@ipk-gatersleben.de (I.K.); hertig@ipk-gatersleben.de (C.H.); hoffier@ipk-gatersleben.de (R.H.)

* Correspondence: kumlehn@ipk-gatersleben.de

† These authors contributed equally to this work.

**Abstract:** Domestication and breeding have created productive crops that are adapted to the climatic conditions of their growing regions. Initially, this process solely relied on the frequent occurrence of spontaneous mutations and the recombination of resultant gene variants. Later, treatments with ionizing radiation or mutagenic chemicals facilitated dramatically increased mutation rates, which remarkably extended the genetic diversity of crop plants. However, a major drawback of conventionally induced mutagenesis is that genetic alterations occur simultaneously across the whole genome and at very high numbers per individual plant. By contrast, the newly emerging Cas endonuclease technology allows for the induction of mutations at user-defined positions in the plant genome. In fundamental and breeding-oriented research, this opens up unprecedented opportunities for the elucidation of gene functions and the targeted improvement of plant performance. This review covers historical aspects of the development of customizable endonucleases, information on the mechanisms of targeted genome modification, as well as hitherto reported applications of Cas endonuclease technology in barley and wheat that are the agronomically most important members of the temperate cereals. Finally, current trends in the further development of this technology and some ensuing future opportunities for research and biotechnological application are presented.

**Keywords:** cereals; CRISPR; crops; genetic engineering; genome editing; plant; *Triticeae*

---

## 1. Introduction

### *1.1. Historical View of Genetic Modification in Crop Plants*

For more than 10,000 years, plants have been cultivated to feed humans and their livestock, as a source of raw materials and to generate energy. Over time, the driving forces of evolution, domestication, and breeding of cultivated plants have changed according to environmental conditions and societal demands. To cope with the growth of the world's population and the consequent increase in demand for food, the total production needs to be substantially increased during the next decades [1,2]. Not only is the growing world population a challenge, but also the expected climatic changes [3,4]. Thus, the plants of the future also have to be better adapted to harsh weather conditions [5,6]. In addition, the demands of modern society are increasing; food should be of better and consistent quality, healthier, and more diverse. At the same time, society calls for a substantially reduced application of fertilizers and pesticides to render agriculture more sustainable [2].

In nature, mutations occur spontaneously and alterations that provide an advantage under the given conditions are likely to prevail over time. For modern plant breeding, however, the rate of spontaneous mutations is too low to keep up with the demands for crop improvements. Moreover, it is in the nature of spontaneous mutations that they are unpredictable in terms of both position and resultant nucleobase sequence. In the 1930s, mutation breeding emerged, in which mutations are deliberately induced with chemicals such as ethylmethanesulfonate (EMS) or with ionizing radiation. Induced mutagenesis breeding has so far produced over 3000 approved varieties [7] and for many further cultivars it has not been documented which induced mutations they have inherited from germplasms they derive from. Using this technology, however, thousands of mutations occur at different sites in a single plant's genome at a time. Therefore, a vast number of undesirable mutations have to be eliminated via cumbersome and time-consuming back-crossing procedures [5]. The method of Targeted Induced Local Lesions in Genomes (TILLING) represented a significant progress in the detection of mutations. With TILLING or Eco-TILLING, respectively, it is possible to readily identify induced and spontaneous mutations in known genes of interest [8,9].

The method of gene transfer using *Agrobacterium tumefaciens* was introduced in the 1980s [10]. Ever since, genes derived from unrelated organisms or genetically modified gene variants can be introduced into plant genomes. This technique makes it possible to modify or introduce performance-determining genes into cultivated plants more quickly and in a more targeted manner. However, there are some limitations, such as the fact that the transferred gene is integrated at a random position in the genome and that changes to the plant's own genes are only possible to a very limited extent. The insertion site in the genome then also determines whether and how effectively the introduced gene is expressed, and thus to what extent the respective trait will be modified. In addition, not all plants can be transformed equally well by the use of *Agrobacterium*, with cereal species being a particular challenge in this respect [11].

With the new methods for targeted genome modification that are based upon customizable endonucleases, it is possible to modify DNA sequences at a previously defined site of choice in the host genome. The standard application of this technology is site-directed mutagenesis, in which the location of the mutation can be precisely determined, whereas the resulting nucleobase sequence is random [12]. However, the precision can be increased by a variety of approaches. The most sophisticated one involves the use of an artificial DNA repair template implemented via homology-directed DNA repair. By this still very challenging principle, any sequence of choice can be created at any predefined genomic locus.

Wheat (*Triticum aestivum*) and barley (*Hordeum vulgare*) are among the most important cereals in the world. Among all crops, wheat occupies the largest cultivated area in the world, and barley is one of the oldest domesticated crops, currently being the fourth most frequently cultivated cereal (http://faostat/fao.org). With a size of about 17 Gbp, wheat has a very large and complex genome. Only recently has the wheat genome been almost completely read out and this data has been made publicly available [13]. Hexaploid bread wheat evolved via hybridizations from several ancestors. This process involved allopolyploidization, resulting in a total of three diploid subgenomes [14]. In contrast to wheat, barley has a genome with a size of 5.1 Gbp. Already in 2012, a detailed draft of the barley genome was published [15], which was recently complemented by much improved data, including genomic sequences of a large number of representative gene bank accessions [16,17].

### 1.2. Platforms of Customizable Endonucleases

In plant research and biotechnology, four platforms of customizable endonucleases have been used so far; meganucleases, zinc-finger nucleases (ZFNs), transcription activator-like effector nucleases (TALENs), and the RNA-guided, clustered, regularly interspaced, short palindromic repeats (CRISPR)-associated (Cas) endonucleases.

Meganucleases are naturally occurring endonucleases that recognize and cut comparatively long (>12 bp) DNA sequences. The most commonly used meganuclease is the I-*Sce*I from *Saccharomyces cerevisiae* [18,19]. The ability of this endonuclease to induce double-strand breaks in plants was first demonstrated by Puchta et al. [20] using *Nicotiana* protoplasts. Meganucleases are very specific and efficient. It is, however, very difficult to reengineer these enzymes for target sequences other than their native ones [21]. Hence, their use is fairly restricted as compared to the alternative endonuclease platforms [22,23]. As a consequence, meganucleases have been used in plants almost exclusively for basic research, in particular for DNA repair mechanisms [24].

Zinc-finger nucleases are hybrid proteins with a DNA recognition domain consisting of at least three zinc-fingers combined with a *Fok*I restriction endonuclease domain [25]. Each zinc-finger specifically interacts with three base pairs (bp) of the genomic target sequence and several zinc-fingers can be consecutively assembled to recognize and bind a total of 9 to 12 bp of DNA [26]. Zinc-finger nucleases must always be used in pairs, since their *Fok*I endonuclease domain is only catalytically active if it is present as a dimer [25]. The target motifs on the DNA are selected in such a way that the two zinc-finger nuclease monomers bind to the target DNA in antiparallel orientation and at a suitable distance to each other. A DNA double-strand break (DSB) is then catalyzed in the interspace between the two binding sites [27,28]. In addition to the particularly complex production process of zinc-finger nucleases, there are limitations in the selection of possible binding sites, as well as unpredictable neighbor effects between adjacent zinc-fingers on the DNA binding specificity [22].

In 1989, it was shown for the first time how transcription activator-like effector (TALE) proteins are transferred from a pathogenic bacterium of the genus *Xanthomonas* into a host plant [29]. The TALE-based nucleases (TALENs) act, similarly to zinc-finger nucleases, as dimers [30]. The DNA-binding domains consist of up to 30 copies of highly conserved repeats of typically 34 amino acids. Only the amino acid positions 12 and 13 (called repeat-variable di-residues) of each repeat are not uniform, because they specify the different DNA base pairs to bind [31]. A *Fok*I endonuclease bound to such a TALE binding domain induces the double-strand break in the target motif [30]. Since each repeat recognizes just one nucleobase of the DNA target, the design and assembly of the binding domain is more straightforward and versatile than with zinc-finger nucleases [31].

The platform of RNA-guided Cas endonucleases is derived from the CRISPR/Cas adaptive immune system of microbes. The most commonly used Cas endonuclease is from *Streptococcus pyogenes* (Sp). The two-component system consists of the Cas restriction enzyme and an artificial guide RNA (gRNA) that navigates the Cas protein to a cognate DNA sequence motif [12]. The gRNA consists of a structurally functional and a variable part. The 3′-tail forms a spatial structure required for binding with Cas to form a ribonucleoprotein complex. The variable part of the gRNA located at the 5′-end usually comprises 20 nucleotides and defines the DNA-binding specificity of the gRNA according to the principle of complementary base pairing. The genomic target motif of such ribonucleoprotein complex includes the same nucleotide sequence as the variable part of the gRNA, which enables the gRNA to specifically bind the opposite DNA strand. This part of the target DNA is often referred to as the protospacer, a term adopted from the microbial immune system these molecules originate from. In addition, this major part of the target motif is complemented by the so-called protospacer-adjacent motif (PAM), which is recognized and bound by the Cas protein. In the case of SpCas9, the nucleobase triplet NGG, in which N stands for any of the four nucleobases, represents the PAM site [12]. The high versatility of the gRNA 5′-end allows a wide variety of target sequences of choice to be addressed. The comparatively simple application and the efficiency and reliability achievable in higher organisms have rendered this platform the most popular and frequently used tool for site-directed genome modification today [32,33].

## 2. Methodological Aspects of Cas Endonuclease Technology

### *2.1. System Components*

When employing Cas endonuclease technology in plants, there are a number of opportunities and some specific requirements in terms of the construction of transformation vectors. Cas9 endonucleases have been modified in various ways for use in plants. Importantly, the coding sequences were complemented by one or two nuclear localization signals (NLSs) and the codon usage of the protein biosynthesis was optimized for use in plants in general, for monocots and dicots, or even more specifically for individual species, such as wheat and barley [34–36].

Depending on the host organism, also various promoters have been used to drive endonuclease expression. In crop plants, the doubled enhanced cauliflower mosaic virus (2x35S) promoter has been mainly used for test systems [34,37,38] and UBIQUITIN promoters for the generation of heritable modifications. Accordingly, the maize POLYUBIQUITIN1 promoter (ZmUBI1) has been commonly used for cas9 expression to generate heritable changes in barley and wheat [37,39,40]. Alternatively, Cas endonuclease-encoding genes can also be expressed by self-replicating virus particles in the plant cell [41]. This expression system ensures a comparatively high gene dosage, which may be particularly useful for homology-based approaches that are described in more detail below. Gil-Humanes et al. [42] have shown that the Wheat Dwarf Virus, a replicon-based geminivirus, can be modified to express gRNA and cas9 in wheat, albeit this approach has not yet been shown to be applicable for the generation of plants with heritable modifications.

The expression cassette for a gRNA typically consists of plant-derived RNA polymerase III (Pol III)-processed promoters and terminators derived from small nuclear RNA (snRNA)-encoding genes. Somehow surprisingly, it also proved to be sufficient in plants to use the transcriptional termination signal of bacterial CRISPR RNAs, which solely consists of a stretch of five to seven thymidines [34,43]. A comparative test in maize protoplasts had shown that wheat and rice U3 promoters were more efficient than the *Arabidopsis* U6 promoter that has been preferentially used in dicots [44]. The wheat U6 promoter has hitherto been mostly used for barley and wheat [37,39,40]. More recently, a study of Kumar et al. [45] indicated that the barley U3 promoter might be more efficient in site-directed mutagenesis of barley than the frequently used rice U3 promoter.

The preference of the U3 and U6 promoters for A and G as the first transcribed nucleotides limits the selection of target motifs to GN20GG and AN20GG, respectively [43]. However, there are thus far no examples where gRNAs were directly driven by Polymerase II-compatible promoters. The mRNAs otherwise expressed by Pol II promoters are processed at both ends and these modifications significantly alter the structure of the gRNA, and thus compromise the ability to bind both Cas9 and the target motif that is to be processed with sufficient efficiency [43].

Several systems have been developed for the expression of multiple gRNAs. Mostly, each gRNA is expressed by a separate Pol III promoter [44,46,47]. Alternatively, multiple gRNAs can be located on only one transcript and post-transcriptionally separated, as has already been demonstrated with wheat [48]. However, such expression systems have not yet led to improved efficiency of the technology with regard to individual target motifs in plant genomes [43,49,50].

### *2.2. Criteria for Target Motif Selection and in Silico gRNA Design*

Thanks to the opportunity to equip the gRNA with any target-specific 5′-sequence, the plant genomic target site of the gRNA/Cas complex can, in principle, be chosen at will. However, the Cas9 endonuclease requires the aforementioned two guanines that are part of the protospacer-adjacent motif (PAM) at the 3′-end of the target sequence. In addition, there are further preferences of the target-specific part of the gRNA, some of which can have a considerable impact on the functionality of the gRNA/Cas9 complex. For instance, it was determined for the so-called seed region, that is, the six nucleotides residing immediately upstream of the PAM site, that a GC content above 50 percent increases the probability of sufficient gRNA functionality [51]. With regards to the entire target motif, an enrichment of guanines and low adenine content was shown to cause increased binding stability and activity of the gRNA/Cas9 complex [52].

The target sequence-specific part of the gRNA usually has a length of 20 nucleotides [12,53]. As previously stated, the transcriptional start defined by the U3 and U6 promoters, which are preferred to drive gRNAs in plants, are A and G, respectively, which requires the selection of genomic targets having the same nucleobases at their 5′-end [38]. In human cell lines and zebrafish, it has been shown that the length of the target-specific part of the gRNA can be shortened to 18 nucleotides without a noticeable effect on mutation efficiency. Surprisingly, such shorter gRNAs even had increased specificity for the on-target sequence as compared with off-targets [52,54,55]. Guide-RNAs with less than 20 target-specific nucleobases also showed high performance in *Arabidopsis* and barley [56,57]. On the other hand, an extension of the gRNA 5′-part beyond 20 nucleotides was reported to cause reduced cleavage efficiency [58].

Quite a number of online platforms have been developed for the selection of target motifs and corresponding gRNAs. For instance, CRISPR-Plant was especially developed for plants [59,60]. While a whole array of plant reference genomes can be directly screened in this platform, the barley and wheat genomes are not implemented. However, for the selection of gRNAs in cereals, DeskGen is a highly instrumental alternative, as these cereal genomes are available for target validation [61,62]. This platform offers a comparatively large scope of options, including various Cas endonucleases and a useful range of lengths of the target-specific gRNA 5′-part. Also, the CRISPOR online tool can be specifically used for barley and wheat. It comprises a comprehensive choice of different Cas enzymes [63,64]. In addition to the selection of gRNAs predicted to be well-performing at their target motif, all above online tools are also capable of indicating potential off-targets in the chosen reference genome. WU-CRISPR is another useful platform, as its algorithm is comparatively strict in selecting particularly useful target motifs. However, this tool is confined to SpCas9 and does not offer off-target screens in reference genomes [65]. A general disadvantage of all these platforms is that their algorithms have been established using data from other organisms than plants, which is thought to be one of the reasons that the reliability of their results is still fairly limited.

Liang et al. [66] have focused on the preservation of the gRNA secondary structure. They determined that three of the stem loops frequently occurring in the gRNA 3′-part are essential to appropriately bind to the Cas9 protein, and hence are also of decisive importance to the overall functionality of the gRNA/Cas complex. These authors also found out that 98 percent of well-performing gRNAs have no more than a total of 12 base pairs formed between their target-specific and 3′-parts and no more than 7 target-specific nucleobases consecutively involved in intra-molecular base pairs. In addition, no more than 6 base pairs should be formed within the target-specific part of the gRNA. In order to ensure these gRNA features, it is recommended to thoroughly investigate the secondary structure of candidate gRNAs. For the prediction of RNA secondary structures, online platforms such as mfold [67–69] or RNAfold [70,71] are available.

In cases where genetic modifications are intended to be performed in genotypes other than those with reference genomes, it is recommended to countercheck pre-selected target motifs for their presence and integrity, because even single nucleotide polymorphisms typically result in a dramatic drop in gRNA/Cas efficiency.

### *2.3. Delivery of Cas Endonucleases and Associated Reagents into Plant Cells*

A number of methods are available for the transfer of DNA, RNA, and proteins into plant cells, which in principle can also be used for Cas endonucleases, customized gRNAs, DNA repair templates, and any further reagents that may be used to support site-directed genome modification. Nowadays, the most widely used approach is based on the genomic integration of expression units with gRNA and Cas-encoding DNA sequences. However, the production of stably transgenic plants is a particular challenge for cereals, because agrobacteria, which are mostly used for plant transformation, have a very limited compatibility with these non-host plants. Moreover, the formation of adventitious shoots from leaf or shoot explants, which is readily achieved in the context of DNA transfer methods applicable to most dicotyledonous species, has only been successful in exceptional, hardly reproducible cases in cereals [72]. On the basis of special methodological approaches, routine genetic transformation of barley and wheat is nevertheless possible, at least by using selected accessions that are comparatively amenable to methods of DNA transfer. Such experimental model genotypes are, for example, the barley cultivar Golden Promise and the Mexican wheat breeding line Bobwhite SH9826 [11]. For stable DNA transfer in cereal crops, immature zygotic embryos are most widely used. Importantly, such embryos must be of a developmental stage at which highly totipotent cells are still abundantly present. In immature embryos of barley and wheat, such cells are preferentially residing in the area that surrounds the shoot apical meristem and the scutellum, respectively. Stable transgenic wheat and barley plants were already produced by ballistic transfer of plasmid-coated gold particles to immature embryos in the early 1990s [73,74]. Using hypervirulent bacterial strains, which had initially been used for the transformation of rice [75], thereafter it was soon also possible to generate transgenic barley and wheat plants using this principle [76,77]. Today, it is routinely possible to generate stable transgenic plants from over 10% of inoculated immature embryos of barley [78], while in wheat, efficiencies of about 5 percent have been achieved both via *Agrobacterium*-mediated and ballistic DNA transfer [79,80]. However, a largely improved protocol was more recently demonstrated to allow for much improved transformation efficiency in wheat ([81,82]. The comparative examination of a variety of genotypes has shown that although some can indeed be transformed in addition to the model lines, the efficiency, however, is much reduced [83,84]. Previous studies on Cas9-induced mutagenesis of barley have been based without exception on *Agrobacterium*-mediated DNA transfer. In contrast to this, site-directed mutagenesis in wheat has mostly relied on ballistic gene transfer [85].

Another particularly interesting principle of DNA transfer into cereal cells is the use of immature, single-celled pollen (microspores), which can undergo cell proliferation and embryogenic development under suitable culture conditions [86]. Since pollen consists of haploid cells, it is possible that homozygous transgenic plants can be directly produced via this developmental pathway in association with DNA transfer and whole genome duplication. *Agrobacterium*-mediated DNA transfer in embryogenic pollen cultures of barley has the further advantage that a winter barley variety can be used, which is comparatively difficult via DNA transfer to immature embryos [87,88]. A similar transformation method has been reported also for wheat, which is, however, based on ballistic DNA transfer [89]. In the authors' laboratory, barley mutants have already been produced via the pollen embryogenesis pathway using TALENs [90], as well as by Cas endonucleases. Furthermore, the production of stable transgenic barley and wheat by ballistic DNA transfer into meristematic tissue from the shoot apex of embryos prepared from mature grains was also successful [91,92]. This method has potential because of particularly low genotype dependence, since plant regeneration is based on conventional germination of the embryos, and therefore no formation of adventitious shoots is necessary. More recently, Hamada et al. [93] have also demonstrated that the production of plants with site-directed modifications using gRNA/cas9 constructs is possible using this method.

After alteration of the genomic target motif, the presence of transgenes coding for gRNA and endonuclease is not only unnecessary but also undesirable, as off-targets can still be mutated, even if they are not identical with the on-target. In this context, an advantage of the application of Cas endonucleases compared to conventional genetic engineering is that the integration site of the transgenes is mostly not coupled with the site of the desired genetic modification. Thus, the elimination of these transgenes by genetic segregation of progeny is comparatively straightforward while maintaining the desired modification. Endonuclease-triggered genetic alterations in primary transgenic plants are often heterozygous and restricted to tissue sectors. Therefore, progeny needs to be screened in order to select homozygously mutated individuals. Consequently, no extra time is required to obtain transgene-free mutants. Haploid technology is a particularly elegant solution for the separation of achieved genetic modifications from unnecessary transgene insertions or for the segregation of several mutations present in chimeric plants of the M1 generation. With this technology, populations of completely homozygous recombinants can be obtained, in which desired genotypes are present at a much higher frequency than among offspring derived from conventional selfing [94,95].

In addition to the various possibilities of genomic integration of gRNA and cas endonuclease genes, methods for transient expression have also been established, which is of considerable value for the preliminary validation of target-specific gRNAs, Cas endonuclease variants, and any further components used in this context. The most widely used transient expression method is based on the transfection of isolated mesophyll protoplasts, whose plasma membrane is rendered porous by application of polyethylene glycol. This enables transgenic plasmid DNA, in vitro transcribed RNA, bacterially pre-produced Cas protein, or ribonucleoprotein complexes to be taken up by the protoplasts. This has been exemplified, amongst others, in wheat and barley [39,57,96]. The activity of the transferred components can then be checked after amplification of the genomic target regions using the T7E1 assay, by Sanger or deep sequencing, as is described in more detail further down below. Another method for the functional validation of gRNA/cas constructs by transient expression is based on the ballistic transformation of leaf epidermis. Epidermal cells are comparatively well suited to be screened microscopically, for instance in regards to Cas-induced restoration of a reporter gene construct [97]. Barley leaf tissue has been used to demonstrate precise editing via homology-based DNA repair using a synthetic DNA repair template [98].

As an alternative to the expression of gRNA- and Cas-encoding DNA, in vitro transcribed RNA or Cas protein can be transferred to plant cells in order to make specific modifications to the genome. Guide RNA and Cas endonuclease can also be pre-assembled to form gRNA/Cas9 ribonucleoprotein complexes prior to their transfer into plant cells. The use of pre-produced RNA and protein molecules or complexes thereof has the advantage over DNA that their effective amount is independent of the expression profile and strength of plant promoters. In addition, the activity of these components is limited in time, as they are not continuously delivered by gene expression but are subject to cellular degradation. Accordingly, fewer mutations will occur in off-target motifs during the subsequent development of the plants. A further advantage is that mutated progeny do not have to be examined for the loss of integrated gRNA- and Cas9-encoding DNA sequences, as all mutated individuals can be used without restriction due to their transgene-free nature. In wheat, the GW2 gene was used as an example to show that the transfer of in vitro transcribed RNA coding for both gRNA and Cas9 leads to mutations in one percent of ballistically transformed cells [99]. About one third of the resulting plants were mutated in all six copies of the hexaploid genome. Compared to the transfer of gRNA and cas9 transgenes, however, this method had a mutation rate that was about 60 percent lower. Liang et al. [96] achieved an improvement in mutation efficiency by assembling gRNA and Cas9 into ribonucleoprotein complexes before ballistic transfer into immature embryos. After selection-free plant regeneration, mutants were identified with a frequency of more than four percent, which is on a par with the efficiency of conventional wheat transformation.

### *2.4. From Site-Directed Mutagenesis to Precise Genome Editing*

Current utilization of Cas endonuclease technology is still largely limited to random alterations of the DNA sequence at the user-defined genomic sites. This comparatively simple approach of site-directed mutagenesis relies on the error rate of the DNA repair mechanism, called non-homologous end-joining (NHEJ), which is predominant in plant cells. In this process, the two DNA ends resulting from a double-strand break are recognized and relegated irrespective of their nucleobase sequence.

Increased predictability of site-directed modifications was shown to result, for example, from two simultaneously induced DNA breaks, which can lead to the precise deletion of the interjacent fragment [22]. In addition, microhomology-based DNA repair produces predictable deletions in a comparatively simple way, provided that identical sequence repeats are present at both DNA ends that are to be relegated [100].

A more ambitious approach is the so-called base editing, in which a single nucleotide is specifically converted into another so that no more than one amino acid of the encoded protein is altered at a time [101]. For this purpose, a Cas-based nickase, that is, a Cas endonuclease derivative that cuts only one strand of DNA due to the mutative alteration of one of its two nucleolytic domains, is bound to a natural or artificial nucleobase deaminase enzyme [102]. Cytidine deaminases can convert C/G basepairs to T/A in the target region, whereas adenosine deaminases induce A/T to G/C conversions [102,103]. The functionality of cytidine and adenosine deaminases has already been demonstrated in several plant species, including wheat [104,105]. The effective base editing area ranges in dependence of the base editor used, e.g., the cytidine deaminase used by Zong et al. [104] can convert positions 1 to 17 within the target site and the adenosine deaminase used by Li et al. [105] is capable of converting positions 4 to 8.

In addition to the aforementioned error-prone non-homologous end-joining (NHEJ) DNA repair mechanism that predominates in plant cells, homology-directed repair (HDR) can also be used to generate precise genome alterations, albeit this process is much less active in somatic cells. Homology-directed repair naturally involves the sister chromatid of the same chromosome or the homologous chromosome as the correct sequence template, allowing the original sequence to be restored, even in cases of comparatively severe DNA damage. By using artificial DNA repair templates that are partially complementary to the site of the plant genome to be modified, it is possible to make even fairly large modifications precisely, as specified by the experimenter [106]. However, due to the methodological challenges, there are very few examples of this precise editing in plants published thus far. A first experimental approach involved the stable integration of the repair template together with the gRNA- and Cas-encoding expression units into the plant genome. The integrated repair template is flanked by the target sequences of the gRNA-mediated endonuclease used, so that it is cut out by the Cas restriction enzyme. This principle has proven to be sufficient in *Arabidopsis* [107]. The use of paired Cas9 nickases, which generate two single-strand breaks on the two opposite DNA strands, has increased the efficiency of homology-directed DNA repair [50,108,109]. With geminivirus replicons as carriers of the repair template, the dose per cell is increased. In this case, a strain of the Bean Yellow Dwarf Virus was modified so that only the elements essential for the method remained and the artificial repair template was amplified with high frequency [110]. In barley and wheat, precise editing has been achieved either only at the cellular level [98] or has the limiting prerequisite that the obtained genetic modification leads to an in vitro selectable trait, e.g., an herbicide resistance [111].

### 2.5. Identification and Characterization of Site-Specifically Modified Plants

The presence of gRNA- and Cas-encoding expression units is usually tested by PCR amplification [112,113]. In some studies, the expression of gRNA and cas9 was additionally assessed by RT-PCR, which can be very informative, especially in the process of method establishment [37]. For the detection of site-directed modifications, PCR products derived from the genomic target region can be analyzed using various methods. Digestion of the PCR product using conventional restriction enzymes is the most straightforward approach. However, the prerequisite for this is that the restriction site of the Cas endonuclease lies within a recognition sequence of the conventional restriction enzyme that then can be used for the test. In contrast to the wild-type sequence, PCR products of mutated target motifs cannot be digested by this enzyme [34]. A disadvantage of this method is that the choice of target motifs is substantially limited. This can be circumvented using the T7E1 assay, in which a mixture of PCR products of mutant and wild-type alleles is digested by the mismatch-sensitive T7 endonuclease I [39,114]. Another disadvantage these methods have in common is that their sensitivity is low, i.e., if mutated alleles of a plant are limited to small tissue sectors, detection may be not possible. This leads to a certain number of mutants not being identified.

To avoid this problem, Liang et al. [115] used target-specific synthetic gRNA and Cas9 protein for in vitro digestion of PCR products from genomic targets. This method is independent of conventional restriction sites in the target motif and was reported to be extremely sensitive. However, a large-scale application of this approach is hampered by the fact that recombinant Cas protein is fairly expensive.

As compared with the aforementioned restriction-based assays, sequencing of PCR product derived from the genomic target region is more informative. In heterozygous or chimeric mutated plants, the sequencing chromatogram shows double and multiple peaks at the individual base positions, which usually starts at the restriction site of the Cas endonuclease and continues in the same direction as the sequencing was conducted. Various bioinformatic tools, such as Tracking of Indels by DEcomposition (TIDE) [116] and DSDecode [117], provide the opportunity to read out such peaks, thereby determining the different allele variants present in the analyzed plant sample. A more reliable approach involves the analysis of several single clones via Sanger sequencing, as this provides a detailed breakdown of the mutations present in the analyzed sample. In addition, more conclusive information on the proportions of the mutated alleles can be obtained. In this context, it is important to consider that Cas endonuclease-triggered mutations are usually heterozygous and often limited to sectors of the plant. Moreover, leaf samples are not necessarily representative for the whole plant. This phenomenon is reflected by the observation that not all mutations detected in primary mutants are inherited by subsequent generations. In addition to Sanger sequencing, deep sequencing methods can be used to analyze mutations in individual plants, which is particularly informative owing to the obtained amount of data [48,118]. In practice, it is typically necessary to find a reasonable compromise between workload, conclusiveness of assays, and costs. Therefore, a combination of different methods is often the most suitable solution [40,42].

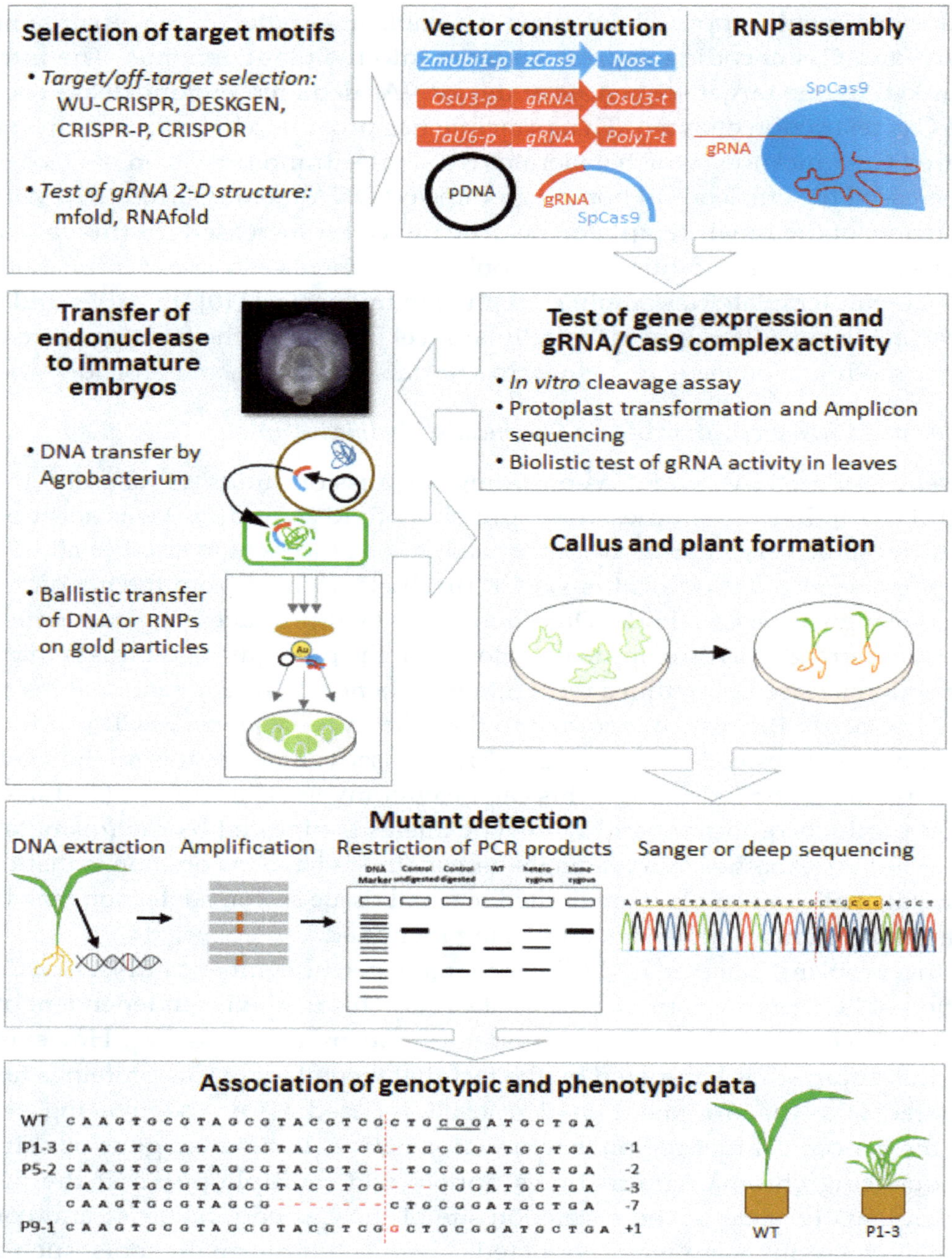


**Figure 1.** General workflow of targeted genome modification in barley and wheat, including target motif selection, vector construction, vector tests, gRNA and Cas9 delivery, plant regeneration, mutant detection, and genotypic and phenotypic analyses.

The molecular analysis of candidate plants can also address accidental modification of so-called off-target sites, which can cause undesired side effects and also influence the efficiency of on-target mutagenesis. In plants, the proportion of off-target mutations is typically below 1% [119]. In contrast to human cells, this does not pose a serious problem. These mutations are often confined to small sectors that are usually not passed on to offspring. Furthermore, due to the high specificity of the gRNA/Cas complex, possible off-target sites are well predictable so that progeny can be readily tested to eliminate segregants with unwanted mutations.

Since primary mutants are often chimeric or heterozygous and integrated cas9 and gRNA transgenes can be inherited, the analysis of progeny is essential to identify homozygous mutants that are null segregants in regards to the transgenes [35,112].

It is not recommended to spend too much effort for phenotypic analyses of primary mutants, because in addition to their uncertain genetic homogeneity and zygosity, the general fitness of these individuals is typically strongly affected owing to the manipulation and culture procedure they have gone through [120]. Importantly, it is advisable to use non-transgenic, non-mutant segregants as wild-type controls for the analysis of plant traits, because the comparison with other plants bears a fairly high risk of obtaining misleading results. A survey of the general workflow of targeted genome modification in barley and wheat is depicted in Figure 1.

## 3. Applications

While in this section only some representative examples for applications of Cas endonuclease technology in barley and wheat are presented in more detail, Table A1 provides a comprehensive overview of studies hitherto published on these crops.

In 2013, Upadhyay et al. [37] were the first to show that Cas endonucleases are in principle applicable in a *Triticeae* species and especially in the large and complex genome of wheat. However, this study was still confined to site-directed mutagenesis in a cell suspension, from which no plants can be regenerated. Mutations were detected in 18 to 22 percent of the sequenced amplicons derived from target regions of the INOSITOL OXYGENASE (TaINOX), and PHYTOENE DESATURASE (TaPDS) genes. The first plants carrying Cas endonuclease-induced mutations were presented by Wang et al. [39]. While heritable loss-of-function mutations were induced in all three homoeologues of the MILDEW-RESISTANCE LOCUS O (TaMLO) gene by the use of TALENs, Cas9-triggered mutations were still confined to the A subgenome. Using two wheat backgrounds, the authors produced a total of 687 gRNA/cas9 primary transgenic plants, of which about 5 percent proved to carry mutations in the target motif. However, in contrast to the TALEN-induced ones, the heritability of these mutations was not shown.

The first published use of Cas endonucleases in barley aimed to induce mutations in HvPM19 [35], which encodes an ABA-induced plasma membrane protein previously described in wheat as a positive regulator of dormancy. Four copies of this gene are present in barley. For two of those, knockout mutants were generated. Amongst 13 primary transgenic plants, three carried mutations in HvPM19-1, whereas one HvPM19-3 mutant was found amongst ten T0 plants. While this study was the first to provide evidence for the heritability of Cas9-induced mutations in a *Triticeae* crop, a resultant phenotype was not described.

Once the applicability of Cas endonuclease technology to achieve heritable mutations was shown for barley and wheat, this principle was used for many further approaches. Holme et al. [40] used Cas endonuclease-induced mutations to investigate the function of the PHYTASE GENE A of barley. HvPAPhy_a acts as the main regulator of phytase content in the barley grain. Phytases are important sources of bioavailable phosphorus, and therefore particularly important for germination. To reduce the activity of HvPAPhy_a independently of HvPAPhy_b, which is very similar in its gene sequence, Holme et al. specifically mutated the promoter of HvPAPhy_a. While small insertions and deletions in homozygous mutant progeny had only little effect, larger deletions, which also affected exon 1 of the target gene, caused a significant reduction of phytase activity.

Taking a Cas9-mediated knockout approach, Wang et al. [118] investigated the function of the GW2 gene of wheat. The rice orthologue of TaGW2 had been previously described as a negative regulator of grain size and thousand-grain weight. Wang et al. produced knockout mutants for all three subgenomes of wheat by RNA-guided Cas9 and generated all possible combinations of mutated homeoalleles by crossing. The knockout of individual TaGW2 homeologues in any of the subgemomes increased the size of the wheat grain, as well as the thousand-grain weight. Consequently, the authors concluded an additive effect of all three TaGW2 homoeologues [118].

Gluten in the wheat grain is relevant for the baking properties of the flour. This protein fraction is encoded by about 100 genes. Gluten intolerance is mainly based on the immune response to a particular peptide within the $\alpha$-gliadins, which are encoded by 45 genes. Sánchez-León et al. [113] designed two guide-RNAs specific to the gene region encoding the hyperallergenic epitopes and produced bread and durum wheat lines with knockout mutations in up to 33 of the 45 $\alpha$-gliadin genes. Thus, immunoreactivity was reduced by as much as 85 percent. The inheritance of the mutations and their effect on the $\alpha$-gliadin content were followed over three generations. In order to further reduce the immunogenic $\alpha$-gliadins in wheat, further work is still necessary [113].

In barley and wheat, Cas endonucleases have been mainly used for the investigation of gene functions by knockout, for the modification of metabolite contents and to increase resistance to fungal pathogens (Table A1). These studies show the great potential of this technology for research and breeding. In addition to the generation of random sequences at the target site, more recent work has demonstrated predictable nucleobase exchanges in wheat by the use of a chimeric Cas9 derivative, which consists of a nickase and an adenosine deaminase [105]. This base editing approach opens up further possibilities of generating functional gene variants rather than knockout alleles.

## 4. Regulation

Parallel to the development of the Cas endonuclease technology, science organizations, politicians, and society are dealing with the effects of the technology. The following section will briefly address some related aspects, in particular the regulation of plants that carry site-directed mutations. The focus here is on the USA and Europe, as these two have very contrasting approaches.

For the United States, Wolt and Wolf [121] summarized the regulation of the so-called green biotechnology, and in particular the use of customizable endonucleases. Commencing at the Asilomar conference on biosafety in 1975, a regulatory framework for plants with recombinant DNA has been developed in the USA since the mid-1980s. Over the years, this framework became more and more complex due to more accompanying research and the increasingly lengthy periods needed for decision-making. Although the focus of regulation in the United States is on the product, the procedure also includes process-based facets. For example, genetically engineered plants have to undergo special approval procedures. However, these differ between plants that have been transformed by use of the plant pathogen *Agrobacterium* and those that have been ballistically transformed. If plants with an identical trait are produced without genetic engineering, no special regulation takes place. For plants bred by use of customized endonuclease technology, the portal "Am I regulated?" has been established, in which the USDA made case-by-case statements at short notice as to whether a plant needs to be regulated or not. In 2017, work began on revising the regulations on biotechnology. The major trigger for regulation was shifted towards the process by a redefinition of "genetic engineering" as "mean techniques that use recombinant or synthetic nucleic acids with the intent to create or alter a genome". This is not applied to processes of targeted genome modification that cause deletions or base edits or result in the targeted insertion of a DNA fragment, which would also be possible by conventional breeding, conventionally induced mutagenesis or even without any human intervention [121].

In a detailed report in 2017 [122], the European Academies' Science Advisory Council (EASAC) stressed the possibilities that targeted genome modification offers. In plant breeding, improved precision compared to undirected mutagenesis methods represents a significant advance. At the same time, some representatives of non-governmental organizations and political parties are viewing biotechnological breeding methods critically. According to EASAC, this gives science a special responsibility to explain its work. In regards to plant breeding, EASAC recommends to politicians not to regulate transgene-free plants carrying site-directed modifications as being "genetically engineered" in the sense of the EU Directive 2001/18/EC on the deliberate release into the environment of genetically modified organisms. Additionally, it advises a revision of the European genetic engineering legislation with a stronger focus on the product instead of the breeding process. The need for international compatibility of regulation is emphasized and reference is also made to best practices from other countries, such as the United States, where site-directed mutagenesis is not regulated as genetic engineering [122].

While some countries, including the United States, Argentina, Brazil, Chile, and more recently Australia, have already made decisions to regulate targeted mutagenesis on a case-by-case or general basis not differently than conventional methods of plant breeding, in Europe, those varieties are the subject of the Directive 2001/18/EC. According to that, field trials with plants carrying targeted mutations need special permission on application, and placing respective varieties on the market needs tedious and very costly approval procedures. The European plant research community has responded to this with demands for a revision of European genetic engineering legislation. For more details see Sprink et al. [123] or the position paper of 95 European plant research institutions (http://www.vib.be/en/news/Documents/Position%20paper%20on%20the%20ECJ%20ruling%20on%20CRISPR%2012%20Nov%202018.pdf).

In many countries in the world, decisions on how to regulate plants generated using customized endonucleases are currently pending. They will play a key role in deciding how and by whom these technologies and their possibilities can be used.

## 5. Perspective

The new possibility of utilizing customized endonucleases in barley and wheat opens up a broad spectrum of opportunities for applications. A generally limiting aspect, however, is that the genetic transformation of cereals poses a particular challenge. In the context of the question for new or improved methods of transferring DNA into regenerable cells of cereals, it has to be considered that a number of methods have already been developed, but have not yet been utilized for the application of Cas endonucleases. A remarkable example of this is the use of isolated microspores or embryogenic pollen cultures generated from them for DNA transfer using *Agrobacterium* [87], ballistics [89] or electroporation [124]. When it comes to the development of new methods, particular attention must also be paid to the fact that they should be applicable to a broad spectrum of genetic backgrounds, which unfortunately does not apply to any of the methods used to date. On the other hand, the particular advantage of Cas endonuclease technology will only be fully exploited if it is possible to directly modify any plant genotype of choice. This is the only way to completely overcome the cumbersome introgression of advantageous gene variants into elite germplasms, and above all, the genetic coupling of desired alleles with undesired ones. In this context, the dissection and culture of ovules is another remarkable possibility for the in vitro regeneration of wheat and barley [125,126]. For the *Agrobacterium*-mediated DNA transfer into cultivated ovules, in which the T-DNA is transferred into few-celled proembryos, it was shown that this method has comparatively low genotype dependence [127].

In all studies published to date in which heritable site-directed modifications have been achieved in the barley and wheat genomes using Cas endonuclease technology, the POLYUBIQUITIN1 promoter from maize was used to express the cas9 gene. For the stable expression of protein-encoding transgenes in the *Triticeae*, however, a vast number of further promoters with different activity profiles are available, some of which may also be used for Cas endonucleases [128–130]. In addition, there is, thus far, unexploited potential for the expression of gRNAs in barley and wheat. For instance in rice, a t-RNA processing-based system was established, in which multiple gRNAs are released from a complex transcript by endogenous RNases [50,131].

To keep pace with the frequent emergence of novel Cas and gRNA formats, as well as regulatory and further functional elements, several modular vector systems have been developed that allow for a rapid and versatile assembly of any combination of components of choice. Some of these systems may also be useful for the application in the *Triticeae* cereals [45,50].

Whereas site-directed mutagenesis represents the currently well-established state of the art of Cas endonuclease technology, more precise techniques of targeted genome modification have unfortunately not yet become routine, and are at best exemplary in plants. A portfolio of chimeric Cas derivatives is now available for the so-called base editing using nucleoside deaminases, which allows conversions from G to A, C to T, T to C, and A to G. Thus, the vast majority of spontaneous point mutations occurring in nature can be specifically generated, or if desired, the corresponding wild-type alleles can be restored as well [103]. A broad application of precise genome editing in association with DNA repair templates, on the other hand, still poses particularly great methodological challenges. Homology-based repair, for example, is comparatively rarely occurring, which entails a correspondingly low efficiency of conventional selection marker genes whose genomic integration and expression is too poorly associated with template-mediated repair events. In addition, the artificial repair templates must compete with the naturally recruited homologous sequences present in the cell on sister chromatid and homologous chromosome(s), which further reduces the efficiency of targeted modification.

A further limitation of Cas endonuclease technology is that the binding site of the gRNA/Cas complex cannot be predefined entirely at will. The most commonly used enzyme for target sequence-specific genome modification is Cas9 from *Streptococcus pyogenes*. This enzyme requires the aforementioned protospacer-adjacent motif (PAM) NGG for binding to the genomic target. However, there are several other Cas9 variants that require other cognate DNA motifs. Such Cas9-orthologous enzymes are known from various representatives of the genera *Streptococcus*, *Staphylococcus*, *Campylobacter*, and *Neisseria* [132–134]. In *Arabidopsis*, for example, the *Staphylococcus aureus* and *Streptococcus thermophilus* Cas9 variants have already been used in addition to the standard SpCas9 [135]. Furthermore, in order to extend the flexibility in the choice of targets beyond the naturally occurring PAMs, Hu et al. [136] developed an artificial Cas9 variant. The advantage of this "expanded" Cas9 (xCas9) is that almost any nucleobase triplet is accepted for its binding to the genomic target. The synthetic Cas9-NG, which was developed in a different way and published shortly thereafter, proved to be even more efficient than xCas9 in many PAM variants [137].

While Cas9 endonucleases are widely employed in plants, Zetsche et al. [138] described another Cas endonuclease, namely Cas12a (Cpf1) from *Francisella novicida*, that can also be used to induce targeted alterations in genomes. Later, Cas12a-orthologous enzymes from other bacteria were also successfully used [139]. Besides some other peculiarities, Cas12a, in contrast to Cas9, produces 5′- overhangs rather than blunt DNA ends. This provides a particularly attractive option for the establishment of robust methods of precise genome editing using homology-directed DNA repair.

Cas endonucleases have also been modified in such a way that they induce single-strand breaks in double-stranded DNA. Among such nickases, there are specific variants for each of the two DNA strands [12,140]. In *Arabidopsis*, paired nickases derived from Cas9 were used to demonstrate that both the specificity of the binding to the genomic DNA and the efficiency of homology-directed DNA repair can be increased [50,108,109].

Whereas knockouts are the standard outcome of site-directed mutagenesis, the possibility to achieve quantitative changes in the functionality of target genes has not yet been exploited in cereals. Using tobacco as experimental model, Schedel et al. [141] have shown that in-frame mutations can maintain gene functionality to a reduced extent. This can be of particular value for the targeted modification of genes that are essential for the plant or the generation of new allelic diversity for the use in plant breeding. In a particularly elegant approach, Rodriguez-Leal et al. [142] produced tomato lines with various CLAVATA3 alleles by simultaneous expression of several gRNAs specified for different target motifs in the promoter of this gene. The resultant gene variants led to a whole series of lines with different fruit sizes according to the individual strength of CLV3 expression.

In other crop plants, there are various application examples of customizable endonucleases, which may also be considered for breeding programs in the *Triticeae* cereals. Since every characteristic of plants is largely genetically determined, it must in fact be possible to improve every trait that is relevant for the utility value of crops by employing Cas endonuclease technology. Due to these extensive possibilities, only a few of the promising options can be mentioned here as examples. A potential field of application for Cas endonucleases is the establishment of resistance to various potyviruses, which has already been achieved, for example, in cucumber. The established resistance relies on the knockout of the host plant's EIF4E gene, which is essential for the translation of viral mRNAs [143]. Resistance-conferring alleles are also known from a related barley gene [144,145]. In addition, there are a number of susceptibility genes for diseases caused by other pathogens, which is similarly suitable for resistance breeding.

The ability of Cas endonucleases to completely eliminate the function of a gene seems particularly attractive when it comes to freeing crops from toxic, carcinogenic, allergenic, or bad-tasting gene products or metabolites. The aforementioned study by Sanchez-Léon et al. [113] on the production of wheat with gluten, whose fraction of immunodominant gliadins is drastically reduced, is a promising example in this respect. Another potential use of Cas endonuclease technology results from the fact that most modern durum wheat varieties carry a mutation in a heavy metal transporter gene, which entails a reduced sequestration of cadmium in the roots so that this toxic element accumulates in the grain to an extent that is of concern in terms of human health [146]. The restoration of the corresponding wild-type allele would, therefore, be a particularly promising way of improving the quality of pasta, which is widely consumed across the entire globe.

As has already been demonstrated in potato [147] and rice [148], it is also possible to modify the starch quality by specifically modulating the biosynthesis of the major starch components amylose and amylopectin. While amylose-free starch is of particularly high value for the paper and chemical industries, starch with reduced amyplopectin content is referred to as resistant starch, which, due to its fiber-like properties, has the potential to counteract type-2 diabetes, a civilization disease that is now increasingly spreading even in developing countries.

For future approaches, it is also conceivable to modify many different genes in one step, for instance to influence complex processes such as photosynthesis [149]. Cas endonuclease technology offers unprecedented prerequisites for this due to the option of simultaneous use of numerous gRNAs.

In summary, it can be stated that the establishment of Cas endonuclease technology represents a quantum leap for plant research and breeding. Even with regard to major methodological challenges, the current pace of research and development in this field means that further very useful solutions are highly probable in the near future. The implementation of this technology, especially in the agronomically important cereals of the *Triticeae* cereals, will likely continue to gain momentum, and is thus expected to increasingly contribute to the effective production of sufficient and high-quality food, feed, and industrial raw materials, while ensuring largely improved environmental compatibility of agriculture.

**Author Contributions:** Conceptualization, J.K.; collection of information, I.K., C.H., R.H., J.K.; writing—original draft preparation, I.K., C.H., R.H., J.K.; writing—review and editing, I.K., C.H., R.H., J.K.; supervision, J.K.; funding acquisition, J.K.

# Appendix A

**Table A1.** Survey of studies employing Cas endonuclease technology in wheat and barley. If not stated otherwise, gRNA and Cas9 were delivered as transgenes, the outcome was site-directed mutagenesis (i.e., random InDels at the target sites), and genetic modifications were shown to be heritable. IEs: immature embryos, RNP: gRNA/Cas9 ribonucleoprotein.

| Target Gene | Reagent Delivery, Recipient Cells, Genotype, Notes | Aim of the Study | Efficiency of Targeted Mutagenesis | Observed Phenotype | Reference |
|---|---|---|---|---|---|
| *Wheat* | | | | | |
| INOSITOL OXYGENASE (INOX), PHYTOENE DESATURASE (PDS) | *Agrobacterium*, callus-derived cell suspension, genotype not reported | method establishment | mutations detected in cell suspension, no plants generated | not reported | [37] |
| MILDEW RESISTANCE LOCUS O (MLO) | ballistic, IEs of "Kenong199" and "Bobwhite" | resistance to powdery mildew | 4 out of 72 T0 plants mutated in A-genome, heritability not shown | not reported | [39] |
| GRAIN WIDTH 2 (GW2) | ballistic transfer of RNP, IEs of "Kenong199", selection-free regeneration | method establishment | 28 primary mutants from 640 IEs, heritability not shown | not reported | [96] |
| α-GLIADIN gene family | ballistic, IEs of "BW208", "THA53", "Don Pedro" | reduced α-gliadin content in grains | Mutation rate in T0 not reported, 17 out of 17 analyzed T1 plants mutated, heritability of genotype and phenotype shown up to T3 | reduced α-gliadin content in grains | [113] |
| ENHANCED DISEASE RESISTANCE 1 (EDR1) | ballistic, IEs of "KN199" | reduced susceptibility to Blumeria graminis | 5 mutant T0 plants identified, 5 out of 207 analyzed T1 plants homozygously mutated in all 3 copies | resistance shown in homozygous T1 mutants | [150] |
| PHYTOENE DESATURASE (PDS) | *Agrobacterium*, IEs of "Fielder" | method establishment | up to 13 out of 73 T0 plants mutated, homozygous mutations in individual subgenomes shown in T2 | not reported (no homozygous triple mutants) | [85] |
| DEP1, GRAIN WIDTH 2 (GW2) | PEG-mediated transfection, protoplasts of "Kenong199", ballistic, IEs of "Kenong199" | establishment of base editing using adenosine deaminase fusion to nCas9 | up to 7.5% of NGS reads from target amplicons showed A -> G conversion, 5 (DEP1) and 2 (GW2) regenerated T0 plants showed heterozygous mutations, no heritability reported | not reported | [105] |
| ZIP4-B2 (homoeolog of B genome only) | *Agrobacterium*, IEs of "Fielder" | increase of crossover frequency | 4 out of 81 analyzed T0 plants mutated, 24 T1 plants (progeny of 1 T0 plant) analyzed | increase of homologous crossovers in KO-mutants under presence of magnesium | [151] |
| GRAIN WIDTH 2 (GW2-ABD), 9-LIPOXYGENASE (LPX-1-BD), MILDEW RESISTANCE LOCUS O (MLO) | Ballistic, IEs of "Bobwhite" | GW2: increased grain size and weight, LPX-1 and TaMLO: resistance to fungi | 20 out of 102 T0 plants with triple gRNAs mutated, 6 out of 61 T0 plants with GW2-gRNA mutated, heritability analyzed by amplicon NGS in progeny of 2 T0 plants, several T1 plants heterozygous for GW2 or MLO, homozygous T1 for GW2 | GW2-KO: increased thousand-grain weight by 27% | [48] |

**Table A1.** *Cont.*

| Target Gene | Reagent Delivery, Recipient Cells, Genotype, Notes | Aim of the Study | Efficiency of Targeted Mutagenesis | Observed Phenotype | Reference |
|---|---|---|---|---|---|
| MALE STERILITY 45 (MS45) | *Agrobacterium*, IEs of "Fielder" and 'SBC0456D' | male sterility | 25 out of 181 analyzed T1 plants mutated in different combinations (single, double, triple KOs) | triple mutants (ABD-KO) proved to be male sterile | [152] |
| GRAIN WIDTH 2 (GW2) | ballistic, IEs of "Bobwhite" | increased grain size and weight | T1, T2, T3 mutants derived from Wang et al. (2018a) used to generate single, double, and triple KO-mutants | thousand-grain weight increased, dosage effect of homeoalleles shown | [118] |
| PUROINDOLINE b (PINb), GRANULE-BOUND STARCH SYNTHASE (WAXY), DA1 | *Agrobacterium*, IEs of "Fielder" (DA1 only) | PINb: grain hardness, WAXY: starch composition, DA1: grain size | 13 out of 24 T0 plants DA1-target mutated in A or B genome, heritability not shown | not reported | [153] |
| ***Barley*** | | | | | |
| PM19-1 and PM19-3 | *Agrobacterium*, IEs of 'Golden Promise' | seed dormancy | 3 out of 13 analyzed T0 plants mutated (PM19-1), 1 out of 10 analyzed T0 plants mutated (PM19-3), heritability of mutations shown in progeny of 4 identified mutants in T2, off-target mutations found in other copies of PM-19 | not reported | [35] |
| Promoter of PHYTASE GENE A (PAPhy_a) | *Agrobacterium*, IEs of "Golden Promise" | reduced phytase content in the grain | 28 out of 64 T0 plants 28 mutated, heritability shown in T2 | homozygous T1 mutants showed reduced phytase activity | 40] |
| ENDO-N-ACETYL-β-D-GLUCOSAMINIDASE (ENGase) | ballistic and *Agrobacterium*, IEs of "Golden Promise", co-transformation for fragment deletion | N-glycan modification in the grain | ballistic: 7 out of 8 T0 plants mutated, Agrobacterium: 15 out of 23 T0 plants mutated, 6 plants with targeted deletion between two targeted positions, heritability shown in T1, homozygous mutants | N-glycan modification in the grain | [112] |
| CYTOKININ DEHYDROGENASE (CKX) | *Agrobacterium*, IEs of "Golden Promise" | gene function in cytokinin metabolism | 15 out of 23 T0 plants mutated, heritability shown in T1, 4 homozygous KO-mutants selected for further analysis | KO-mutants showed higher cytokinin content and slightly reduced root growth | [154] |
| PHYTOENE DESATURASE (PDS) | *Agrobacterium*, IEs of "Golden Promise" | photo-bleaching | 6 out of 40 T0 plants mutated, heritability not reported | chimeric photobleaching in T0 mutants | [85] |
| Member of MICRORCHIDIA GHKL ATPASE subfamily (MORC1) | *Agrobacterium*, IEs of "Golden Promise" | resistance to fungal pathogens | callus screened for mutations, progeny of randomly selected T0 plants screened for mutations, 12 and 10 T1 families analyzed for two constructs, mutations in 19 of the 22 families, proportion of mutated plants not reported | homozygous morc1-KO T1 plants were less susceptible to fungal pathogens and showed higher expression of transposons | [45] |
| PROTEIN TARGETING TO STARCH 1 (PTST1), GRANULE-BOUND STARCH SYNTHASE (GBSS1a) | *Agrobacterium*, IEs of "Golden Promise" | starch accumulation in the grain | 6 out of 8 (PTST1) and 5 out of 9 (GBSS1a) T0 plants mutated, heritability shown in T1 | ptst1-KO: starch-free grains, no germination, gbss1a-KO: amylose-free grains | [155] |

## References


1. Baulcombe, D.; Crute, I.; Davies, B.; Dunwell, J.; Gale, M.; Jones, J.; Pretty, J.; Sutherland, W.; Toulmin, C. *Reaping the Benefits: Science and the Sustainable Intensification of Global Agriculture*; The Royal Society: London, UK, 2009; p. 72.
2. Ronald, P. Plant genetics, sustainable agriculture and global food security. *Genetics* **2011**, *188*, 11–20. [CrossRef]
3. *Global Warming of 1.5 °C*; IPCC Special Report; IPCC: Geneva, Switzerland, 2018. Available online: https://www.ipcc.ch/sr15/ (accessed on 28 May 2019).
4. Modrzejewski, D.; Hartung, F.; Sprink, T.; Krause, D.; Kohl, C.; Schiemann, J.; Wilhelm, R. What is the available evidence for the application of genome editing as a new tool for plant trait modification and the potential occurrence of associated off-target effects: A systematic map protocol. *Environ. Evid.* **2018**, *7*, 11. [CrossRef]
5. Parry, M.A.J.; Madgwick, P.J.; Bayon, C.; Tearall, K.; Hernandez-Lopez, A.; Baudo, M.; Rakszegi, M.; Hamada, W.; Al-Yassin, A.; Ouabbou, H.; et al. Mutation discovery for crop improvement. *J. Exp. Bot.* **2009**, *60*, 2817–2825. [CrossRef] [PubMed]
6. Reynolds, M.; Foulkes, M.J.; Slafer, G.A.; Berry, P.; Parry, M.A.J.; Snape, J.W.; Angus, W.J. Raising yield potential in wheat. *J. Exp. Bot.* **2009**, *60*, 1899–1918. [CrossRef] [PubMed]
7. Sovová, T.; Kerins, G.; Demnerová, K.; Ovesná, J. Genome Editing with Engineered Nucleases in Economically Important Animals and Plants: State of the Art in the Research Pipeline. *Curr. Issues Mol. Biol.* **2017**, *21*, 41–62.
8. McCallum, C.M.; Comai, L.; Greene, E.A.; Henikoff, S. Targeting Induced LocalLesions IN Genomes (TILLING) for Plant Functional Genomics. *Plant Physiol.* **2000**, *123*, 439–442. [CrossRef] [PubMed]
9. Comai, L.; Young, K.; Till, B.J.; Reynolds, S.H.; Greene, E.A.; Codomo, C.A.; Enns, L.C.; Johnson, J.E.; Burtner, C.; Odden, A.R.; et al. Efficient discovery of DNA polymorphisms in natural populations by Ecotilling. *Plant J.* **2004**, *37*, 778–786. [CrossRef] [PubMed]
10. Herrera-Estrella, L.; Depicker, A.; van Montagu, M.; Schell, J. Expression of chimaeric genes transferred into plant cells using a Ti-plasmid-derived vector. *Nature* **1983**, *303*, 209–213. [CrossRef]
11. Kumlehn, J.; Hensel, G. Genetic transformation technology in the Triticeae. *Breed. Sci.* **2009**, *59*, 553–560. [CrossRef]
12. Jinek, M.; Chylinski, K.; Fonfara, I.; Hauer, M.; Doudna, J.A.; Charpentier, E. A programmable dual-RNA-guided DNA endonuclease in adaptive bacterial immunity. *Science* **2012**, *337*, 816–821. [CrossRef]
13. Ramírez-González, R.H.; Borrill, P.; Lang, D.; Harrington, S.A.; Brinton, J.; Venturini, L.; Davey, M.; Jacobs, J.; van Ex, F.; Pasha, A.; et al. The transcriptional landscape of polyploid wheat. *Science* **2018**, 361. [CrossRef] [PubMed]
14. Brenchley, R.; Spannagl, M.; Pfeifer, M.; Barker, G.L.A.; D'Amore, R.; Allen, A.M.; McKenzie, N.; Kramer, M.; Kerhornou, A.; Bolser, D.; et al. Analysis of the bread wheat genome using whole-genome shotgun sequencing. *Nature* **2012**, *491*, 705–710. [CrossRef]
15. International Barley Genome Sequencing Consortium; Mayer, K.F.; Waugh, R.; Brown, J.W.; Schulman, A.; Langridge, P.; Platzer, M.; Fincher, G.B.; Muehlbauer, G.J.; Sato, K.; et al. A physical, genetic and functional sequence assembly of the barley genome. *Nature* **2012**, *491*, 711–716. [CrossRef] [PubMed]
16. Mascher, M.; Gundlach, H.; Himmelbach, A.; Beier, S.; Twardziok, S.O.; Wicker, T.; Radchuk, V.; Dockter, C.; Hedley, P.E.; Russell, J.; et al. A chromosome conformation capture ordered sequence of the barley genome. *Nature* **2017**, *544*, 427–433. [CrossRef] [PubMed]
17. Milner, S.G.; Jost, M.; Taketa, S.; Mazón, E.R.; Himmelbach, A.; Oppermann, M.; Weise, S.; Knüpffer, H.; Basterrechea, M.; König, P.; et al. Genebank genomics highlights the diversity of a global barley collection. *Nat. Genet.* **2019**, *51*, 319–326. [CrossRef]
18. Plessis, A.; Perrin, A.; Haber, J.E.; Dujon, B. Site-specific recombination determined by I-SceI, a mitochondrial group I intron-encoded endonuclease expressed in the yeast nucleus. *Genetics* **1992**, *130*, 451–460.
19. Pauwels, K.; Podevin, N.; Breyer, D.; Carroll, D.; Herman, P. Engineering nucleases for gene targeting: Safety and regulatory considerations. *New Biotechnol.* **2014**, *31*, 18–27. [CrossRef]

20. Puchta, H.; Dujon, B.; Hohn, B. Homologous recombination in plant cells is enhanced by in vivo induction of double strand breaks into DNA by a site-specific endonuclease. *Nucleic Acids Res.* **1993**, *21*, 5034–5040. [CrossRef]
21. Gao, H.; Smith, J.; Yang, M.; Jones, S.; Djukanovic, V.; Nicholson, M.G.; West, A.; Bidney, D.; Falco, S.C.; Jantz, D.; et al. Heritable targeted mutagenesis in maize using a designed endonuclease. *Plant J.* **2010**, *61*, 176–187. [CrossRef] [PubMed]
22. Puchta, H.; Fauser, F. Synthetic nucleases for genome engineering in plants: Prospects for a bright future. *Plant J.* **2014**, *78*, 727–741. [CrossRef] [PubMed]
23. Kouranova, E.; Forbes, K.; Zhao, G.; Warren, J.; Bartels, A.; Wu, Y.; Cui, X. CRISPRs for Optimal Targeting: Delivery of CRISPR Components as DNA, RNA, and Protein into Cultured Cells and Single-Cell Embryos. *Hum. Gene Ther.* **2016**, *27*, 464–475. [CrossRef] [PubMed]
24. Vu, G.T.H.; Cao, H.X.; Watanabe, K.; Hensel, G.; Blattner, F.R.; Kumlehn, J.; Schubert, I. Repair of Site-Specific DNA Double-Strand Breaks in Barley Occurs via Diverse Pathways Primarily Involving the Sister Chromatid. *Plant Cell* **2014**, *26*, 2156–2167. [CrossRef]
25. Kim, Y.G.; Cha, J.; Chandrasegaran, S. Hybrid restriction enzymes: Zinc finger fusions to Fok I cleavage domain. *Proc. Natl. Acad. Sci. USA* **1996**, *93*, 1156–1160. [CrossRef]
26. Voytas, D.F. Plant genome engineering with sequence-specific nucleases. *Annu. Rev. Plant Biol.* **2013**, *64*, 327–350. [CrossRef]
27. Smith, J.; Berg, J.M.; Chandrasegaran, S. A detailed study of the substrate specificity of a chimeric restriction enzyme. *Nucleic Acids Res.* **1999**, *27*, 674–681. [CrossRef]
28. Doyon, Y.; McCammon, J.M.; Miller, J.C.; Faraji, F.; Ngo, C.; Katibah, G.E.; Amora, R.; Hocking, T.D.; Zhang, L.; Rebar, E.J.; et al. Heritable targeted gene disruption in zebrafish using designed zinc-finger nucleases. *Nat. Biotechnol.* **2008**, *26*, 702–708. [CrossRef]
29. Bonas, U.; Stall, R.E.; Staskawicz, B. Genetic and structural characterization of the avirulence gene avrBs3 from Xanthomonas campestris pv. vesicatoria. *Mol. Gen. Genet. MGG* **1989**, *218*, 127–136. [CrossRef]
30. Christian, M.; Cermak, T.; Doyle, E.L.; Schmidt, C.; Zhang, F.; Hummel, A.; Bogdanove, A.J.; Voytas, D.F. Targeting DNA double-strand breaks with TAL effector nucleases. *Genetics* **2010**, *186*, 757–761. [CrossRef]
31. Boch, J.; Scholze, H.; Schornack, S.; Landgraf, A.; Hahn, S.; Kay, S.; Lahaye, T.; Nickstadt, A.; Bonas, U. Breaking the code of DNA binding specificity of TAL-type III effectors. *Science* **2009**, *326*, 1509–1512. [CrossRef]
32. Doudna, J.A.; Charpentier, E. Genome editing. The new frontier of genome engineering with CRISPR-Cas9. *Science* **2014**, *346*, 1258096. [CrossRef] [PubMed]
33. Kumlehn, J.; Pietralla., J.; Hensel, G.; Pacher, M.; Puchta, H. The CRISPR/Cas revolution continues: From efficient gene editing for crop breeding to plant synthetic biology. *J. Integr. Plant Biol.* **2018**, *60*, 1127–1154. [CrossRef]
34. Shan, Q.; Wang, Y.; Li, J.; Zhang, Y.; Chen, K.; Liang, Z.; Zhang, K.; Liu, J.; Xi, J.J.; Qiu, J.-L.; et al. Targeted genome modification of crop plants using a CRISPR-Cas system. *Nat. Biotechnol.* **2013**, *31*, 686–688. [CrossRef] [PubMed]
35. Lawrenson, T.; Shorinola, O.; Stacey, N.; Li, C.; Østergaard, L.; Patron, N.; Uauy, C.; Harwood, W. Induction of targeted, heritable mutations in barley and Brassica oleracea using RNA-guided Cas9 nuclease. *Genome Biol.* **2015**, *16*, 258. [CrossRef] [PubMed]
36. Liu, X.; Wu, S.; Xu, J.; Sui, C.; Wei, J. Application of CRISPR/Cas9 in plant biology. *Acta Pharm. Sin. B* **2017**, *7*, 292–302. [CrossRef] [PubMed]
37. Upadhyay, S.K.; Kumar, J.; Alok, A.; Tuli, R. RNA-guided genome editing for target gene mutations in wheat. *G3 Genes Genome Genet.* **2013**, *3*, 2233–2238. [CrossRef] [PubMed]
38. Shan, Q.; Wang, Y.; Li, J.; Gao, C. Genome editing in rice and wheat using the CRISPR/Cas system. *Nat. Protoc.* **2014**, *9*, 2395–2410. [CrossRef]
39. Wang, Y.; Cheng, X.; Shan, Q.; Zhang, Y.; Liu, J.; Gao, C.; Qiu, J.-L. Simultaneous editing of three homoeoalleles in hexaploid bread wheat confers heritable resistance to powdery mildew. *Nat. Biotechnol.* **2014**, *32*, 947–951. [CrossRef]
40. Holme, I.B.; Wendt, T.; Gil-Humanes, J.; Deleuran, L.C.; Starker, C.G.; Voytas, D.F.; Brinch-Pedersen, H. Evaluation of the mature grain phytase candidate HvPAPhy_a gene in barley (*Hordeum vulgare* L.) using CRISPR/Cas9 and TALENs. *Plant Mol. Biol.* **2017**, *95*, 111–121. [CrossRef]

41. Čermák, T.; Baltes, N.J.; Čegan, R.; Zhang, Y.; Voytas, D.F. High-frequency, precise modification of the tomato genome. *Genome Biol.* **2015**, *16*, 232. [CrossRef]
42. Gil-Humanes, J.; Wang, Y.; Liang, Z.; Shan, Q.; Ozuna, C.V.; Sánchez-León, S.; Baltes, N.J.; Starker, C.; Barro, F.; Gao, C.; et al. High-efficiency gene targeting in hexaploid wheat using DNA replicons and CRISPR/Cas9. *Plant J.* **2017**, *89*, 1251–1262. [CrossRef]
43. Gao, Y.; Zhao, Y. Self-processing of ribozyme-flanked RNAs into guide RNAs in vitro and in vivo for CRISPR-mediated genome editing. *J. Integr. Plant Biol.* **2014**, *56*, 343–349. [CrossRef] [PubMed]
44. Xing, H.-L.; Dong, L.; Wang, Z.-P.; Zhang, H.-Y.; Han, C.-Y.; Liu, B.; Wang, X.-C.; Chen, Q.-J. A CRISPR/Cas9 toolkit for multiplex genome editing in plants. *BMC Plant Biol.* **2014**, *14*, 327. [CrossRef]
45. Kumar, N.; Galli, M.; Ordon, J.; Stuttmann, J.; Kogel, K.-H.; Imani, J. Further analysis of barley MORC1 using a highly efficient RNA-guided Cas9 gene-editing system. *Plant Biotechnol. J.* **2018**, *16*, 1892–1903. [CrossRef] [PubMed]
46. Ma, X.; Zhang, Q.; Zhu, Q.; Liu, W.; Chen, Y.; Qiu, R.; Wang, B.; Yang, Z.; Li, H.; Lin, Y.; et al. A Robust CRISPR/Cas9 System for Convenient, High-Efficiency Multiplex Genome Editing in Monocot and Dicot Plants. *Mol. Plant* **2015**, *8*, 1274–1284. [CrossRef]
47. Lowder, L.G.; Zhang, D.; Baltes, N.J.; Paul, J.W.; Tang, X.; Zheng, X.; Voytas, D.F.; Hsieh, T.-F.; Zhang, Y.; Qi, Y. A CRISPR/Cas9 Toolbox for Multiplexed Plant Genome Editing and Transcriptional Regulation. *Plant Physiol.* **2015**, *169*, 971–985. [CrossRef] [PubMed]
48. Wang, W.; Pan, Q.; He, F.; Akhunova, A.; Chao, S.; Trick, H.; Akhunov, E. Transgenerational CRISPR-Cas9 Activity Facilitates Multiplex Gene Editing in Allopolyploid Wheat. *Cris. J.* **2018**, *1*, 65–74. [CrossRef]
49. Tang, X.; Zheng, X.; Qi, Y.; Zhang, D.; Cheng, Y.; Tang, A.; Voytas, D.F.; Zhang, Y. A Single Transcript CRISPR-Cas9 System for Efficient Genome Editing in Plants. *Mol. Plant* **2016**, *9*, 1088–1091. [CrossRef] [PubMed]
50. Čermák, T.; Curtin, S.J.; Gil-Humanes, J.; Čegan, R.; Kono, T.J.Y.; Konečná, E.; Belanto, J.J.; Starker, C.G.; Mathre, J.W.; Greenstein, R.L.; et al. A Multipurpose Toolkit to Enable Advanced Genome Engineering in Plants. *Plant Cell* **2017**, *29*, 1196–1217. [CrossRef]
51. Ren, X.; Yang, Z.; Xu, J.; Sun, J.; Mao, D.; Hu, Y.; Yang, S.J.; Qiao, H.H.; Wang, X.; Hu, Q.; et al. Enhanced specificity and efficiency of the CRISPR/Cas9 system with optimized sgRNA parameters in Drosophila. *Cell Rep.* **2014**, *9*, 1151–1162. [CrossRef]
52. Moreno-Mateos, M.A.; Vejnar, C.E.; Beaudoin, J.D.; Fernandez, J.P.; Mis, E.K.; Khokha, M.K.; Giraldez, A.J. CRISPRscan: Designing highly efficient sgRNAs for CRISPR-Cas9 targeting in vivo. *Nat. Methods* **2015**, *12*, 982–988. [CrossRef]
53. Cong, L.; Ran, F.A.; Cox, D.; Lin, S.; Barretto, R.; Habib, N.; Hsu, P.D.; Wu, X.; Jiang, W.; Marraffini, L.A.; et al. Multiplex genome engineering using CRISPR/Cas systems. *Science* **2013**, *339*, 819–823. [CrossRef] [PubMed]
54. Fu, Y.; Sander, J.D.; Reyon, D.; Cascio, V.M.; Joung, J.K. Improving CRISPR-Cas nuclease specificity using truncated guide RNAs. *Nat. Biotechnol.* **2014**, *32*, 279–284. [CrossRef]
55. Zhang, J.P.; Li, X.L.; Neises, A.; Chen, W.; Hu, L.P.; Ji, G.Z.; Yu, J.Y.; Xu, J.; Yuan, W.P.; Cheng, T.; et al. Different Effects of sgRNA Length on CRISPR-mediated Gene Knockout Efficiency. *Sci. Rep.* **2016**, *6*, 28566. [CrossRef] [PubMed]
56. Osakabe, Y.; Watanabe, T.; Sugano, S.S.; Ueta, R.; Ishihara, R.; Shinozaki, K.; Osakabe, K. Optimization of CRISPR/Cas9 genome editing to modify abiotic stress responses in plants. *Sci. Rep.* **2016**, *6*, 26685. [CrossRef] [PubMed]
57. Gerasimova, S.V.; Korotkova, A.M.; Hertig, C.; Hiekel, S.; Hoffie, R.; Budhagatapalli, N.; Otto, I.; Hensel, G.; Shumny, V.K.; Kochetov, A.V.; et al. Targeted genome modifcation in protoplasts of a highly regenerable Siberian barley cultivar using RNA-guided Cas9 endonuclease. *Vavilovskii Zhurnal Genet. I Selektsii* **2019**, *22*, 1033–1039. [CrossRef]
58. Cho, S.W.; Kim, S.; Kim, Y.; Kweon, J.; Kim, H.S.; Bae, S.; Kim, J.S. Analysis of off-target effects of CRISPR/Cas-derived RNA-guided endonucleases and nickases. *Genome Res.* **2014**, *24*, 132–141. [CrossRef]
59. Lei, Y.; Lu, L.; Liu, H.Y.; Li, S.; Xing, F.; Chen, L.L. CRISPR-P: A web tool for synthetic single-guide RNA design of CRISPR-system in plants. *Mol. Plant* **2014**, *7*, 1494–1496. [CrossRef] [PubMed]
60. Liu, H.; Ding, Y.; Zhou, Y.; Jin, W.; Xie, K.; Chen, L.L. CRISPR-P 2.0: An Improved CRISPR-Cas9 Tool for Genome Editing in Plants. *Mol. Plant* **2017**, *10*, 530–532. [CrossRef] [PubMed]

61. Doench, J.G.; Hartenian, E.; Graham, D.B.; Tothova, Z.; Hegde, M.; Smith, I.; Sullender, M.; Ebert, B.L.; Xavier, R.J.; Root, D.E. Rational design of highly active sgRNAs for CRISPR-Cas9-mediated gene inactivation. *Nat. Biotechnol.* **2014**, *32*, 1262–1267. [CrossRef]
62. Doench, J.G.; Fusi, N.; Sullender, M.; Hegde, M.; Vaimberg, E.W.; Donovan, K.F.; Smith, I.; Tothova, Z.; Wilen, C.; Orchard, R.; et al. Optimized sgRNA design to maximize activity and minimize off-target effects of CRISPR-Cas9. *Nat. Biotechnol.* **2016**, *34*, 184–191. [CrossRef] [PubMed]
63. Haeussler, M.; Schonig, K.; Eckert, H.; Eschstruth, A.; Mianne, J.; Renaud, J.B.; Schneider-Maunoury, S.; Shkumatava, A.; Teboul, L.; Kent, J.; et al. Evaluation of off-target and on-target scoring algorithms and integration into the guide RNA selection tool CRISPOR. *Genome Biol.* **2016**, *17*, 148. [CrossRef]
64. Concordet, J.P.; Haeussler, M. CRISPOR: Intuitive guide selection for CRISPR/Cas9 genome editing experiments and screens. *Nucleic Acids Res.* **2018**, *46*, W242–W245. [CrossRef] [PubMed]
65. Wong, N.; Liu, W.; Wang, X. WU-CRISPR: Characteristics of functional guide RNAs for the CRISPR/Cas9 system. *Genome Biol.* **2015**, *16*, 218. [CrossRef]
66. Liang, G.; Zhang, H.; Lou, D.; Yu, D. Selection of highly efficient sgRNAs for CRISPR/Cas9-based plant genome editing. *Sci. Rep.* **2016**, *6*, 21451. [CrossRef]
67. Zuker, M.; Jacobson, A.B. Using reliability information to annotate RNA secondary structures. *RNA* **1998**, *4*, 669–679. [CrossRef]
68. Waugh, A.; Gendron, P.; Altman, R.; Brown, J.W.; Case, D.; Gautheret, D.; Harvey, S.C.; Leontis, N.; Westbrook, J.; Westhof, E.; et al. RNAML: A standard syntax for exchanging RNA information. *RNA* **2002**, *8*, 707–717. [CrossRef]
69. Zuker, M. Mfold web server for nucleic acid folding and hybridization prediction. *Nucleic Acids Res.* **2003**, *31*, 3406–3415. [CrossRef]
70. Gruber, A.R.; Lorenz, R.; Bernhart, S.H.; Neubock, R.; Hofacker, I.L. The Vienna RNA websuite. *Nucleic Acids Res.* **2008**, *36*, W70–W74. [CrossRef] [PubMed]
71. Lorenz, R.; Bernhart, S.H.; Honer Zu Siederdissen, C.; Tafer, H.; Flamm, C.; Stadler, P.F.; Hofacker, I.L. ViennaRNA Package 2.0. *Algorithms Mol. Biol.* **2011**, *6*, 26. [CrossRef] [PubMed]
72. Pasternak, T.P.; Rudas, V.A.; Lörz, H.; Kumlehn, J. Embryogenic Callus Formation and Plant Regeneration from Leaf Base Segments of Barley (*Hordeum vulgare* L.). *J. Plant Physiol.* **1999**, *155*, 371–375. [CrossRef]
73. Vasil, V.; Castillo, A.M.; Fromm, M.E.; Vasil, I.K. Herbicide Resistant Fertile Transgenic Wheat Plants Obtained by Microprojectile Bombardment of Regenerable Embryogenic Callus. *Bio-Technol.* **1992**, *10*, 667–674. [CrossRef]
74. Wan, Y.; Lemaux, P.G. Generation of Large Numbers of Independently Transformed Fertile Barley Plants. *Plant Physiol.* **1994**, *104*, 37–48. [CrossRef]
75. Hiei, Y.; Ohta, S.; Komari, T.; Kumashiro, T. Efficient transformation of rice (*Oryza sativa* L.) mediated by Agrobacterium and sequence analysis of the boundaries of the T-DNA. *Plant J.* **1994**, *6*, 271–282. [CrossRef] [PubMed]
76. Tingay, S.; McElroy, D.; Kalla, R.; Fieg, S.; Wang, M.; Thornton, S.; Brettell, R. Agrobacterium tumefaciens-mediated barley transformation. *Plant J.* **1997**, *11*, 1369–1376. [CrossRef]
77. Cheng, M.; Fry, J.E.; Pang, S.; Zhou, H.; Hironaka, C.M.; Duncan, D.R.; Conner, T.W.; Wan, Y. Genetic Transformation of Wheat Mediated by Agrobacterium tumefaciens. *Plant Physiol.* **1997**, *115*, 971–980. [CrossRef]
78. Hensel, G.; Kastner, C.; Oleszczuk, S.; Riechen, J.; Kumlehn, J. Agrobacterium-mediated gene transfer to cereal crop plants: Current protocols for barley, wheat, triticale, and maize. *Int. J. Plant Genom.* **2009**, *2009*, 835608. [CrossRef]
79. Rasco-Gaunt, S.; Riley, A.; Cannell, M.; Barcelo, P.; Lazzeri, P.A. Procedures allowing the transformation of a range of European elite wheat (*Triticum aestivum* L.) varieties via particle bombardment. *J. Exp. Bot.* **2001**, *52*, 865–874. [CrossRef] [PubMed]
80. Wu, H.; Sparks, C.; Amoah, B.; Jones, H.D. Factors influencing successful Agrobacterium-mediated genetic transformation of wheat. *Plant Cell Rep.* **2003**, *21*, 659–668. [CrossRef]
81. Richardson, T.; Thistleton, J.; Higgins, T.J.; Howitt, C.; Ayliffe, M. Efficient Agrobacterium transformation of elite wheat germplasm without selection. *Plant Cell Tissue Organ Cult.* **2014**, *119*, 647–659. [CrossRef]
82. Ishida, Y.; Tsunashima, M.; Hiei, Y.; Komari, T. Wheat (*Triticum aestivum* L.) transformation using immature embryos. *Methods Mol. Biol.* **2015**, *1223*, 189–198. [CrossRef] [PubMed]

83. Hensel, G.; Valkov, V.; Middlefell-Williams, J.; Kumlehn, J. Efficient generation of transgenic barley: The way forward to modulate plant-microbe interactions. *J. Plant Physiol.* **2008**, *165*, 71–82. [CrossRef]
84. Yeo, F.K.S.; Hensel, G.; Vozábová, T.; Martin-Sanz, A.; Marcel, T.C.; Kumlehn, J.; Niks, R.E. Golden SusPtrit—A genetically well transformable barley line for studies on the resistance to rust fungi. *Theor. Appl. Genet.* **2014**, *127*, 325–337. [CrossRef] [PubMed]
85. Howells, R.M.; Craze, M.; Bowden, S.; Wallington, E.J. Efficient generation of stable, heritable gene edits in wheat using CRISPR/Cas9. *BMC Plant Biol.* **2018**, *18*, 215. [CrossRef] [PubMed]
86. Daghma, D.E.S.; Hensel, G.; Rutten, T.; Melzer, M.; Kumlehn, J. Cellular dynamics during early barley pollen embryogenesis revealed by time-lapse imaging. *Front. Plant Sci.* **2014**, *5*, 675. [CrossRef]
87. Kumlehn, J.; Serazetdinova, L.; Hensel, G.; Becker, D.; Loerz, H. Genetic transformation of barley (*Hordeum vulgare* L.) via infection of androgenetic pollen cultures with Agrobacterium tumefaciens. *Plant Biotechnol. J.* **2006**, *4*, 251–261. [CrossRef]
88. Otto, I.; Muller, A.; Kumlehn, J. Barley (*Hordeum vulgare* L.) transformation using embryogenic pollen cultures. *Methods Mol. Biol.* **2015**, *1223*, 85–99. [CrossRef] [PubMed]
89. Shim, Y.S.; Pauls, K.P.; Kasha, K.J. Transformation of isolated barley (*Hordeum vulgare* L.) microspores: II. Timing of pretreatment and temperatures relative to results of bombardment. *Genome* **2009**, *52*, 175–190. [CrossRef]
90. Gurushidze, M.; Hensel, G.; Hiekel, S.; Schedel, S.; Valkov, V.; Kumlehn, J. True-breeding targeted gene knock-out in barley using designer TALE-nuclease in haploid cells. *PLoS ONE* **2014**, *9*, e92046. [CrossRef] [PubMed]
91. Zhang, S.; Cho, M.J.; Koprek, T.; Yun, R.; Bregitzer, P.; Lemaux, P.G. Genetic transformation of commercial cultivars of oat (*Avena sativa* L.) and barley (*Hordeum vulgare* L.) using in vitro shoot meristematic cultures derived from germinated seedlings. *Plant Cell Rep.* **1999**, *18*, 959–966. [CrossRef]
92. Hamada, H.; Linghu, Q.; Nagira, Y.; Miki, R.; Taoka, N.; Imai, R. An in planta biolistic method for stable wheat transformation. *Sci. Rep.* **2017**, *7*, 11443. [CrossRef]
93. Hamada, H.; Liu, Y.; Nagira, Y.; Miki, R.; Taoka, N.; Imai, R. Biolistic-delivery-based transient CRISPR/Cas9 expression enables in planta genome editing in wheat. *Sci. Rep.* **2018**, *8*, 14422. [CrossRef]
94. Kapusi, E.; Hensel, G.; Coronado, M.J.; Broeders, S.; Marthe, C.; Otto, I.; Kumlehn, J. The elimination of a selectable marker gene in the doubled haploid progeny of co-transformed barley plants. *Plant Mol. Biol.* **2013**, *81*, 149–160. [CrossRef]
95. Gurushidze, M.; Trautwein, H.; Hoffmeister, P.; Otto, I.; Müller, A.; Kumlehn, J. Doubled Haploidy as a Tool for Chimaera Dissolution of TALEN-Induced Mutations in Barley. In *Biotechnologies for Plant Mutation Breeding*; Springer: Cham, Switzerland, 2017; Chapter 8; pp. 129–141. [CrossRef]
96. Liang, Z.; Chen, K.; Li, T.; Zhang, Y.; Wang, Y.; Zhao, Q.; Liu, J.; Zhang, H.; Liu, C.; Ran, Y.; et al. Efficient DNA-free genome editing of bread wheat using CRISPR/Cas9 ribonucleoprotein complexes. *Nat. Commun.* **2017**, *8*, 14261. [CrossRef] [PubMed]
97. Budhagatapalli, N.; Schedel, S.; Gurushidze, M.; Pencs, S.; Hiekel, S.; Rutten, T.; Kusch, S.; Morbitzer, R.; Lahaye, T.; Panstruga, R.; et al. A simple test for the cleavage activity of customized endonucleases in plants. *Plant Methods* **2016**, *12*, 18. [CrossRef]
98. Budhagatapalli, N.; Rutten, T.; Gurushidze, M.; Kumlehn, J.; Hensel, G. Targeted Modification of Gene Function Exploiting Homology-Directed Repair of TALEN-Mediated Double-Strand Breaks in Barley. *G3 Genes Genome Genet.* **2015**, *5*, 1857–1863. [CrossRef]
99. Zhang, Y.; Liang, Z.; Zong, Y.; Wang, Y.; Liu, J.; Chen, K.; Qiu, J.L.; Gao, C. Efficient and transgene-free genome editing in wheat through transient expression of CRISPR/Cas9 DNA or RNA. *Nat. Commun.* **2016**, *7*, 12617. [CrossRef] [PubMed]
100. Bae, S.; Kweon, J.; Kim, H.S.; Kim, J.-S. Microhomology-based choice of Cas9 nuclease target sites. *Nat. Methods* **2014**, *11*, 705–706. [CrossRef]
101. Zong, Y.; Wang, Y.; Li, C.; Zhang, R.; Chen, K.; Ran, Y.; Qiu, J.-L.; Wang, D.; Gao, C. Precise base editing in rice, wheat and maize with a Cas9-cytidine deaminase fusion. *Nat. Biotechnol.* **2017**, *35*, 438–440. [CrossRef]
102. Komor, A.C.; Kim, Y.B.; Packer, M.S.; Zuris, J.A.; Liu, D.R. Programmable editing of a target base in genomic DNA without double-stranded DNA cleavage. *Nature* **2016**, *533*, 420–424. [CrossRef] [PubMed]

103. Gaudelli, N.M.; Komor, A.C.; Rees, H.A.; Packer, M.S.; Badran, A.H.; Bryson, D.I.; Liu, D.R. Programmable base editing of A•T to G•C in genomic DNA without DNA cleavage. *Nature* **2017**, *551*, 464–471. [CrossRef] [PubMed]
104. Zong, Y.; Song, Q.; Li, C.; Jin, S.; Zhang, D.; Wang, Y.; Qiu, J.-L.; Gao, C. Efficient C-to-T base editing in plants using a fusion of nCas9 and human APOBEC3A. *Nat. Biotechnol.* **2018**, *36*, 950–953. [CrossRef]
105. Li, C.; Zong, Y.; Wang, Y.; Jin, S.; Zhang, D.; Song, Q.; Zhang, R.; Gao, C. Expanded base editing in rice and wheat using a Cas9-adenosine deaminase fusion. *Genome Biol.* **2018**, *19*, 59. [CrossRef]
106. Puchta, H.; Dujon, B.; Hohn, B. Two different but related mechanisms are used in plants for the repair of genomic double-strand breaks by homologous recombination. *Proc. Natl. Acad. Sci. USA* **1996**, *93*, 5055–5060. [CrossRef]
107. Fauser, F.; Roth, N.; Pacher, M.; Ilg, G.; Sánchez-Fernández, R.; Biesgen, C.; Puchta, H. In planta gene targeting. *Proc. Natl. Acad. Sci. USA* **2012**, *109*, 7535–7540. [CrossRef]
108. Ran, F.A.; Hsu, P.D.; Lin, C.-Y.; Gootenberg, J.S.; Konermann, S.; Trevino, A.E.; Scott, D.A.; Inoue, A.; Matoba, S.; Zhang, Y.; et al. Double nicking by RNA-guided CRISPR Cas9 for enhanced genome editing specificity. *Cell* **2013**, *154*, 1380–1389. [CrossRef]
109. Mikami, M.; Toki, S.; Endo, M. Precision Targeted Mutagenesis via Cas9 Paired Nickases in Rice. *Plant Cell Physiol.* **2016**, *57*, 1058–1068. [CrossRef] [PubMed]
110. Baltes, N.J.; Gil-Humanes, J.; Cermak, T.; Atkins, P.A.; Voytas, D.F. DNA replicons for plant genome engineering. *Plant Cell* **2014**, *26*, 151–163. [CrossRef]
111. Ran, Y.; Patron, N.; Kay, P.; Wong, D.; Buchanan, M.; Cao, Y.Y.; Sawbridge, T.; Davies, J.P.; Mason, J.; Webb, S.R.; et al. Zinc finger nuclease-mediated precision genome editing of an endogenous gene in hexaploid bread wheat (Triticum aestivum) using a DNA repair template. *Plant Biotechnol. J.* **2018**, *16*, 2088–2101. [CrossRef] [PubMed]
112. Kapusi, E.; Corcuera-Gomez, M.; Melnik, S.; Stoger, E. Heritable Genomic Fragment Deletions and Small Indels in the Putative ENGase Gene Induced by CRISPR/Cas9 in Barley. *Front. Plant Sci.* **2017**, *8*, 540. [CrossRef] [PubMed]
113. Sánchez-León, S.; Gil-Humanes, J.; Ozuna, C.V.; Gimenez, M.J.; Sousa, C.; Voytas, D.F.; Barro, F. Low-gluten, nontransgenic wheat engineered with CRISPR/Cas9. *Plant Biotechnol. J.* **2018**, *16*, 902–910. [CrossRef]
114. Xie, K.; Yang, Y. RNA-guided genome editing in plants using a CRISPR-Cas system. *Mol. Plant* **2013**, *6*, 1975–1983. [CrossRef]
115. Liang, Z.; Chen, K.; Yan, Y.; Zhang, Y.; Gao, C. Genotyping genome-edited mutations in plants using CRISPR ribonucleoprotein complexes. *Plant Biotechnol. J.* **2018**, *16*, 2053–2062. [CrossRef]
116. Brinkman, E.K.; Chen, T.; Amendola, M.; van Steensel, B. Easy quantitative assessment of genome editing by sequence trace decomposition. *Nucleic Acids Res.* **2014**, *42*, e168. [CrossRef]
117. Liu, W.; Xie, X.; Ma, X.; Li, J.; Chen, J.; Liu, Y.G. DSDecode: A Web-Based Tool for Decoding of Sequencing Chromatograms for Genotyping of Targeted Mutations. *Mol. Plant* **2015**, *8*, 1431–1433. [CrossRef]
118. Wang, W.; Simmonds, J.; Pan, Q.; Davidson, D.; He, F.; Battal, A.; Akhunova, A.; Trick, H.N.; Uauy, C.; Akhunov, E. Gene editing and mutagenesis reveal inter-cultivar differences and additivity in the contribution of TaGW2 homoeologues to grain size and weight in wheat. *Theor. Appl. Genet.* **2018**, *131*, 2463–2475. [CrossRef]
119. Kim, D.; Alptekin, B.; Budak, H. CRISPR/Cas9 genome editing in wheat. *Funct. Integr. Genome* **2018**, *18*, 31–41. [CrossRef]
120. Hensel, G.; Marthe, C.; Kumlehn, J. Agrobacterium-Mediated Transformation of Wheat Using Immature Embryos. *Methods Mol. Biol.* **2017**, *1679*, 129–139. [CrossRef] [PubMed]
121. Wolt, J.D.; Wolf, C. Policy and Governance Perspectives for Regulation of Genome Edited Crops in the United States. *Front. Plant Sci.* **2018**, *9*, 1606. [CrossRef]
122. E.A.S.A. *Council Genome Editing: Scientific Opportunities, Public Interests and Policy Options in the European Union*; EASAC Secretariat Deutsche Akademie der Naturforscher Leopoldina German National Academy of Sciences: Halle, Germany, 2017; p. 34.
123. Sprink, T.; Eriksson, D.; Schiemann, J.; Hartung, F. Regulatory hurdles for genome editing: Process-vs. product-based approaches in different regulatory contexts. *Plant Cell Rep.* **2016**, *35*, 1493–1506. [CrossRef]
124. Bhowmik, P.; Ellison, E.; Polley, B.; Bollina, V.; Kulkarni, M.; Ghanbarnia, K.; Song, H.; Gao, C.; Voytas, D.F.; Kagale, S. Targeted mutagenesis in wheat microspores using CRISPR/Cas9. *Sci. Rep.* **2018**, *8*, 6502. [CrossRef]

125. Kumlehn, J.; Schieder, O.; Lorz, H. In vitro development of wheat (*Triticum aestivum* L.) from zygote to plant via ovule culture. *Plant Cell Rep.* **1997**, *16*, 663–667. [CrossRef]
126. Holme, I.B.; Brinch-Pedersen, H.; Lange, M.; Holm, P.B. Transformation of barley (*Hordeum vulgare* L.) by Agrobacterium tumefaciens infection of in vitro cultured ovules. *Plant Cell Rep.* **2006**, *25*, 1325–1335. [CrossRef] [PubMed]
127. Holme, I.B.; Brinch-Pedersen, H.; Lange, M.; Holm, P.B. Transformation of different barley (*Hordeum vulgare* L.) cultivars by Agrobacterium tumefaciens infection of in vitro cultured ovules. *Plant Cell Rep.* **2008**, *27*, 1833–1840. [CrossRef]
128. Hensel, G.; Himmelbach, A.; Chen, W.; Douchkov, D.K.; Kumlehn, J. Transgene expression systems in the Triticeae cereals. *J. Plant Physiol.* **2011**, *168*, 30–44. [CrossRef]
129. Freeman, J.; Sparks, C.A.; West, J.; Shewry, P.R.; Jones, H.D. Temporal and spatial control of transgene expression using a heat-inducible promoter in transgenic wheat. *Plant Biotechnol. J.* **2011**, *9*, 788–796. [CrossRef] [PubMed]
130. Jöst, M.; Hensel, G.; Kappel, C.; Druka, A.; Sicard, A.; Hohmann, U.; Beier, S.; Himmelbach, A.; Waugh, R.; Kumlehn, J.; et al. The INDETERMINATE DOMAIN Protein BROAD LEAF1 Limits Barley Leaf Width by Restricting Lateral Proliferation. *Curr. Biol.* **2016**, *26*, 903–909. [CrossRef]
131. Xie, K.; Minkenberg, B.; Yang, Y. Boosting CRISPR/Cas9 multiplex editing capability with the endogenous tRNA-processing system. *Proc. Natl. Acad. Sci. USA* **2015**, *112*, 3570–3575. [CrossRef] [PubMed]
132. Esvelt, K.M.; Mali, P.; Braff, J.L.; Moosburner, M.; Yaung, S.J.; Church, G.M. Orthogonal Cas9 proteins for RNA-guided gene regulation and editing. *Nat. Methods* **2013**, *10*, 1116–1121. [CrossRef] [PubMed]
133. Ran, F.A.; Cong, L.; Yan, W.X.; Scott, D.A.; Gootenberg, J.S.; Kriz, A.J.; Zetsche, B.; Shalem, O.; Wu, X.; Makarova, K.S.; et al. In vivo genome editing using Staphylococcus aureus Cas9. *Nature* **2015**, *520*, 186–191. [CrossRef]
134. Kim, H.; Kim, S.T.; Ryu, J.; Kang, B.C.; Kim, J.S.; Kim, S.G. CRISPR/Cpf1-mediated DNA-free plant genome editing. *Nat. Commun.* **2017**, *8*, 14406. [CrossRef]
135. Steinert, J.; Schiml, S.; Fauser, F.; Puchta, H. Highly efficient heritable plant genome engineering using Cas9 orthologues from Streptococcus thermophilus and Staphylococcus aureus. *Plant J.* **2015**, *84*, 1295–1305. [CrossRef] [PubMed]
136. Hu, J.H.; Miller, S.M.; Geurts, M.H.; Tang, W.; Chen, L.; Sun, N.; Zeina, C.M.; Gao, X.; Rees, H.A.; Lin, Z.; et al. Evolved Cas9 variants with broad PAM compatibility and high DNA specificity. *Nature* **2018**, *556*, 57–63. [CrossRef]
137. Nishimasu, H.; Shi, X.; Ishiguro, S.; Gao, L.; Hirano, S.; Okazaki, S.; Noda, T.; Abudayyeh, O.O.; Gootenberg, J.S.; Mori, H.; et al. Engineered CRISPR-Cas9 nuclease with expanded targeting space. *Science* **2018**, *361*, 1259–1262. [CrossRef] [PubMed]
138. Zetsche, B.; Gootenberg, J.S.; Abudayyeh, O.O.; Slaymaker, I.M.; Makarova, K.S.; Essletzbichler, P.; Volz, S.E.; Joung, J.; van der Oost, J.; Regev, A.; et al. Cpf1 is a single RNA-guided endonuclease of a class 2 CRISPR-Cas system. *Cell* **2015**, *163*, 759–771. [CrossRef] [PubMed]
139. Tang, X.; Lowder, LG.; Zhang, T.; Malzahn, AA.; Zheng, X.; Voytas, DF.; Zhong, Z.; Chen, Y.; Ren, Q.; Li, Q.; Kirkland, ER.; Zhang, Y.; Qi, Y. A CRISPR-Cpf1 system for efficient genome editing and transcriptional repression in plants. *Nat. Plants* **2017**, *3*, 17018. [CrossRef] [PubMed]
140. Fauser, F.; Schiml, S.; Puchta, H. Both CRISPR/Cas-based nucleases and nickases can be used efficiently for genome engineering in Arabidopsis thaliana. *Plant J.* **2014**, *79*, 348–359. [CrossRef]
141. Schedel, S.; Pencs, S.; Hensel, G.; Muller, A.; Rutten, T.; Kumlehn, J. RNA-Guided Cas9-Induced Mutagenesis in Tobacco Followed by Efficient Genetic Fixation in Doubled Haploid Plants. *Front. Plant Sci.* **2017**, *7*, 1995. [CrossRef]
142. Rodriguez-Leal, D.; Lemmon, Z.H.; Man, J.; Bartlett, M.E.; Lippman, Z.B. Engineering Quantitative Trait Variation for Crop Improvement by Genome Editing. *Cell* **2017**, *171*, 470–480. [CrossRef] [PubMed]
143. Chandrasekaran, J.; Brumin, M.; Wolf, D.; Leibman, D.; Klap, C.; Pearlsman, M.; Sherman, A.; Arazi, T.; Gal-On, A. Development of broad virus resistance in non-transgenic cucumber using CRISPR/Cas9 technology. *Mol. Plant Pathol.* **2016**, *17*, 1140–1153. [CrossRef] [PubMed]
144. Stein, N.; Perovic, D.; Kumlehn, J.; Pellio, B.; Stracke, S.; Streng, S.; Ordon, F.; Graner, A. The eukaryotic translation initiation factor 4E confers multiallelic recessive Bymovirus resistance in *Hordeum vulgare* (L.). *Plant J.* **2005**, *42*, 912–922. [CrossRef]

145. Yang, P.; Lupken, T.; Habekuss, A.; Hensel, G.; Steuernagel, B.; Kilian, B.; Ariyadasa, R.; Himmelbach, A.; Kumlehn, J.; Scholz, U.; et al. PROTEIN DISULFIDE ISOMERASE LIKE 5-1 is a susceptibility factor to plant viruses. *Proc. Natl. Acad. Sci. USA* **2014**, *111*, 2104–2109. [CrossRef]
146. Maccaferri, M.; Harris, N.S.; Twardziok, S.O.; Pasam, R.K.; Gundlach, H.; Spannagl, M.; Ormanbekova, D.; Lux, T.; Prade, V.M.; Milner, S.G.; et al. Durum wheat genome highlights past domestication signatures and future improvement targets. *Nat. Genet.* **2019**, *51*, 885–895. [CrossRef]
147. Andersson, M.; Turesson, H.; Nicolia, A.; Falt, A.S.; Samuelsson, M.; Hofvander, P. Efficient targeted multiallelic mutagenesis in tetraploid potato (*Solanum tuberosum*) by transient CRISPR-Cas9 expression in protoplasts. *Plant Cell Rep.* **2017**, *36*, 117–128. [CrossRef] [PubMed]
148. Sun, Y.; Jiao, G.; Liu, Z.; Zhang, X.; Li, J.; Guo, X.; Du, W.; Du, J.; Francis, F.; Zhao, Y.; et al. Generation of High-Amylose Rice through CRISPR/Cas9-Mediated Targeted Mutagenesis of Starch Branching Enzymes. *Front. Plant Sci.* **2017**, *8*, 298. [CrossRef]
149. Simkin, A.J.; Lopez-Calcagno, P.E.; Raines, C.A. Feeding the world: Improving photosynthetic efficiency for sustainable crop production. *J. Exp. Bot.* **2019**, *70*, 1119–1140. [CrossRef]
150. Zhang, Y.; Bai, Y.; Wu, G.; Zou, S.; Chen, Y.; Gao, C.; Tang, D. Simultaneous modification of three homoeologs of TaEDR1 by genome editing enhances powdery mildew resistance in wheat. *Plant J.* **2017**, *91*, 714–724. [CrossRef] [PubMed]
151. Rey, M.-D.; Martín, A.C.; Smedley, M.; Hayta, S.; Harwood, W.; Shaw, P.; Moore, G. Magnesium Increases Homoeologous Crossover Frequency During Meiosis in ZIP4 (Ph1 Gene) Mutant Wheat-Wild Relative Hybrids. *Front. Plant Sci.* **2018**, *9*, 509. [CrossRef] [PubMed]
152. Singh, M.; Kumar, M.; Albertsen, M.C.; Young, J.K.; Cigan, A.M. Concurrent modifications in the three homeologs of Ms45 gene with CRISPR-Cas9 lead to rapid generation of male sterile bread wheat (*Triticum aestivum* L.). *Plant Mol. Biol.* **2018**, *97*, 371–383. [CrossRef]
153. Zhang, S.; Zhang, R.; Song, G.; Gao, J.; Li, W.; Han, X.; Chen, M.; Li, Y.; Li, G. Targeted mutagenesis using the Agrobacterium tumefaciens-mediated CRISPR-Cas9 system in common wheat. *BMC Plant Biol.* **2018**, *18*, 302. [CrossRef] [PubMed]
154. Holubová, K.; Hensel, G.; Vojta, P.; Tarkowski, P.; Bergougnoux, V.; Galuszka, P. Modification of Barley Plant Productivity Through Regulation of Cytokinin Content by Reverse-Genetics Approaches. *Front. Plant Sci.* **2018**, *9*, 1676. [CrossRef]
155. Zhong, Y.; Blennow, A.; Kofoed-Enevoldsen, O.; Jiang, D.; Hebelstrup, K.H. Protein Targeting to Starch 1 is essential for starchy endosperm development in barley. *J. Exp. Bot.* **2019**, *70*, 485–496. [CrossRef] [PubMed]

# 11

# Instability of Alien Chromosome Introgressions in Wheat Associated with Improper Positioning in the Nucleus

**Kateřina Perničková [1], Veronika Koláčková [1], Adam J. Lukaszewski [2], Chaolan Fan [2], Jan Vrána [1], Martin Duchoslav [3], Glyn Jenkins [4], Dylan Phillips [4], Olga Šamajová [5], Michaela Sedlářová [3], Jozef Šamaj [5], Jaroslav Doležel [1] and David Kopecký [1,*]**

1 Institute of Experimental Botany, Centre of the Region Haná for Biotechnological and Agricultural Research, Šlechtitelů 31, 78371 Olomouc, Czech Republic; pernickova@ueb.cas.cz (K.P.); kolackova@ueb.cas.cz (V.K.); vrana@ueb.cas.cz (J.V.); dolezel@ueb.cas.cz (J.D.)
2 Department of Botany and Plant Sciences, University of California, Riverside, CA 92521, USA; adam.lukaszewski@ucr.edu (A.J.L.); chaolanf@ucr.edu (C.F.)
3 Department of Botany, Faculty of Science, Palacký University Olomouc, Šlechtitelů 27, 783 71 Olomouc, Czech Republic; martin.duchoslav@upol.cz (M.D.); michaela.sedlarova@upol.cz (M.S.)
4 Institute of Biological, Environmental and Rural Sciences, Aberystwyth University, Aberystwyth, Ceredigion, Wales SY23 3DA, UK; gmj@aber.ac.uk (G.J.); dwp@aber.ac.uk (D.P.)
5 Department of Cell Biology, Centre of the Region Haná for Biotechnological and Agricultural Research, Faculty of Science, Palacký University Olomouc, Šlechtitelů 27, 783 71 Olomouc, Czech Republic; olga.samajova@upol.cz (O.Š.); jozef.samaj@upol.cz (J.Š.)
* Correspondence: kopecky@ueb.cas.cz.

**Abstract:** Alien introgressions introduce beneficial alleles into existing crops and hence, are widely used in plant breeding. Generally, introgressed alien chromosomes show reduced meiotic pairing relative to the host genome, and may be eliminated over generations. Reduced pairing appears to result from a failure of some telomeres of alien chromosomes to incorporate into the leptotene bouquet at the onset of meiosis, thereby preventing chiasmate pairing. In this study, we analysed somatic nuclei of rye introgressions in wheat using 3D-FISH and found that while introgressed rye chromosomes or chromosome arms occupied discrete positions in the Rabl's orientation similar to chromosomes of the wheat host, their telomeres frequently occupied positions away from the nuclear periphery. The frequencies of such abnormal telomere positioning were similar to the frequencies of out-of-bouquet telomere positioning at leptotene, and of pairing failure at metaphase I. This study indicates that improper positioning of alien chromosomes that leads to reduced pairing is not a strictly meiotic event but rather a consequence of a more systemic problem. Improper positioning in the nuclei probably impacts the ability of introgressed chromosomes to migrate into the telomere bouquet at the onset of meiosis, preventing synapsis and chiasma establishment, and leading to their gradual elimination over generations.

**Keywords:** chromatin; 3D-FISH; nucleus; introgression; rye; hybrid; wheat; genome stability

---

## 1. Introduction

During interphase, chromosomes are decondensed and occupy distinct regions of the nucleus, called chromosome domains or chromosome territories [1,2]. In flowering plants, chromosome territories show two predominant configurations, known as the Rabl's orientation and Rosette [3,4] with the former being more frequent. The Rabl's configuration [5] reflects the orientation of chromosomes from the preceding anaphase: all centromeres are grouped at or close to the nuclear periphery at one pole of the nucleus while telomeres are dispersed toward the opposite pole [6]. Such an orientation presumably simplifies homologue search and the initiation of chromosome pairing in early meiosis. Chromosome pairing is initiated during leptotene (or telomere) bouquet, where all telomeres are located close to each other at one pole of the nucleus. The association of the telomeres into the bouquet at the onset of the first meiotic division facilitates the initiation of synapsis of homologous chromosomes, which in turn is a prerequisite for crossing over and chiasmate pairing at metaphase I [7], both of which are critical for the success of meiosis. It has been clearly demonstrated that misalignment of telomeres of homologues during the bouquet formation restricts synapsis and drastically reduces metaphase I pairing [8,9].

Many crops are allopolyploids, that is, products of wide hybridization, a natural process commandeered frequently by plant breeders to widen the gene pool of a crop by introduction of agronomically important alleles. Such introgressions may take the form of new amphiploids, such as triticale (X Triticosecale Wittmack), introgressions of single chromosomes, chromosome arms or even smaller chromosome segments [10]. One of the most successful intergeneric introgressions is the 1RS.1BL centric wheat-rye translocation in wheat (*Triticum aestivum* L.), where the short arm of wheat chromosome 1B is replaced by its counterpart from the rye genome (*Secale cereale* L.). Many other introgressions have been released in wheat breeding and research programs involving rye, barley (*Hordeum* sp.), *Aegilops*, *Agropyron* and *Haynaldia* species [11,12].

As a general rule, early generations of interspecific hybrids suffer irregular chromosome pairing which may affect different genomes in various ways. This hampers wider commercial utilization of amphiploids and introgression lines in agriculture. Nishiyama [13] described reduction of chromosome number in successive generations of synthetic decaploid interspecific hybrids of oats (*Avena*) from 2n = 70 to 2n = 42–58. Similarly, reversion to bread wheat (loss of the entire rye genome) has been reported in octoploid triticale (amphiploids of bread wheat with rye) [14]. In tetraploid and hexaploid triticales, rye chromosomes fail to pair more frequently than wheat chromosomes [15]. Among disomic additions and substitutions of individual rye chromosomes in bread wheat, Orellana et al. [16] found reduced metaphase I pairing of rye chromosomes and significantly higher numbers of univalents compared to wheat chromosomes.

The problem of reduced chromosome pairing in amphiploids has been a subject of much debate over the decades. In wheat-rye hybrids, five different concepts have been presented, but none has been proven to be correct (reviewed in [17]). After several recent studies, a relationship between the behaviour of telomeres and the success of chromosome pairing has gained credence. Murphy and Bass [18] have shown that the desynaptic (*dy*) mutant of maize displays multiple defects in telomere-nuclear envelope interactions, homologous chromosome synapsis, recombination and chromosome segregation. Similarly, Naranjo [19] reported that reduced pairing of rye chromosomes in wheat appears to be a consequence of disturbed migration of rye telomeres into the leptotene bouquet. Telomere positioning and migration are preferentially studied in meiosis. However, our previous study suggested that the problem may be systemic in nature, and aberrant arrangement of telomeres in pollen mother cells (PMCs) is only an extension of their erratic behaviour in other (somatic) tissues. We observed that the frequency of out-of-bouquet rye telomere position at leptotene was virtually identical to that in the nuclei of somatic cells, and was similar to the frequency of synapsis of the normal and inverted chromosome arms in a heterozygote for an inversion of a rye chromosome arm in wheat [20].

In this study, we analysed wheat lines with introgressed chromatin from rye, involving disomic whole chromosome substitutions, ditelosomic line and centric (whole arm) translocations in the 3D space of wheat nuclear volume of somatic tissue in order further investigate the possible link between telomere positioning in somatic tissues, chromosome pairing and stability of introgressed alien chromatin. Special attention was paid to the distribution of the telomeres and centromeres.

## 2. Results

### *2.1. Morphometrical Characteristics of G1 Interphase Nuclei of Wheat-Rye Introgression Lines*

In total, we analysed 315 nuclei with most of the parameters described in Material and Methods. The morphology of G1 nuclei flow sorted into a polyacrylamide gel ranged from spherical and ellipsoidal to irregular shapes with varying degrees of contortion. Only nuclei with spherical and slightly ellipsoidal shapes were selected for analyses. Nuclear volumes ranged from 753 to 2996 $\mu m^3$ (mean $\pm$ SD 1677 $\pm$ 506 $\mu m^3$). 3D-GISH analysis showed that the chromosome territories (CTs) of rye chromosome arms appeared as compact structures of regular shapes spanning the entire nucleus and arranged in a typical Rabl's orientation (Figure 1). Centromeres were generally close to each other and located on the nuclear periphery at one pole, while telomeres were at the periphery of the opposite pole. In a majority of the analysed nuclei, centromeres of rye chromosome arms were closer to each other than their telomeres (Table 1).

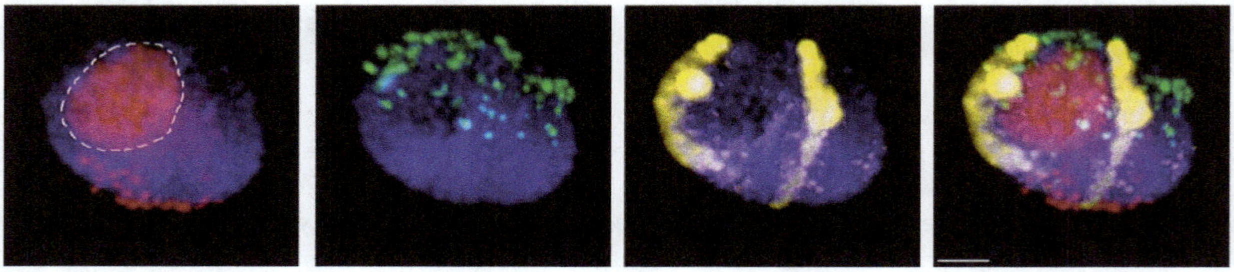


**Figure 1.** Rabl's orientation of rye chromosomes. Nucleus with a pair of homologous rye $_{del}$1RS.1RL chromosomes in the proper Rabl's orientation. Total genomic DNA of rye was labelled with TRITC using Nick translation (yellow color), centromeres of both wheat and rye chromosomes were visualized using oligonucleotide probe (red color), and telomere-specific sequence was PCR-labelled with FITC (green color). Nuclear DNA was counterstained with DAPI (blue color). Note the difference in signal intensity along the chromosome arms with subtelomeric heterochromatin labelled dark yellow and the remaining rye arm labelled light yellow. Nucleoli are indicated by white dashed lines. Scale bar 5 μm.

**Table 1.** Morphometrical characteristics of rye chromosome arms in wheat-rye introgression lines. The values (Mean $\pm$ SD) of arm length, distances between centromere to centromere (C-C) and telomere to telomere (T-T) (given in μm) were normalized according to the volume of the nucleus (absolute distance in μm/nuclear volume $\times$ 1000).

| Introgression | Nuclear Volume ($\mu m^3$) | Arm Length (μm) | C-C (μm) | T-T (μm) |
|---|---|---|---|---|
| 1AS.1RL | 1545 $\pm$ 224 | 7.73 $\pm$ 1.48 | 4.01 $\pm$ 1.98 | 4.88 $\pm$ 1.71 |
| 1RS.1BL | 1501 $\pm$ 362 | 8.21 $\pm$ 1.69 | 3.49 $\pm$ 1.50 | 4.18 $\pm$ 1.66 |
| 1RS.1DL | 1998 $\pm$ 285 | 6.86 $\pm$ 1.31 | 2.83 $\pm$ 1.15 | 3.71 $\pm$ 1.74 |
| 2RS.2BL | 1597 $\pm$ 490 | 8.51 $\pm$ 2.58 | 3.38 $\pm$ 1.70 | 4.09 $\pm$ 1.95 |
| 2BS.2RL | 1510 $\pm$ 357 | 8.96 $\pm$ 1.62 | 4.05 $\pm$ 1.80 | 4.99 $\pm$ 2.25 |
| 5RS.5BL | 1656 $\pm$ 329 | 6.95 $\pm$ 1.54 | 3.06 $\pm$ 1.29 | 3.97 $\pm$ 1.48 |
| $_{del}$1RS.1RL | 1986 $\pm$ 390 | S [1]: 4.04 $\pm$ 1.03 | 2.88 $\pm$ 1.37 | 3.86 $\pm$ 1.79 |
| | | L [2]: 7.56 $\pm$ 1.67 | | 3.56 $\pm$ 1.87 |
| 1RS.$_{del}$1RL | 2006 $\pm$ 500 | S [1]: 6.18 $\pm$ 1.58 | 3.57 $\pm$ 1.69 | 4.20 $\pm$ 1.92 |
| | | L [2]: 5.21 $\pm$ 1.41 | | 4.12 $\pm$ 1.53 |

[1] S: short arm [2] L: long arm.

### *2.2. Positions of Rye Telomeres and Centromeres Relative to the Nuclear Envelope and Positioning of Rye Telomeres in the Telomere Cluster*

In the majority of nuclei, telomeres and centromeres of rye chromosomes/chromosome arms were positioned at the nuclear periphery (NP). As we did not perform any tests for the attachment of telomeres to the inner nuclear envelope, we cannot be certain that the positioning of telomeres or centromeres at the nuclear periphery reflects their attachment to the nuclear envelope. However, it is a safe assumption that the positioning of a telomere away from the nuclear periphery indicates an absence of such attachment. It also needs to be stressed that the position of the nuclear envelope was inferred from the edge of the DAPI-stained chromatin. Therefore, in the absence of definitive proof, all telomeres located away from the edge of nuclear periphery are assumed to be unattached to the nuclear envelope; those at the nuclear periphery are labelled below as "in contact with NP".

In a small proportion of nuclei across most of the genotypes, a centromere or a telomere of one rye chromosome arm was positioned away from the NP (Table 2). In control diploid rye, essentially all chromosomes were in Rabl's orientation with all centromeres positioned at the centromere pole and 99.3% of telomeres located at the telomere pole. Similarly, 94.3% of the telomeres were in contact with the NP. Rye chromosome arms in the introgression lines differed in the proportions of the telomere (log-linear models using likelihood-ratio chi-square test, $\chi^2 = 54.36$, DF = 12, $p < 0.001$) and centromere ($\chi^2 = 25.61$, DF = 12, $p = 0.012$) positions relative to the NP, while only minor differences were observed among genotypes in proportions of telomeres positioned at the telomere pole ($\chi^2 = 22.08$, DF = 12, $p = 0.036$).

**Table 2.** Frequencies of proper positioning of telomeres of rye chromosome arms (at the telomere pole) and rye telomeres and centromeres in contact with the nuclear periphery (NP).

| Introgression | Rye Chromosome arm Length (Mb) [3] | Chromosome Length (Mb) | Arm Ratio | Number of Nuclei | Telomere Proper Positioning (%) | Telomere in Contact with the NP (%) | Centromere in Contact with the NP (%) |
|---|---|---|---|---|---|---|---|
| 1AS.1RL | 626 | 902 | 2.27 | 25 | 98.00 | 98.00 | 84.00 |
| 1RS.1BL | 423 | 959 | 1.27 | 40 | 98.75 | 98.75 | 100.00 |
| 1RS.1DL | 423 | 804 | 0.90 | 40 | 97.50 | 95.00 | 98.75 |
| 2RS.2BL | 595 | 1102 | 0.85 | 40 | 98.75 | 96.25 | 98.75 |
| 2BS.2RL | 693 | 1116 | 1.64 | 40 | 98.75 | 100.00 | 100.00 |
| 5RS.5BL | 346 | 928 | 1.68 | 40 | 93.75 | 87.50 | 96.25 |
| $_{del}$1RS.1RL | short arm: 271 [1]<br>long arm: 626 | 897 | 2.31 | 25 | 1RS: 82.00<br>1RL: 100.00 | 1RS: 46.00<br>1RL: 100.00 | 90.00 |
| 1RS.$_{del}$1RL | short arm: 423<br>long arm: 438 [2] | 861 | 1.04 | 21 | 1RS: 92.90;<br>1RL: 100.00 | 1RS: 90.50<br>1RL: 97.60 | 88.10 |
| 1R(1A) | short arm: 423<br>long arm: 626 | 1049 | 1.48 | 22 | 1RS: 100.00<br>1RL: 100.00 | 1RS: 93.20;<br>1RL: 100.00 | 90.90 |
| t1RS | 423 | 423 | - | 22 | 100.00 | 93.20 | 79.50 |

[1] deletion of about 36% of 1RS arm (proximal part); [2] deletion of about 30% of 1RL arm (proximal part); [3] estimated values of chromosome and chromosome arm length and arm ratio have been calculated from karyotypes of Schlegel et al. [21] and Naranjo [22] for rye and Gill et al. [23] for wheat and genome size estimations [24].

In the majority of cases, centromeres were tightly clustered at one pole of the nucleus; telomeres were located in the about one third of the nuclear volume opposite to the centromere pole (see measurements of C-C and T-T in Table 1). Rarely, a telomere was located away from the telomere pole (Figure 2, Figure A2 and Video S1) and/or away from the nuclear envelope (NE) (Figure 3; Figures A1 and A2 and Video S2). The frequencies of telomere proper positioning and telomere contact with the nuclear periphery (NP) were correlated (Spearman correlation coefficient, $r_s = 0.57$, $p = 0.043$), and they appeared to correlate with the length of a chromosome arm (telomere proper positioning: $r_s = 0.57$, $p = 0.040$, telomere in contact with NP: $r_s = 0.53$, $p = 0.063$): shorter arms were less likely to be in contact with the NP and more frequently were out-of-position (Table 2). Chromosome length and arm ratio had no effect on the proper positioning of telomeres in Rabl's orientation and the proportions of centromeres and telomeres in contact with the NP. Shortening of 1RS by a deletion reduced the in-contact with the NP frequency from 93.2% (in 1R) to 46.0% (in $_{del}$1RS.1RL) and proper positioning from 100% to 82%. In all cases, telomeres of long arms were in contact with the NP much more frequently than those of the short arms. Even the reduction in length of 1RL did not change this pattern. Telomeres of the 1RS arm were less likely to be in contact with the NP relative to those of the $_{del}$1RL arm (90.5% vs. 97.6%, respectively). Telocentric chromosome t1RS had its telomere at the nuclear periphery with the same frequency as the short arm of 1R (93.2%) and was always properly arranged in the Rabl's configuration (the same as 1RS in normal 1R). Interestingly, the telomere located at the centromere of t1RS was positioned at the centromere pole but less frequently in contact with the NP than the centromere of 1R (79.5% vs. 90.9%).

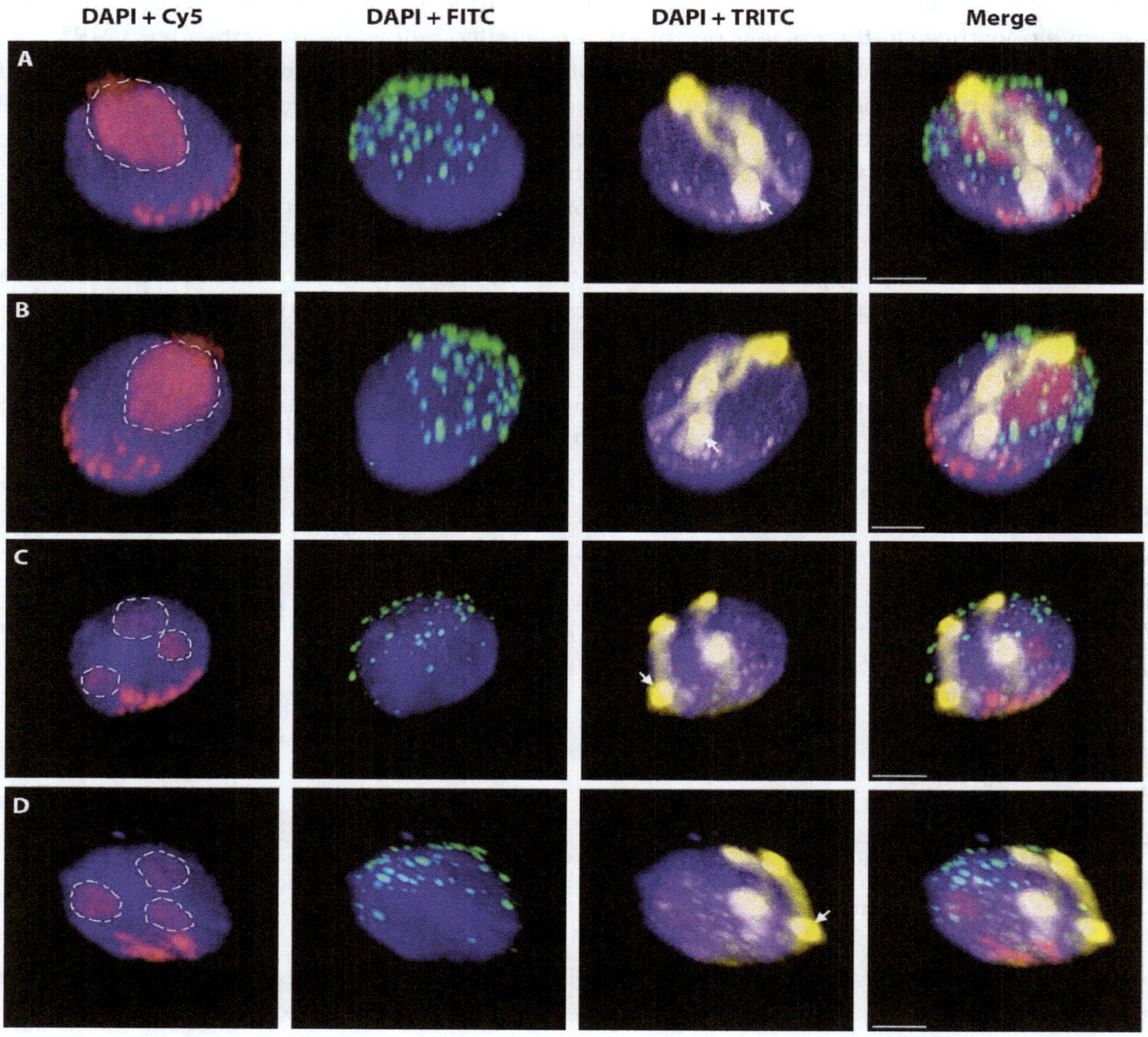


**Figure 2.** Telomere out-positioning of rye chromosome arms. Two nclei ((**A**,**B**) is the same nucleus from different angle; the same for (**C**,**D**)) with a pair of homologous rye $_{del}$1RS.1RL chromosomes after 3D-FISH. Total genomic DNA of rye was labelled with TRITC using Nick translation (yellow colour), centromeres of both wheat and rye chromosomes were visualized using oligonucleotide probe (red colour), and telomere-specific sequence was PCR-labelled with FITC (green colour). Nuclear DNA was counterstained with DAPI (blue colour). Rye telomeres positioned out of the telomere pole are indicated by arrows. Nucleoli are indicated by white dashed lines. Scale bar 5 μm.

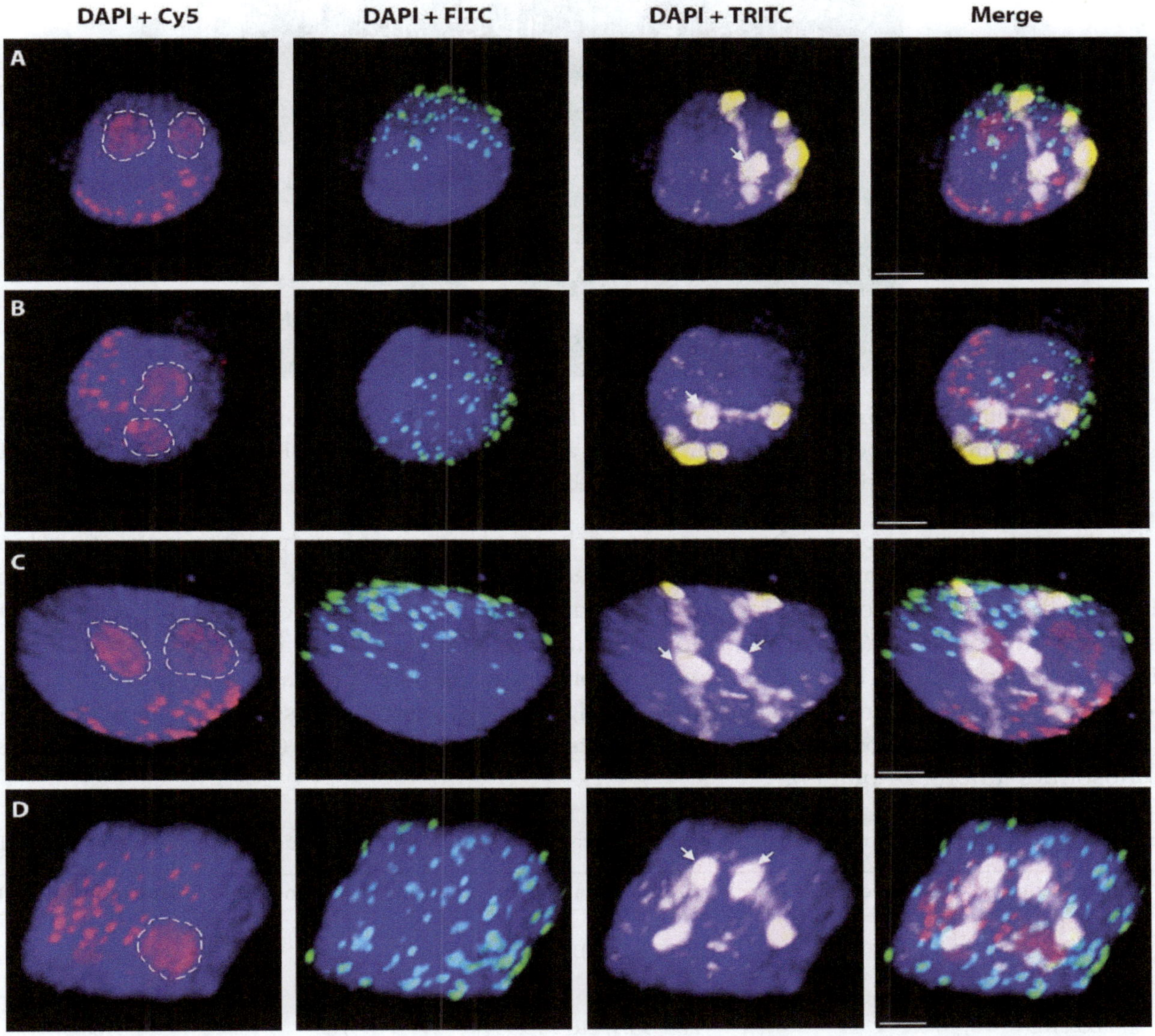


**Figure 3.** Non-attachment of rye telomeres. Two nuclei ((**A**,**B**) is the same nucleus from different angles; same for (**C**,**D**)) with a pair of homologous rye $_{del}$1RS.1RL chromosomes after 3D-FISH. Total genomic DNA of rye was labelled with TRITC using Nick translation (yellow colour), centromeres of both wheat and rye chromosomes were visualized using oligonucleotide probe (red colour), and telomere-specific sequence was PCR-labelled with FITC (green colour). Nuclear DNA was counterstained with DAPI (blue color). Rye telomeres without visual contact to nuclear envelope are indicated by arrows. Nucleoli are indicated by white dashed lines. Scale bar 5 μm.

### 2.3. *Chromosome Pairing and Transmission Rate*

The pairing frequencies of rye chromosome arms, whether in chromosome 1R or in centric translocations, were high, but varied for different chromosome arms and even for the same arms but in different translocations or configurations. In centric translocations 1RS.1BL and 2BS.2RL, rye chromosome arms paired less frequently than the wheat arms present (90.0% vs. 94.0% and 93.9% vs. 95.9%, respectively). In the disomic substitution and deletion lines of chromosome 1R, the short arm (1RS) always paired less frequently than the long one. Deletion of a large portion of an arm, whether in the short or the long arm, did not have any major effect on pairing frequency of the arm with the deletion, or the other arm of the chromosome (Figure 4).

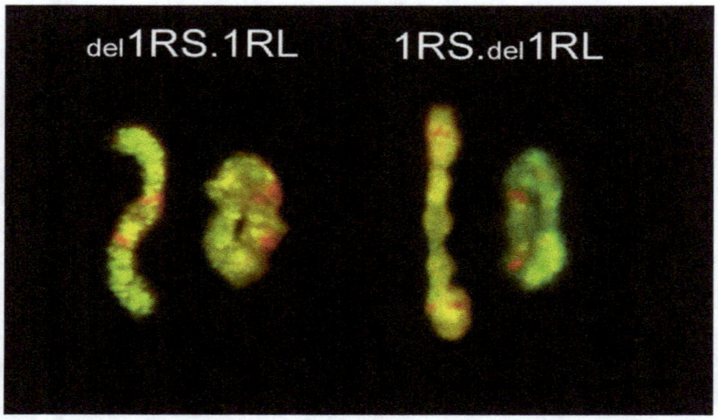


**Figure 4.** Metaphase I pairing of chromosomes 1R with deletions. From left to right: chromosome $_{del}$1RS.1RL paired as a rod in the short arm and as a ring and chromosome 1RS. $_{del}$1RL paired as a rod in the long arm and as a ring. Short arms are marked by non-hybridized (red) bands of the NOR region.

The transmission rate of rye chromosomes or chromosome arms in wheat-rye centric translocation was also high, but with some variation among the lines (Table 3). Centric translocations were highly stable with the transmission rate in most cases at 100%. The only exception was 5RS.5BL with a transmission rate of 98.6% (one out of 71 plants was nullisomic for this translocation). Similarly, high stability was observed in 1R (99.4%; one monosomic among 80 plants) and $_{del}$1RS.1RL (98.7%; two monosomics among 80 plants). On the other hand, much lower transmission rates were observed for 1RS.$_{del}$1RL (91.9%; 13 monosomics among 80 plants) and ditelosomic 1RS (87.2%; two nullisomics and 15 monosomics among 74 plants).

**Table 3.** Pairing of rye chromosome arms during metaphase I and their transmission to the next generation.

| Introgression | Number of PMC | Chromosome Pairing | Number of Progeny | Transmission (%) |
|---|---|---|---|---|
| 1AS.1RL | | | 63 | 100.0 |
| 1RS.1BL | 50 | 1RS: 90.0; 1BL: 94.0 | 32 | 100.0 |
| 1RS.1DL | | | 30 | 100.0 |
| 2RS.2BL | | | 38 | 100.0 |
| 2BS.2RL | 49 | 2BS: 95.9; 2RL: 93.9 | 76 | 100.0 |
| 5RS.5BL | | | 71 | 98.6 |
| $_{del}$1RS.1RL | 60 | 1RS: 90.0; 1RL: 100.0 | 80 | 98.7 |
| 1RS.$_{del}$1RL | 59 | 1RS: 89.8; 1RL: 94.9 | 80 | 91.9 |
| 1R(1A) | 289 [1] | 1RS: 80.8; 1RL: 94.4 | 80 | 99.4 |
| t1RS | | | 74 | 87.2 |

[1] [25].

## 3. Discussion

Interspecific hybrids and allopolyploids frequently display genome instability. This instability may take different forms, such as parental chromosome competition. In *Lolium* × *Festuca* hybrids, the *Lolium* genome predominates: chromosomes of *Festuca* are slowly, but continuously replaced by those of *Lolium* in consecutive generations [26,27]. The mechanism underlying this phenomenon is not yet clear. Alternatively, genome instability may manifest itself as reduced meiotic pairing of one of the parental genomes, causing the higher univalency rate for the low pairing genome, and its gradual elimination from the hybrid. Wheat-rye hybrids are perhaps the best studied in this respect, with known cases of complete elimination of the rye genome, especially from octoploid triticales [14]. This phenomenon is also well documented in single chromosome wheat-rye addition lines [16]; as a general rule rye chromosomes pair poorly and are eliminated over time, which requires careful control of such lines to maintain their status. It has been argued that meiotic instability of wheat-rye hybrids is a consequence of mis-matched chromosome pairing control systems of the two parental species, and, especially, the effect of the *Ph1* system of wheat [17]. However, the *Ph1* system in wheat appears to impose strict stringency levels for crossover formation, and rye homologues in established triticale lines and wheat-rye addition lines are virtually the same. In stocks analysed in this study, rye chromosome arms always paired less frequently than their wheat counterparts. This fits well with the general pattern of rye chromosome behaviour when introgressed into wheat. As shown earlier by [16], rye chromosomes introgressed into wheat as disomic substitution or addition lines generate univalents in frequencies ranging from 1 to 19% (6% on average; Figure 5). Similarly, rye univalents are twice as frequent as those of wheat in tetraploid triticale, despite equal proportions of wheat and rye genomes [15].

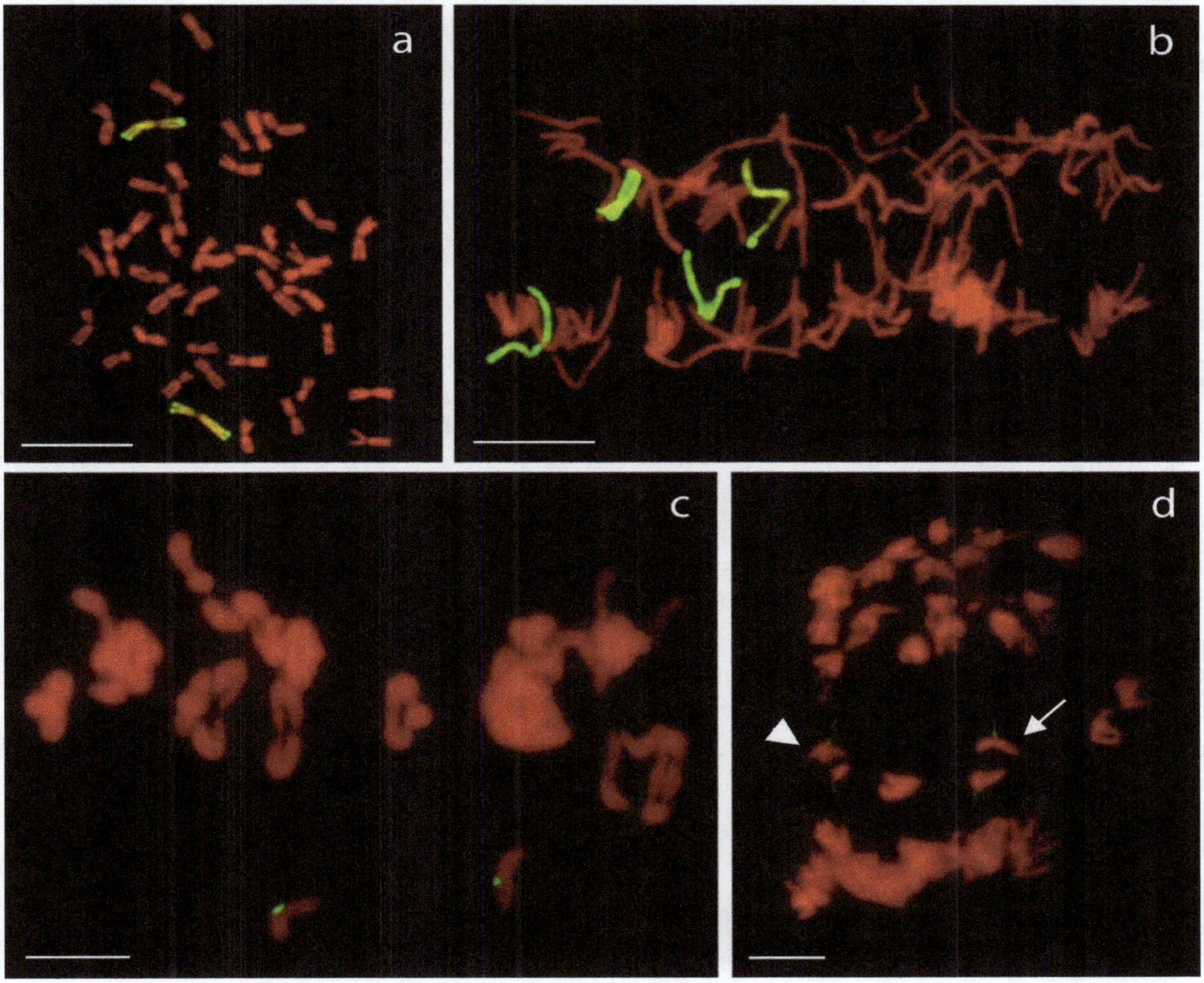


**Figure 5.** Mitotic (**a**,**b**) and meiotic (**c**,**d**) behavior of rye chromosome pair introgressed into wheat background. Rye chromosomes usually behaves perfectly normal during mitotic metaphase (**a**) and anaphase (**b**), while reduced pairing was observed during meiotic metaphase I (**c**). This may result in the separation of sister chromatids (arrow) or misdivision (in this case one arm of one chromatid is being separated from the rest of the chromosome; arrowhead) during anaphase I (**d**). Rye mitotic chromosomes were visualized using labelled genomic DNA of rye (labelled with FITC; green color), while rye-specific centromeric probe was used for meiotic preparations (labelled with FITC; green color). Scale bar 10 μm.

Naranjo [22] indicated that reduced metaphase I pairing of rye chromosomes in wheat might be a consequence of a lower probability of rye telomeres clustering into the leptotene bouquet at the onset of meiosis. Proper positioning of telomeres in the bouquet is believed to be prerequisite for the initiation of synapsis, and thus, regular chromosome pairing [7]. In our previous work [20], we observed a surprising agreement between the frequency of out-of-position telomeres at leptotene and in somatic nuclei on the one hand, and the failure of metaphase I pairing on the other. However, the situation is probably much more complex. There is chromosome movement at the onset of meiosis mediated by protein complexes of the inner nuclear envelope [28]. Such movement may be less frequent, or perhaps impossible, if the telomeres are not attached to the nuclear envelope (NE) already in the premeiotic interphase. The frequencies of non-attached rye telomeres observed in this study (0–12.5% for normal length chromosome arms) are again surprisingly similar to the frequencies of their out-of-bouquet positioning at the leptotene-zygotene transition as shown by Naranjo [19,22] and the frequencies of their pairing failure ([16] and our results). Interestingly, the transmission rate of rye introgressions in disomic centric wheat-rye translocations was high. This is because wheat chromosome arms in such translocations pair with normal frequencies assuring normal disjunction of the translocated chromosomes, their regular inclusion into the products of meiosis and then normal transmission of rye chromatin into subsequent generations. This is despite reduced (relative to diploid rye) pairing of rye chromosome arms. On the other hand, general instability of ditelosomics (such as t1RS analysed here) appears to be associated with lower frequencies of telomeres attached to the NE (here detected as telomeres in contact with NP). Correlation also exists between the out-positioning of rye telomeres in diploid rye nuclei observed here (0.66%) and average pairing failure per arm in a population of rye (0.21%) [29,30]. Thus, what appears to emerge from rather fragmentary data is that there is a direct link between somatic arrangement of chromosome arms, the leptotene bouquet and metaphase I pairing success.

The study of Naranjo [19] and our results indicate that there may be a relationship between chromosome arm length and the positioning of that arm's telomere at the telomere cluster/pole. Naranjo found that only 83.3% and 73.5% of telomeres of the short arms of chromosomes 1R and 6R, respectively, were located in the telomere cluster at the leptotene-zygotene transition, while none of 1RL telomeres and only 0.5% of 6RL telomeres were out of the telomere clusters [19]. Similarly, the lowest frequency of rye telomere clustering was for chromosome arm 5RS, the shortest rye chromosome arm, and the highest observed frequencies were for the arm 6RL, the longest in the karyotype [19,22,31]. This correlates well with our results, where telomeres of shorter chromosome arms were attached to the NE (or positioned at the telomere cluster/pole) less frequently than the telomeres of longer chromosome arms and significantly higher frequencies of pairing of long arms compared to short arms of chromosomes 1R, 2R and 5R in disomic addition lines (84.5 vs. 91.0, 64.8 vs. 96.8% and 31.0 vs. 79.0, respectively) [16]. Similarly, reduction in length of 1RS arm by deletion in the 1R introgression line caused reduction of its telomere attachment from 93.2% (in regular 1R introgression) to 46%.

## 4. Materials and Methods

### 4.1. Plant Material

The plant material consisted of a set of lines of hexaploid bread wheat (*Triticum aestivum* L., 2n = 6× = 42) cv. 'Pavon 76' with disomic (homozygous) introgressions of rye chromosomes or chromosome arms: substitution of rye chromosome 1R for wheat chromosome 1A [1R(1A)], ditelosomic addition line 1RS, a deletion line $_{del}$1RS.1RL where ca. proximal 36% of the short arm is missing, a deletion line 1RS.$_{del}$1RL where proximal ca. 30% of the long arm is missing (Figure 6), and centric wheat-rye chromosome translocations 1RS.1BL, 1RS.1DL, 1AS.1RL, 2RS.2BL, 2BS.2RL and 5RS.5BL. The telosomic line and all centric translocation lines were created by centric misdivision of complete normal chromosomes of rye and their wheat homoeologues; the deletion chromosomes were identified during selection of centric translocations [32]. As a control, we used a population of rye (*Secale cereale* L.) cv. Dankowskie Zlote. About 25 seedlings were used for each line.

Note on terminology: this manuscript uses the original terminology of Bridges for chromosome aberrations where "deficiency" is a loss of a terminal chromosome segment and "deletion" indicates a loss of an intercalary segment [33].

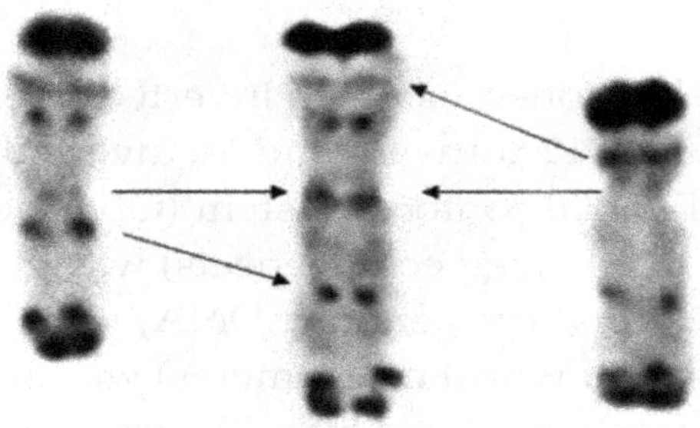

**Figure 6.** Three 1R chromosomes differing in the length of their arms. From left to right: $1RS._{del}1RL$ with deletion of about 30% of the 1RL arm (proximal part), regular 1R chromosome and $_{del}1RS.1RL$ with deletion of about 36% of 1RS arm (proximal part).

### *4.2. Isolation of Nuclei and Flow Sorting*

Seeds of the introgression lines were germinated in Petri dishes on moist filter paper at 25 °C in the dark. Root tips from young seedlings were collected and fixed in 2% (*v*/*v*) freshly prepared formaldehyde in meiocyte buffer A (15 mM PIPES, pH 6.8, 80 mM KCl, 20 mM NaCl, 0.5 mM EGTA, 2 mM EDTA, 0.15 mM spermine tetra-HCl, 0.05 mM spermidine tri-HCl, 1 mM dithiothreitol, 0.32 M sorbitol) for 20 min at 5 °C. After fixation, root tips were washed three times for 15 min at room temperature in meiocyte buffer A. Meristem tissue of root tips was cut and transferred into a 5 mL sample tube containing 400 μL of meiocytes buffer A, and homogenized using a Polytron PT1200 homogenizer (20,000 rpm/13 sec). The homogenate was filtered through 20 μm nylon mesh into a 5 mL polystyrene tube and stored on ice until used. The nuclear suspension was stained with 2 μg/mL DAPI (4′,6-diamidino-2-phenylindole). Nuclei in G1 phase of the cell cycle were identified and sorted using a FACSAria II SORP flow cytometer (BD Biosciences, San Jose, CA, USA) into a sample tube containing 10 μl of meiocyte buffer A [34]. About 50,000 nuclei at G1 were obtained from one sample prepared from 50 root tips. Using flow cytometry enabled us to have a sample of the nuclei exclusively from one phase of cell cycle (G1), rather than a mixture of cells from different phases, which could bias the results from different lines.

### *4.3. Probe Preparation and 3D-FISH*

Total genomic DNA of *S. cereale* L. was labelled with Texas Red or TRITC using a Nick Translation Kit (Roche Applied Science, Penzberg, Germany) according to manufacturer's instructions, and applied as a probe. Total genomic DNA of rye labelled the rye chromosome arms yellow, and their subtelomeric heterochromatin dark yellow. Both rye and wheat centromeres were visualized by an oligonucleotide probe based on the sequence of clone pHind258 [35] and directly labelled with Cy5. The telomeric probe was prepared using PCR and FITC-directly labelled nucleotides. Total genomic DNA of wheat was sheared to 200–500 bp fragments by boiling and used as blocking DNA at a ratio of 1:150 (probe/blocking DNA). As a control, we used rye nuclei with probes specific for telomeres and centromeres.

3D-FISH experiments were performed according to [36,37] with minor modifications. The suspension of flow-sorted nuclei was dripped onto a 22 × 22 mm coverslip and mixed with acrylamide solution (30% Acrylamide/Bis-acrylamide Mix Solution, ratio 29:1, Sigma-Aldrich, St. Louis, USA) in meiocyte buffer A, ammonium persulfate (20%) and sodium sulphate anhydrous (20%). The drop of solidifying nuclear suspension was covered with another coverslip and the "sandwich" was placed in an oven at 37 °C for 1 h for polymerization. Thereafter, the coverslip "sandwich" was gently separated by a razor blade. The coverslip with polymerized nuclei was washed in meiocyte buffer A in a small Petri dish for 3 × 5 min at room temperature. The coverslip was then placed on a new slide with a drop of hybridization mixture for FISH. The FISH mixture included probes, blocking DNA, 70% formamide and 2× saline sodium citrate (SSC). Hybridization was conducted according to [37]. The nuclei were counterstained with 1.5 μg/mL DAPI in Vectashield antifade mounting medium (Vector Laboratories, Burlingame, USA).

### *4.4. Image Acquisition and Analysis*

Probed nuclei were optically sectioned using an inverted laser spinning disk microscope (Axio Observer Z1, ZEISS) and ZEN Blue 2012 software and an inverted motorized microscope Olympus IX81 equipped with a Fluoview FV1000 confocal system (Olympus, Tokyo, Japan) and FV10-ASW software. Fluorescein (binding to telomere-specific repeats) was excited by a 488 nm line of an argon laser, while Texas Red and TRITC (total rye genomic DNA) were excited by He-Ne laser at 543 nm or 561 nm. Cy5 (labelling both rye and wheat centromeres) was excited at 639 nm. The excitation of DAPI (to visualize nuclear DNA) was performed by a 405 nm diode laser.

For each nucleus, 80–120 optical sections in 160 – 200 nm steps were taken and merged into a 3D model. Subsequent analyses were performed using Imaris 9.2 software (Bitplane, Oxford Instruments, Zurich, Switzerland). Imaris applications 'Contour Surface', 'Spot Detection' and '3D Measurement' were used for manual analysis of each nucleus. The volume and the centre of the nucleus (CN) were determined from the rendering of primary intensity of DAPI staining using the function 'Surfaces'. The 'Spot' function was used to mark the positions of centromeres (C) and telomeres (T). Distances between the centromeres of rye homologues (C-C) and between their telomeres (T-T) were measured using the 'Line' function.

Special attention was paid to the positioning of rye telomeres and centromeres relative to the nuclear envelope and to the (wheat) telomere cluster. No additional steps were taken to visualize the nuclear envelope: its position was inferred from the edge of the DAPI-stained chromatin. 'Display Adjustment' was used to adjust the channel contrast and thus, to improve the visualization of all analysed objects. Between 21 and 40 nuclei were analysed per genotype.

### *4.5. Chromosome Pairing and Transmission Rate*

Meiotic pairing of selected chromosomes was analysed using standard protocols for material collection, fixation and genomic in situ hybridization (GISH) described in detail previously [34]. Briefly, from each sampled flower, a portion of one anther was removed, fresh-squashed in a drop of acetocarmine and if the desired meiotic stage was present, the remaining anthers from the flower were fixed in a mixture of absolute ethyl alcohol and glacial acetic acid in proportion 3:1 at 37 °C for a week and stored at −20 °C until used. Chromosome pairing was analysed on squashed preparations using GISH. Rye chromosomes and chromosome arms were visualized with DIG-labelled total genomic rye DNA and anti-DIG FITC, with wheat DNA sheared to ca. 200–500 bp fragments as a blocking DNA at 1:150 ratio (probe: blocking DNA). Following hybridization, preparations were counterstained with 1.5% propidium iodide (PI) in VectaShield antifade (BioRad, Hercules, CA, USA), mounted and observed under a microscope. Individual arms of rye chromosome 1R were identified by the presence of an unlabelled NOR band on the short arm. Pairing frequencies of individual arms of chromosome 1R were taken from a previous study [25]. The same protocol was used on root-tip meristems to estimate transmission rates to further generations. Seeds were germinated on wet filter paper in Petri dishes, root tips were collected to ice water for 26–30 h and fixed in a mixture of absolute alcohol and glacial acetic acid (3:1) at 37 °C for seven days. GISH was done the same way as for meiotic preparations with a probe prepared from total genomic DNA of rye and unlabelled genomic DNA of wheat as a blocking DNA.

## 5. Conclusions

To conclude, our study indicates that reduced chromosome pairing of introgressed alien chromosomes (or chromosome arms) may be predetermined already in somatic cells, and is more systemic in nature: telomeres of rye chromosomes in wheat often fail to attach to the nuclear envelope, and less frequently assume proper positions within the telomere cluster. This may hamper their migration to the telomere bouquet at the onset of meiosis resulting in reduced synapsis and reduced metaphase I pairing. Consequently, reduced metaphase I pairing lowers the transmission rate of such chromosomes to successive generations, and thus, destabilizes the integrity of a hybrid genome.

**Supplementary Materials:**
**Video S1 (.mp4).** Visualization of nucleus of disomic deletion line $1RS._{del}1RL$ after 3D-FISH. Total genomic DNA of rye was labelled with TRITC using Nick translation (yellow color), centromeres of both wheat and rye chromosomes were visualized using oligonucleotide probe (red color), and telomere-specific sequence was PCR-labelled with FITC (green color). Nuclear DNA was counterstained with DAPI (blue color). Out-positioning of one rye chromosome arm is indicated by arrow. **Video S2 (.mp4).** Visualization of nucleus of disomic deletion line $_{del}1RS$. 1RL after 3D-FISH. Total genomic DNA of rye was labelled with TRITC using Nick translation (yellow color), centromeres of both wheat and rye chromosomes were visualized using oligonucleotide probe (red color), and telomere-specific sequence was PCR-labelled with FITC (green color). Nuclear DNA was counterstained with DAPI (blue color). Non-attachment of telomeres of both rye $_{del}1RS$ chromosome arms is indicated by arrows.

**Author Contributions:** Conceptualization, A.J.L. and D.K.; Methodology, K.P., V.K., A.J.L. and D.K.; Formal Analysis, K.P., V.K., J.V., M.D., D.P., O.Š. and D.K.; Investigation, K.P., V.K., A.J.L., C.F. and D.K.; Resources, A.J.L., J.V., M.S., G.J., O.Š., J.Š. and J.D.; Data Curation, M.D. and D.K.; Writing—Original Draft Preparation, K.P., V.K., A.J.L. and D.K.; Writing— Review & Editing, G.J., J.Š., M.S., O.Š. and J.D.; Visualization, K.P., A.J.L. and D.K.; Supervision, D.K.; Project Administration, D.K.; Funding Acquisition, J.Š., J.D. and D.K.

**Acknowledgments:** We would like to thank Radka Tušková for excellent technical support.

## Abbreviations

| | |
|---|---|
| 3D-FISH | Three-dimensional fluorescent in situ hybridization |
| CN | Centre of the nucleus |
| CT | Chromosome territory |
| GISH | Genomic in situ hybridization |
| NE | Nuclear envelope |
| NOR | Nucleolar organizing region |
| NP | Nuclear periphery |
| PMC | Pollen mother cell |

## Appendix A

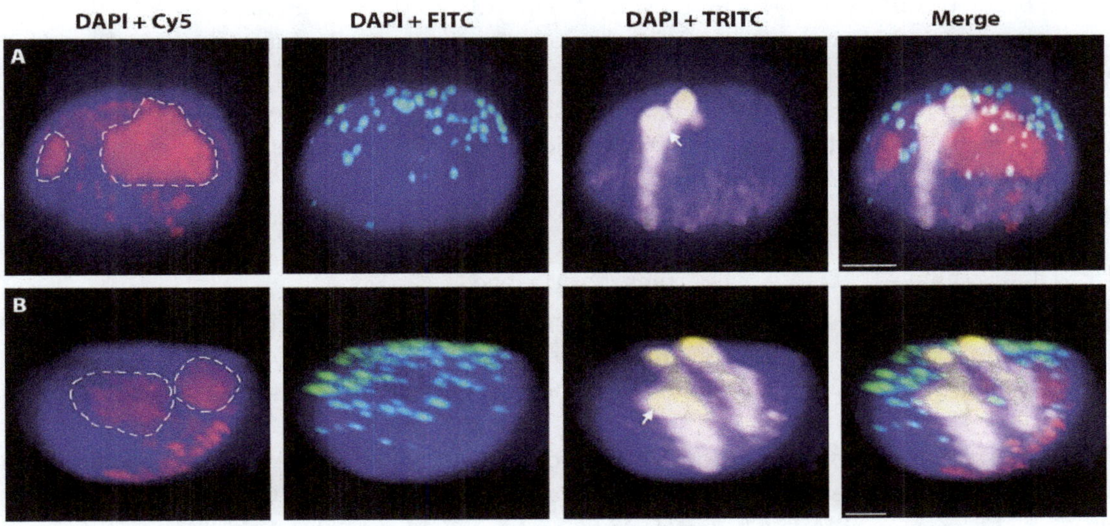


**Figure A1.** Non-attachment of rye telomeres. Nuclei with a pair of telocentric 1RS rye chromosomes (**A**) and a pair of 1R rye chromosomes (**B**) after 3D-FISH. Total genomic DNA of rye was labelled with TRITC using Nick translation (yellow color), centromeres of both wheat and rye chromosomes were visualized using oligonucleotide probe (red color), and telomere-specific sequence was PCR-labelled with FITC (green color). Nuclear DNA was counterstained with DAPI (blue color). Rye telomeres without visual contact to nuclear envelope are indicated by arrows. Nucleoli are indicated by white dashed lines. Scale bar 5 μm.

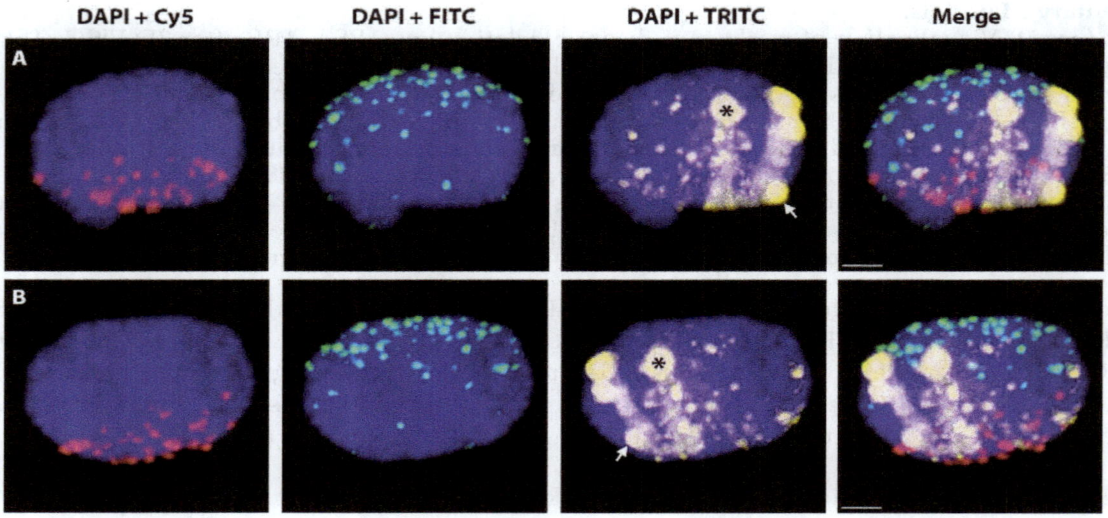


**Figure A2.** Combination of non-attachment and out-positioning of rye telomeres. A nucleus ((**A**,**B**) is the same nucleus from different angles) with a pair of homologous rye 1RS. $_{del}$1RL chromosomes after 3D-FISH. Total genomic DNA of rye was labelled with TRITC using Nick translation (yellow color), centromeres of both wheat and rye chromosomes were visualized using oligonucleotide probe (red color), and telomere-specific sequence was PCR-labelled with FITC (green color). Nuclear DNA was counterstained with DAPI (blue color). Rye telomere without visual contact to nuclear envelope is indicated by asterisk, rye telomere positioned out of the telomere pole (at the centromere pole) is indicated by arrow. Nucleoli are indicated by white dashed lines. Scale bar 5 μm.

## References


1. Cremer, T.; Cremer, C. Chromosome territories, nuclear architecture and gene regulation in mammalian cells. *Nat. Rev. Genet.* **2001**, *2*, 292–301. [CrossRef] [PubMed]
2. Fritz, A.J.; Barutcu, A.R.; Martin-Buley, L.; van Wijnen, A.J.; Zaidi, S.K.; Imbalzano, A.N.; Lian, J.B.; Stein, J.L. Chromosomes at Work: Organization of Chromosome Territories in the Interphase Nucleus. *J. Cell. Biochem.* **2016**, *117*, 9–19. [CrossRef] [PubMed]
3. Fransz, P.; de Jong, J.H.; Lysak, M.; Castiglione, M.R.; Schubert, I. Interphase chromosomes in Arabidopsis are organized as well defined chromocenters from which euchromatin loops emanate. *Proc. Natl. Acad. Sci. USA* **2002**, *9*, 14584–14589. [CrossRef] [PubMed]
4. Tiang, C.L.; He, Y.; Pawlowski, W.P. Chromosome Organization and Dynamics during Interphase, Mitosis, and Meiosis in Plants. *Plant Physiol.* **2012**, *158*, 26–34. [CrossRef]
5. Rabl, C. Über Zellteilung. *Morph. Jahrb.* **1885**, *10*, 214–330.
6. Dong, F.G.; Jiang, J.M. Non-Rabl patterns of centromere and telomere distribution in the interphase nuclei of plant cells. *Chromosome Res.* **1998**, *6*, 551–558. [CrossRef] [PubMed]
7. Dawe, R.K. Meiotic chromosome organization and segregation in plants. *Annu. Rev. Plant Physiol. Plant Mol. Biol.* **1998**, *49*, 371–395. [CrossRef] [PubMed]
8. Moens, P.B.; Bernei-Moens, C.; Spyropoulos, B. Chromosome core attachment to the meiotic nuclear envelope regulates synapsis in Chloealtis (Orthoptera). *Genome* **1989**, *32*, 601–610. [CrossRef]
9. Curtis, C.A.; Lukaszewski, A.J.; Chrzastek, M. Metaphase I pairing of deficient chromosomes and genetic mapping of deficiency breakpoints in common wheat. *Genome* **1991**, *34*, 553–560. [CrossRef]
10. Mujeeb-Kazi, A. Intergeneric crosses: Hybrid production and utilization. In *Utilizing Wild Grass Biodiversity in Wheat Improvement, 15 Years of Wide cross Research at CIMMYT*; Mujeeb-Kazi, A., Hettel, G.P., Eds.; CIMMYT: Texcoco, Mexico, 1995; p. 140.
11. Friebe, B.; Jiang, J.; Raupp, J.W.; Mclntosh, R.A.; Gill, B.S. Characterization of wheat-alien translocations conferring resistance to diseases and pests: Current status. *Euphytica* **1996**, *91*, 59–87. [CrossRef]
12. Molnár-Láng, M. The crossability of wheat with rye and other related species. In *Alien Introgression in Wheat: Cytogenetics, Molecular Biology, and Genomics*; Molnár-Láng, M., Ceoloni, C., Doležel, J., Eds.; Springer International Publishing: Chem, Switzerland, 2015; pp. 103–120.

13. Nishiyama, I. Cytogenetic studies in Avena, IX New synthetic oats in the progenies of induced decaploid interspecific hybrids. *Jpn. J. Genet.* **1962**, *37*, 118–130. [CrossRef]
14. Tsunewaki, K. Genetic studies on a 6x derivative from an 8x triticale. *Can. J. Genet. Cytol.* **1964**, *6*, 1–11. [CrossRef]
15. Lukaszewski, A.J.; Apolinarska, B.; Gustafson, J.P.; Krolow, K.D. Chromosome-pairing and aneuploidy in tetraploid triticale 1. Stabilized karyotypes. *Genome* **1987**, *29*, 554–561. [CrossRef]
16. Orellana, J.; Cermeno, M.C.; Lacadena, J.R. Meiotic pairing in wheat-rye addition and substitution lines. *Can. J. Genet. Cytol.* **1984**, *26*, 25–33. [CrossRef]
17. Lukaszewski, A.J.; Gustafson, J.P. Cytogenetics of Triticale. *Plant Breed. Rev.* **1987**, *5*, 41–93.
18. Murphy, S.P.; Bass, H.W. The maize (*Zea mays*) *desynaptic* (*dy*) mutation defines a pathway for meiotic chromosome segregation, linking nuclear morphology, telomere distribution and synapsis. *J. Cell Sci.* **2012**, *125*, 3681–3690. [CrossRef] [PubMed]
19. Naranjo, T. Dynamics of Rye Telomeres in a Wheat Background during Early Meiosis. *Cytogenet. Genome Res.* **2014**, *143*, 60–68. [CrossRef]
20. Pernickova, K.; Linc, G.; Gaal, E.; Kopecky, D.; Samajova, O.; Lukaszewski, A.J. Out-of-position telomeres in meiotic leptotene appear responsible for chiasmate pairing in an inversion heterozygote in wheat (*Triticum aestivum* L.). *Chromosoma* **2019**, *128*, 31–39. [CrossRef]
21. Schlegel, R.; Melz, G.; Nestrowicz, R. A universal reference karyotype in rye, *Secale cereale* L. *Theor. Appl. Genet.* **1987**, *74*, 820–826. [CrossRef]
22. Naranjo, T. Variable Patterning of Chromatin Remodeling, Telomere Positioning, Synapsis, and Chiasma Formation of Individual Rye Chromosomes in Meiosis of Wheat-Rye Additions. *Front. Plant Sci.* **2018**, *9*. [CrossRef]
23. Gill, B.S.; Friebe, B.; Endo, T.R. Standard karyotype and nomenclature system for description of chromosome bands and structural abberations in wheat (*Triticum aestivum*). *Genome* **1991**, *34*, 830–839. [CrossRef]
24. Doležel, J.; Greilhuber, J.; Lucretti, S.; Meister, A.; Lysák, M.A.; Nardi, L.; Obermayer, R. Plant genome size estimation by flow cytometry: Inter-laboratory comparison. *Ann. Bot.* **1998**, *82*, 17–26. [CrossRef]
25. Lukaszewski, A.J. Unexpected behavior of an inverted rye chromosome arm in wheat. *Chromosoma* **2008**, *117*, 569–578. [CrossRef]
26. Kopecky, D.; Loureiro, J.; Zwierzykowski, Z.; Ghesquiere, M.; Dolezel, J. Genome constitution and evolution in *Lolium x Festuca* hybrid cultivars (Festulolium). *Theor. Appl. Genet.* **2006**, *113*, 731–742. [CrossRef]
27. Zwierzykowski, Z.; Kosmala, A.; Zwierzykowska, E.; Jones, N.; Joks, W.; Bocianowski, J. Genome balance in six successive generations of the allotetraploid *Festuca pratensis x Lolium perenne*. *Theor. Appl. Genet.* **2006**, *113*, 539–547. [CrossRef] [PubMed]
28. Alleva, B.; Smolikove, S. Moving and stopping: Regulation of chromosome movement to promote meiotic chromosome pairing and synapsis. *Cell* **2017**, *8*, 613–624. [CrossRef] [PubMed]
29. Schlegel, R.; Mettin, D. Studies on intraindividual and interindividual variation of chromosome-pairing in diploid and tetraploid populations. 2. interindividual variation. *Biol. Zbl.* **1975**, *94*, 703–715.
30. Schlegel, R. *Rye: Genetics, Breeding, and Cultivation*; CRC Press: Boca Raton, FL, USA, 2014; p. 359.
31. Naranjo, T.; Valenzuela, N.T.; Perera, E. Chiasma Frequency Is Region Specific and Chromosome Conformation Dependent in a Rye Chromosome Added to Wheat. *Cytogenet. Genome Res.* **2010**, *129*, 133–143. [CrossRef] [PubMed]
32. Lukaszewski, A.J. Behavior of centromeres in univalents and centric misdivision in wheat. *Cytogenet. Genome Res.* **2010**, *129*, 97–109. [CrossRef]
33. Bridges, C.B. Deficiency. *Genetics* **1917**, *2*, 445–465.
34. Vrana, J.; Simkova, H.; Kubalakova, M.; Cihalikova, J.; Dolezel, J. Flow cytometric chromosome sorting in plants: The next generation. *Methods* **2012**, *57*, 331–337. [CrossRef] [PubMed]
35. Ito, H.; Nasuda, S.; Endo, T.R. A direct repeat sequence associated with the centromeric retrotransposons in wheat. *Genome* **2004**, *47*, 747–756. [CrossRef] [PubMed]
36. Phillips, D.; Nibau, C.; Ramsay, L.; Waugh, R.; Jenkins, G. Development of a Molecular Cytogenetic Recombination Assay for Barley. *Cytogenet. Genome Res.* **2010**, *129*, 154–161. [CrossRef] [PubMed]
37. Howe, E.S.; Murphy, S.P.; Bass, H.W. Three-Dimensional Acrylamide Fluorescence in Situ Hybridization for Plant Cells. In *Plant Meiosis. Methods and Protocols*; Pawlowski, W., Grelon, M., Armstrong, S., Eds.; Springer International Publishing AG: Dordrecht, The Netherlands, 2014; pp. 53–66.

# 12

# Transcriptional Dynamics of Grain Development in Barley (*Hordeum vulgare* L.)

**Jianxin Bian [1,†], Pingchuan Deng [1,†], Haoshuang Zhan [1], Xiaotong Wu [1], Mutthanthirige D. L. C. Nishantha [1], Zhaogui Yan [2], Xianghong Du [1], Xiaojun Nie [1,*] and Weining Song [1,3,*]**

1 State Key Laboratory of Crop Stress Biology in Arid Areas, College of Agronomy and Yangling Branch of China Wheat Improvement Center, Northwest A&F University, Yangling, Shaanxi 712100, China; brian1791@nwafu.edu.cn (J.B.); dengpingchuan@nwsuaf.edu.cn (P.D.); zhanhaoshuang@nwsuaf.edu.cn (H.Z.); wuxiaotong@nwsuaf.edu.cn (X.W.); mdlcnishantha@gmail.com (M.D.L.C.N.); xianghongdu@nwsuaf.edu.cn (X.D.)

2 Huazhong Agricultural University, Wuhan 430070, China; gyan@mail.hzau.edu.cn

3 Joint Research Center for Agriculture Research in Arid Areas, Yangling, Shaanxi 712100, China

* Correspondence: small@nwsuaf.edu.cn (X.N.); sweining2002@yahoo.com (W.S.).

† These authors contributed equally to this work.

**Abstract:** Grain development, as a vital process in the crop's life cycle, is crucial for determining crop quality and yield. However, the molecular basis and regulatory network of barley grain development is not well understood at present. Here, we investigated the transcriptional dynamics of barley grain development through RNA sequencing at four developmental phases, including early prestorage phase (3 days post anthesis (DPA)), late prestorage or transition phase (8 DPA), early storage phase (13 DPA), and levels off stages (18 DPA). Transcriptome profiling found that pronounced shifts occurred in the abundance of transcripts involved in both primary and secondary metabolism during grain development. The transcripts' activity was decreased during maturation while the largest divergence was observed between the transitions from prestorage phase to storage phase, which coincided with the physiological changes. Furthermore, the transcription factors, hormone signal transduction-related as well as sugar-metabolism-related genes, were found to play a crucial role in barley grain development. Finally, 4771 RNA editing events were identified in these four development stages, and most of the RNA editing genes were preferentially expressed at the prestore stage rather than in the store stage, which was significantly enriched in "essential" genes and plant hormone signal transduction pathway. These results suggested that RNA editing might act as a 'regulator' to control grain development. This study systematically dissected the gene expression atlas of barley grain development through transcriptome analysis, which not only provided the potential targets for further functional studies, but also provided insights into the dynamics of gene regulation underlying grain development in barley and beyond.

**Keywords:** Barley; Grain development; Transcriptional dynamics; RNA editing; RNA-seq

## 1. Introduction

Grain development is one of the most essential processes for the life cycle of plants, and especially for crops, which is not only critical for colonization of the environment for plant survival, but also provides the main food source for human beings [1–4]. It is a very complex biological process involving multiple metabolic regulation pathways, which can be further divided into two major phases, embryogenesis and maturation. During the embryogenesis process, embryonic cells divide and go through characteristic stages of development, while maturation mainly involves accumulation of organic materials such as protein, lipids, and carbohydrates [4–7]. A better understanding of the molecular mechanisms of grain development will contribute to crop improvement, including seed formation, morphogenesis, and storage reserve accumulation. During the last few years, a larger number of studies have been conducted to investigate the specific genes or pathways controlling grain development in different species, particularly in model organisms such as rice [8] and Arabidopsis [3,5,9]. These studies have provided important information on both of the regulators of transcription and the genes involved in grain development.

Barley (*Hordeum vulgare*), a member of the Poaceae family, was domesticated thousands of years ago and has been utilized mainly for animal feed, malting, and the brewing industry, and as a human food source [1,10]. Because of its diploid nature and rich genetic diversity, barley is also a well-studied crop regardless of genetics or genomics and once acted as a model plant for Triticeae research. Since the grain is the agriculturally most important organ of crops, many studies have been carried out on barley grain development [1,6,7,11–13]. Preliminary transcriptomic analyses of barley seeds have provided important information regarding tissue-specific metabolic pathways and the regulatory network controlling grain development, including storage reserve accumulation [6], photosynthesis [7], hormone biosynthesis [6,14], programmed cell death [12], sucrose transport [13], and zinc trafficking [15]. Particularly, a comparative study of transcript and metabolite profiling at both temporal and spatial levels dissected that glumes acted as important transitory resource buffers during barley endosperm filling [16]. Furthermore, the differentially expressed genes (DEGs) involved in barley grain development through RNA-seq are also reported and available in barleyGenes—Barley RNA-seq Database (https://ics.hutton.ac.uk/barleyGenes/about_material.html). However, the transcriptomic dynamics of barley grain development are not well understood, and in particular, there is little known about the regulatory network at present. Recently, the high-quality reference sequence of the barley genome was completed, which provides the unprecedented opportunity to investigate the dynamic of the barley grain development at the whole transcriptome level [17]. Here, we performed a study of mRNA-seq analysis at four stages spanning important developmental phases of seed filling with the purpose to the transcriptional dynamics and regulatory process of grain development in barley.

## 2. Results

### 2.1. RNA-Seq Analysis of Grain Development in Barley

In this study, the dynamics of mRNA abundance were systematically investigated at four important stages (3, 8, 13, and 18 DPA) of grain development in barley (Figure S1). Totally, 52.69 Gb of raw data was obtained for all of the samples, with the average of about 65 million pair-end reads with 100 bp in size for each sample (Table S1). After removing adaptor sequences and low-quality reads (reads with N and empty reads), there were around 521 million clean reads remaining, which represented 98.97% of the raw data. After quality filter, all of the clean reads were mapped to the barley reference genome [17]. Results showed that on average, 55.10 million paired-end reads (84.53%) could be uniquely mapped to the reference genome [17], with the range from 53.86 (82.75%) to 56.28 million reads (86.29%) (Figure S2A). Out of 39,734 annotated protein-coding genes in the barley genome, 19,165 (48.23%) genes were found to be expressed during grain development (Table S1), which was consistent with a previous study in Arabidopsis (~44%) and hexaploid wheat (~55%) [2,18].

To determine whether this gene expression correlated with developmental stages, reads numbers were firstly normalized to RPKM value, and then were subjected to the usual correlation coefficient ($R^2$) and hierarchical clustering analysis [19]. We found that two biological replicates of all samples showed consistent determinations of transcript abundance with a coefficient ($R^2$) greater than 0.93 (Figure 1A). In addition, gene expression correlations between different stages were compared. Results found that the coefficients of 3 DPA (stage01) and 8 DPA (stage02) were 0.73 ($R^2$ > 0.73) and those of 13 DPA (stage03) and 18 DPA (stage04) were higher with the value of 0.89. However, the coefficient between 13 DPA (stage03) and 8 DPA (stage02) was only 0.10, indicating a large change in gene expression pattern occurred during the process. It may be the transition from prestorage phase to storage phase, which was consistent with a previous study finding that the most dramatic transcriptional and physiological changes occurred during transition from the prestorage to the storage phase [6].

Similar stage expression patterns were supported by correlation dendrogram analysis (Figure 1B). The correlation dendrogram demonstrated that the four stages were clearly clustered into two groups, namely 3 DPA (stage01) and 8 DPA (stage02) grouped together, and 13 DPA (Stage03) and 18 DPA (Stage04) clustered into another group. Then, we also used the publicly available barley grain RNA-seq data (5 DPA and 15 DPA) to perform correlation analysis. Results found that 3 DPA and 8 DPA showed high correlation coefficient with 5 DPA and 13 DPA and 18 DPA showed high correlation coefficient with 15 DPA, which also could be clustered into two groups (Figure S3). Overall, the correlation analyses results were in agreement with the developmental order as expected, which were consistent with the previous study [7].

Furthermore, we compared the changes of gene expression abundance at different development stages. The total number of expressed genes in each sample varied from 13,290 (stage04) to 16,532 (stage01) (Table S2) and approximately 63.79% (11,826/18,540) of them were expressed across all of these four stages (Figure 1C). Interestingly, the majority of the genes (80.35%) exhibited peak expression at the prestorage phase (stage01 and stage02) rather than the storage phase (stage03 and stage04) (Figure 1D). GO (Gene ontology)analysis identified the peak expressed genes were significantly enriched in molecular functions at each stage. For example, at the prestorage phase, the peak genes were enriched in terms of cell cycle, DNA, RNA, peptidase biosynthesis, and modification such as cell cycle process (GO: 0022402), cell division (GO:0051301), nuclease activity (GO:0004518), DNA metabolic process (GO: 0006259), tRNA processing (GO: 0008033), double-stranded DNA binding (GO:0003690), microtubule motor activity(GO:0003777), lipoprotein biosynthetic process (GO:0042158), fatty acid biosynthetic process (GO:0006633) (Figure 1E) (Table S3). However, at the storage phase, gene sets were enriched in terms of matter accumulation, such as nutrient reservoir activity (GO: 0045735), catalytic activity (GO: 0003824), 6-phosphofructokinase activity (GO: 0003872), fructose 6-phosphate metabolic process (GO: 0006002), response to temperature stimulus (GO:0009266), response to heat(GO:0009408), disaccharide metabolic process (GO:0005984), protein transporter activity (GO:0008565), carbohydrate derivative catabolic process (GO:1901136) (Figure 1F) (Table S4).

Genes displaying enriched expression patterns in a given developmental stage or tissue are important for understanding the specialized processes within these stages or tissues. As described previously, an empirical cutoff value for positively expressed genes (FPKM $\geq$ 1) was used to detect stage-specific expression candidates [20].

In our study, we identified 2158 stage-enriched genes, of which stage01 possessed the highest number (1183), followed by stage02 (538), stage04 (254), and stage03 (183) (Figure 1C). GO enrichment results showed the stage-specific expression genes involved in DNA metabolic process (GO:0006259), DNA repair (GO:0006281), reproductive process (GO:0022414), RNA processing (GO:0006396) for stage01, transport (GO:0006810), single-organism transport (GO:0044765), lipid metabolic process (GO:0006629), carbohydrate binding (GO:0030246) for stage02, and aromatic compound biosynthetic process (GO:0019438), organic cyclic compound biosynthetic process (GO:1901362), biosynthetic process (GO:0009058), organic substance biosynthetic process (GO:1901576) for stage03, as well as serine-type endopeptidase inhibitor activity (GO:0004867), transferase activity, transferring hexosyl groups (GO:0016758) for stage04 (Table S4), which represented the grain developmental program and biological process.

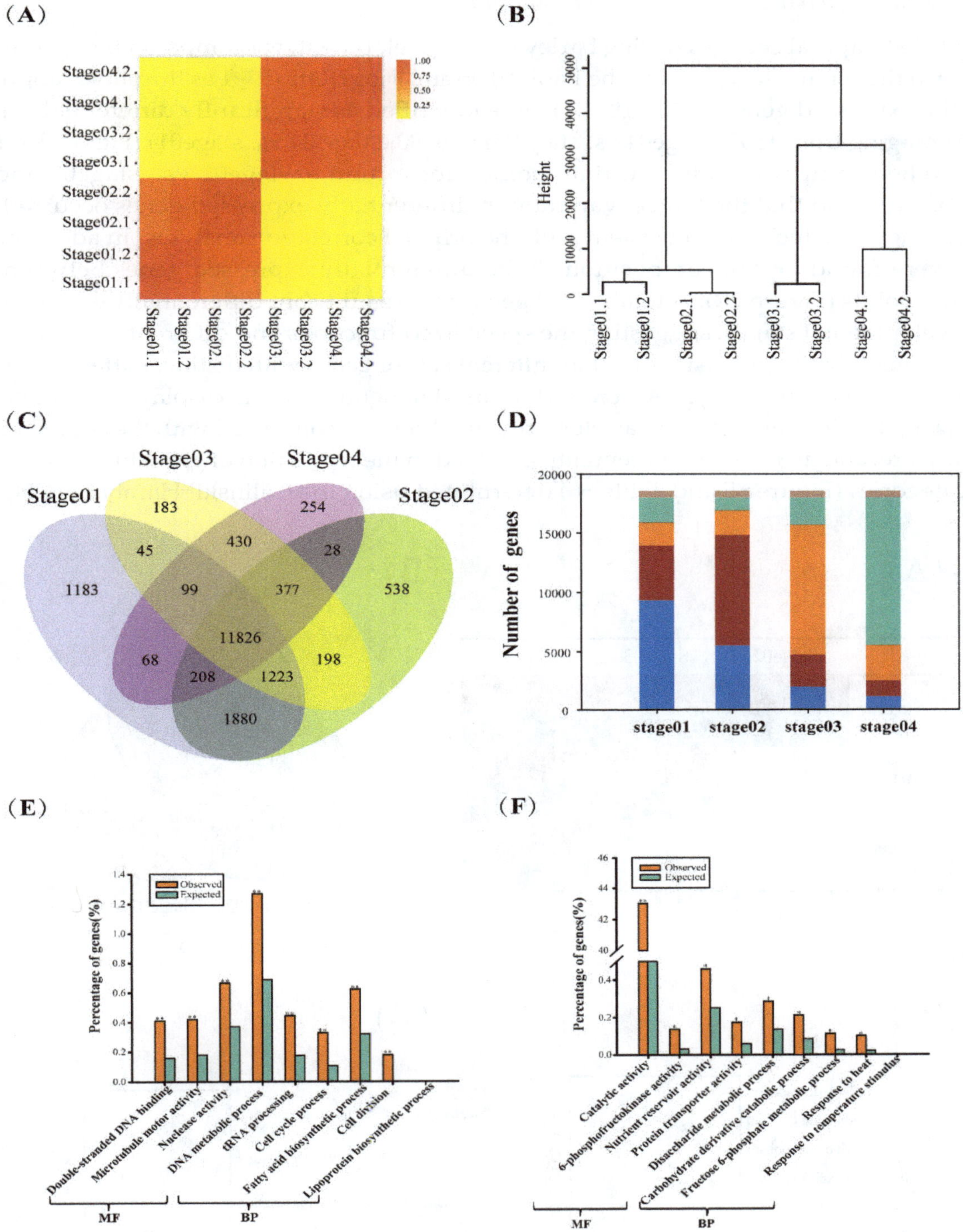


**Figure 1.** Global characterization of gene expression patterns during the four stages of barley grain development. (**A**) Correlation coefficients between gene expression data sets from two biological duplicates and stages. (**B**) Cluster dendrogram showing global relationships of gene expression in different stages. The branch length indicates the degree of variance. (**C**) Venn diagram analyses of stage-specific genes in barley. (**D**) Overview of genes with different activity degree among the four stages. The different number suggests different activity ranking at each stage and the activity degree decreases as the number increases. For example, '1' indicates gene showed the most activity at this stage and '4' suggests the most inactive stage. (**E**) to (**F**) Functional categories of genes showing peak expression at the prestorage phase (stage01 and stage02) (**E**) and storage phase (stage03 and stage04). (**F**) Padj-adjusted $p$ values, * $p < 0.05$ and ** $p < 0.01$. Observed, numbers of genes observed in this study; Expected, numbers of genes in this same category of annotated barley gene models.

*2.2. Transcriptome Dynamics of Grain Development in Barley*

To capture temporal changes during barley grain development, we compared the gene expression levels between these four stages. Using the Padj < 0.05 and | log2Ratio | ≥1 as threshold, approximately 51.38% of the expressed genes (9527/18,541) were identified as significantly differential expressions, with values ranging from 1532 (stage04 vs. stage03) to 6380 (stage 04 vs. stage01) (Figure 2A, Figure S4, Table S5). When comparing adjacent time points (for example, stage02 vs. stage01 and stage03 vs. stage02), we found that the largest variation in differentially expressed genes occurred between stage03 and stage02, which was consistent with the original correlation analyses. In addition, obvious variations were found on the composition of the differentially expressed genes between any two adjacent time points (Figure 2B). A total of 312 genes were significant differentially expressed at all of the four developmental stages, suggesting the specialized functions and differentiation during barley grain development that might be involved in different sets of genes with distinct patterns of expression. All the differentially expressed genes were further used to define clusters displaying distinct temporal patterns during development. A K-mean clustering method was conducted with the squared Euclidean distance measure, and all the genes potentially involved in the regulation of grain filling were classified into nine categories (Figure S5 and Table S6) determined using the Calinski–Harabasz (CH) index.

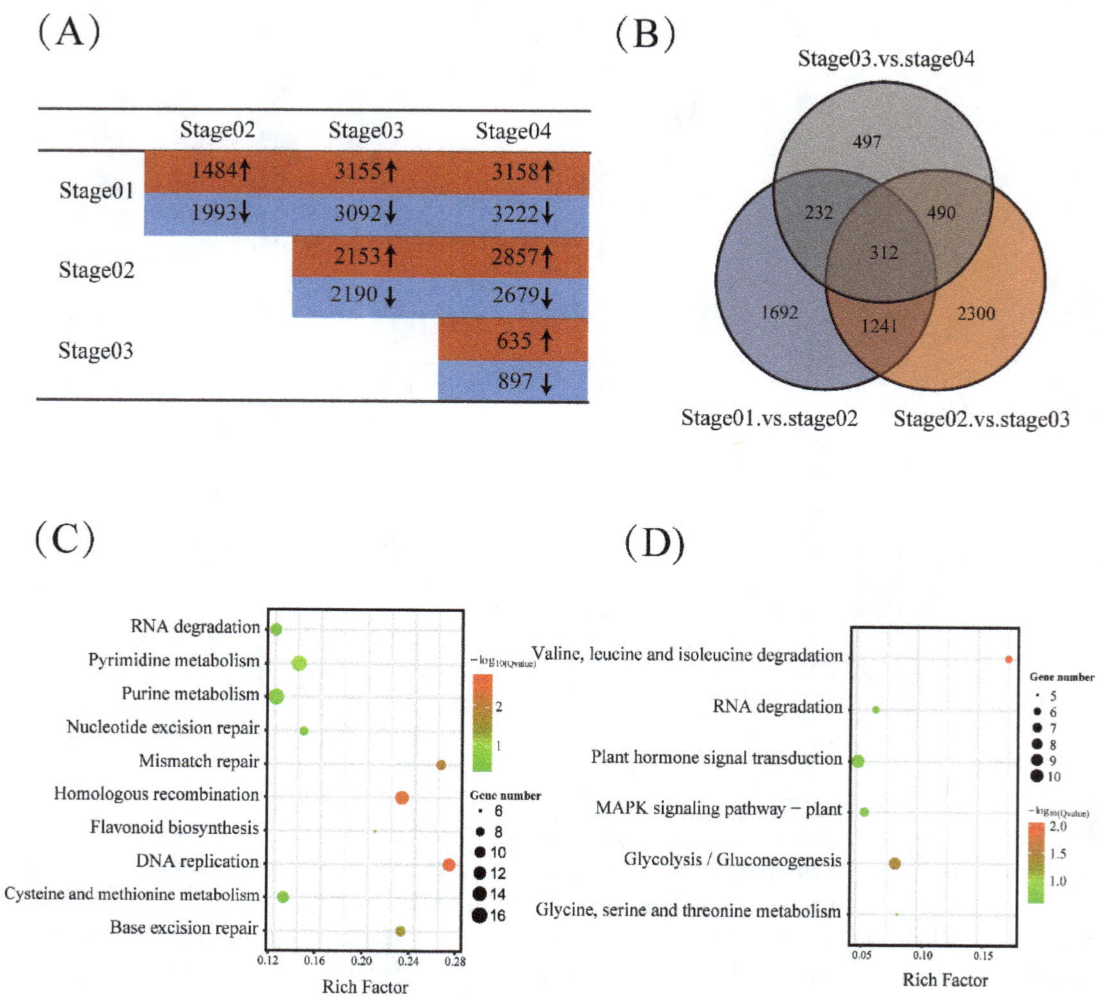


**Figure 2.** Differential expression genes during barley grain development. (**A**) Number of genes showing up- or downregulated expression during barley grain development (Padj < 0.05 and | log2Ratio | ≥1). (**B**) Venn diagram analyses of genes that were differentially expressed between any two consecutive stages in barley. (**C**) KEGG enrichment analysis of DEGs in cluster V during barley grain development. (**D**) KEGG enrichment analysis of DEGs in cluster VII.

Based on the growth characteristics, starch accumulation patterns, and metabolite profiles, the major events of barley endosperm development were classified into four main stages: prestorage phase (0–5 DPA), late prestorage or transition phase (6–10 DPA), early storage phase (11–15 DPA), and levels off stages (16–20 DPA) [6]. The first four clusters (I–IV) of differentially expressed genes were preferentially expressed at these four development stages, respectively (Table S6). Cluster I comprised 1823 genes with the peak expression at the first stage and then declined sharply at later stages. This cluster showed a clear enrichment of genes involved in protein, chromosome, and cell wall pathways such as protein dimerization activity (GO:0046983), chromosome organization (GO:0051276), cell wall organization (GO:0071555). The transition phase is characterized mainly by an initial accumulation of starch. Expecting for some pathways involved in preparing for starch accumulation such as protein and RNA, genes preferentially expressed at this phase (cluster II, 1311 genes) were also enrichment for some carbohydrate metabolism pathways such as photosynthesis (GO:0015979), transporter activity (GO:0005215), and carbohydrate binding (GO:0030246). Expression of genes in cluster III group (472 genes) activated significant enrichment of genes involved in intracellular transport (GO:0046907), macromolecule localization (GO:0033036), phosphorelay signal transduction system (GO:0000160), pyrophosphatase activity (GO:0016462), and nutrient reservoir activity (GO:0045735). At the levels off stage, there was a significant increase in expression (cluster IV, 746 genes) of genes related to multicellular organism development (GO:0007275) and developmental process (GO:0032502). Cluster V group contained the largest number of differentially expressed genes (2,217 genes), which showed an increased expression at the prestorage phase (stage01 and stage02). In preparation for seed filling, genes related to protein, RNA, cell organization, and cell division were activated and significantly enriched for cell cycle (GO:0007049), cytokinesis (GO:0000910), lipid metabolic process (GO:0006629), cellulose synthase activity (GO:0016759), RNA processing (GO:0006396), nucleosome organization (GO:0034728), chromosome organization (GO:0051276), cellular component biogenesis (GO:004085), protein metabolic process (GO:0019538), organic acid metabolic process (GO:0006082). The KEGG pathway analysis found they were significantly enriched in the DNA replication, mismatch repair pathway (Figure 2C). Transcripts belonging to cluster VII with a peak of expression at the storage phases (stage03 and stage04) were mainly involved in nutrient reservoir activity (GO:0045735), enzyme inhibitor activity (GO:0004857), peptidase inhibitor activity (GO:0030414). Compared to cluster V, genes in this cluster were significantly enriched in glycolysis/gluconeogenesis and plant hormone signal transduction pathway (Figure 2D). Other functional classes, such as stress, transport, cell wall, protein, and RNA, were also over-represented in cluster VI, cluster VIII, and cluster IX (Table S6).

### 2.3. *Functional Analyses of Differentially Expressed Transcription Factors during Barley Grain Development*

Transcription factors (TFs) are one of the main factors for signal transduction and regulating gene expression, which play important roles in controlling the growth, development, and stress response in plants. We further identified the transcription factor gene among the identified differentially expressed genes. Results found that there were 606 transcription factors in these DEGs, which could be assigned into 61 families based on the presence of conserved domains. Among them, MYB family members were the most prominent (50), followed by AP2/ERF-ERF (44), NAC (42), bHLH (38), bZIP (35), and MYB-related (30) (Figure 3A and Table S7). MYB proteins are key factors in regulatory networks controlling development and metabolism under biotic and abiotic stress [21]. In this dataset, 17 (34%) MYB TFs members abundantly expressed at the prestorage phase, while another 10 genes (20%) were preferentially expressed at the storage phase (Figure 3B). The differential function of MYBs contributed to controlling development and stress response, which might be the reason the MYB family holds the largest TFs family during grain development. In addition, MADS transcription factor family, known to play crucial roles in flower organ development [22], were found to be abundantly expressed (20 members), implying an unexplored distinct role of MADS box TFs during barley grain development. On the other hand, only 22 TFs (36.73%) were expressed at the level of FPKM $\geq$50, indicating that most TFs were not highly expressed in grain (Table S7). Interestingly, two groups of differently expressed TFs with the highest number exhibited opposite dynamic expression trends from early to late stages: expression levels of 150 gene (cluster V, 24.75%) were highly expressed at prestage (stage01 and stage02), while in contrast, 72 genes in cluster VIII (11.88%) showed low expression at the prestage but highly expressed at the storage phases (stage03 and stage04). Some transcription factors related to meristem function and cell division such as HB (Homeobox), CPP, FAR1 were found in prestage (Figure 3B). Furthermore, 44.44% (4) of the ARFs (auxin response factors), responding to phytohormone auxin and important in determining plant architecture, were identified and their expression levels decreased gradually from early to late stages. This suggests that ARFs play important roles in early stages of barley grain development.

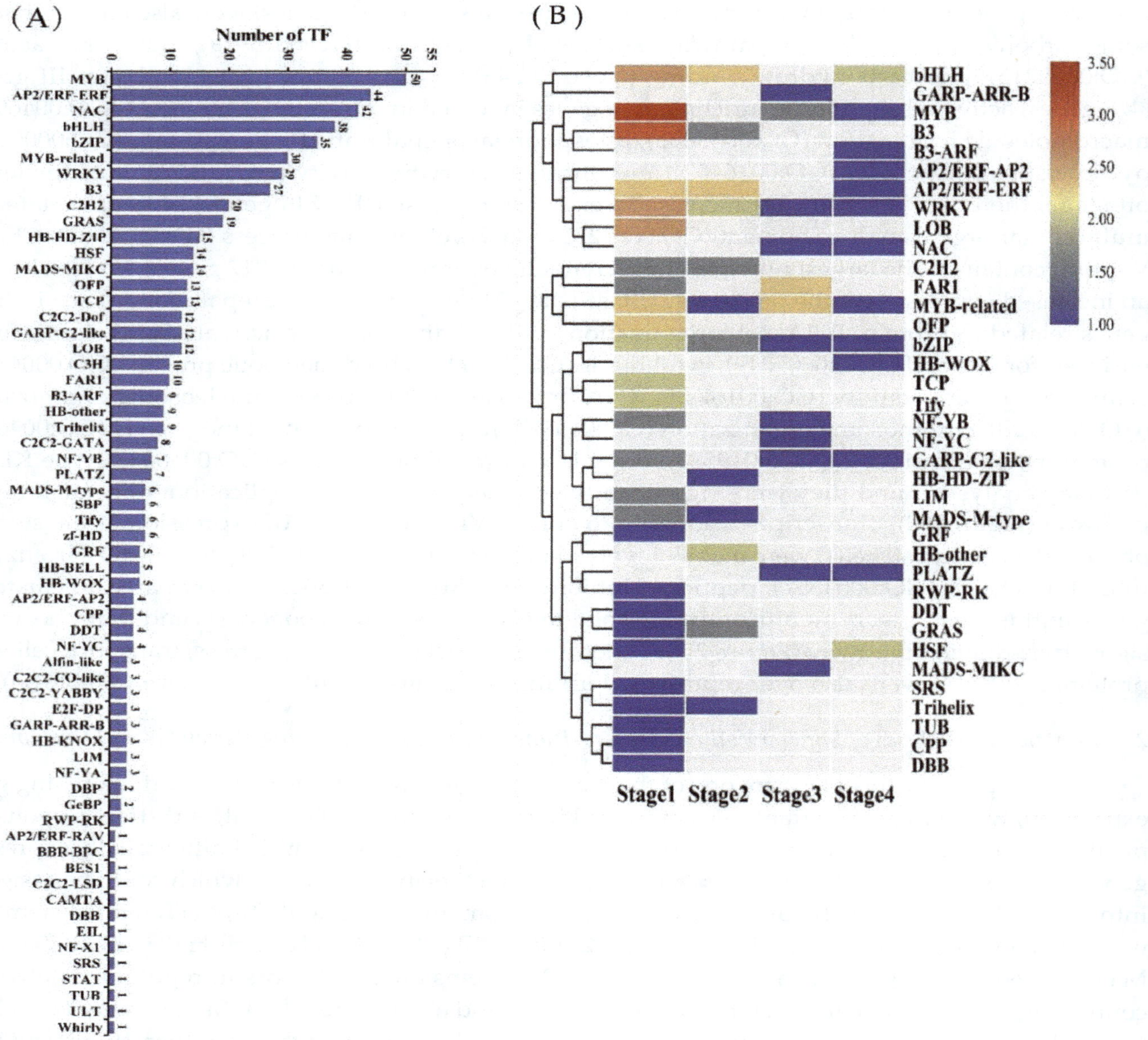


**Figure 3.** Characterization of transcription factors (TFs) differentially expressed during barley grain development. (**A**) Number of transcription factors (TFs) differentially expressed between any two stages of barley grain development. (**B**) Characterization of TF in each expression pattern.

*2.4. Prediction of RNA Editing Sites*

RNA editing is an important post-transcription mechanism to enrich genetic information and regulate diverse biological processes. After filtering the lower coverage sites as described in Method section, a total of 4711 editing sites were identified during the barley grain development, with 12 base change types, of which four editing types (A->G; C->T; G->A; T->C) were significantly higher than the other eight types (Figure 4A, Table S8). This result showed the transition occurred more abundantly than transversion in RNA editing, which was consistent with previous studies [23]. Then, the stage-specific and overlapped editing sites of these four stages were investigated. Results showed that stage01 owned the highest unique sites, with the number of 1048, followed by stage02, stage03, and stage04, with the numbers of 847, 707, and 356, respectively (Figure 4B). Among these editing events, a total of 3410 sites were found in 1795 protein-coding genes (Table S8), of which 2412 were found in CDS region, and 356 and 642 sites were located in 5′ and 3′ UTR regions (Figure 4C), respectively, suggesting that most editing could change the genetic information to result in the change of the encoded protein sequence. The gene HORVU7Hr1G000580 showed the most abundant editing sites with the total value of 15, and most of the other genes were observed to have one editing site (Figure 5A). Totally, 783 editing sites in 542 genes were observed to result in the amino acid change, of which 139 editing events (25.65%) led to a conversion from hydrophilic amino acid to hydrophobic amino acid, and 102 (18.82%) changing from hydrophobic amino acid to hydrophilic amino acid (Figure S6). A series of genes related to starch and sugar metabolic processes were found to be edited, such as beta-amylase 5 (HORVU4Hr1G089510), fructose-1,6-bisphosphatase class 1 (HORVU4Hr1G060630, HORVU1Hr1G074330) and low-molecular-weight glutenin subunit (HORVU1Hr1G001350, HORVU1Hr1G001420, HORVU1Hr1G001120, HORVU1Hr1G001120, HORVU1Hr1G000930, HORVU1Hr1G000930, HORVU1Hr1G001080) (Table S9). What is more, 4 terminators were found to be converted to 1 hydrophobic and 3 hydrophilic amino acids, and 1 hydrophobic and 11 hydrophilic amino acids were converted to terminators due to RNA editing. The embryo-defective related gene (HORVU2Hr1G037470) had 4 C to T editing events occur during barley grain development, and two of them caused the amino acid change (H>Y, A>V), one site with no amino acid change (C>C), and one resulted in the termination of the coding transformation (Q>*) (Figure S7). The stop code of this gene caused by editing may lead to the defect development of the embryo (Figure 5B). Furthermore, combined with the expression patterns, 308 DEGs were found to have RNA editing occur and the edited genes were more abundant in cluster I, cluster II, and cluster V, compared to cluster III, cluster IV, and cluster VII (Figure 5C). Then, KEGG analysis of them showed that they were enriched in metabolic pathways, biosynthesis of secondary metabolites, carbon fixation in photosynthetic organisms, carbon metabolism, base excision repair, pentose phosphate pathway, and glycolysis/gluconeogenesis (Figure 5D). These results demonstrated that RNA editing may function as the crucial regulator to control the complex process of barley grain development.

(A)

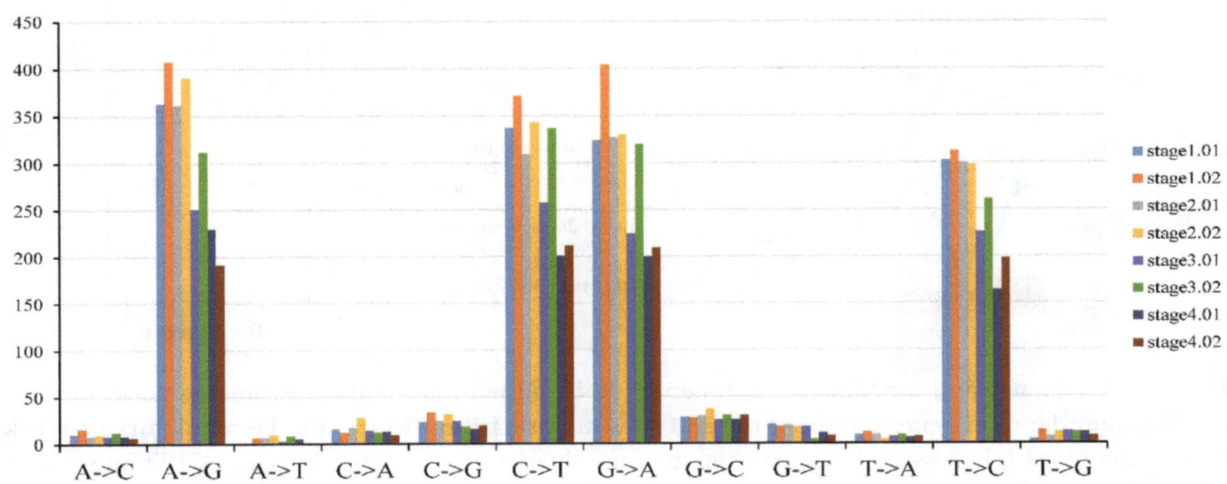


**Figure 4.** *Cont.*

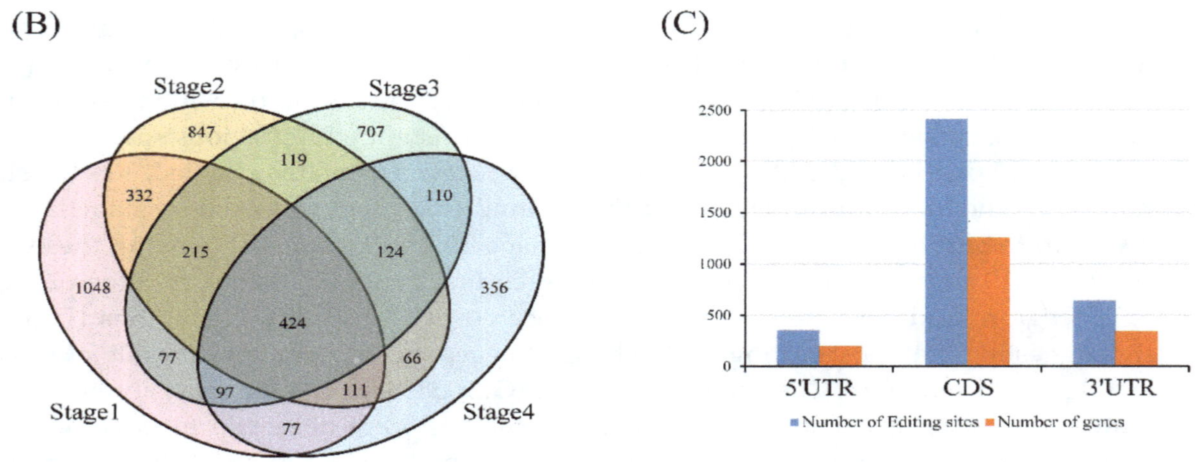


**Figure 4.** Identification of the RNA editing events during barley grain development. (**A**) The frequency of different types of nucleotide editing at four stages of barley grain development. (**B**) Venn diagram analysis of the abundance of RNA editing events among four stages of barley grain development. (**C**) The localization of RNA editing events in genic regions.

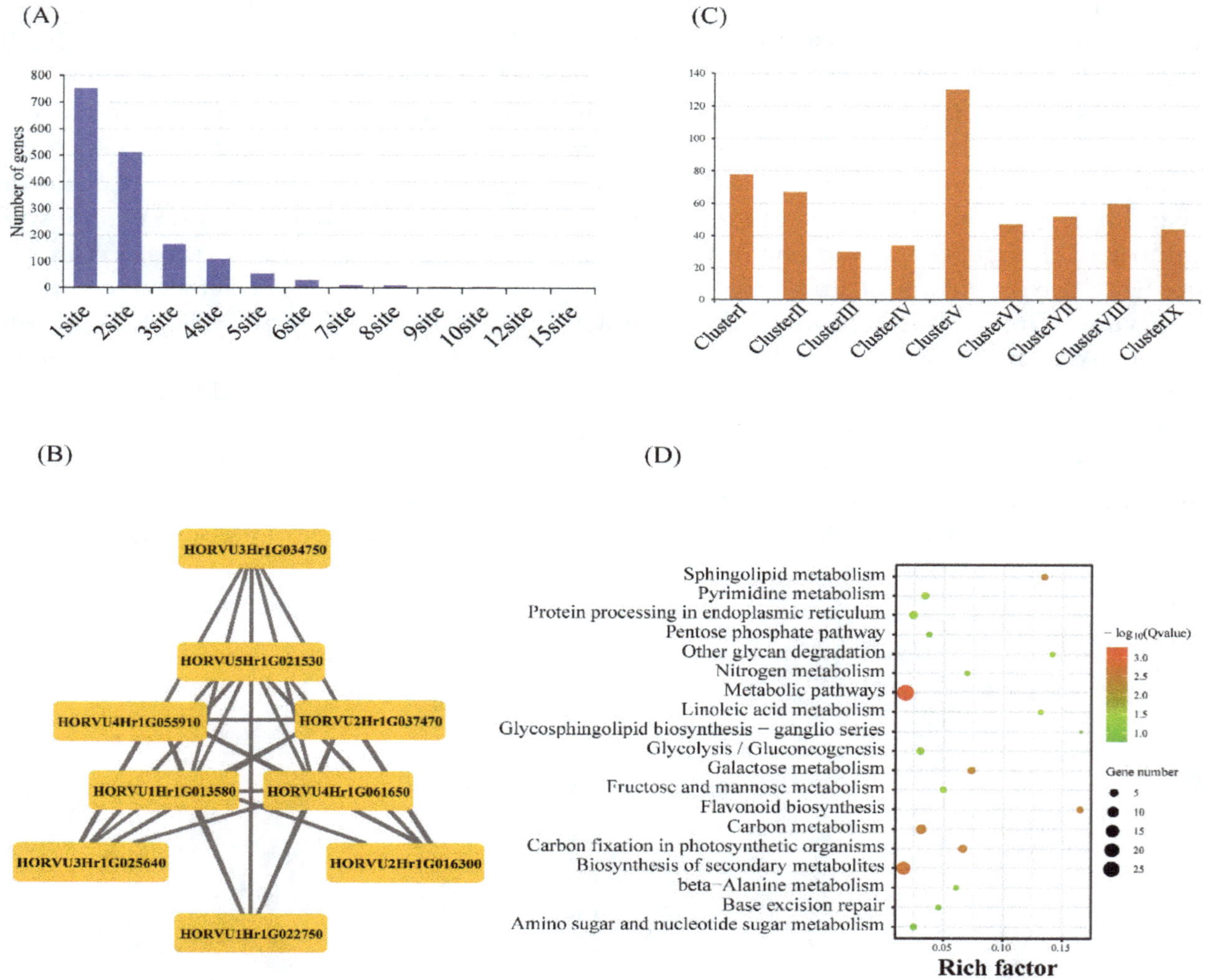


**Figure 5.** Frequency and functional enrichment of RNA-edited differentially expressed genes (DEGs). (**A**) The abundance of genes harbored the edit sites ranging from 1 to 15. (**B**) The regulatory network of the identified DEGs with RNA editing sites involved in embryonic defects. (**C**) Frequency of RNA-edited DEGs in nine expression clusters. (**D**) KEGG enrichment analysis of the RNA-edited DEGs identified in this study.

## 3. Discussion

### *3.1. mRNA Reveals Insights into Dynamic Regulation of Grain Development in Barley*

Grain development represents an elaborate and complex phase in the plant life cycle and has been extensively studied in recent years [6,7,13]. It has been suggested that differences in the molecular regulation of grain development may arise from spatial and temporal variation in gene expression [1,2,18]. Hence, interpretation of transcriptomic data providing some clues for the development of grain in plants, and barley in particular, becomes imperative. In this study, we performed a comprehensive analysis of transcriptional dynamics during barley grain development using a transcriptomic sequencing strategy. By pairwise comparisons of the data between four stages, differential transcript abundance and sequence variation were found. The levels of transcripts showed close correlation with the physiological changes during grain development. The majority of the genes exhibited peak expression at the prestorage phase (stage01 and stage02), and relatively low expression at the storage and mature phases (stage03 and stage04). We demonstrated that a dramatic change took place during the transition of the seeds from a prestorage phase to maturation, which enhanced our understanding of the developmental shifts during the filling process of barley grain. Previous studies have identified several pathways or genes involved in regulating barley grain development based on microarray hybridization [1,7,12,14]. In contrast to cDNA chip hybridization, the RNA-seq technology has obvious advantages for transcriptome dynamics analysis, which could generate millions of sequence reads with high reproducibility, especially for the lowly expressed transcripts. In the current study, 48.23% (19,165/39,734) of the annotated barley genes were detected to have the RPKM value more than 1 in at least one stage. These results were similar with that of hexaploid wheat (~55%) [2] and Arabidopsis (~44%) [18], but significantly higher than previous studies in barley [1,6,7,12,14]. Comparison analysis revealed that a total of 9527 genes were considered to be potentially involved in the regulation of barley grain development. These genes were significantly enriched in 16 KEGG pathways (Figure S8A), which provided the clues to explore the global activation of metabolic pathways and gene regulatory networks activated during barley seed development.

### *3.2. Identification of Genes Required for Grain Development in Barley*

Seed development is a complex process in which most of the essential developmental processes are initiated, including embryo and endosperm development, storage essence accumulation, desiccation tolerance forms, and seed dormancy. To obtain the information concerning the genes governing this complex developmental process, several studies have been focused on the identification of "essential" genes (lethal-phenotype genes) for seed development [3,9]. At present, a total of 481 genes were characterized as "essential" genes required for normal embryo development in Arabidopsis and presented in the SeedGenes database (http://seedgenes.org/index.html). The majority (74.43%) of these genes have been confirmed by function analysis [9]. In this study, we identified 3172 seed development-related genes with significant similarity to Arabidopsis based on bidirectional best BLAST hits and an e-value threshold below 1e-30. Among them, 887 genes were significantly differentially expressed during barley grain development and could be considered as potential barley EMB genes (Table S10). Mutations of the majority of these genes (85.34%, 757/887) will cause defects in embryo development. While the other 100 EMB genes are essential to seed pigmentation, mutation of these genes will lead to morphologically normal seeds with color changes such as albino and pale green [3,9]. To define their detailed functional categories, KEGG enrichment analysis of the potential EMB genes for barley used KOBAS (version 3.0) software [24]. Enrichment analysis found that the nucleotide, protein, lipid, and carbohydrate metabolism pathways were significantly enriched (Figure S8B). While these results demonstrate that the majority of genes related to seed development of Arabidopsis [3,9] are also present in barley grains, the definitive functions of these genes in barley seed development remain to be determined.

Developing barley grains are green and contain functional chloroplasts capable of photosynthesis during seed filling in barley. The prestorage phase (0–5 DAP) and transition phase (6–10 DAP) are characterized mainly by cell division and absence (or initial accumulation) of starch in the endosperm [6]. Previous studies in Arabidopsis indicated that the level of photosynthesis was high in the early and middle stages of grain development, when the majority of photosynthesis-related genes showed peak expression [5]. Our findings showed that a total of 29 genes were clustered as photosynthesis-related genes and 68.97% (20/29) of them peaked at stage02, represented by photosynthesis, photosystem I, and photosystem II. For example, several genes encoding photosynthetic light reaction (such as light-harvesting complex II, including Lhca1, Lhca2, Lhca3, Lhca4, Lhcb1, Lhcb2, Lhcb3, Lhcb5, Lhcb6, and Lhcb7) were low in the youngest seeds, increased sharply by 8 DAP (stage02), and then decreased afterwards.

### *3.3. Plant Hormone Signal Transduction and Sugar Signaling Interaction Networks During the Transition from Pre-Storage to Storage Phase*

Plant hormone plays a critical role in the entire growth period in plants. For instance, auxin contributes mainly to cell enlargement and division, and ABA (Abscisic acid) is believed to be involved in the initiation of storage compounds synthesis and linked to seed desiccation tolerance level and seed dormancy [1,4,5,7,14,25,26]. The plant hormone signal transduction pathway bridges the hormone and its response genes. Our transcriptome analysis indicated that auxin signal transduction-related genes were related to cell enlargement and division and their expression showed peak at prestore stage (Figure S9A), but the ABA signal transduction pathway showed a peak at store stage (Figure S9B).

Extensive studies have elucidated that ABA has an important role in the genetic control of seed development. There are four TFs known as main regulators of late embryogenesis in A. thaliana. These regulators are ABA-dependent genes, including LEC1 (HAP3 subunit), LEC2 (B3-domain), ABI3 (ABA-insensitive 3), and FUSCA/FUS3 (fused cotyledon3) [4,14,26]. FUS3 regulates endogenous ABA synthesis, which will activate the ABA-insensitive protein phosphatases 2C (PP2C) and sucrose nonfermenting 1 (Snf1)-related protein kinases 2 (SnRK2). SnRK2 subsequently initiates downstream targets, including ABF, AREB, and ABI5 [4,26]. In this study, we found several ABA signal transduction-related genes were differentially expressed in barley during grain development, and HvPP2C, HvSnRK, and HvABI were preferentially expressed at the storage phase (Figure S9B).

It has been demonstrated that both sugars and ABA played critical roles in controlling the transition from the cell division phase to the cell enlargement phase as storage reserves were accumulated [4,26]. SnRK, known for the regulation of key starch biosynthesis genes, such as sucrose synthase (SUS) and ADP-glucose pyrophosphorylase (AGPase), has been shown to act as a bridge between ABA and sugar signaling [4,26]. During legume seed development, glucose and sucrose acted in almost opposite fashions, with glucose promoting cell division and sucrose being associated with cell expansion [4]. Importantly, previous studies regarding metabolite contents in developing barley endosperm showed that hexoses were highest during early development whereas sucrose peaked at 8 DPA [13,16]. In the current study, our data showed some sugar-related genes were significantly expressed to maintain different contents of glucose and sucrose in distinct stages, thus medicating the phase transition from prestorage to storage phase. In potato tubers, for instance, the expression of invertase or sucrose phosphorylase bypasses sucrose synthase and decreases sucrose levels [4]. Two genes encoding sucrose phosphorylase were significantly expressed in barley and all of them showed a peak expression level at the prestorage phase (stage01 and stage02). It appears that the critical role these sugar-related genes played was to maintain a relatively low content of sucrose at the early stages (Figure S9C). In addition, one sucrose synthase gene, together with an enzyme that hydrolyzes sucrose to glucose and fructose, was preferentially expressed at the early stage, contributing to the high content of glucose at this stage [14] (Figure S9C). Starch biosynthesis was detected at high levels at the storage phase (stage03, stage04) (Figure S9D). A similar pattern was found for one sucrose phosphate and two sucrose synthases genes, indicating their important roles in starch biosynthesis at the storage phase. To conclude, we might hypothesize that the extensive interactions between sugar and ABA signaling have critical roles in mediating the phase transition of barley grain development according to the results obtained in this study.

### *3.4. RNA Editing Regulated Barley Grain Development*

RNA editing is an important post-transcription mechanism to enrich genetic information and regulate diverse biological processes by altering the genetic identity between RNA and genomic DNA. Although the RNA editing also occurs in the noncoding regions, such as 3′-UTR and 5′-UTR, rRNA and tRNAs, it mostly occurs in protein-coding regions, resulting in the change of encoded protein as well as protein function [27–29]. A series of studies have evidenced that the RNA-editing events have important roles in regulating organ formation, architecture growth, and development [30–33]. In this study, during grain development, the editing DEGS were more observed in the prestore stages (cluster V) than store stages (cluster VII) (Figure 5C). In addition, RNA editing was found to occur in some of the genes involved in the embryonic defects development and breakdown of the macromolecule organic matter, and editing has caused the stop coding or amino acid change (Figure S7; Tables S8 and S9), consistent with the empirical evidence that mostly nucleic acid activity and cell cycle occur in embryonic development stages. Combined with the DEGs, KEGG pathway enrichment analysis showed that the editing genes were significantly enriched in metabolic pathways, carbon fixation in photosynthetic organisms, base excision repair, and glycolysis/gluconeogenesis, which is consistent with the physiological process of spike development (Figure 5D). Thus, the RNA editing may function as a 'regulator' to ensure the plant's normal growth and development.

## 4. Materials and Methods

### *4.1. Sample Preparation and Sequencing Library Construction*

Barley plants (cv. Clipper) were grown in field trial of Northwest A&F University (Yangling, Shaanxi, China) under normal agricultural management. The time of anthesis and developmental stage of caryopses were determined for each head based on the dissection of the middle spikelet as previously described [13]. Individual spikes were tagged at the time of anthesis and harvested in the morning (08:30–09:30 h) at 3, 8, 13, and 18 DPA [16]. Four biological replicates were sampled. For each sample, tissues were collected directly into liquid nitrogen by pooling the middle parts of the spike from 10 to 15 individual plants for the purposes of homogeneity.

Then, two of the four replicates were randomly selected to isolate total RNAs using TRIzol reagent (Invitrogen, Waltham, MA, USA) and were treated with RNase-free DNase I to remove any contaminating genomic DNA. The purified RNAs were used to construct the RNA sequencing library following the Illumina's standard pipeline (Illumina, San Diego, CA, USA). In brief, mRNAs were extracted from purified RNA by Dynabeads oligo (dT) (Dynal, Invitrogen). Double-stranded cDNAs were synthesized using reverse transcriptase (Superscript II, Invitrogen) and random hexamer primers. After size selection and purification through agarose gel electrophoresis, the selected fragments with the length of 500 bp were ligated to sequencing adaptors. Then, the ligated RNAs were reverse transcribed and amplified for Illumina sequencing. The high-throughput sequencing was performed by pair-end (PE) approach with the read length of 100 bp on Illumina HiSeq2000 platform following the standard protocol at BGI Comp (Shenzhen, China). Each sample was sequenced with no less than 60 million reads. All the high-throughput sequencing data were deposited into the Sequence Read Archive (SRA) database in NCBI with the accession number of SUB2126402.

### *4.2. Expression Profiles Analysis*

After removing the adaptor sequences, empty reads, and low-quality reads ($Q < 20$) with FASTX-toolkit (FASTX-toolkit 0.0.14, http://hannonlab.cshl.edu/fastx_toolkit/), all the clean reads were aligned to barley genome [17] using the default parameters for HISAT2 (version 2.0.5) [34]. Reads counts statistic was calculated using HTSeq (Version 0.7.1) [35] with htseq-count command. Then, gene expression levels were estimated using RPKM values (Reads Per Kilobase transcriptome per million reads). Each sample was treated individually. Pearson correlations between biological replicates were conducted based on the RPKM values of all expressed genes using the function cor in R, which were further used to generate a dendrogram of samples.

*4.3. Identification of Differentially Expressed Genes (DEGs)*

Differential expression analysis was performed using R packages of DESeq2 (version 1.20.0) between any two stages with two biological replicates [8]. Genes with an adjusted $p$-value < 0.05 and at least twofold changes ($|\text{Log2 (treatment/control)}| \geq 1$) were considered as differentially expressed. Then, a K-means clustering was used to extract the fundamental patterns of gene expression inherent in the development of barley caryopses [36]. The following three additional steps were performed to pretreat the dataset using the standard tools in cluster 1.52 [37]: (i) Log-transform the data and all data values (FPKM) were replaced by log2(FPKM); (ii) for center arrays [mean], subtract the row-wise mean from the values in each row of data; (iii) normalize arrays: specifies to normalize each row in the data so that the sum of the squares of the values in each row is 1.0. The Venn diagrams in this study were prepared using the Vennerable package (version 3.0) in R based on the gene list for each tissue type.

*4.4. Identification of the Transcript Factors in DEGs*

All the nucleotide sequences for DEGs were extracted from the barley genome [17] and then used for prediction of the potential transcript factors by the iTAK (version 17.09) online software [38].

*4.5. RNA Editing Analysis*

All potential RNA editing sites were predicted using SPRINT (SnP-free Rna editing IdeNtification Toolkit) [39] software with default parameters. To support the prediction, we also identified the SNP between RNA reads and reference genomic DNA using SNP calling tools Samtools (version 1.3.1) (http://www.htslib.org) to analyze the mapping BAM files. The overlapped results obtained by both methods remained for further analysis. Then, the editing sites were further filtered using the following parameters: (1) the edited sites had more than five mapped reads; (2) the ratio of editing reads and total mapped reads more than 50%; (3) the editing sites were found in both biological replications.

*4.6. Gene Ontology (GO) and KEGG Enrichment Analysis*

GO terms that are significantly overrepresented in each cluster were determined by the AgriGO (version 2.0) [40]. Singular Enrichment Analysis (SEA) in AgriGO was used to detect over-represented GO categories in each cluster compared to the whole genes. GO terms with corrected Padj less than 0.05 were taken as significant ones. KEGG was used to identify processes and pathways of specific gene sets [41]. The annotations on genes were carried out using a simple unidirectional BLAST (blastx) search against already classified proteins from Arabidopsis thaliana (thale cress), Oryza sativa japonica (Japanese rice) (RefSeq), and Zea mays (maize) by KAAS (KEGG Automatic Annotation Server) software [41]. Then, KOBAS (version 3.0) software was used to enrich the pathways of all the genes [24].

## 5. Conclusions

In this study, we systematically investigated the gene expression atlas and RNA editome during barley grain development. Results showed that the expression profiles of transcripts involved in both primary and secondary metabolism were significantly changed between the transitions from prestorage phase to storage phase, which could reflect the physiological changes of grain development. Furthermore, the essential gene and regulatory networks involved in this process were investigated. Results showed that plant hormone signal transduction and sugar signaling interaction networks could play a crucial role in regulating grain development. Finally, 4771 RNA editing sites were detected in these four development stages, and most RNA editing genes were preferentially expressed at the prestore stage compared to store stage, suggesting that RNA editing might act as a 'regulator' to control grain development. This is the first study to report RNA edit events involved in grain development in barley at the whole transcriptome level. Our study systematically dissected the transcriptional dynamics of barley grain development, which will contribute to improved understanding of the molecular mechanism and dynamics of gene regulation underlying grain development in barley and beyond.

**Supplementary Materials:** Figure S1: The photos of four stages during barley grain development; Figure S2: RNA-seq read mapping counts against the barley reference genome; Figure S3: The correlation relationship of the gene expression level between four stages reported in this study and two stages in barley RNA-seq database; Figure S4: The volcano map of the number of differently expressed genes between two adjacent stages in barley grain development. A: stage 1 vs. stage 2; B: stage 2 vs. stage 3; C: stage 3 vs. stage; Figure S5: Clusters analysis of genes showing representative expression patterns during barley grain development; Figure S6: The abundance of change type among hydrophobic and hydrophilic amino acid as well as terminator caused by RNA editing; Figure S7: The aligment of the normal and edited protein sequences of the embryonic defect-related gene HORVU2Hr1G037470; Figure S8: Functional characterization of the differentially expressed genes during barley grain development; Figure S9: Identification of the DEGs involved in hormone signal transduction and sugar metabolism processes; Table S1: RNA-seq read mapping counts against the barley reference genome; Table S2: The expression of genes of all the stages(FPKM>=1); Table S3: The GO enrichment of pre-store stage genes; Table S4: The GO enrichment of store stage genes; Table S5: The set of genes showing stage-specific expression during barley grain development (FPKM < 1 in all other stages); Table S6: Go enrichment of the DEGs clustered in different clusters; Table S7: The detail list of transcription factors (TFs) differentially expressed during barley grain development; Table S8: RNA editing sites in coding regions of morex genome predicted by SPRINT tool; Table S9: The function annotation and expression level of editing DEGs; Table S10: List of putative EMB genes for barley grain development.

**Author Contributions:** Data curation, J.B., X.D., and X.N.; Formal analysis, J.B., P.D., H.Z., X.W., N.M.D.L.C., and X.N.; Funding acquisition, W.S.; Writing—original draft, J.B. and P.D.; Writing—review & editing, Z.Y., X.N., and W.S.

**Acknowledgments:** We thank Kewei Feng and Licao Cui for technical assistance on RNA editing analysis. We are also grateful to the High-Performance Computing (HPC) platform of Northwest A&F University for providing computing resources.

## References

1. Druka, A.; Muehlbauer, G.; Druka, I.; Caldo, R.; Baumann, U.; Rostoks, N.; Schreiber, A.; Wise, R.; Close, T.; Kleinhofs, A.; et al. An atlas of gene expression from seed to seed through barley development. *Funct. Integr. Genomics* **2006**, *6*, 202–211. [CrossRef] [PubMed]
2. Pfeifer, M.; Kugler, K.G.; Sandve, S.R.; Zhan, B.; Rudi, H.; Hvidsten, T.R.; Mayer, K.F.X.; Olsen, O.-A. Genome interplay in the grain transcriptome of hexaploid bread wheat. *Science* **2014**, *345*, 1250091. [CrossRef] [PubMed]
3. Tzafrir, I.; Pena-Muralla, R.; Dickerman, A.; Berg, M.; Rogers, R.; Hutchens, S.; Sweeney, T.C.; McElver, J.; Aux, G.; Patton, D.; et al. Identification of Genes Required for Embryo Development in Arabidopsis. *Plant Physiol.* **2004**, *135*, 1206–1220. [CrossRef]
4. Weber, H.; Borisjuk, L.; Wobus, U. Molecular physiology of legume seed development. *Annu. Rev. Plant Biol.* **2005**, *56*, 253–279. [CrossRef] [PubMed]
5. Ruuska, S.A.; Girke, T.; Benning, C.; Ohlrogge, J.B. Contrapuntal networks of gene expression during Arabidopsis seed filling. *Plant Cell* **2002**, *14*, 1191–1206. [CrossRef] [PubMed]
6. Sreenivasulu, N.; Altschmied, L.; Radchuk, V.; Gubatz, S.; Wobus, U.; Weschke, W. Transcript profiles and deduced changes of metabolic pathways in maternal and filial tissues of developing barley grains. *Plant J.* **2004**, *37*, 539–553. [CrossRef] [PubMed]
7. Sreenivasulu, N.; Usadel, B.; Winter, A.; Radchuk, V.; Scholz, U.; Stein, N.; Weschke, W.; Strickert, M.; Close, T.J.; Stitt, M.; et al. Barley grain maturation and germination: Metabolic pathway and regulatory

network commonalities and differences highlighted by new MapMan/PageMan profiling tools. *Plant Physiol.* **2008**, *146*, 1738–1758. [CrossRef] [PubMed]

8. Wang, L.; Xie, W.B.; Chen, Y.; Tang, W.J.; Yang, J.Y.; Ye, R.J.; Liu, L.; Lin, Y.J.; Xu, C.G.; Xiao, J.H.; et al. A dynamic gene expression atlas covering the entire life cycle of rice. *Plant J.* **2010**, *61*, 752–766. [CrossRef] [PubMed]
9. Devic, M. The importance of being essential: EMBRYO-DEFECTIVE genes in Arabidopsis. *C R Biol.* **2008**, *331*, 726–736. [CrossRef] [PubMed]
10. Feuillet, C.; Langridge, P.; Waugh, R. Cereal breeding takes a walk on the wild side. *Trends Genet.* **2008**, *24*, 24–32. [CrossRef] [PubMed]
11. Hansen, M.; Friis, C.; Bowra, S.; Holm, P.B.; Vincze, E. A pathway-specific microarray analysis highlights the complex and co-ordinated transcriptional networks of the developing grain of field-grown barley. *J. Exp. Bot.* **2009**, *60*, 153–167. [CrossRef] [PubMed]
12. Sreenivasulu, N.; Radchuk, V.; Strickert, M.; Miersch, O.; Weschke, W.; Wobus, U. Gene expression patterns reveal tissue-specific signalling networks controlling programmed cell death and ABA-regulated maturation in developing barley seeds (vol 47, pg 310, 2006). *Plant J.* **2006**, *47*, 987. [CrossRef] [PubMed]
13. Weschke, W.; Panitz, R.; Sauer, N.; Wang, Q.; Neubohn, B.; Weber, H.; Wobus, U. Sucrose transport into barley seeds: Molecular characterization of two transporters and implications for seed development and starch accumulation. *Plant J.* **2000**, *21*, 455–467. [CrossRef] [PubMed]
14. Seiler, C.; Harshavardhan, V.T.; Rajesh, K.; Reddy, P.S.; Strickert, M.; Rolletschek, H.; Scholz, U.; Wobus, U.; Sreenivasulu, N. ABA biosynthesis and degradation contributing to ABA homeostasis during barley seed development under control and terminal drought-stress conditions. *J. Exp. Bot.* **2011**, *62*, 2615–2632. [CrossRef] [PubMed]
15. Tauris, B.; Borg, S.; Gregersen, P.L.; Holm, P.B. A roadmap for zinc trafficking in the developing barley grain based on laser capture microdissection and gene expression profiling. *J. Exp. Bot.* **2009**, *60*, 1333–1347. [CrossRef] [PubMed]
16. Kohl, S.; Hollmann, J.; Erban, A.; Kopka, J.; Riewe, D.; Weschke, W.; Weber, H. Metabolic and transcriptional transitions in barley glumes reveal a role as transitory resource buffers during endosperm filling. *J. Exp. Bot.* **2015**, *66*, 1397–1411. [CrossRef] [PubMed]
17. Mascher, M.; Gundlach, H.; Himmelbach, A.; Beier, S.; Twardziok, S.O.; Wicker, T.; Radchuk, V.; Dockter, C.; Hedley, P.E.; Russell, J.; et al. A chromosome conformation capture ordered sequence of the barley genome. *Nature* **2017**, *544*, 427–433. [CrossRef] [PubMed]
18. Schmid, M.; Davison, T.S.; Henz, S.R.; Pape, U.J.; Demar, M.; Vingron, M.; Scholkopf, B.; Weigel, D.; Lohmann, J.U. A gene expression map of Arabidopsis thaliana development. *Nat. Genet.* **2005**, *37*, 501–506. [CrossRef] [PubMed]
19. Mortazavi, A.; Williams, B.A.; Mccue, K.; Schaeffer, L.; Wold, B. Mapping and quantifying mammalian transcriptomes by RNA-Seq. *Nat. Methods* **2008**, *5*, 621–628. [CrossRef] [PubMed]
20. Gan, Q.A.; Schones, D.E.; Eun, S.H.; Wei, G.; Cui, K.R.; Zhao, K.J.; Chen, X. Monovalent and unpoised status of most genes in undifferentiated cell-enriched Drosophila testis. *Genome Biol.* **2010**, *11*, R42. [CrossRef]
21. Dubos, C.; Stracke, R.; Grotewold, E.; Weisshaar, B.; Martin, C.; Lepiniec, L. MYB transcription factors in Arabidopsis. *Trends Plant Sci.* **2010**, *15*, 573–581. [CrossRef] [PubMed]
22. De Folter, S.; Immink, R.G.H.; Kieffer, M.; Parenicova, L.; Henz, S.R.; Weigel, D.; Busscher, M.; Kooiker, M.; Colombo, L.; Kater, M.M.; et al. Comprehensive interaction map of the Arabidopsis MADS box transcription factors. *Plant Cell* **2005**, *17*, 1424–1433. [CrossRef] [PubMed]
23. Levanon, E.Y.; Eisenberg, E.; Yelin, R.; Nemzer, S.; Hallegger, M.; Shemesh, R.; Fligelman, Z.Y.; Shoshan, A.; Pollock, S.R.; Sztybel, D.; et al. Systematic identification of abundant A-to-I editing sites in the human transcriptome. *Nat. Biotechnol.* **2004**, *22*, 1001–1005. [CrossRef] [PubMed]
24. Wu, J.M.; Mao, X.Z.; Cai, T.; Luo, J.C.; Wei, L.P. KOBAS server: A web-based platform for automated annotation and pathway identification. *Nucleic Acids Res.* **2006**, *34*, W720–W724. [CrossRef] [PubMed]
25. Locascio, A.; Roig-Villanova, I.; Bernardi, J.; Varotto, S. Current perspectives on the hormonal control of seed development in Arabidopsis and maize: A focus on auxin. *Front. Plant Sci.* **2014**, *5*, 412. [CrossRef] [PubMed]

26. Trontin, J.-F.; Klimaszewska, K.; Morel, A.; Hargreaves, C.; Lelu-Walter, M.-A. Molecular Aspects of Conifer Zygotic and Somatic Embryo Development: A Review of Genome-Wide Approaches and Recent Insights. In *In Vitro Embryogenesis in Higher Plants*; Germana, M.A., Lambardi, M., Eds.; Springer: New York, NY, USA, 2016; pp. 167–207. ISBN 978-1-4939-3061-6.
27. Maier, R.M.; Zeitz, P.; Kössel, H.; Bonnard, G.; Gualberto, J.M.; Grienenberger, J.M. RNA editing in plant mitochondria and chloroplasts. In *Post-Transcriptional Control of Gene Expression in Plants*; Filipowicz, W., Hohn, T., Eds.; Springer: Dordrecht, The Netherlands, 1996; pp. 343–365. ISBN 978-94-009-0353-1.
28. Steinhauser, S.; Beckert, S.; Capesius, I.; Malek, O.; Knoop, V. Plant mitochondrial RNA editing. *J. Mol. Evol.* **1999**, *48*, 303–312. [CrossRef] [PubMed]
29. Ichinose, M.; Sugita, M. RNA Editing and Its Molecular Mechanism in Plant Organelles. *Genes* **2017**, *8*, 5. [CrossRef]
30. Marechaldrouard, L.; Ramamonjisoa, D.; Cosset, A.; Weil, J.H.; Dietrich, A. Editing Corrects Mispairing in the Acceptor Stem of Bean and Potato Mitochondrial Phenylalanine Transfer-Rnas. *Nucleic Acids Res.* **1993**, *21*, 4909–4914. [CrossRef]
31. MarechalDrouard, L.; Kumar, R.; Remacle, C.; Small, I. RNA editing of larch mitochondrial tRNA(His) precursors is a prerequisite for processing. *Nucleic Acids Res.* **1996**, *24*, 3229–3234. [CrossRef]
32. Fey, J.; Weil, J.H.; Tomita, K.; Cosset, A.; Dietrich, A.; Small, I.; Marechal-Drouard, L. Role of editing in plant mitochondrial transfer RNAs. *Gene* **2002**, *286*, 21–24. [CrossRef]
33. Grewe, F.; Herres, S.; Viehover, P.; Polsakiewicz, M.; Weisshaar, B.; Knoop, V. A unique transcriptome: 1782 positions of RNA editing alter 1406 codon identities in mitochondrial mRNAs of the lycophyte Isoetes engelmannii. *Nucleic Acids Res.* **2011**, *39*, 2890–2902. [CrossRef] [PubMed]
34. Kim, D.; Landmead, B.; Salzberg, S.L. HISAT: A fast spliced aligner with low memory requirements. *Nat. Methods* **2015**, *12*, 357–360. [CrossRef] [PubMed]
35. Anders, S.; Pyl, P.T.; Huber, W. HTSeq-a Python framework to work with high-throughput sequencing data. *Bioinformatics* **2015**, *31*, 166–169. [CrossRef] [PubMed]
36. Brock, G.; Datta, S.; Pihur, V.; Datta, S. clValid: An R package for cluster validation. *J. Stat. Softw.* **2008**, *25*, 1–22. [CrossRef]
37. de Hoon, M.J.L.; Imoto, S.; Miyano, S. Statistical analysis of a small set of time-ordered gene expression data using linear splines. *Bioinformatics* **2002**, *18*, 1477–1485. [CrossRef] [PubMed]
38. Zheng, Y.; Jiao, C.; Sun, H.H.; Rosli, H.G.; Pombo, M.A.; Zhang, P.F.; Banf, M.; Dai, X.B.; Martin, G.B.; Giovannoni, J.J.; et al. iTAK: A Program for Genome-wide Prediction and Classification of Plant Transcription Factors, Transcriptional Regulators, and Protein Kinases. *Mol. Plant* **2016**, *9*, 1667–1670. [CrossRef] [PubMed]
39. Zhang, F.; Lu, Y.L.; Yan, S.J.; Xing, Q.H.; Tian, W.D. SPRINT: An SNP-free toolkit for identifying RNA editing sites. *Bioinformatics* **2017**, *33*, 3538–3548. [CrossRef]
40. Tian, T.; Liu, Y.; Yan, H.Y.; You, Q.; Yi, X.; Du, Z.; Xu, W.Y.; Su, Z. agriGO v2.0: A GO analysis toolkit for the agricultural community, 2017 update. *Nucleic Acids Res.* **2017**, *45*, W122–W129. [CrossRef] [PubMed]
41. Ogata, H.; Goto, S.; Sato, K.; Fujibuchi, W.; Bono, H.; Kanehisa, M. KEGG: Kyoto Encyclopedia of Genes and Genomes. *Nucleic Acids Res.* **1999**, *27*, 29–34. [CrossRef]

# 13

# Overexpression of a Wheat Aquaporin Gene, *Td*PIP2;1, Enhances Salt and Drought Tolerance in Transgenic Durum Wheat cv. Maali

**Malika Ayadi, Faiçal Brini and Khaled Masmoudi** *,†

Biotechnology and Plant Improvement Laboratory, Centre of Biotechnology of Sfax (CBS), University of Sfax, B.P 1177, 3018 Sfax, Tunisia; malika.ayadi@cbs.rnrt.tn (M.A.); faical.brini@cbs.rnrt.tn (F.B.)

* Correspondence: khaledmasmoudi@uaeu.ac.ae

† Present address: Arid Land Department, College of Food and Agriculture, United Arab Emirates University, P.O. Box 15551, Al Ain, UAE.

**Abstract:** In this study, we generated transgenic durum wheat cv. Maali overexpressing the wheat plasma membrane aquaporin *Td*PIP2;1 gene under the control of Pr*Td*PIP2;1 promoter or under the constitutive PrCaMV35S promoter. Histochemical analysis of the fusion Pr*Td*PIP2;1::*Td*PIP2;1::GusA in wheat plants showed that the β-glucuronidase (GUS) activity was detected in the leaves, stems and roots of stably transformed wheat T3 plants. Our results showed that transgenic wheat lines overexpressing the *Td*PIP2;1 gene exhibited improved germination rates and biomass production and retained low $Na^+$ and high $K^+$ concentrations in their shoots under high salt and osmotic stress conditions. In a long-term study under greenhouse conditions on salt or drought stress, transgenic *Td*PIP2;1 lines produced filled grains, whereas wild-type (WT) plants either died at the vegetative stage under salt stress or showed drastically reduced grain filling under drought stress. Performing real time RT-PCR experiments on wheat plants transformed with the fusion Pr*Td*PIP2;1::GusA, we showed an increase in the accumulation of GusA transcripts in the roots of plants challenged with salt and drought stress. Study of the antioxidant defence system in transgenic wheat *Td*PIP2;1 lines showed that these lines induced the antioxidative enzymes Catalase (CAT) and Superoxide dismutase (SOD) activities more efficiently than the WT plants, which is associated with lower malondialdehyde and hydrogen peroxide contents. Taken together, these results indicate the high potential of the *Td*PIP2;1 gene for reducing water evaporation from leaves (water loss) in response to water deficit through the lowering of transpiration per unit leaf area (stomatal conductance) and engineering effective drought and salt tolerance in transgenic *Td*PIP2;1 lines.

**Keywords:** abiotic stress; antioxidant enzymes; aquaporin; *Td*PIP2;1; histochemical analysis; transgenic wheat; transpiration

---

## 1. Introduction

Wheat constitutes the most widely grown and consumed cereal in the world. There is a growing imbalance between its supply and demand. The loss of fertile lands for cereal production due to climate change and water scarcity is considered major obstacles against further increases in yield. Drought stress induces different types of responses in plants [1]. It promotes oxidative damage in chloroplasts [2,3], decreases photosynthesis and metabolic reactions [4–7], induces osmotically active compound (i.e., soluble sugars and glycine betaine) and signal molecules [8–10] and changes in cellular lipid composition [11]. To tolerate drought stress, plants have generated different strategies such as the formation of larger and deeper root systems [12], adjustment of stomatal closure by controlling turgor pressure changes in guard cells to reduce water loss [13], accumulation of organic metabolites of low molecular weight known as compatible solutes and protective proteins [14] and enhancement of antioxidative systems [15].

Water transport across the plasma membrane is very important for the plant cell. Many reports on the relationship between plant aquaporins (AQPs) and plant water have been published [16]. Earlier research has shown that AQPs play a major role in the transportation through the membrane of different physiologically important molecules such as water, glycerol, $CO_2$ and $H_2O_2$, as well as assisting with various physiological processes such as drought, salt and chilling responses [17–20]. This suggests that AQPs play an important role in transporting a large quantity of water with the minimum of energy expenditure and seem to regulate the trans-cellular transport of water [17].

AQPs are proteins belonging to the conserved family of major intrinsic proteins (MIPs) [16,21]. There are five subfamilies of plant MIPs and among them the Plasma Membrane Intrinsic Proteins, (PIPs) are the most represented. PIP proteins can be divided into two groups, PIP1 and PIP2. These proteins, when expressed in *Xenopus laevis* oocytes, exhibit different water channel activities and can interact physically, leading to an increase of the osmotic water permeability coefficient (*Pf*) of the oocyte membrane [22–24]. Previous reports on AQPs have demonstrated their contribution to the water permeability of root cortex cells and their involvement in osmotic water transport in the entire root system [25]. AQPs are key players in the transport of sap-assimilated elements into phloem, stomatal closure, control of cellular homeostasis and leaf movement. [21,24,26,27]. During transpiration, AQPs control the transport of water from the roots to the leaves [21,24,26,27]. Transgenic plants overexpressing AQP proteins provide promising strategies to explore the hydraulic conductance in roots and leaves, plant transpiration, stomatal aperture and gas exchange under water deficit.

Plant AQPs play an important role in water relations. Their activities regulation and gene expression, which depend on complex processes at the transcriptional, post-transcriptional, translational and post-translational levels, are considered part of the adaptation mechanisms to environmental constraints. Over the last decade, the physiological contributions of AQPs have been investigated in planta by reverse genetics. Assuming that overexpression or down-regulation of a gene can help identify its function, this kind of approach is of key importance in the field. Such studies have already provided results demonstrating the central role of AQPs in plant physiology [28]. Under abiotic stresses or other physiological or developmental changes, differential expression of AQP transcripts or proteins was observed and reported. In rice, under osmotic stress the *Os*PIP2 transcripts in roots increased substantially [29]. In contrast, under salinity stress, expression of *Os*PIP1;1 in rice increased in leaves but was reduced in roots [30]. In barley and under drought stress, the expression of *Hv*PIP2;1 was reduced in roots and enhanced in shoots [31]. Numerous studies have confirmed that modulating AQP expression in transgenic plants can increase resistance to stresses. Indeed, expression of AQP PIP1 of *Vicia faba* (*Vf*PIP1) in transgenic *Arabidopsis thaliana* enhanced drought resistance by the reduction of transpiration rates through stomatal closure [32]. Recently, it was shown that overexpression of a barley aquaporin gene, *Hv*PIP2;5, confers salt and osmotic stress tolerance in yeast and plants [33]. Moreover, transgenic banana plants overexpressing a native plasma membrane aquaporin *Musa*PIP1;2 present high levels of tolerance to different abiotic stresses [34]. In addition, it was reported that overexpression of *Mf*PIP2-7 from *Medicago falcata* promotes cold tolerance and growth under $NO^{-3}$ deficiency in transgenic tobacco plants [35]. Conversely, the heterologous expression of *Ta*TIP2;2 in *Arabidopsis thaliana*, compromised its drought and salinity tolerance, suggesting that *Ta*TIP2;2 may be a negative regulator to abiotic stress. This was correlated with all down-regulated stress tolerance related genes acting in an ABA-independent manner, such as SOS1, SOS2, SOS3, CBF3 and DREB2A [36]. Moreover, it was reported that plasma membrane aquaporin overexpressed in transgenic tobacco increases plant vigour under favourable growth conditions but not under drought or salt stress [37]. The regulation of AQP activity and gene expression by various developmental and environmental factors, such as salinity and drought stress, relies on complex processes and signalling pathways [38].

In a previous work, we have isolated a durum wheat PIP2 gene, named *Td*PIP2;1 and have characterized its expression in *Xenopus* oocytes [39]. The generated transgenic tobacco plants overexpressing the *Td*PIP2;1 gene showed a phenotype of tolerance towards drought and salt stress [39]. To understand this tolerance mechanism at the transcriptional level, we isolated and characterized the promoter region of the *Td*PIP2;1 gene. When challenged with drought stress, the transgenic rice plants overexpressing Pr*Td*PIP2;1 in fusion with the *Td*PIP2;1 gene, showed enhanced drought stress tolerance, while WT plants were more sensitive and exhibited symptoms of wilting and chlorosis. These results suggest that expression of the *Td*PIP2;1 gene regulated by its own promoter achieves enhanced drought tolerance in transgenic rice plants [40].

In this study, we ectopically overexpressed the *Td*PIP2;1 gene under its own promoter and demonstrated that the promoter region contained all the regulatory elements required to mediate expression in transgenic durum wheat plants cv. Maali, resulting in improved tolerance to salt and drought stresses. We confirmed the role of the *Td*PIP2;1 gene in developing tolerant-crops for specific abiotic stress without any penalty phenotype.

## 2. Results

### 2.1. Production of Transgenic Wheat TdPIP2;1 Plants

Transgenic wheat plants expressing the *Td*PIP2;1 gene under a strong constitutive promoter (SP construct) or the native core promoter Pr*Td*PIP2;1 (PR construct), were generated. Eight transgenic lines for each different construct were checked and approved by PCR for their transgenic status. For each construct, three transgenic T3 homozygous lines were selected for the further evaluation of GusA transcript accumulation and the analysis of phenotypic and physiological parameters for drought and salinity stress responses in durum wheat.

### 2.2. Analysis of GUS Activity in Relation to Stress Treatment

The histochemical staining of GUS activity for whole plants at an early developmental stage (7 day-old seedlings), grown under control conditions, uncovered a coloured product in the leaves and in roots (Figure 1a). To validate the status of stress-inducible expression of the Pr*Td*PIP2;1, GusA transcript accumulation was examined by quantitative real-time qPCR in WT, SP3, PR3 and PR6 lines grown under control conditions or subjected to salt stress treatment (150 mM NaCl) and osmotic stress (20% PEG). The changes in GusA gene expression levels, across multiple samples and stress, were carried out by comparative quantification based on the housekeeping Actin gene as a reference gene. After 48 h of stress treatment, the level of GusA transcripts in the roots was higher in transgenic lines expressing the GusA gene under the native promoter of the *Td*PIP2;1 gene (PR6 and PR3), compared to the transgenic SP3 line expressing the GUS gene under the 35S constitutive promoter (Figure 1c). In leaves, the expression level of GusA transcripts was higher in the transgenic PR6 line compared to the two other transgenic PR3 and SP3 lines (Figure 1b). For each of the two responses (relative root expression and relative leaf expression), a Two-way analysis of variance with factors stress type and genotype was carried out. The results show that the genotype effect was highly significant ($p$-value $< 0.001$). There was a significant interaction between genotype and stress factors. The Tukey multiple comparisons procedure shows that there was a significant difference between the three genotypes, with the highest expression in roots for PR6, followed by PR3 and then SP3. For salt stress, there was no significant difference in gene expression in roots between the PR6 and the PR3 lines but both of them differed from the SP3 line. Similar results were obtained for drought stress. However, there was a significant difference for PR6 between salt and drought stress (Figure 1c). All these results suggest that the Pr*Td*PIP2;1 promoter is an abiotic and a tissue stress inducible promoter.

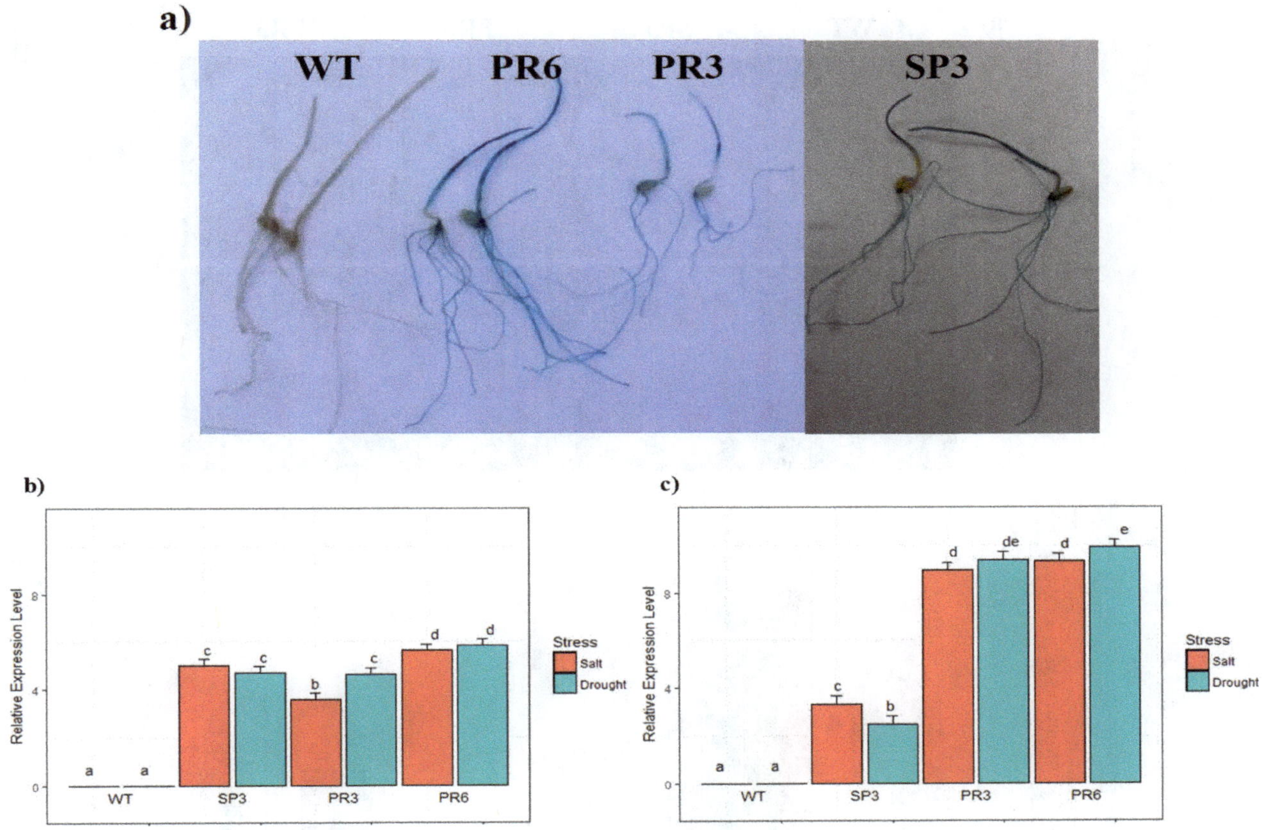


**Figure 1.** Histochemical analysis and GusA expression level of wheat transgenic lines overexpressing the *Td*PIP2;1 gene. (**a**) Histochemical GUS staining of 7 day-old wildtype (WT) and transgenic SP3, PR3 and PR6 wheat seedlings grown on MS medium. (**b**,**c**) qRT-PCR analysis of GusA gene expression in leaves and roots, respectively, of 15 day-old plantlets of WT, SP3, PR3 and PR6 lines subjected to salt (150 mM NaCl) and drought (20% PEG 6000) treatments over 48 h. Expression level of the GusA gene in control conditions was used as a reference. Bars show standard deviations of the replicates. Values represent the means of 5 replicates ($n = 5 \pm$ SD). Treatments with different letters have significant differences.

### 2.3. *Evaluation of the Wheat TdPIP2;1 Gene for in Vitro Stress Tolerance*

#### 2.3.1. Effect of Salt Stress on Seed Germination

In a first trial, the homozygous T3 seedlings were tested in vitro for their tolerance to salt and osmotic (mannitol) stresses. Wheat seeds from WT and three homozygous transgenic lines, SP3, PR6 and PR3, were germinated in petri-dishes with two sheets of pre-wetted filter paper supplemented with 0 or 150 mM NaCl. Under a control condition, the germination rate increased from 4 to 7 days of culture for all tested lines (Figure 2), indicating that aquaporin gene expression has no adverse effect on seed germination. However, germination of WT and SP3 seeds was significantly reduced under stress conditions when compared to transgenic PR3 and PR6 seeds. In fact, when challenged with 150 mM NaCl, the WT and SP3 seeds did not germinate, whereas after 4 days, seeds of the PR3 and PR6 lines had a germination rate of 13.3 and 20%, respectively (Figure 2). Moreover, after 7 days of salt stress treatment, seeds of the PR3 and PR6 lines showed a germination rate of 56% and 63%, respectively (Figure 2). The Two-way analysis of variance shows that the effect of stress and genotype were highly significant. PR6 lines showed the highest germination rate, followed by PR3, then WT and SP3. From the germination experiments, we noticed that the transgenic seedlings had shorter leaves and roots when grown for 7 days under salt stress.

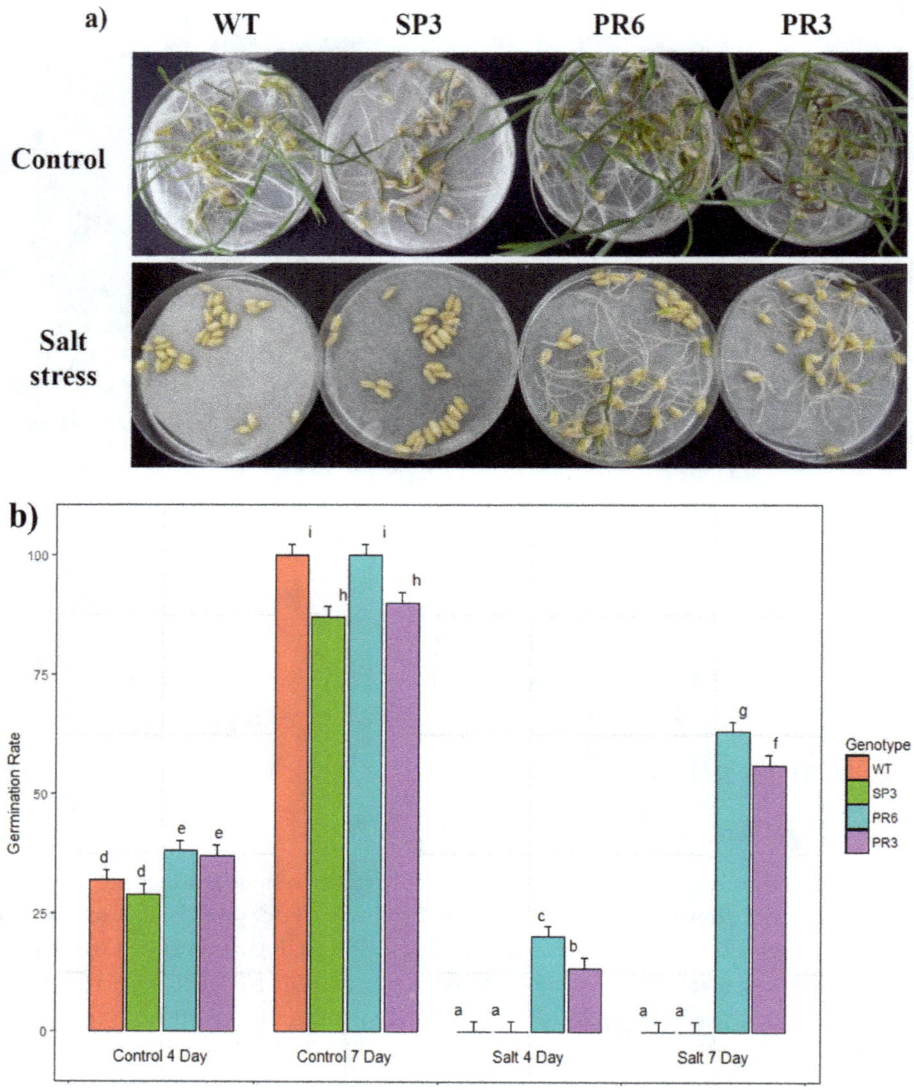


**Figure 2.** Effect of salt (NaCl 150 mM) on seed germination of transgenic wheat plants overexpressing *Td*PIP2;1 gene. (**a**) Photographs taken one week after seed germination. (**b**) Percentage of seed germination in the presence of 0 and 150 mM NaCl at 4 and 7 days of WT and T3 homozygous transgenic plants (SP3, PR3 and PR6). Values are means of 3 replicates ± SD ($n$ = 3) (20 seeds per repetition). Treatments with different letters have significant differences.

### 2.3.2. Effect of Salt and Osmotic Stresses on Root and Leaf Length

In order to further explore the consequences of salt and osmotic stress on the growth rate, 3 day-old WT and transgenic (SP3 and PR6) seedlings were transferred to MS, MSs (NaCl 150 mM) or MSm (mannitol 300 mM). After an additional 10 days, WT seedlings recorded a shorter leaf and root length compared to transgenic lines in control condition (Figure 3a,d,e). Under the same condition, WT plants showed severe root length reduction of 71.3% and 58% when compared to transgenic lines PR6 and SP3, respectively, while leaf length reduction in WT plants was only 34% and 38.5% when compared to transgenic lines SP3 and PR6, respectively (Figure 3d,e). Interestingly, inhibition of root elongation under salinity stress was 40% in WT, 23% and 34% in transgenic lines SP3 and PR6, respectively. Under NaCl treatment, leaf length showed non-significant difference among the three genotypes but when compared to the control MS media, the three genotypes showed significant leaf length reduction (Figure 3d). When grown in mannitol-supplemented culture media, WT and PR6 transgenic line did not show significant leaf length reduction when compared to control MS culture media. In contrast, the SP3 transgenic line showed significant leaf length reduction in comparison with the control. Under salinity treatment, root length differed significantly among the three genotypes and when compared to the control condition, they showed significant root length reduction (Figure 3e). Seedlings conducted with mannitol-supplemented culture showed a significant increase in root length for WT and SP3 lines, while the PR6 line showed a significant decrease when compared to control condition (Figure 3e). The reduction in root length in WT was 42% when compared to transgenic lines SP3 and PR6 (Figure 3a–c). It is noteworthy to notice that reduction in root length observed in transgenic plants challenged with NaCl or mannitol stress was compensated by an augmentation in the root number when compared to control conditions (Figure 3a–c). When challenged with NaCl or mannitol stress, transgenic seedlings produced two leaves, whereas WT seedlings showed inhibition and produced only one short leaf. Leaf length was inhibited by 30% under salinity stress in WT plants when compared to control condition, while no inhibition was observed in leaf length under mannitol stress (Figure 3d). Similarly, in transgenic line PR6, leaf length inhibition under salinity stress was about 45% but no inhibition was observed under mannitol stress (Figure 3b–e). The Two-way analysis of variance shows that the effect of stress and genotype were highly significant. PR6 lines showed the highest root and leaf length under control, salt and mannitol stress conditions, followed by SP3 and then WT. The difference in leaf or root length reduction or inhibition observed between WT and transgenic wheat plants overexpressing the *Td*PIP2;1 gene under its own promoter, shows the ability of the transgenic plants to tolerate in vitro salinity and osmotic stresses.

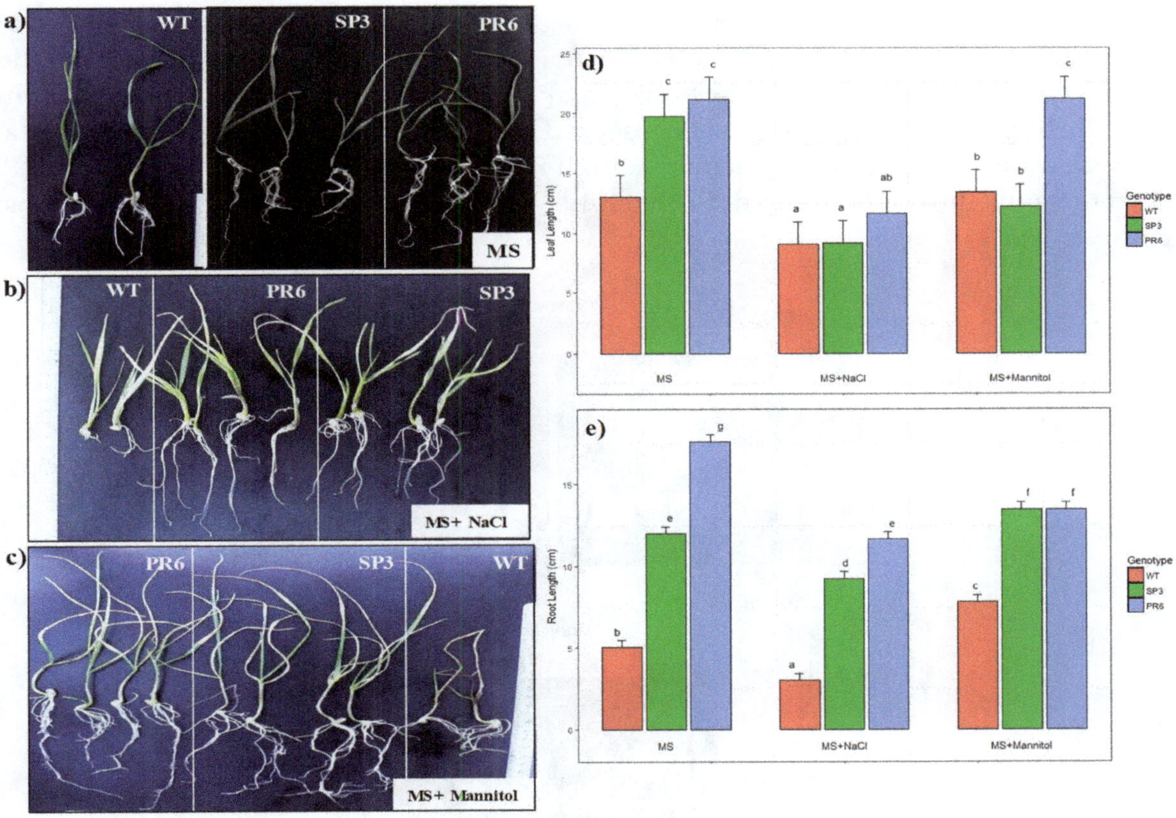


**Figure 3.** Effect of stress on growth rate of wild-type and *Td*PIP12;1 transgenic lines (SP3, PR3 and PR6). (**a**) MS medium. (**b**): MS medium supplemented with 150 mM NaCl. (**c**): MS medium supplemented with 300 mM Mannitol. (**d**,**e**) Analysis of root and leaf length of WT, SP3 and PR6 transgenic plants cultured on MS, MS + 150 mM NaCl and MS + 300 mM Mannitol, respectively. The photographs were taken 15 days after stress application. The results are expressed as the means SE of measurements from three different experiments. Values are means SD ($n = 5$). Treatments with different letters have significant differences.

### 2.4. *Evaluation of the Wheat TdPIP2;1 Gene for Stress Tolerance under Greenhouse Conditions*

In vitro growth of the transgenic wheat plants overexpressing the *Td*PIP2;1 gene showed a promising level of tolerance under salt and osmotic stresses. It was worth studying the effect of this gene on biomass and grain production. Accordingly, we carried out salt and drought stress experiments with WT and the two T3 generation of transgenic lines (PR3, PR6 and SP3) grown in soil in a greenhouse.

#### 2.4.1. Fraction of Transpirable Soil Water (FTSW)

As the soil dried progressively for about 20 days, the transpiration rate (TR) was recorded for the generated transgenic wheat plants overexpressing the Pr*Td*PIP2;1 promoter in fusion with the *Td*PIP2;1 gene. For each construct of the generated transgenic T3 homozygous lines (SP3, PR3, PR6), plants were evaluated in vitro and under greenhouse conditions for their growth (Figure 4). The Normalized transpiration rate (NTR) (transpiration of stressed plants /average transpiration of control plants = Gs/Gsmax) was calculated to reflect the daily transpiration rate and FTSW was calculated to reflect the soil-water content. The relationship between NTR and FTSW correlated perfectly to the plateau regression function (Figure 5). For the NTR decline, the FTSW threshold values were close for both transgenic wheat lines, SP3 and PR6. In contrast, the FTSW thresholds for the NTR decline were higher for the transgenic PR3 lines compared to PR6 lines. The threshold for the decline of transpiration (NTR) occurred when FTSW values of about 0.363, 0.909 and 0.983 were obtained for WT, SP3 and PR3, respectively. The FTSW threshold wherein transpiration rates began to decline was calculated for each line tested using a plateau regression procedure. The determined FTSW threshold values ranged from 0.363 to 0.909 for WT and PR3 (Table 1). There was no significant trend of change of the FTSW threshold for transpiration rate decrease between SP3 and PR6 lines. However, a significant linear ($r^2 = 0.656$) decline of FTSW threshold was observed with PR3. In this line, a preventive strategy to skip excessive water loss by stomata closure at high FTSW was observed. Accordingly, the PR3 line was shown to be the most tolerant to drought stress, whereas the WT line was the most sensitive to drought stress.

**Figure 4.** Phenotype and growth rate of WT and Pr*Td*PIP2;1 transgenic wheat lines PR6 and SP3 grown under (**a**) control conditions or subjected (**b**) to an average of 15 days of water withholding (FTSW = 0.1). Transgenic and WT wheat plants were gown in soil during 20 days in controlled greenhouse conditions. The photographs were taken 15 days after water stress application.

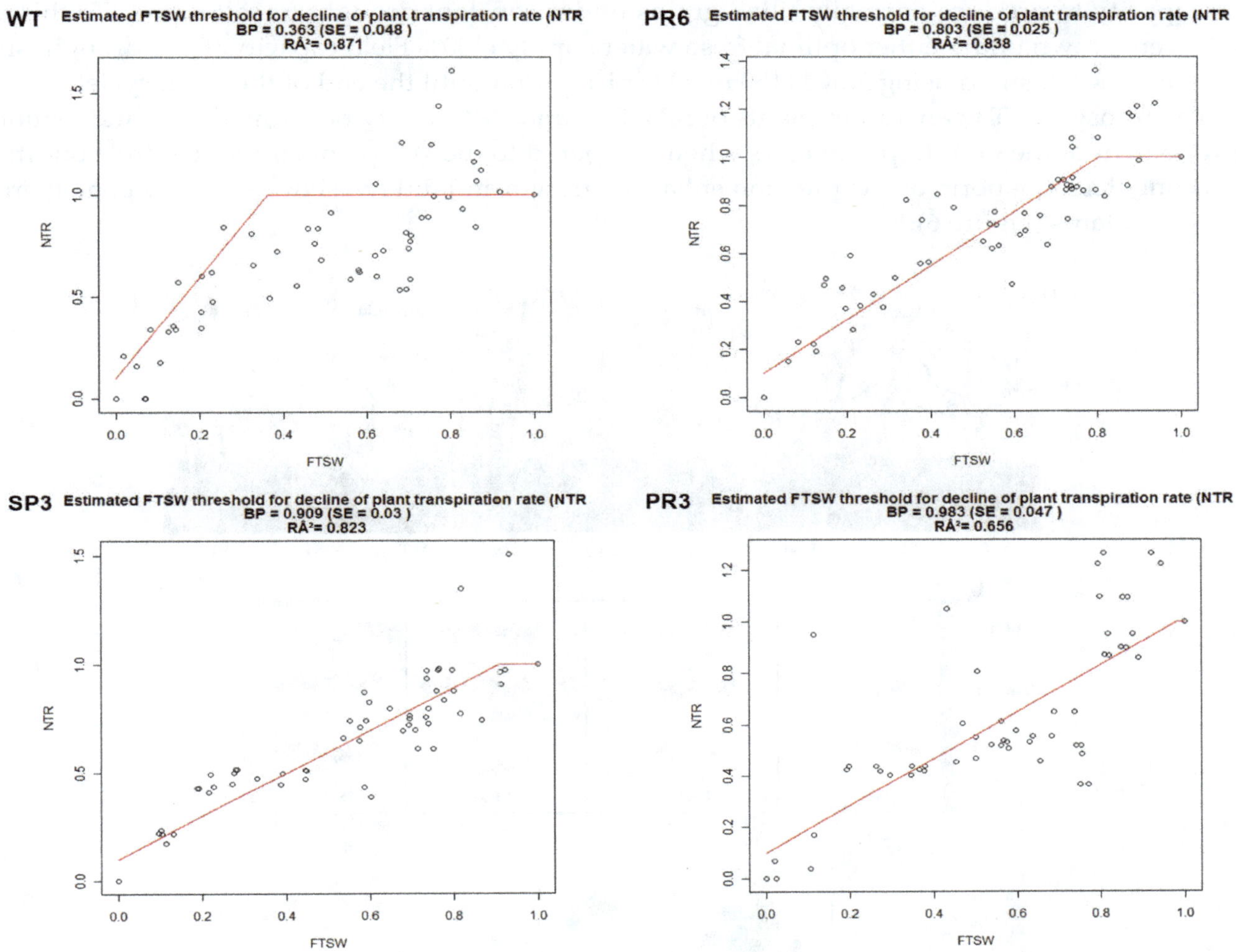


**Figure 5.** Variability for normalized transpiration rate (NTR) control under drought stress among the WT and transgenic wheat lines tested. Classification of wheat lines are from least sensitive WT, PR6 and SP3 to the most sensitive (stomata closure for a high FTSW) (PR3).

**Table 1.** FTSW threshold values.

| | BP | ($R^2$) |
|---|---|---|
| WT | 0.363 | 0.871 |
| SP3 | 0.909 | 0.823 |
| PR3 | 0.983 | 0.656 |
| PR6 | 0.803 | 0.838 |

The same results were observed when these lines were subjected to salt stress.. When challenged with salt or drought stress, these lines exhibited a high level of tolerance. In fact, in response to water deficit, minimization of water loss constitutes a major aspect of drought tolerance and this can be reached by lowering the transpiration per unit leaf area (stomatal conductance). In this context, it is worthwhile elucidating the contribution of this protein in the tolerance mechanism to salt and drought stress by overexpressing the isolated *Td*PIP2;1 gene under its own promoter in wheat plants.

### 2.4.2. Effect on Biomass and Grain Production

The increase in yield stability under harsh environmental and growth conditions with improved varieties adapted to salt and drought stresses would constitute a major breakthrough for farmers. To achieve this goal, we tested the generated transgenic wheat lines under greenhouse conditions for their capacity to produce normally filled grains under constant drought or salt stress. To this end, plants were grown under either optimal fresh water supply (100% Field capacity (FC)), drought stress or continuous salt stress using NaCl (150 mM) for irrigation until the end of the plant cycle.

The transgenic T3 generation plants of SP3, PR3 and PR6 clearly performed well and exhibited improvement of the growth parameters when compared to the WT plants under control conditions, confirming that aquaporin overexpression enhances growth and did not show any yield penalty in the transgenic plants (Figure 6).

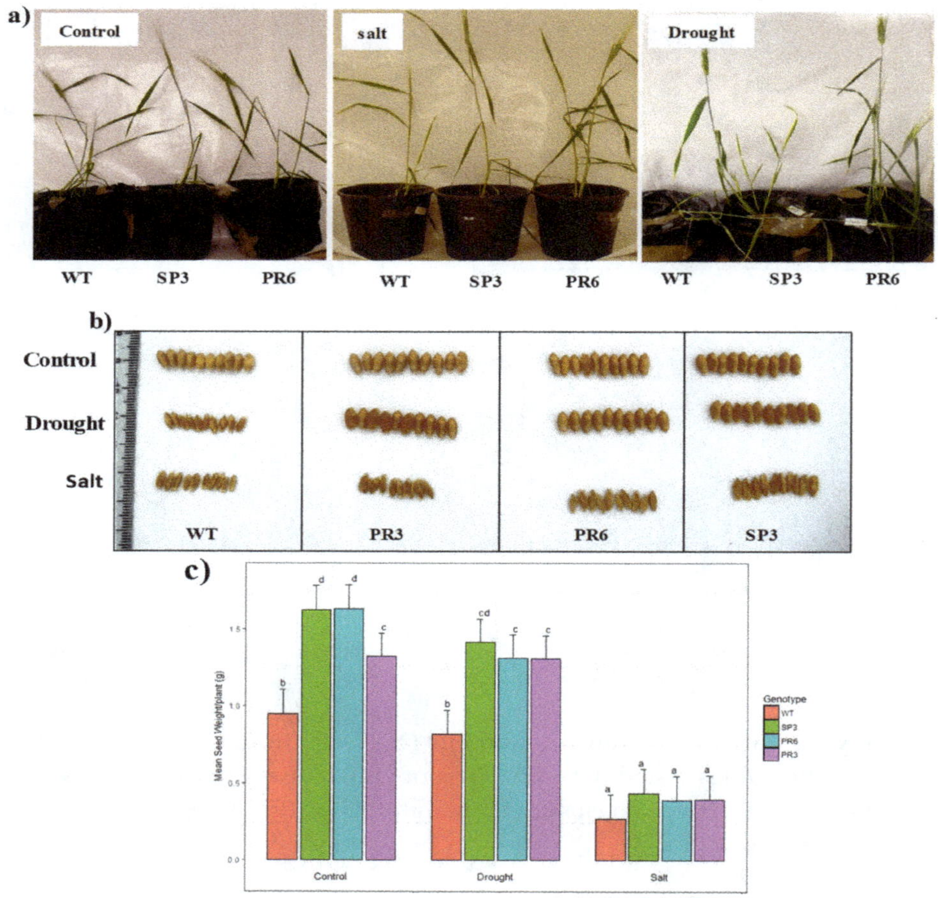


**Figure 6.** Effect of salt and drought stresses on growth and size of grains of wild-type and transgenic *Td*PIP2;1 wheat lines (SP3 and PR6). (**a**) Photographs show WT and transgenic plants grown under either normal conditions (100% FC), a continuous presence of 150 mM NaCl or a continuous drought stress (40% FC) until the end of the plant cycle. (**b**) Grain aspects of WT and transgenic (SP3, PR3 and PR6) grown under control, drought and salinity conditions. (**c**) Mean weight of seeds produced per plant under control and stress conditions. Treatments with different letters have significant differences.

When challenged with drought stress, the transgenic wheat plants were able to continue to grow, reached maturity, flowered and set grains; whereas the WT plants' growth was strongly inhibited and the produced grains were poorly filled. Under salt stress, the WT plants were highly affected and showed chlorosis and a stunted phenotype, were unable to produce viable grains and ultimately died (Figure 6a,b). Under drought stress and when compared to the control condition, the decrease in the weight of 30 grains showed a non-significant difference for the WT, SP3 and PR3 lines, while the PR6 line showed a significant difference when compared to the control condition (Figure 6c). The recorded reduction in the weight of 30 grains was about 13% and 20% in transgenic lines SP3 and PR6, respectively in comparison to the obtained weight under control condition. In contrast, the PR3 line showed no reduction in the weight of 30 grains compared to the control condition. Under salinity stress, the weight of 30 grains was drastically affected in all genotypes and showed a significant difference when compared to the control condition. The reduction in the weight of 30 grains was above 70% in all genotypes compared to the control condition (Figure 6a–c).

### 2.5. *$Na^+$ and $K^+$ Accumulation in Transgenic Wheat TdPIP2;1 Plants under Salt Stress Treatment*

To understand the basic mechanism of salt and/or drought-tolerance, the endogenous $Na^+$ and $K^+$ contents in leaves of transgenic wheat lines overexpressing the *Td*PIP2;1 gene were monitored. Under the control condition, there was no significant difference observed in $Na^+$ or $K^+$ accumulation or partitioning in leaves of WT and transgenic plants, except for PR6 plants where the accumulation was slightly increased for $Na^+$ (Figure 7a,b). Nevertheless, under drought or salt stress conditions, the increase of $Na^+$ accumulation in leaves showed a significant difference among all genotypes when compared to control condition. Hence, the increase of $Na^+$ concentration in leaves of the transgenic plants was approximately threefold that of WT (Figure 7a). Furthermore, $K^+$ accumulation was higher in control condition for all genotypes compared to stress conditions and the decrease in $K^+$ accumulation observed in all genotypes under drought or salt stress was highly significant compared to the control condition (Figure 7b). On the other hand, under drought or salt stress conditions, $K^+$ accumulation was greater in transgenic lines than in WT (range between 1.8 to 4.13 fold) (Figure 7b). The two-way analysis of variance shows that the effect of stress and genotype were highly significant. Overall, these results suggest that transgenic plants retained selectively more $K^+$ than $Na^+$ in their leaves and the $Na^+/K^+$ ratio, which is an important stress tolerance trait, decreased from 3.9 in WT plants to 2.85 in PR3 and PR6 transgenic lines and to 2.5 in transgenic SP3 lines, thus preventing the young photosynthetically active organs from $Na^+$ accumulation and toxicity.

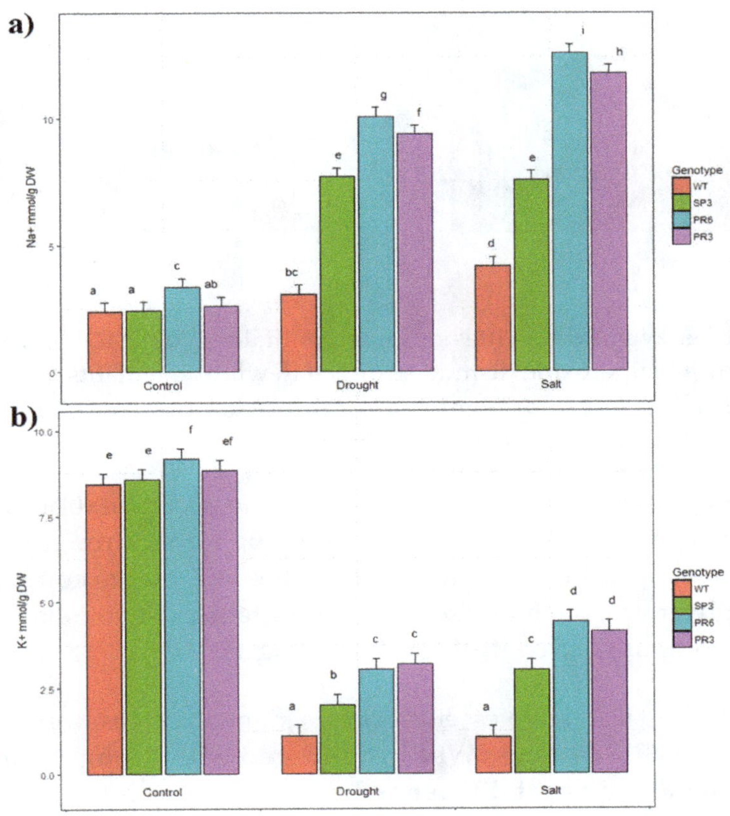


**Figure 7.** (**a**) $Na^+$ and (**b**) $K^+$ accumulation in leaves of the wild-type and transgenic lines overexpressing the *Td*PIP2;1 gene and grown under control or continuous presence of 150 mM NaCl or drought stress (40% FC). Values are the mean ± standard deviation (SD) ($n = 3$). Treatments with different letters have significant differences.

### 2.6. Oxidative Stress Evaluation of Transgenic Wheat TdPIP2;1 Plants

In the present study, Malondialdehyde (MDA) content in the leaves of WT and transgenic wheat lines correlated with drought stress induced growth inhibition. Moreover, when compared to the transgenic lines, the MDA content increased significantly in WT plants under drought stress (Figure 8a). In contrast, MDA content decreased significantly in the two transgenic lines PR6 and PR3 compared to WT plants under drought stress conditions (Figure 8a). The decrease in MDA content under drought stress was significantly different for PR3 and PR6 lines compared to control condition, while for the SP3 line the difference was not significant. Hence, transgenic wheat *Td*PIP2;1 lines with lower MDA content were shown to be more tolerant to drought.

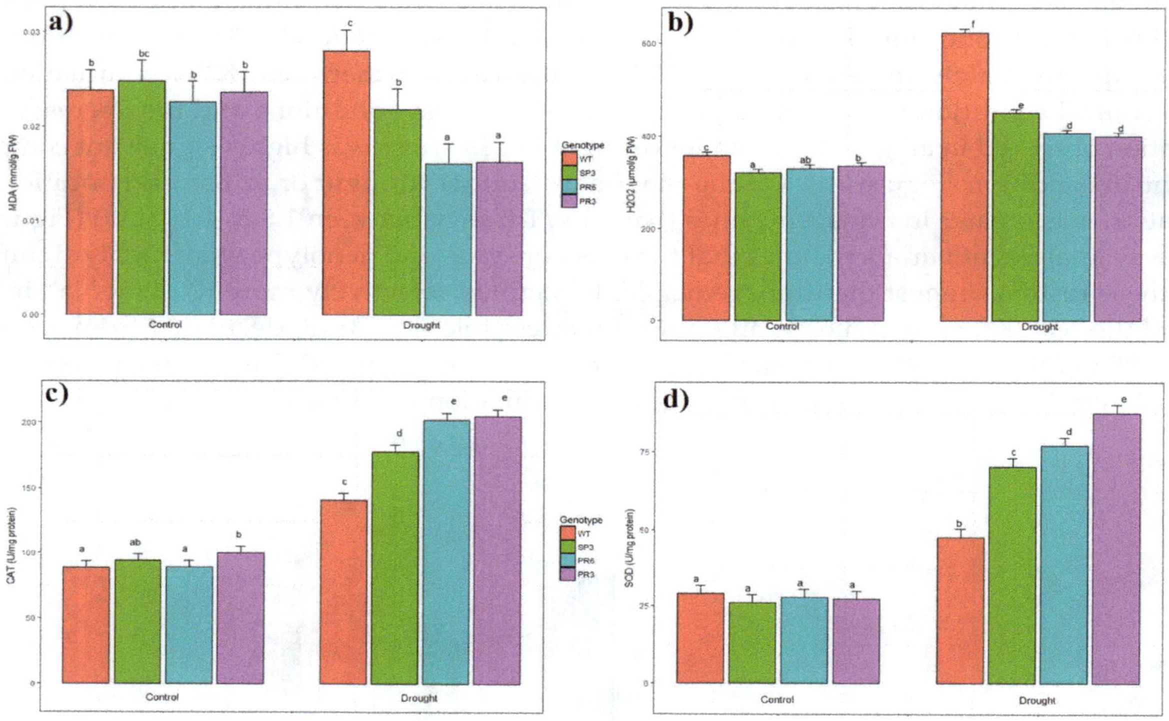


**Figure 8.** Analysis of enzymatic activities of (**a**) malondialdehyde (MDA), (**b**) hydrogen peroxide, (**c**) catalase (CAT) and (**d**) superoxide dismutase (SOD) in wild-type plants and transgenic lines (SP3, PR3 and PR6) subjected or not to drought stress. Values are means of 5 replicates ± SD ($n$ = 5). Treatments with different letters have significant differences.

The oxidative damage was checked by measuring the accumulation of $H_2O_2$ in leaves of the control and stressed transgenic wheat lines. As shown in Figure 8b, drought stress treatment induced significant accumulation of $H_2O_2$ content in leaves of the WT plants compared to transgenic lines (1.53 fold). Under drought stress, the increase in $H_2O_2$ content in leaves showed significant difference for all genotypes when compared to control condition (Figure 8b).

The activities of SOD and CAT, two essential antioxidant enzymes, were analysed in the leaves of WT and transgenic wheat *Td*PIP2;1 lines. When challenged with drought stress, the increase in CAT and SOD activities showed a significant difference for all genotypes compared to the control condition (Figure 8c,d). CAT and SOD activities increased significantly in transgenic lines, reaching about 2.2 and 3.2 fold, respectively relative to the non-treated plants (Figure 8c,d). However, this increase in CAT and SOD activities is lower in WT when comparing control and stress conditions (Figure 8c,d). The Two-way analysis of variance shows that the effect of stress and genotype were highly significant.

## 3. Discussion

In a large number of plant species, constitutive promoters were used to study the expression of transgenes. However, in some specific cases, constitutive expression could be damaging to the recipient plant, generating sterility, development retardation, atypical morphology, yield penalty, modified grain composition or transgene silencing [41]. To overcome the negative effects of the constitutive promoter, the use of a tissue-specific promoter is expected to be the solution for driving candidate gene expression restricted to tissues of interest and at given developmental stages. It is expected that utilization of stress inducible promoters will have a great impact on improving plant tolerance to either abiotic or biotic stress. Aquaporins (AQP) are ubiquitous membrane proteins with members exhibiting both tissue specific and inducible regulation. It has been shown that the *Nt*AQP1 gene is specifically expressed in the mesophyll tissue of *Arabidopsis thaliana* and plays an important role in increasing both net photosynthesis and mesophyll $CO_2$ conductance. Moreover, the *Nt*AQP1 gene targeted at the cells of the vascular envelope significantly improved the plants' stress response [42].

In this report, we demonstrated that the overexpression of the *Td*PIP2;1 gene in transgenic durum wheat homozygous lines significantly enhanced tolerance under constant salt and drought stresses (Figure 4). These results affirm our previous findings in rice [40]. In fact, transgenic rice plants transformed with the Pr*Td*PIP2;1 fused to the *Td*PIP2;1 gene and treated with either salt or drought stress, exhibited growth and vigour enhancement when compared to wild-type plants and therefore showed higher tolerance to abiotic stresses [40]. Our results were similar to those observed in soybean, where the overexpression of *Gm*PIP2;9 under a native promoter increased tolerance to drought stress in both solution and soil plots. This was correlated with expanded net $CO_2$ assimilation of photosynthesis, stomata conductance and transpiration rate [43].

Characterization of transgenic (SP3) wheat plants transformed with the 35S promoter fused to the *Td*PIP2;1 gene and cultivated in vitro under salt and drought stress showed enhanced growth and vigour compared to wild-type plants (Figure 3b,c). In our preceding work on the *Td*PIP2;1 gene, similar results were reported where the expression of the *Td*PIP2;1 gene was positively correlated with stress response pathways in transgenic Tobacco plants [39] and rice plants [40].

Our analysis using transcriptional (Pr*Td*PIP2;1::GUS) and translational (Pr*Td*PIP2;1::*Td*PIP2;1::GUS) fusions demonstrated that the *Td*PIP2;1 gene is expressed in the homologous transgenic wheat system. Using a histochemical staining technique, we were able to show that the Pr*Td*PIP2;1 promoter can drive GusA reporter gene expression with inducible and tissue-specific patterns (Figure 1a). The differential expression of AQP transcripts or proteins in response to abiotic stress or other physiological or developmental changes was reported in a number of studies (reviewed in Reference [44]). However, the transcriptional regulation of these genes is still elusive. The analysis of the promoter sequences of a variety of PIP and TIP genes from rice, maize and *A. thaliana* has shown the presence of putative regulatory elements such as DREs (drought responsive elements), LTREs (low temperature responsive elements) and ABREs (ABA responsive elements) [44,45]. In wheat, the *Td*PIP2;1 gene sequence revealed the presence of three highly conserved motifs in positions (261(+) CNGTTR; 435(+) CNGTTR; and 526(+) CNGTTR or CANNTG); identified at MYBCORE transcription factor binding sites responsive to water stress and induced by dehydration [36]. In addition, the sequence analysis of pro*Td*PIP2;1 revealed the presence of WRKY transcription factor required for positive and negative regulation of abscisic acid signalling [40]. This indicates the regulation of these genes at a transcriptional level in response to abiotic stresses [44,45].

To confirm the results obtained in vitro for the transgenic wheat plants transformed with the *Td*PIP2;1 gene driven by its own promoter (PR3 and PR6) or the constitutive 35S promoter (SP3) under salt and osmotic (mannitol) stress, we designed an experiment under greenhouse conditions and tested these transgenic plants for abiotic stress tolerance. Our results showed that transpiration rates were affected in control wheat plants under drought stress by closure of stomata and changes in leaf morphology. If drought stress was initiated at the vegetative phase, the first perceived response was a relatively quick decrease in leaf expansion, leaf rolling and stomata closure (Figure 4). Interestingly, the PR3 line, which is the most sensitive to stomata closure for a high FTSW, was the most tolerant line to drought stress compared to SP3 and PR6 (Figure 5). In this case, there was a significant linear decline of the FTSW threshold ($r^2 = 0.656$).

Under abiotic stress, oxidative damage can be induced in many plants such as pea, rice and tomato [46,47]. ROS degrades polyunsaturated lipids forming MDA which is considered a biomarker of the level of lipid peroxidation. Accordingly, our results indicated that drought stress induced oxidative stress in transgenic *Td*PIP2;1 wheat lines manifested by low MDA content and high $H_2O_2$ contents (Figure 8a,b). Nevertheless, under drought stress, the lipid peroxidation level and $H_2O_2$ content were higher in WT than in transgenic lines SP3, PR3 and PR6. These data suggest that the transgenic wheat lines were preserved against oxidative damage under drought stress.

For the antioxidant enzymes activities, drought stress induced up-regulation of the activities in WT and transgenic wheat lines. When comparing the enzymatic activity of these genotypes, higher activities were obtained in the transgenic lines than in WT (Figure 8c,d). This suggested that transgenic lines present a higher capacity for scavenging ROS than WT under stress conditions. Consequently, the reduction in the content of $H_2O_2$ in transgenic lines is the result of the SOD reaction, which is followed by an increased enzymatic activity to catalyse the dismutation of the superoxide radical. The correlation between abiotic stress tolerance and an increase in SOD activity has been reported in many previous works [48–52]. Overall, it seems that the salt tolerance phenotype of transgenic wheat lines could be the result of either a low production of ROS or a better capacity to prevent ROS than the WT plants.

## 4. Materials and methods

### 4.1. Plant Material

Wheat seeds (*Triticum turgidum* L. subsp. *durum*), cv. Maali (New seed lot) were supplied by INRAT, Laboratoire de Physiologie Végétale (Tunis, Tunisia) and were used for genetic transformation.

### 4.2. Construction of the Binary Vector and Wheat Transformation Procedure

To perform gene expression and histochemical analysis, we used translational fusion with the reporter GUS gene, Pr*Td*PIP2;1::*Td*PIP2;1::GUS (named PR) and CaMV35S::*Td*PIP2;1::GUS (named SP) [40]. The resulting constructs were then transferred into *Agrobacterium tumefaciens* hyper-virulent strain EHA105 [53] using the freeze–thaw method [54] and finally used for durum wheat transformation experiments. The transformation method employed in this study was slightly modified from the applied protocol to bread wheat [55]. *A. tumefaciens* strains, harbouring the resulting constructs (PR and SP), were cultured on MG/L agar medium [56] containing Hygromycin (50 μg/mL) and Rifampicin (10 μg/mL) for two days at 28 °C. Bacteria were harvested when they reached an $OD_{600} = 1$ and were suspended in an $R_2$ co-cultivation liquid medium ($R_2$CL, pH 5.2) containing $R_2$ Basic medium [57], glucose 10 g/L and 100 μM acetosyringone. An aliquot of 5 mL from this culture was diluted five times with the same $R_2$CL medium containing the appropriate antibiotics and kept growing at 28 °C with shaking until $OD_{600}$ reached 0.4–0.6. Durum wheat seeds cv. Maali were soaked first in 70% (*v/v*) ethanol for 3 min, then in sodium hypochlorite 1% (*v/v*) with gentle shaking for 20 min for sterilization. After that, seeds were washed five times with sterile distilled water. Seeds were additionally soaked at 4 °C in 0.1% (*v/v*) Pelt 44 (thiophanate-methyl at 450 g/L) fungicide (Bayer CropScience) and then germinated on wet filter paper with R2 basic medium at 25 °C for 2 days in the dark. A needle (Terumo Syringe 1 mL Insulin (U100) with 29 g × 0.5 needle ×100: BS-N1H2913) that had been previously filled with *A. tumefaciens* inoculum solution, was used to pierce twice, to a depth of approximately 1 mm, the embryonic apical meristem from the 2-d-old seedlings. The inoculated seeds were then co-cultivated on wet filter paper at 25 °C for an additional two days in the dark to avoid overgrowth of the Agrobacterium on the seeds. To eliminate the inoculum of Agrobacterium, seeds were dipped overnight at 4 °C in an aqueous solution containing Cefotaxime (1 g/L). Eventually, seeds were immersed in sodium hypochlorite 0.1% (*v*/*v*) for 15 min and then rinsed with sterile distilled water for five times. Ultimately, the inoculated wheat seeds were sown in pots filled with a mixture of sand (50%) and peat (50%), allowed to grow to maturity and to produce seeds (T1) under greenhouse conditions. The regenerated hygromycin-resistant wheat plants (T1) were numbered and named as indicated above (SP and PR). For the control, we used the WT wheat plants.

### 4.3. Histochemical GUS Staining

To help understand the presence of staining in tissues for GUS activity, a histochemical assay was carried out by incubating wheat seedlings under vacuum infiltration in 50 mM $Na_2HPO_4$ buffer (pH 7.0), 0.5 mM $K_3(Fe[CN]_6)$, 0.5 mM $K_4(Fe[CN]_6)$, 0.1% Triton X-100 and 1 mg/L X-Gluc (5-bromo-4-chloro-3-indolyl β-D-glucuronide cyclohexyl ammonium salt) for several minutes and then keeping them overnight at 37 °C. The chlorophyll and pigments were eliminated by soaking the stained seedlings tissues for several hours in 70% ethanol. Finally, the destained tissues were photographed. Transgenic plants overexpressing the *Td*PIP2;1 gene driven by the CaMV35S promoter and WT wheat plants were considered the positive and negative controls, respectively.

### 4.4. GusA Expression Analysis by the Quantitative Real-Time qPCR

Transcription levels of the *GusA* gene under the control of the Pr*Td*PIP2;1 or CaMV35S promoters were assessed by real-time qPCR in T3 homozygous transgenic wheat plants (leaves and roots of two-week-old plants) treated with 150 mM NaCl or 20% PEG 6000 for 48 h. The Trizol method (Invitrogen) was employed to extract total RNA from 100 mg plant tissues in accordance with the manufacturer's recommendations [39]. To check the concentration and the quality of the extracted RNA, agarose gel (1% RNase-free) and spectrophotometer measurements (260/280 nm) were performed. To synthesize first strand cDNA, 3 μg of total RNA was treated with DNase I (Promega, USA) and then reverse transcribed using oligo (dT)18 and SuperScript II reverse transcriptase (Invitrogen, Canada), per the supplier's recommendations. Primer 3 software was used to identify the primer pairs that amplify fragments of GusA and the housekeeping actin genes. The primers used for real-time qPCR were: qGusAF, qGusAR, q-Act-F and q-Act-R, (Table 2). PCR reactions were performed in a 10 μL final volume containing 4 μL cDNA (obtained from 40 ng of DNase-treated RNA), 0.5 μL of each primer (at 10 μM), 5 μL 2× SYBR Green I master mix and 1 μL of RNase-free water (Sigma, Canada). The reaction consisted of an initial denaturation step at 94 °C for 10 min followed by 45 cycles composed of 10 s at 94 °C, 10 s at 60 °C and 15 s at 72 °C, then a melting curve (5 s at 95 °C, 1 min at 65 °C and 5 min with temperature increasing from 65 °C to 97 °C). For each experimental condition three biological repetitions were performed, with three technical repetitions for each sample. The threshold cycle (CT) values of the triplicate PCRs were averaged and used for transcripts quantification at the end of the reaction. The relative expression ratio of the GusA gene was calculated by using the comparative CT method with the actin gene as an internal expression standard [58]. The relative expression level was calculated from triplicate measurements based on the 2-ΔΔCT, where ΔΔCT = (CT, Target gene—CT, Actin) stressed—(CT, Target gene—CT, Actin) control. Relative expression ratios from three independent experiments (three biological repetitions) were reported.

**Table 2.** List of primers used in real-time PCR.

| Primers | Nucleotides Sequences (5′-3′) |
|---|---|
| Act F: | TGCATAGAGGGAAAGCACG |
| Act R: | AACCCAAAAGCCAACAACAGAGA |
| q-GusF: | CACGCCGTATGTTATTGCCG |
| q-GusR: | TCTTGCCGTTTTCGTGGGTA |

### 4.5. *Evaluation of Transgenic Wheat Plants for Abiotic Stress Tolerance*

#### 4.5.1. In Vitro Assays

Seeds of wild type and transgenic homozygous T3 generation of durum wheat cv. Maali were employed for abiotic stress assays. For three experimental repetitions, thirty seeds from WT and transgenic plants were surface-sterilized by soaking in 70% ethanol for 1 min, washed with sterile distilled water and treated with 40% solution of sodium hypochlorite for 30 min. Finally, seeds were rinsed five times with sterile distilled water. Seeds were then incubated in sterile distilled water and maintained in a growth chamber (16 h of light per day, 500 $\mu E\ m^{-2}\ S^{-1}$, 28 °C/25 °C day/night) for germination and growth rate parameters. After 2 days, seeds from T3 transgenic lines (PR and SP) and WT plant were sown in Petri dishes on two layers of filter paper saturated with distilled water, supplemented with 0 or 150 mM NaCl. Plates were incubated at 25 °C in a growth chamber under a 16 h light/8 h dark photoperiod. Germination rates were recorded after one week of incubation. The seed germination experiment was repeated three times. Young seedlings were transplanted to a new medium supplemented with 150 mM NaCl or 300 mM Mannitol to estimate the growth rate under osmotic or salt stress conditions. These dishes were removed and placed vertically in a growth chamber. After 15 days of culture, root/shoot lengths were determined for WT and transgenic seedlings using the UTHSCSA Image Tool, a free image processing and analysis program used to acquire, display, edit and analyse images (available online: http://www.ddsdx.uthscsa.edu/dig/iTdesc.html).

#### 4.5.2. Greenhouse Assays

To assess the effect of salt and drought treatments on biomass production, seeds from WT and transgenic lines (SP and PR) were germinated under greenhouse conditions. To initiate salt treatments, pots were irrigated with water supplemented with 150 mM NaCl. Soil moisture regimes were monitored daily by weighing the pots. For both salt and drought treatments, seeds were allowed to germinate and to grow for additional 20 days before harvesting the plants to measure their fresh weight, dry weight and plant transpiration rate. Seeds from T3 transgenic lines (PR and SP) and WT plants were sown in pots filled with peat moss and grown for 2 weeks before challenging them with stress treatments. Under control conditions, pots were irrigated weekly with fresh water to maintain plants at 100% FC. For salt stress treatment, the same irrigation program was used for watering the plants with addition of 150 mM NaCl. This NaCl concentration was maintained until the end of the plant cycle. Young leaves (top) and old leaves (bottom), were collected from salt-stressed and control plants and then dried at 80 °C for 24 h. Finally, the dried material was incubated in 0.5% $HNO_3$ for a week. The $Na^+$, $Ca^{2+}$ and $K^+$ contents were analysed in the filtrate using atomic absorption spectrophotometry. For drought stress, plants were maintained at 40% FC until the end of plant cycle. This was achieved by refiling the exact amount of water lost by evapotranspiration.

#### 4.5.3. Water Treatments

To monitor the soil water status, we used the fraction of transpirable soil water (FTSW). According to Sinclair et al. [59], when water stress is expressed as FTSW, plants respond to the progressive drying of soil in a similar manner.

Calculation of each pot's FTSW value was carried out according to Sinclair and Ludlow's method [60]. First, pots were fully watered daily and drained overnight until a day prior to initiating measurements. After one day of drainage, all pots were weighted to assess their initial water holding capacity. The total transpirable soil water (TTSW) was obtained by calculating the difference between initial pot capacities (Wi) and final pot weight after soil desiccation (Wf). The ratio of actual transpirable soil water (ATSW) to TTSW allowed the estimation of FTSW, where ATSW was the mass difference between daily (Wt) and final pot weight. A single drought cycle was implemented in half of the pots, when the plants had developed 8 leaves (five pots for each genotype).

$$FTSW = ATSW/TTSW = (Wt - Wf)/(Wi - Wf) \quad (1)$$

TTSW: total transpirable soil water; ATSW: actual transpirable soil water.

To restore the daily water loss, the remaining five control pots were fully watered in late afternoon. In contrast, for the stressed pots, water deficit was maintained by withholding irrigation and covering the pots with a plastic bag in order to prevent soil evaporation. Original FTSW values were assigned a value of 1. When a transpiration rate of less than 10% was reached for each stressed pot in comparison to the control, the experiment was terminated [60].

4.5.4. Plant Transpiration Rate

Plant transpiration (TRj) per unit leaf area (mmol·m$^{-2}$·s$^{-1}$) was determined every day for each pot as the mass difference between weights every 24 h (including the watering for unstressed pots) and were divided by plant total leaf area on the previous day. At the same time, stomatal conductance (mmol·m$^{-2}$·s$^{-1}$) was measured each morning between 9:00 a.m. and 12:00 p.m. on the last panicle leaf per plant. These measures were acquired with a porometer (type AP4-UM-3) on the control and treated plants. Conductance measurements were performed on leaves and on its abaxial face, where stomatal density is greater.

$$\text{Normalized transpiration rate (NTR)} = \text{transpiration of stressed plants/average transpiration of control plants} = (Gs/Gsmax) \quad (2)$$

Porometry was also used to determine the last day of measurements (when gs(stressed)/gs(control) was less than 0.1), allowing us to estimate the value of the TTSW of each pot. The mean TTSW was remarkably stable.

### *4.6. Whole Plant Response Modelling to Increasing Water Deficit*

To monitor the daily transpiration rate fluctuation due to evaporative demand changes, the drought treatment values were divided by the corresponding mean values on the well-watered (control) treatment in order to calculate the relative TR values daily. When the soil was still moist, daily TR ratio values for FTSW > 0.6 were adjusted to a mean value of 1, in order to minimize the effect of variation in initial plant size variation. It was essential to normalize the calculation of plant transpiration (NTR) for the accuracy of a further linear model. A two-slope linear relation with one parameter (FTSWt) was used to meet the plant responses to water deficit, indicating the FTSW threshold below which conductance starts decreasing. By applying a two-slope linear model to the experimental data, the parameter FTSWt was thus estimated (Equation (3)):

$$\begin{gathered}\text{if } FTSW \geq FTSWt \\ y = 1; \\ \text{else } y = 1/FTSWt \times FTSW; \\ y \text{ being NTR}\end{gathered} \quad (3)$$

### *4.7. Lipid Peroxidation*

By using the method of Ben Amor et al. [49] to calculate the amount of MDA in the leaves, the extent of lipid peroxidation was estimated. 1.0 mL of 0.5% (*w*/*v*) thiobarbituric acid (TBA) in 20% (*w*/*v*) TCA solution was added to an aliquot of 0.5 mL from the homogenate of fresh shoots in 0.1% (*w*/*v*) trichloroacetic acid (TCA) solution. Subsequently, the mixture was incubated at 90 °C for 30 min and then cooled on ice. By measuring absorbance at 532 and 600 nm in reference to an MDA standard curve, the equivalent MDA was calculated.

### 4.8. *Quantitative $H_2O_2$ Measurement*

The concentration of $H_2O_2$ was estimated according to the method described by Velikova et al. [61]. Fresh shoots tissue (0.5 g) was homogenized with 5 mL of 0.1% (*w*/*v*) TCA. This homogenate was then centrifuged at 12,000× *g* for 15 min. 10 mM phosphate buffer (pH 7.0) and 1 M potassium iodide were added to 0.5 mL of the supernatant. The mixture was then vortexed and its absorbance at 390 nm was recorded and $H_2O_2$ content was calculated using a standard curve with concentration ranging from 0.05 to 0.1 mM.

### 4.9. *Enzyme Assays*

Aliquots of frozen fresh shoot material (0.5 g) were ground to a fine powder with liquid nitrogen and homogenized in a cold solution containing 100 mM Tris–HCl buffer (pH 8.0), 10 mM EDTA (Ethylenediaminetetraacetic acid), 50 mM KCl, 20 mM $MgCl_2$, 0.5 mM PMSF (Phenyl methyl sulfonyl fluoride) and 2% (*w*/*v*) PVP (Polyvinylpyrolidone). The homogenate was centrifuged at 14,000× *g* for 30 min at 4 °C and the supernatant was used for determination of the antioxidative enzyme activities.

Total catalase (CAT) activity was measured according to the method of Aebi [62], by monitoring the decline in absorbance at 240 nm as $H_2O_2$ was consumed. An aliquot of crude enzyme extract was added to the reaction mixtures containing 50 mM phosphate buffer (pH 7), 30 mM $H_2O_2$. One unit of CAT was defined as 1 μmol $mL^{-1}$ $H_2O_2$ decomposed per minute.

Total superoxide dismutase (SOD) activity was determined by measuring the percentage of inhibition of the pyrogallol autoxidation [63]. A crude enzyme extract aliquot was added to the reaction mixture containing 10 mM pyrogallol in Tris-cacodylic acidediethylene triamine penta acetic acid buffer (pH 7.4–8). One unit of SOD was defined as the enzyme quantity required to inhibit 50% of the pyrogallol autoxidation.

### 4.10. *Statistical Analysis*

All data analysis was carried out using the Two-way Analysis of Variance (ANOVA) with SPSS package. Means were compared using Tukey's (HSD) multiple comparison procedure.

## 5. Conclusions

In our previously published works, we reported that the expression of the *Td*PIP2;1 gene was strongly associated with abiotic stress response pathways in Tobacco [39] and rice [40]. In the present study, we found similar results, confirming that the expression of the *Td*PIP2;1 gene was closely related to abiotic stress response pathways in durum wheat, a staple crop, without causing any undesirable growth phenotypes or yield penalty. In addition, we found that mannitol, a major photosynthetic product in many higher plants, enhanced the activity of the *Td*PIP2;1 promoter, suggesting that Pr*Td*PIP2;1 may be involved in dehydration response. Our results provide insights for the *Td*PIP2;1 gene regulated by its own promoter which generates enhanced drought tolerance in transgenic wheat. In general, regulation of the activity and gene expression of aquaporins are considered part of the adaptation mechanisms to stress conditions through complex processes. Further studies are required to gain a deeper understanding of the function of this gene, particularly through transcriptome profiling and identification of *Td*PIP2;1 interacting proteins in plant PIPs. Recent advances in genome editing with CRISPR/Cas9 was efficiently used in combination with haploid induction (HI) to induce edits in nascent seeds of diverse monocot species [64]. HI enables widespread application of genome editing technology for crop improvement and offers a new path to understand the stress resistance mechanisms, when generating knockout of down-regulated AQP genes in wheat.

**Author Contributions:** K.M. conceived and designed the experiments. M.A., K.M. and F.B. performed the experiments, analysed the data and wrote the paper. All authors read and approved the manuscript.

**Acknowledgments:** We are thankful to the Plant Physiology Laboratory, INRAT, Tunisia for the provision of durum wheat seed lots of cv. Maali. We are thankful for the help with the editing of this manuscript by Elizabeth Whitehouse from the writing centre at UAEU. We are also thankful to Kilani Ghoudi from the Statistics Department at UAEU for his help with the analysis.

## References

1. Chaves, M.M.; Maroco, J.P.; Pereira, J.S. Understanding plant response to drought-from genes to the whole plant. *Funct. Plant Biol.* **2003**, *30*, 239–264. [CrossRef]
2. Munne-Bosch, S.; Jubany-Mari, T.; Alegre, L. Drought-induced senescence is characterized by a loss of antioxidant defences in chloroplasts. *Plant Cell Environ.* **2001**, *24*, 1319–1327. [CrossRef]
3. Jubany-Mari, T.; Munne-Bosch, S.; Alegre, L. Redox regulation of water stress responses in field-grown plants. Role of hydrogen peroxide and ascorbate. *Plant Physiol. Biochem.* **2010**, *48*, 351–358. [CrossRef]
4. Kyparissis, A.; Drilias, P.; Manetas, Y. Seasonal fluctuations in photoprotective (xanthophyll cycle) and photoselective (chlorophylls) capacity in eight Mediterranean plant species belonging to two different growth forms. *Funct. Plant Biol.* **2000**, *27*, 265–272. [CrossRef]
5. Kyparissis, A.; Petropoulou, Y.; Manetas, Y. Summer survival of leaves in a soft-leaved shrub (*Phlomis fruticosa* L., Labiatae) under Mediterranean field conditions: Avoidance of photoinhibitory damage through decreased chlorophyll contents. *J. Exp. Bot.* **2000**, *46*, 1825–1831. [CrossRef]
6. Havaux, M.; Tardy, F. Loss of chlorophyll with limited reduction of photosynthesis as an adaptive response of Syrian barley landraces to high-light and heat stress. *Aust. J. Plant Physiol.* **1999**, *26*, 569–578. [CrossRef]
7. Beck, P.A.; Hutchison, S.; Gunter, A.S.; Losi, C.T.; Stewart, B.C.; Capps, K.P.; Phillips, J.M. Chemical composition and in situ dry matter and fiber disappearance of sorghum x Sudangrass hybrids. *J. Anim. Sci.* **2007**, *85*, 545–555. [CrossRef] [PubMed]
8. Pelleschi, S.; Rocher, J.P.; Prioul, J.L. Effect of water restriction on carbohydrate metabolism and photosynthesis in mature maize leaves. *Plant Cell Environ.* **1997**, *20*, 493–503. [CrossRef]
9. Pinheiro, C.; Chaves, M.M.; Ricardo, C.P. Alterations in carbon and nitrogen metabolism induced by water deficit in the stems and leaves of *Lupinus albus* L. *J. Exp. Bot.* **2001**, *52*, 1063–1070. [CrossRef]
10. Yang, J.; Zhang, J.; Wang, Z.; Zhu, Q.; Wang, W. Remobilization of carbon reserves in response to water-deficit during grain filling of rice. *Field Crops Res.* **2001**, *71*, 47–55. [CrossRef]
11. Toumi, I.; Gargouri, M.; Nouairi, I.; Moschou, P.; Salem-Fnayou, A.; Mliki, A.; Zarrouk, M.; Ghorbel, A. Water stress induced changes in the leaf lipid composition of four grapevine genotypes with different drought tolerance. *Biol. Plant.* **2008**, *52*, 161–164. [CrossRef]
12. Passioura, J. Increasing crop productivity when water is scarce- from breeding to field management. In Proceedings of the 4th International Crop Science Congress: New Directions for a Diverse Planet, Brisbane, Australia, 26 September–1 October 2004.
13. Cornic, G. Drought stress inhibits photosynthesis by decreasing stomatal aperture–not by affecting ATP synthesis. *Trends Plant Sci.* **2000**, *5*, 187–188. [CrossRef]
14. Chen, T.H.; Murata, N. Enhancement of tolerance of abiotic stress by metabolic engineering of betaines and other compatible solutes. *Curr. Opin. Plant Biol.* **2002**, *5*, 250–257. [CrossRef]
15. Zhang, J.; Kirkham, B. Antioxidant responses to drought in sunflower and sorghum seedlings. *New Phytol.* **1996**, *132*, 361–373. [CrossRef]
16. Chaumont, F.; Tyerman, S.D. Aquaporins: Highly regulated channels controlling plant water relations. *Plant Physiol.* **2014**, *164*, 1600–1618. [CrossRef]
17. Pandey, B.; Sharma, P.; Pandey, D.M.; Sharma, I.; Chatrath, R. Identification of New Aquaporin Genes and Single Nucleotide Polymorphism in Bread Wheat. *Evol. Bioinform.* **2013**, *9*, 437–452. [CrossRef]
18. Zhuang, L.; Liu, M.; Yuan, X.; Yang, Z.; Huang, B. Physiological Effects of Aquaporin in Regulating Drought Tolerance through Overexpressing of *Festuca arundinacea* Aquaporin Gene *FaPIP2;1*. *J. Am. Soc. Hort. Sci.* **2015**, *140*, 404–412. [CrossRef]
19. Kaldenhoff, R.; Kei, L.; Uehlein, N. Aquaporin and membrane diffusion of $CO_2$ in living organisms. *Biochim. Biophys. Acta-Gen. Subj.* **2013**, *1840*, 1592–1595. [CrossRef] [PubMed]
20. Bienert, G.P.; Schjoerring, J.K.; Jahn, T.P. Membrane transport of hydrogen peroxide. *Biochim. Biophys. Acta-Biomembr.* **2006** *1758*, 994–1003. [CrossRef]
21. Maurel, C.; Verdoucq, L.; Luu, D.T.; Santoni, V. Plant aquaporins: Membrane channels with multiple

integrated functions. *Ann. Rev. Plant Biol.* **2008**, *59*, 595–624. [CrossRef] [PubMed]
22. Chaumont, F.; Barrieu, F.; Jung, R.; Chrispeels, M.J. Plasma membrane intrinsic proteins from maize cluster in two sequence subgroups with differential aquaporin activity. *Plant Physiol.* **2000**, *122*, 1025–1034. [CrossRef] [PubMed]
23. Fetter, K.; van Wilder, V.; Moshelion, M.; Chaumont, F. Interactions between plasma membrane Aquaporins modulate their water channel activity. *Plant Cell* **2004**, *16*, 215–228. [CrossRef] [PubMed]
24. Chaumont, F.; Moshelion, M.; Daniels, M.J. Regulation of plant aquaporin activity. *Biol. Cell* **2005**, *97*, 749–764. [CrossRef]
25. Javot, H.; Maurel, C. The role of aquaporins in root water uptake. *Ann. Bot.* **2002**, *90*, 301–313. [CrossRef] [PubMed]
26. Maurel, C.; Javot, H.; Lauvergeat, V.; Gerbeau, P.; Tournaire, C.; Santoni, V. Molecular physiology of aquaporins in plants. *Int. Rev. Cytol.* **2002**, *215*, 105–148.
27. Heinen, R.B.; Ye, Q.; Chaumont, F. Role of aquaporins in leaf physiology. *J. Exp. Bot.* **2009**, *60*, 2971–2985. [CrossRef] [PubMed]
28. Hachez, C.; Zelazny, E.; Chaumont, F. Modulating the expression of aquaporin genes in planta: A key to understand their physiological functions? *Biochim. Biophys. Acta (BBA)-Biomembr.* **2006**, *1758*, 1142–1156. [CrossRef]
29. Guo, L.; Wang, Z.Y.; Lin, H.; Cui, W.E.; Chen, J.; Liu, M.H.; Chen, Z.L.; Qu, L.J.; Gu, H.Y. Expression and functional analysis of the rice plasma-membrane intrinsic protein gene family. *Cell Res.* **2006**, *16*, 277–286. [CrossRef]
30. Liu, C.; Fukumoto, T.; Matsumoto, T.; Gena, P.; Frascaria, D.; Kaneko, T.; Katsuhara, M.; Kitagawa, Y. Aquaporin *Os*PIP1;1 promotes rice salt resistance and seed germination. *Plant Physiol. Biochem.* **2013**, *63*, 151–158. [CrossRef]
31. Katsuhara, M.; Koshio, K.; Shibasaka, M.; Hayashi, Y.; Hayakawa, T.; Kasamo, K. Overexpression of a barley aquaporin increased the shoot/root ratio and raised salt sensitivity in transgenic rice plants. *Plant Cell Physiol.* **2003**, *44*, 1378–1383. [CrossRef]
32. Cui, X.H.; Hao, F.S.; Chen, H.; Chen, J.; Wang, X.C. Expression of the *Vicia faba VfPIP1* gene in Arabidopsis thaliana plants improves drought resistance. *J. Plant Res.* **2003**, *121*, 207–214. [CrossRef] [PubMed]
33. Alavilli, H.; Awasthi, J.P.; Rout, G.R.; Sahoo, L.; Lee, B.H.; Panda, S.K. Overexpression of a Barley Aquaporin gene, HvPIP2;5 confers salt and osmotic stress tolerance in yeast and plants. *Front. Plant Sci.* **2016**, *7*, 1566. [CrossRef]
34. Sreedharm, S.; Shekhawat, U.K.; Gamapathi, T.R. Transgenic Banana plants overexpressing a native plasma membrane aquaporin MusaPIP1;2 display high tolerance levels to different abiotic stresses. *Plant Biotechnol. J.* **2013**, *8*, 942–952. [CrossRef] [PubMed]
35. Zhuo, C.; Wang, T.; Guo, Z.; Lu, S. Overexpression of MfPIP2-7 from *Medicago falcata* promotes cold tolerance and growth under $NO^{-}_{3}$ deficiency in transgenic tobacco plants. *BMC Plant Biol.* **2016**, *16*, 138. [CrossRef] [PubMed]
36. Xu, C.; Wang, M.; Zhou, L.; Quan, T.; Xia, G. Heterologous expression of the wheat Aquaporin gene *Ta*PIP2;2 compromises the abiotic stress tolerance of Arabidopsis thaliana. *PLoS ONE* **2013**, *8*, e79618.
37. Aharon, R.; Shahak, Y.; Winniger, S.; Bendov, R.; Kapulink, Y.; Galili, G. Overexpression of a plasma membrane aquaporin in transgenic tobacco improves plant vigour under favourable growth conditions but not under drought or salt stress. *Plant Cell* **2003**, *15*, 439–447. [CrossRef]
38. Kapilan, R.; Vaziri, M.; Zwiazek, J.J. Regulation of aquaporins in plants under stress. *Biol. Res.* **2018**, *51*, 4. [CrossRef] [PubMed]
39. Ayadi, M.; Cavez, D.; Miled, N.; Chaumont, F.; Masmoudi, K. Identification and characterization of two plasma membrane aquaporins in durum wheat (*Triticum turgidum* L. *subsp. durum*) and their role in abiotic stress tolerance. *Plant Physiol. Biochem.* **2011**, *49*, 1029–1039. [CrossRef] [PubMed]
40. Ayadi, M.; Mieulet, D.; Fabre, D.; Verdeil, J.L.; Vernet, A.; Guiderdoni, E.; Masmoudi, K. Functional analysis of the durum wheat gene *Td*PIP2; 1 and its promoter region in response to abiotic stress in rice. *Plant Physiol. Biochem.* **2014**, *79*, 98–108. [CrossRef]
41. Gago, J.; Grima-Pettenati, J.; Gallego, P.P. Vascular-specific expression of GUS and GFP reporter genes in transgenic grapevine (*Vitis vinifera* L. cv. Albariño) conferred by the EgCCR promoter of *Eucalyptus gunnii*. *Plant Physiol. Biochem.* **2011**, *49*, 413–419. [CrossRef] [PubMed]

42. Sade, N.; Galle, A.; Flexas, J.; Lerner, S.; Peleg, G.; Yaaran, A.; Moshelion, M. Differential tissue-specific expression of *Nt*AQP1 in Arabidopsis thaliana reveals a role for this protein in stomatal and mesophyll conductance of CO2 under standard and salt-stress conditions. *Planta* **2013**, *239*, 357–366. [CrossRef]
43. Lu, L.; Dong, C.; Liu, R.; Zhou, B.; Wang, C.; Shou, H. Roles of soybean plasma membrane intrinsic protein GmPIP2;9 in drought tolerance and seed development. *Front. Plant Sci.* **2018**, *9*, 530. [CrossRef]
44. Forrest, K.L.; Bhave, M. The PIP and TIP aquaporins in wheat form a large and diverse family with unique gene structures and functionally important features. *Funct. Integr. Genom.* **2008**, *8*, 115e133. [CrossRef]
45. Siefritz, F.; Tyree, M.T.; Lovisolo, C.; Schubert, A.; Kaldenhoff, R. PIP1 plasma membrane aquaporins in tobacco: From cellular effects to function in plants. *Plant Cell* **2002**, *14*, 869–876. [CrossRef] [PubMed]
46. Mittova, V.; Tal, M.; Volokita, M.; Guy, M. Up-regulation of the leaf mitochondrial and peroxisomal antioxidative systems in response to salt-induced oxidative stress in the wild salt-tolerant tomato species *Lycopersicon pennellii. Plant Cell Environ.* **2003**, *26*, 845–856. [CrossRef]
47. Deng, L.; Chen, F.; Jiang, L.; Lam, H.M.; Xiao, G. Ectopic expression of *GmPAP3* enhances salt tolerance in rice by alleviating oxidative damage. *Plant Breed.* **2014**, *133*, 348–355. [CrossRef]
48. Hernández, I.; Alegrel, L.; Munné-Bosch, S. Drought-induced changes in flavonoids and other low molecular weight antioxidants in *Cistus clusii* grown under Mediterranean field conditions. *Tree Physiol.* **2004**, *24*, 1303–1311. [CrossRef]
49. Ben Amor, N.; Ben Hamed, K.; Debez, A.; Grignon, C.; Abdelly, C. Physiological and antioxidant responses of the perennial halophyte *Crithmum maritimum* to salinity. *Plant Sci.* **2005**, *168*, 889–899. [CrossRef]
50. Mandhania, S.; Madan, S.; Sawhney, V. Antioxidant defense mechanism under salt stress in wheat seedling. *Biol. Plant.* **2006**, *227*, 227–231. [CrossRef]
51. Koca, H.; Bor, M.; Ozdemir, F.; Turkan, I. Effect of salt stress on lipid peroxidation, antioxidative enzymes and proline content of sesame cultivars. *Environ. Exp. Bot.* **2007**, *60*, 344–351. [CrossRef]
52. Ellouzi, H.; Ben Hamed, K.; Cela, J.; Munne'-Bosch, S.; Abdelly, C. Early effects of salt stress on the physiological and oxidative status of *Cakile maritima* (halophyte) and *Arabidopsis thaliana* (glycophyte). *Physiol. Plant.* **2011**, *142*, 128–143. [CrossRef] [PubMed]
53. Hood, E.E.; Gelvin, S.B.; Melchers, S.; Hoekema, A. New Agrobacterium helper plasmids for gene transfer to plant. *Transgenic Res.* **1993**, *2*, 208–218. [CrossRef]
54. Chen, H.; Nelson, R.S.; Sherwood, J.L. Enhanced recovery of transformants of *Agrobacterium tumifaciens* after freeze–thaw transformation and drug selection. *Biotechniques* **1994**, *16*, 664–668. [PubMed]
55. Supartana, P.; Shimizu, T.; Nogawa, M.; Shioiri, H.; Nakajima, T.; Haramoto, N.; Nozue, M.; Kojima, M. Development of simple and efficient in-planta transformation method for wheat (*Triticum aestivum* L.) using *Agrobacterium tumefaciens*. *J. Biosci Bioeng.* **2006**, *102*, 162–170. [CrossRef] [PubMed]
56. Garfinkel, D.J.; Nester, E.W. *Agrobacterium tumefaciens* mutants affected in crown gall tumorigenesis and octopine catabolism. *J. Bacteriol.* **1980**, *144*, 732–774. [PubMed]
57. Ohira, K.; Ojima, K.; Fujiwara, A. Studies on the nutrition of rice cell culture 1: A simple defined medium for rapid growth in suspension culture. *Plant Cell Physiol.* **1973**, *14*, 1113–1121.
58. Livak, K.J.; Schmittgen, T.D. Analysis of relative gene expression data using real-time quantitative PCR and the 2 DDCT method. *Methods* **2001**, *25*, 402–408. [CrossRef]
59. Sinclair, T.R.; Zwieniecki, M.A.; Holbrook, N.M. Low leaf hydraulic conductance associated with drought tolerance in soybean. *Physiol. Plant.* **2008**, *132*, 446–451. [CrossRef]
60. Sinclair, T.R.; Ludlow, M.M. Influence of soil water supply on the plant water balance of four tropical grain legumes. *Aust. J. Plant Physiol.* **1986**, *13*, 329–341. [CrossRef]
61. Velikova, V.; Yordanov, I.; Edreva, A. Oxidative stress and some antioxidant system in acid rain treated bean plants: Protective role of exogenous polyamines. *Plant Sci.* **2000**, *151*, 59–66. [CrossRef]
62. Aebi, H. Catalase in vitro. *Method Enzymol.* **1984**, *105*, 121–126.
63. Marklund, S.; Marklund, G. Involvement of the superoxide anion radical in the autoxidation of pyrogallol and a convenient assay for superoxide dismutase. *Eur. J. Biochem.* **1974**, *47*, 469–474. [CrossRef] [PubMed]
64. Kelliher, T.; Starr, D.; Su, X.; Tang, G.; Chen, Z.; Carter, J.; Wittich, P.E.; Dong, S.; Green, J.; Burch, E.; et al. One-step genome editing of elite crop germplasm during haploid induction. *Nat. Biotechnol.* **2019**, *37*, 287–292. [CrossRef] [PubMed]

# 14

# Using the 6RL$^{Ku}$ Minichromosome of Rye (*Secale cereale* L.) to Create Wheat-Rye 6D/6RL$^{Ku}$ Small Segment Translocation Lines with Powdery Mildew Resistance

**Haimei Du [1,†], Zongxiang Tang [1,2,†], Qiong Duan [1,2], Shuyao Tang [1,2] and Shulan Fu [1,*]**

1 College of Agronomy, Sichuan Agricultural University, Wenjiang, Chengdu 611130, China; 18314441196@163.com (H.D.); zxtang@sicau.edu.cn (Z.T.); 18380443767@163.com (Q.D.); tangshuyao705708@sina.com (S.T.)

2 Institute of Ecological Agriculture, Sichuan Agricultural University, Wenjiang, Chengdu 611130, China

* Correspondence: fushulan@sicau.edu.cn

† These authors contributed equally to this work.

**Abstract:** Long arms of rye (*Secale cereale* L.) chromosome 6 (6RL) carry powdery mildew resistance genes. However, these sources of resistance have not yet been successfully used in commercial wheat cultivars. The development of small segment translocation chromosomes carrying resistance may result in lines carrying the 6R chromosome becoming more commercially acceptable. However, no wheat-rye 6RL small segment translocation line with powdery mildew resistance has been reported. In this study, a wheat-rye 6RL$^{Ku}$ minichromosome addition line with powdery mildew resistance was identified, and this minichromosome was derived from the segment between L2.5 and L2.8 of the 6RL$^{Ku}$ chromosome arm. Following irradiation, the 6RL$^{Ku}$ minichromosome divided into two smaller segments, named 6RL$^{Kumi200}$ and 6RL$^{Kumi119}$, and these fragments participated in the formation of wheat-rye small segment translocation chromosomes 6DS/6RL$^{Kumi200}$ and 6DL/6RL$^{Kumi119}$, respectively. The powdery mildew resistance gene was found to be located on the 6RL$^{Kumi119}$ segment. Sixteen 6RL$^{Kumi119}$-specific markers were developed, and their products were cloned and sequenced. Nucleotide BLAST searches indicated that 14 of the 16 sequences had 91–100% similarity with nine scaffolds derived from 6R chromosome of *S. cereale* L. Lo7. The small segment translocation chromosome 6DL/6RL$^{Kumi119}$ makes the practical utilization in agriculture of powdery mildew resistance gene on 6RL$^{Ku}$ more likely. The nine scaffolds are useful for further studying the structure and function of this small segment.

**Keywords:** wheat; rye; 6R; small segment translocation; powdery mildew

---

## 1. Introduction

It has already been reported that the long arms of rye (*Secale cereale* L.) chromosome 6 (6RL) carry powdery mildew resistance gene *Pm20* [1], and this gene was introduced into wheat background in the form of a 6BS/6RL translocation chromosome [1]. The gene *Pm20* still has a broad spectrum of resistance to *Blumeria graminis* f. sp. *tritici* (Bgt) isolates [2,3]. The 6RL chromosome arm that carries *Pm20* was derived from *S. cereale* L. cv. Prolific [1]. Recently, some reports indicated that 6RL arms derived from *S. cereale* cv. Jingzhouheimai, *S. cereale* cv. German White, and *S. cereale* cv. Kustro also carried powdery mildew resistance genes [3–6]. It has already been established that the powdery mildew resistance gene on 6RL of German White was different from the gene *Pm20* [3], and this indicates that different 6RL arms may also display genetic diversity for powdery mildew resistance genes. However, the powdery mildew resistance genes on 6RL arms have not been successfully used in commercial wheat cultivars because of agronomic disadvantages, possibly caused by non-compensation and linkage drag of the 6RL arm. The development of small segment translocations between wheat chromosomes and 6RL, which causes minimal loss of indispensable wheat genes, may resolve this problem [3,4,7,8]. Only one translocation chromosome carrying a small segment of a 6RL arm, Ti4AS.4AL-6RL-4AL, has been reported. In this case, the 6RL segment was derived from the telomeric region and carried the Hessian fly-resistant gene *H25* [9,10]. So far, no wheat-rye 6RL small segment translocation lines with powdery mildew resistance have been reported. In this study, a wheat-rye $6RL^{Ku}$ minichromosome addition line was developed, and $6D/6RL^{Ku}$ small segment translocation lines were identified from the irradiated seeds of this minichromosome addition line.

## 2. Results

### 2.1. Obtaining Wheat-Rye $6RL^{Ku}$ Minichromosome Addition Line

A $6RL^{Ku}$ minichromosome addition line, $MiA6RL^{Ku}$, was found among the self-pollinated progeny of wheat-rye $6RL^{Ku}$ monotelosomic addition line $MTA6RL^{Ku}$. Line $MiA6RL^{Ku}$ contained one minichromosome derived from the $6RL^{Ku}$ arm (Figure 1). According to the FISH map of $6RL^{Ku}$ arm constructed based on the signal patterns of probes Oligo-pSc200, Oligo-pSc250, and Oligo-pSc119.2-1 [6], the $6RL^{Ku}$ minichromosome was derived from the segment between L2.5 and L2.8 (Figure 1). Through measuring the fraction length of the $6RL^{Ku}$ arm, combined with the fraction length standard of 6RL built by Mukai et al. [11], it can be deduced that the $6RL^{Ku}$ minichromosome comprised about 11% of the original $6RL^{Ku}$ length.

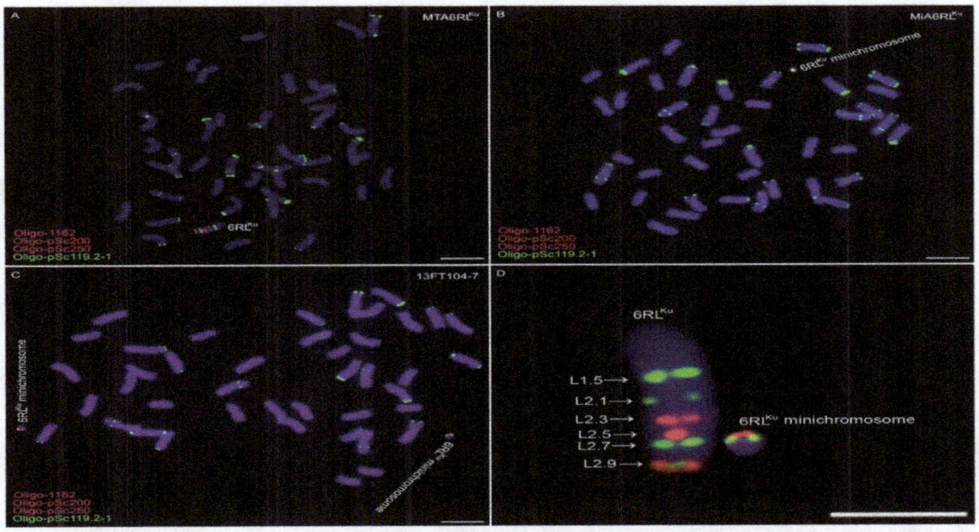


**Figure 1.** ND-FISH analysis of $MTA6RL^{Ku}$ and $MiA6RL^{Ku}$. (**A**,**B**,**C**) Oligo-1162 (**red**), Oligo-pSc200 (**red**), Oligo-pSc250 (**red**), and Oligo-pSc119.2-1 (**green**) were used as probes. (**D**) Cut and pasted $6RL^{Ku}$ from $MTA6RL^{Ku}$ and $6RL^{Ku}$ minichromosome from 13FT104-7. The FISH map of $6RL^{Ku}$ is the same as the one reported by Li et al. [6]. Chromosomes were counterstained with 4′-6-diamidino-2-phenyllindole (DAPI) (**blue**). Scale bar 10 μm.

### 2.2. Transmission of $6RL^{Ku}$ Minichromosome

Thirty-four seeds were randomly selected from the self-fertilized progeny of line $MiA6RL^{Ku}$ for ND-FISH analysis. Among the 34 plants, 24 had no $6RL^{Ku}$ minichromosome; nine contained one $6RL^{Ku}$ minichromosome; and one plant, 13FT104-7, contained a pair of this minichromosome (Figure 1C). From the self-fertilized progeny of line 13FT104-7, 100 seeds were randomly selected for ND-FISH analysis; 25 plants contained two $6RL^{Ku}$ minichromosomes, 62 plants contained one $6RL^{Ku}$ minichromosomes, and the remaining 13 plants had none of this minichromosome. The progeny of

13FT104-7 were named 15T154, and some of these plants were used for developing additional 6RL$^{Ku}$ minichromosome-specific markers and producing wheat-rye small segment translocations.

### *2.3. Development of 6D/6RL$^{Ku}$ Small Segment Translocation Lines*

Some seeds that were derived from line 14T154-35 with two 6RL$^{Ku}$ minichromosomes were exposed to $^{60}$Co-$\gamma$ rays. A total of 1428 M1 seeds were analyzed using ND-FISH, and ten wheat-rye 6D/6RL$^{Ku}$ small segment translocation lines were detected and named 16T379 or 16T380 (Table 1, Figure 2). In these 6D/6RL$^{Ku}$ small segment translocation lines, the 6RL$^{Ku}$ minichromosome divided into two smaller segments. One small segment, with the Oligo-pSc200 and Oligo-pSc250 signals and a strong Oligo-pSc119.2-1 signal, was named 6RL$^{Kumi119}$. The other small segment, with the Oligo-pSc200 and Oligo-pSc250 signals and a weak Oligo-pSc119.2-1 signal, was named 6RL$^{Kumi200}$. The segment 6RL$^{Kumi200}$ had been translocated onto the 6DS arm, while the segment 6RL$^{Kumi119}$ was translocated onto the 6DL arm (Figure 2). The small segment translocation chromosome fused with 6DS was named 6DS/6RL$^{Kumi200}$, and the other translocation attached with 6DL was named 6DL/6RL$^{Kumi119}$ (Figure 2).

**Table 1.** Wheat-rye 6D/6RL$^{Ku}$ small segment translocation lines with 6DS/6RL$^{Kumi200}$ and 6DL/6RL$^{Kumi119}$ translocation chromosomes.

| Small Segment Translocation Lines | Small Segment Translocation Chromosomes |
|---|---|
| 16T379-1 | one 6DS/6RL$^{Kumi200}$ and one 6DL/6RL$^{Kumi119}$ |
| 16T379-4 | two 6DS/6RL$^{Kumi200}$ and two 6DL/6RL$^{Kumi119}$ |
| 16T379-6 | one 6DS/6RL$^{Kumi200}$ and one 6DL/6RL$^{Kumi119}$ |
| 16T379-8 | two 6DS/6RL$^{Kumi200}$ and two 6DL/6RL$^{Kumi119}$ |
| 16T379-9 | one 6DS/6RL$^{Kumi200}$ and one 6DL/6RL$^{Kumi119}$ |
| 16T379-11 | one 6DS/6RL$^{Kumi200}$ and one 6DL/6RL$^{Kumi119}$ |
| 16T379-13 | one 6DS/6RL$^{Kumi200}$ and one 6DL/6RL$^{Kumi119}$ |
| 16T379-14 | two 6DS/6RL$^{Kumi200}$ and two 6DL/6RL$^{Kumi119}$ |
| 16T380-2 | one 6DS/6RL$^{Kumi200}$ and one 6DL/6RL$^{Kumi119}$ |
| 16T380-3 | one 6DS/6RL$^{Kumi200}$ and one 6DL/6RL$^{Kumi119}$ |

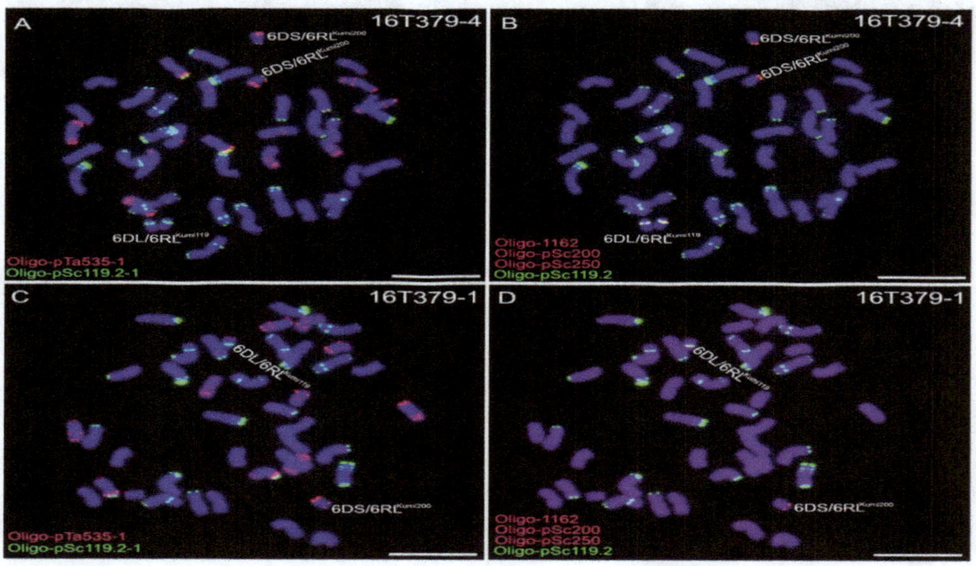


**Figure 2.** Oligo-pTa535-1 (**red**), Oligo-pSc119.2-1 (**green**), Oligo-1162 (**red**), Oligo-pSc200 (**red**), and Oligo-pSc250 (**red**) were used as probes for ND-FISH analysis of wheat-rye 6D/6RL$^{Ku}$ small segment translocation lines. (**A**,**B**) Line 16T379-4 represents lines with two 6DS/6RL$^{Kumi200}$ and two 6DL/6RL$^{Kumi119}$ chromosomes. (**A**) and (**B**) are the same cell. (**C**,**D**) Line 16T379-1 represents lines with one 6DS/6RL$^{Kumi200}$ and one 6DL/6RL$^{Kumi119}$ chromosome. (**C**) and (**D**) are the same cell. Chromosomes were counterstained with DAPI (**blue**). Scale bar 10 μm.

## 2.4. *Transmission Rates of* $6DS/6RL^{Kumi200}$ *and* $6DL/6RL^{Kumi119}$

Among 100 randomly selected seeds from the progeny of 16T379-1, 21 derived seedlings contained no translocation chromosomes, 16 seedlings contained two $6DS/6RL^{Kumi200}$ and two $6DL/6RL^{Kumi119}$ chromosomes, 42 seedlings contained one $6DS/6RL^{Kumi200}$ and one $6DL/6RL^{Kumi119}$ chromosome, eight contained two $6DS/6RL^{Kumi200}$ and one $6DL/6RL^{Kumi119}$ chromosome (Figure 3A), three plants contained one $6DS/6RL^{Kumi200}$ and two $6DL/6RL^{Kumi119}$ chromosomes (Figure 3B), three plants contained only one $6DS/6RL^{Kumi200}$ chromosome (Figure 3C), and seven plants contained only one $6DL/6RL^{Kumi119}$ chromosome (Figure 3D). The progeny of 16T379-1 were named 17T256, and some of these plants were used for further marker development. All of the randomly selected 100 seeds from the progeny of 16T379-4 contained two $6DS/6RL^{Kumi200}$ and two $6DL/6RL^{Kumi119}$ chromosomes. These results indicate that the $6DS/6RL^{Kumi200}$ and $6DL/6RL^{Kumi119}$ chromosomes show a high frequency of transmission to progeny.

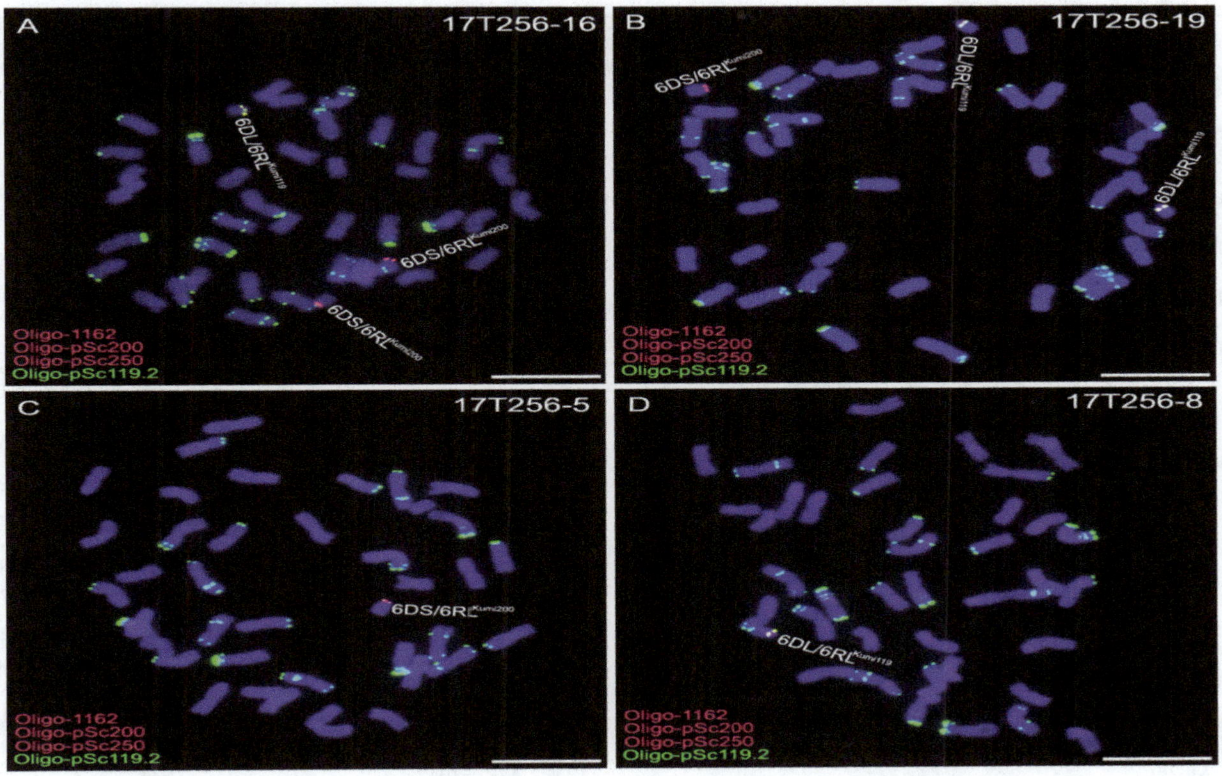


**Figure 3.** Oligo-pSc119.2-1 (**green**), Oligo-1162 (**red**), Oligo-pSc200 (**red**), and Oligo-pSc250 (**red**) were used as probes for ND-FISH analysis of the progeny of line 16T379-1. (**A**) Line 17T256-16 carried two $6DS/6RL^{Kumi200}$ chromosomes and one $6DL/6RL^{Kumi119}$ chromosome. (**B**) Line 17T256-19 carried one $6DS/6RL^{Kumi200}$ and two $6DL/6RL^{Kumi119}$ chromosomes. (**C**) Line 17T256-5 possessed only one $6DS/6RL^{Kumi200}$ chromosome. (**D**) Line 17T256-8 had only one $6DL/6RL^{Kumi119}$ chromosome. Chromosomes were counterstained with DAPI (**blue**). Scale bar 10μm.

## 2.5. *Development of* $6RL^{Ku}$ *Minichromosome-Specific Markers*

Common wheat Chinese Spring (CS), Mianyang 11 (MY11), rye Kustro, line $MTA6RL^{Ku}$, line 14T154-35 with two $6RL^{Ku}$ minichromosomes, and two lines (14T154-31 and 14T154-41) without the minichromosome were used to develop $6RL^{Ku}$ minichromosome-specific markers. One thousand and eighty-six pairs of primers were designed according to the $6RL^{Ku}$ minichromosome-specific pair-end reads. Sixteen of the 1086 primer pairs amplified their target bands from the genomic DNAs of rye Kustro, line $MTA6RL^{Ku}$, and line 14T154-35, and not from the genomic DNA of CS, MY11, and lines 14T154-31 and 14T154-41 (Figure 4). Therefore, the 16 primer pairs were tentatively regarded as 6RL$^{Ku}$ minichromosome-specific markers and are listed in Table 2. From the progeny of 16T379-1, three lines (17T256-7, 17T256-10, and 17T256-17) without 6D/6RL$^{Ku}$ translocation chromosomes, three lines (17T256-8, 17T256-12, and 17T256-14) that only contained one 6DL/6RL$^{Kumi119}$ chromosome, and three lines (17T256-5, 17T256-9, and 17T256-11) that only contained one 6DS/6RL$^{Kumi200}$ chromosome, were used to validate the 16 primer pairs. The 16 primer pairs only amplified target bands from lines 17T256-8, 17T256-12, and 17T256-14 (Figure 5). This indicated that all the 16 6RL$^{Ku}$ minichromosome-specific markers were located on the segment 6RL$^{Kumi119}$. Therefore, the 16 markers were 6RL$^{Kumi119}$-specific.

**Table 2.** 6RL$^{Ku}$ minichromosome-specific markers and the similarity between the sequences amplified by the markers and the *S. cereale* Lo7 scaffolds.

| Marker | Forward (5′-3′) | Reverse (5′-3′) | GenBank Accession Number of Amplified Sequence | Similarity of Amplified Sequence with the *S. cereale* Lo7 Scaffolds |
|---|---|---|---|---|
| 6RL-M8 | CAACCTATTCG GACCAGAGC | GATTAAACCGCT GGTGAGAAAC | MK051036 | 99% similarity with 7794–8206 bp of Lo7_v2_scaffold_453717 6R |
| 6RL-M11 | GGGGGAACTTT GAGTATGCTT | GATCGGATCGGT TGAGTTGT | MK051037 | 99% similarity with 464–1239 bp of Lo7_v2_scaffold_651086 0R |
| 6RL-M55 | TGATGCAAGTT CGTTGGTGT | CGTTGACTCCCTT CCGTTAG | MK051038 | 91% similarity with 1–108 bp of Lo7_v2_scaffold_457844 6R |
| 6RL-M63 | TCGAAATGCAT CGGACAAT | TCCATGGTCTCCT CGAGTGT | MK051039 | 100% similarity with 144–422 bp of Lo7_v2_scaffold_492428 6R |
| 6RL-M102 | CGGGAGAGGA CTGGTTCTT | CATATGTACAAC AGAGGCATCTTC | MK051040 | 98% similarity with 26957–27168 bp, 27841-27881 bp and 27923-28047 bp of Lo7_v2_scaffold_445202 6R |
| 6RL-M118 | TCCCCCTTCTA GGGTTTCAT | ATAGCCCCATCTG CAAACAC | MK051041 | 100% similarity with 488–896 bp of Lo7_v2_scaffold_484582 6R |
| 6RL-M149 | AATGGCTGCAA TTTCTTGGA | AAAAAGCCACA AAACACTGC | MK051042 | 100% similarity with 34805–34422 bp of Lo7_v2_scaffold_445202 6R |
| 6RL-M220 | GCACAAGTCCA TGTCCTTCA | GATCCATCTGGCT GTGTGTG | MK051043 | 99% similarity with 4112–4449 bp of Lo7_v2_scaffold_448816 6R |
| 6RL-M221 | CGCTATATGCA ATGCAGGTG | CTTGCTTGCAAC ACCAAAAA | MK051044 | 98% similarity with 41511–41912 bp of Lo7_v2_scaffold_445202 6R |
| 6RL-M255 | CCTTATGACCA CCCATGCTC | TTCATAGCTGCCT CTTTTAGGTG | MK051045 | 99% similarity with 31812–32230 bp of Lo7_v2_scaffold_445202 6R |
| 6RL-M710 | CAAACTCACAC GAAGCCAAA | CTGATCCAAATTT GCCCAGT | MK051046 | 92% similarity with 3600–3681 bp of Lo7_v2_scaffold_457146 6R |
| 6RL-M828 | TTTGTCGAGAG CAACAATGG | CCCGCTTCTAAGT TCAATCG | MK051047 | 100% similarity with 39318–39668 bp of Lo7_v2_scaffold_445202 6R |
| 6RL-M869 | GGGTCAACCCA TCTTGTTTC | CCTCTTCCACTGC AGAGCTT | MK051048 | 99% similarity with 589–962 bp of Lo7_v2_scaffold_451612 6R |
| 6RL-M896 | GACGAAACAC AACAAATCATTCA | GGGAAAATCGAA AACTGCAA | MK051049 | 100% similarity with 10161–10344 bp of Lo7_v2_scaffold_620512 0R |
| 6RL-M1074 | AAAGCCGATG AAAAATGGTG | GAAGAAGAAGA AGATGGGGTGTT | MK051050 | 100% similarity with 9159–9365 bp of Lo7_v2_scaffold_445253 6R |
| 6RL-M1081 | TTGCATGCTCG CTTTAGTTG | CCACTTGACGTT GCCCTATT | MK051051 | 100% similarity with 8940–9193 bp of Lo7_v2_scaffold_445253 6R |

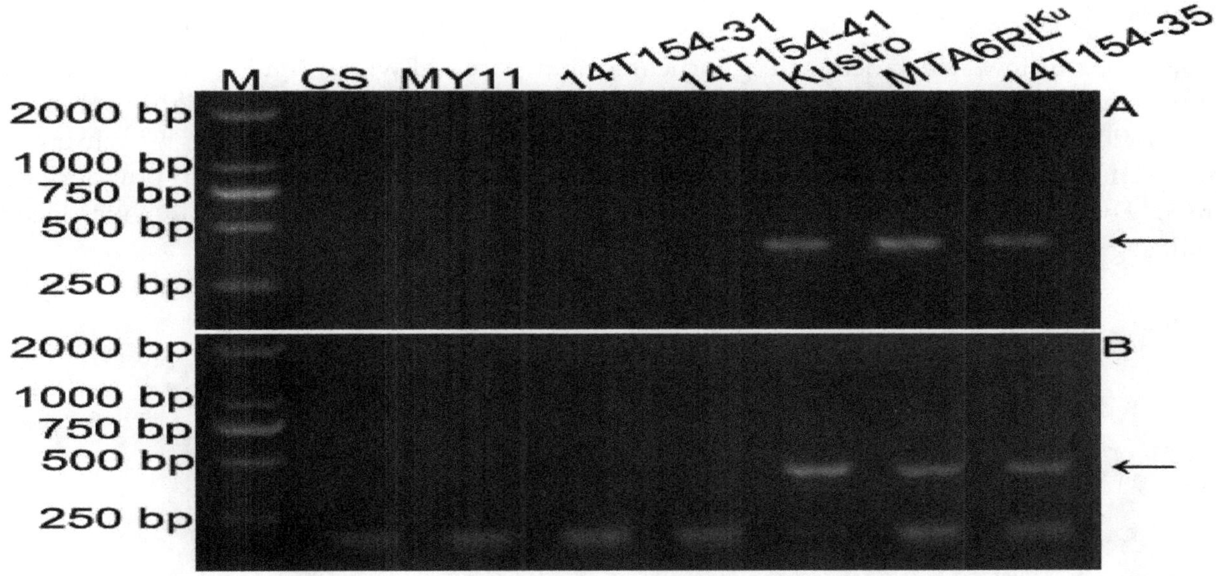


**Figure 4.** Representative results of developing $6RL^{Ku}$ minichromosome-specific markers. (**A**) Products amplified marker 6RL-M55. (**B**) Products amplified marker 6RL-M63. M, DNA marker. Arrows indicate $6RL^{Ku}$ minichromosome-specific bands.

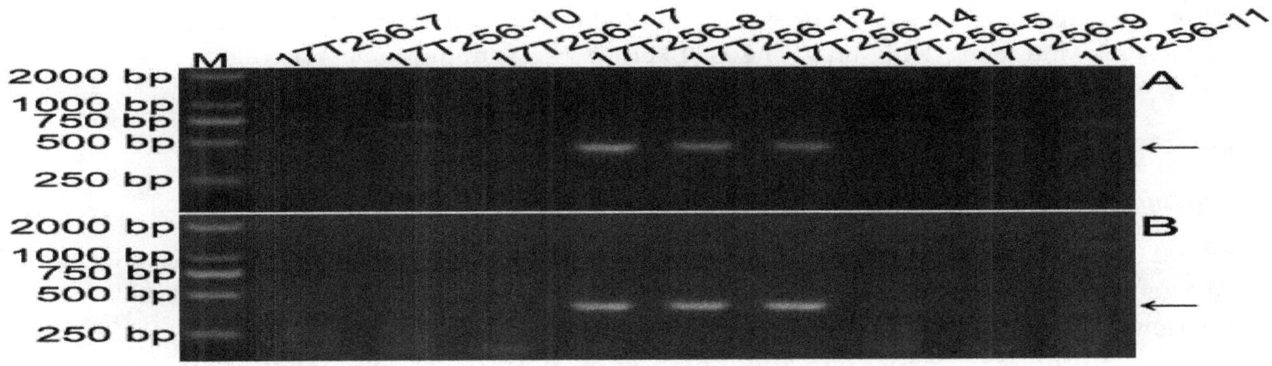


**Figure 5.** Representative results of locating $6RL^{Ku}$ minichromosome-specific markers on $6RL^{Kumi119}$ segment. (**A**) Products amplified by marker 6RL-M11. (**B**) Products amplified by marker 6RL-M102. M, DNA marker. Arrows indicate $6RL^{Kumi119}$-specific bands.

### *2.6. Sequence Characteristics of the Products Amplified by the 16 Markers*

The sequences amplified by the 16 $6RL^{Kumi119}$-specific markers were isolated. These sequences have been deposited in the GenBank Database (GenBank accession numbers: MK051036–MK051051). Sequence alignment using the BLAST tool in NCBI and gene annotation indicated that all the 16 sequences did not belong to any gene sequence; therefore, they are probably not involved in active genes. Analysis using RepeatMasker software (http://repeatmasker.org/) also indicated that these sequences were not involved in repetitive DNA sequences. Nucleotide BLAST searches indicated that 14 of the 16 sequences had 91–100% similarity with the scaffolds from the 6R chromosome of *S. cereale* L. Lo7, and the other two sequences had 99–100% similarity with the scaffolds derived from 0R of *S. cereale* L. Lo7 scaffolds (Table 2). It can be noted that all the five sequences (MK051040, MK051042, MK051044, MK051045, and MK051047) were located in the scaffold Lo7_v2_scaffold_445202 6R, and both MK051050 and MK051051 were located in the scaffold Lo7_v2_scaffold_445253 6R (Table 2). According to the gene annotation of these scaffolds published by Bauer et al. [12], they did not contain genes; however, these scaffolds were still useful for the future study on the structure and function of the $6RL^{Kumi119}$ segment.

### 2.7. Powdery Mildew Resistance

Powdery mildew resistance testing indicated that parental wheat MY11, lines without $6RL^{Ku}$ minichromosome, and lines with only one $6DS/6RL^{Kumi200}$ chromosome were highly susceptible to powdery mildew (Figure 6). Lines with both the $6DS/6RL^{Kumi200}$ and $6DL/6RL^{Kumi119}$ chromosomes and lines with only one $6DL/6RL^{Kumi119}$ chromosome displayed high resistance to powdery mildew (Figure 6). These results indicated that the powdery mildew resistance gene on $6RL^{Ku}$ was located on the small segment $6RL^{Kumi119}$.

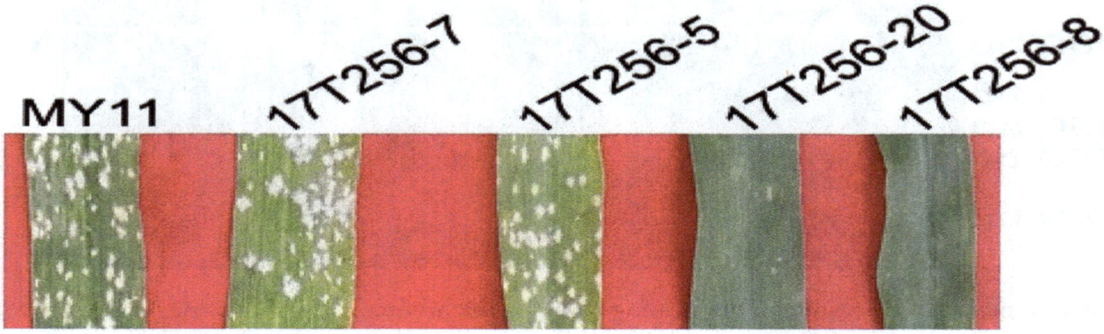


**Figure 6.** Identification of resistance to powdery mildew. Parental wheat and lines without $6D/6RL^{Ku}$ translocation chromosomes and lines with only one $6DS/6RL^{Kumi200}$ chromosome are highly susceptible to powdery mildew. Lines with one $6DS/6RL^{Kumi200}$ and one $6DL/6RL^{Kumi119}$ chromosome and lines with only one $6DL/6RL^{Kumi119}$ chromosome are highly resistant to powdery mildew.

## 3. Discussion

### 3.1. Extending Genetic Basis of Powdery Mildew Resistance Genes

Several reports have highlighted the current narrowness of genetic diversity of powdery mildew resistance genes in wheat breeding programs in China [13–15]. For example, the wheat cultivars (lines) from the Yangtze River region mainly contained genes *Pm4a* and *Pm21* [13]. The gene *Pm21* is widely used in wheat cultivars from Sichuan and Guizhou provinces [14], while the wheat cultivars (lines) from Henan province mainly carry genes *Pm2, Pm4, Pm21*and unknown gene(s) located on a new 1RS.1BL translocation chromosome [15]. It should be noted that gene *Pm21* has played an important role in wheat powdery mildew resistance breeding programs in China, but there is the risk of pathogen directional selection caused by the high frequency of *Pm21* gene in wheat breeding programs [15]. Therefore, extending the genetic basis of powdery mildew resistance genes in wheat breeding programs in China must be an urgent priority. 6R chromosomes of rye (*S. cereale* L.) carry powdery mildew resistance genes, and still display effective resistance to powdery mildew pathotypes in China [1–6,16]. In addition, the powdery mildew resistance genes on 6R chromosomes have genetic diversity [3,16], which may provide recipient wheat lines with resistance to future virulent powdery mildew races that might defeat existing genes. However, these 6R-derived powdery mildew resistance genes have not been successfully used in commercial wheat cultivars because of the loss of indispensable wheat genes and alien chromosome-associated linkage drag. Therefore, the development of new wheat-rye 6RL small segment translocation chromosomes with powdery mildew resistance gene(s) is of pressing concern [3,4,8]. It has been reported that few recombinants (sometime none) can be recovered between 6R and wheat chromosomes, even in a *ph1b* mutant background. Furthermore, it is difficult to produce suitable wheat-rye 6RL small segment translocations [17]. In this study, a wheat-rye small segment translocation chromosome $6DL/6RL^{Kumi119}$ with powdery mildew resistance was developed using a $6RL^{Ku}$ minichromosome. According to the fraction length standard of the 6RL arm constructed by Mukai et al. [11], the size of the $6RL^{Kumi119}$ segment might be slightly larger than the small 6RL segment in the Ti4AS.4AL-6RL-4AL translocation chromosome, in which the 6RL segment occupied 10% of the total length of 6RL [11]. Therefore, the occurrence of possible deleterious genes on the $6RL^{Ku}$ arm may have been greatly reduced. Additionally, the $6DS/6RL^{Kumi200}$ and $6DL/6RL^{Kumi119}$ chromosomes showed high transmission to progeny. These characteristics make the successful utilization of the powdery mildew resistance gene on $6RL^{Ku}$ more likely, although the agronomic traits of the lines with $6DS/6RL^{Kumi200}$ and $6DL/6RL^{Kumi119}$ chromosomes have not been investigated.

### 3.2. $6RL^{Ku}$ Minichromosome Specific Markers

For the effective utilization of elite genes in wild germplasm, it is necessary to increase our understanding of the molecular basis of target alien chromosome regions [18,19]. In this study, the small segment translocation chromosome $6DL/6RL^{Kumi119}$ provides an opportunity to understand the molecular basis of the small segment $6RL^{Kumi119}$ carrying powdery mildew resistance. Additionally, 16 $6RL^{Ku}$ minichromosome-specific markers were developed, and these markers were subsequently located onto the segment $6RL^{Kumi119}$ with powdery mildew resistance. These markers represent convenient tools to use the $6DL/6RL^{Kumi119}$ translocation chromosome in wheat breeding programs. Using the sequences amplified by the 16 markers, 11 scaffolds derived from winter rye inbred line *S. cereale* L. Lo7 [12] were located on the segment $6RL^{Kumi119}$. These scaffolds are useful for the further detailed studying of this segment. However, published data indicated that no active genes could be found among the 11 scaffolds [12]. There are two possible explanations for this phenomenon. First, the $6RL^{Ku}$ minichromosome-specific markers may be insufficient in number. Second, large gaps may exist in these scaffolds. Therefore, a greater number of $6RL^{Ku}$ minichromosome-specific markers are required and more complete rye genomic sequences are needed.

## 4. Materials and Methods

### 4.1. Plant Materials

A monotelosomic addition line $MTA6RL^{Ku}$ was developed according to the methods described by Qiu et al. [20]. From the self-pollinated progeny of $MTA6RL^{Ku}$, a $6RL^{Ku}$ minichromosome addition line $MiA6RL^{Ku}$ was identified. Some seeds selected from the progeny of $MiA6RL^{Ku}$ were irradiated with $^{60}Co$-$\gamma$ rays at the Biotechnology and Nuclear Technology Research Institute, Sichuan Academy of Agricultural Sciences, China. Common wheat (*Triticum aestivum* L.) Mianyang 11 (MY11) and Chinese Spring (CS) were used as controls.

### 4.2. Cytological Analysis

The root-tip metaphase cells were analyzed using non-denaturing fluorescence *in situ* hybridization (ND-FISH) technology. Oligonucleotide (oligo) probes Oligo-1162, Oligo-pSc200, Oligo-pSc250, Oligo-pSc119.2-1, and Oligo-pTa535-1 [21,22] were used. These oligo probes were 5′end-labelled with 6-carboxyfluorescein (6-FAM) or 6-carboxytetramethylrhodamine (Tamra). Root-tip metaphase chromosomes were prepared following the methods described by Han et al. [23]. ND-FISH analysis was performed following the methods described by Fu et al. [22] with the following minor modification. When dropped onto cell spreads, the probe mixture containing Oligo-1162 around the slides was kept above 28 °C, and the slides were immediately placed into a moist box that was incubated at 42 °C in advance for 1–2 h. Then, the slides were washed 15–20 s in 2 × SSC at 42 °C.

### 4.3. Development of $6RL^{Ku}$ Minichromosome-Specific PCR-Based Markers

Genomic DNAs of *S. cereale* L. Kustro and $MiA6RL^{Ku}$ were sequenced by using the Specific Length Amplified Fragment Sequencing (SLAF-seq) technique (Biomarker, Beijing, China). The sequencing procedure followed the methods described by Duan et al. [24]. The pair-end reads derived from Kustro and $MiA6RL^{Ku}$ were compared with the whole genome shotgun assembly sequences (IWGSC WGA v0.4) of common wheat *cv.* Chinese Spring (*Triticum aestivum* L.) (http://www.wheatgenome.org) using SOAP software (Beijing Genomics Institute (BGI), Beijing, China) [25].

The pair-end reads with high wheat homology were discarded. At last, after comparing specific pair-end reads of Kustro and MiA6RL$^{Ku}$, the 6RL$^{Ku}$ minichromosome-specific pair-end reads were obtained. Primers were designed according to the 6RL$^{Ku}$ minichromosome-specific pair-end reads using the software Primer 3 (version 4.0). The optimal melting temperature and size values of primers were set following the methods described by Duan et al. [24].

### 4.4. PCR Analysis and Sequence Cloning

The PCR amplifications were carried out according to the procedure described by Li et al. [6]. Agarose gel (2%) electrophoresis was used to detect the amplicons in 1 $\times$ TAE buffer. The products amplified by the 6RL$^{Kumi119}$-specific primers were recovered using Universal DNA Purification Kit (Tiangen Biotech Co., Ltd, Beijing, China) and cloned using pClone007 Vector Kit (TsingKe Biotech Co., Ltd, Beijing, China). Inserts were sequenced by TsingKe Biotech (Beijing) Co., Ltd. These sequences were deposited in the GenBank Database. Finally, these sequences were used for Nucleotide BLAST searches against the *S. cereale* L. Lo7 scaffolds database using the BLAST tool in GrainGenes (https://wheat.pw.usda.gov/cgi-bin/seqserve/blast_rye.cgi) and gene annotation was carried out according to the data sets related to the publication by Bauer et al. [12]. Additionally, RepeatMasker software (http://repeatmasker.org/) was used to identify whether these sequences amplified by the 6RL$^{Kumi119}$-specific primers are repetitive DNA.

### 4.5. Powdery Mildew Resistance Test

The resistance of line MiA6RL$^{Ku}$, the progeny of MiA6RL$^{Ku}$, 6D/6RL$^{Ku}$ small segment translocation lines, and parental wheat MY11 to powdery mildew were evaluated. Plants were grown in Qionglai, Sichuan, China. The materials were naturally inoculated by locally occurring field derived powdery mildew, and infection types (IT) were scored according to the standard described by Fu et al. [5].

## 5. Conclusions

In conclusion, a new wheat-rye small segment translocation chromosome 6DL/6RL$^{Kumi119}$ with powdery mildew resistance was produced. Sixteen 6RL$^{Kumi119}$ specific markers were developed, and 11 *S. cereale* L. Lo7 scaffolds were located on the small rye segment 6RL$^{Kumi119}$. The small segment translocation chromosome 6DL/6RL$^{Kumi119}$ makes the practical utilization of the powdery mildew resistance gene on 6RL$^{Ku}$ more likely. The 6RL$^{Kumi119}$-specific markers and the *S. cereale* L. Lo7 scaffolds that were located on the 6RL$^{Kumi119}$ segment may find application in future detailed studies of the structure and function of this small segment.

**Author Contributions:** S.F. and Z.T. created the materials, designed the study, analyzed the data, and wrote the manuscript. H.D. and Q.D. did the ND-FISH analysis, the PCR experiments, and cloned the sequences. S.T. did the BLAST searches. S.F. and Z.T. did the powdery mildew resistance test.

**Acknowledgments:** We gratefully acknowledge Ian Dundas, The University of Adelaide, Australia and Zujun Yang, School of Life Science and Technology, University of Electronic Science and Technology of China for discussion and revision of this manuscript.

## References

1. Friebe, B.; Heun, M.; Tuleen, N.; Zeller, F.J.; Gill, B.S. Cytogenetically monitored transfer of powdery mildew resistance from rye into wheat. *Crop Sci.* **1994**, *34*, 621–625. [CrossRef]
2. Zhao, Z.H.; Sun, H.G.; Song, W.; Lu, M.; Huang, J.; Wu, L.F.; Wang, X.M.; Li, H.J. Genetic analysis and detection of the gene MlLX99 on chromosome 2BL conferring resistance to powdery mildew in the wheat cultivar Liangxing 99. *Theor. Appl. Genet.* **2013**, *126*, 3081–3089. [CrossRef]

3. An, D.; Zheng, Q.; Luo, Q.; Ma, P.; Zhang, H.; Li, L.; Han, F.; Xu, H.; Zhang, X.; Zhou, Y. Molecular cytogenetic identification of a new wheat-rye 6R chromosome disomic addition line with powdery mildew resistance. *PLoS ONE* **2015**, *10*, e0134534. [CrossRef] [PubMed]
4. Wang, D.; Zhuang, L.F.; Sun, L.; Feng, Y.G.; Pei, Z.Y.; Qi, Z.J. Allocation of a powdery mildew resistance locus to the chromosome arm 6RL of *Secale cereale* L. cv. 'Jingzhouheimai'. *Euphytica* **2010**, *176*, 157–166. [CrossRef]
5. Fu, S.; Ren, Z.; Chen, X.; Yan, B.; Tan, F.; Fu, T.; Tang, Z. New wheat-rye 5DS-4RS.4RL and 4RS-5DS.5DL translocation lines with powdery mildew resistance. *J. Plant Res.* **2014**, *127*, 743–753. [CrossRef] [PubMed]
6. Li, M.; Tang, Z.X.; Qiu, L.; Wang, Y.Y.; Tang, S.Y.; Fu, S.L. Identification and physical mapping of new PCR-based markers specific for the long arm of rye (*Secale cereale* L.) chromosome 6. *J. Genet. Genomics* **2016**, *43*, 199–206. [CrossRef] [PubMed]
7. Fu, S.; Tang, Z.; Ren, Z.; Zhang, H. 2010 Transfer to wheat (*Triticum aestivum*) of small chromosome segments from rye (*Secale cereale*) carrying disease resistance genes. *J. Appl. Genet.* **2010**, *51*, 115–121. [CrossRef] [PubMed]
8. Lukaszewski, A.J. *Alien Introgression in Wheat*, 1st ed.; Molnár-Láng, M., Ceoloni, C., Doležel, J., Eds.; Springer: Cham, Switzerland, 2015; Charpter 7; pp. 163–189.
9. Friebe, B.; Hatchett, J.H.; Gill, B.S.; Mukai, Y.; Sebeata, E.E. Transfer of Hessian fly resistance from rye to wheat via radiation-induced terminal and intercalary chromosomal translocations. *Theor. Appl. Genet.* **1991**, *83*, 33–40. [CrossRef]
10. Friebe, B.; Jiang, J.; Raupp, W.J.; McIntosh, R.A.; Gill, B.S. Characterization of wheat-alien translocations conferring resistance to diseases and pests: current status. *Euphytica* **1996**, *91*, 59–87. [CrossRef]
11. Mukai, Y.; Friebe, B.; Hatchett, J.H.; Yamamoto, M.; Gill, B.S. Molecular cytogenetic analysis of radiation-induced wheat-rye terminal and intercalary chromosomal translocations and the detection of rye chromatin specifying resistance to Hessian fly. *Chromosoma* **1993**, *102*, 88–95. [CrossRef]
12. Bauer, E.; Schmutzer, T.; Barilar, I.; Mascher, M.; Gundlach, H.; Martis, M.M.; Twardziok, S.O.; Hackauf, B.; Gordillo, A.; Wilde, P.; et al. Towards a whole-genome sequence for rye (*Secale cereale* L.). *Plant J.* **2017**, *89*, 853–869. [CrossRef] [PubMed]
13. Li, H.J.; Wang, X.M.; Song, F.J.; Wu, C.P.; Wu, X.F.; Zhang, N.; Zhou, Y.; Zhang, X.Y. Response to powdery mildew and detection of resistance genes in wheat cultivars from China. *Acta Agron. Sin.* **2011**, *37*, 943–954, (in Chinese with English abstract). [CrossRef]
14. Jiang, Z.; Wang, Q.L.; Wu, J.H.; Xue, W.B.; Zeng, Q.D.; Huang, L.L.; Kang, Z.S.; Han, D.J. Distribution of powdery mildew resistance gene *Pm21* in Chinese winter wheat cultivars and breeding lines based on gene-specific marker. *Scientia Agri. Sin.* **2014**, *47*, 2078–2087, (in Chinese with English abstract). [CrossRef]
15. Wu, Q.H.; Chen, Y.X.; Li, D.; Wang, Z.Z.; Zhang, Y.; Yuan, C.G.; Wang, X.C.; Zhao, H.; Cao, T.J.; Liu, Z.Y. Large scale detection of powdery mildew resistance genes in wheat via SNP and bulked segregate analysis. *Acta Agron. Sin.* **2018**, *44*, 1–14. [CrossRef]
16. Hao, M.; Luo, J.; Fan, C.; Yi, Y.; Zhang, L.; Yuan, Z.; Ning, S.; Zheng, Y.; Liu, D. Introgression of powdery mildew resistance gene *Pm56* on rye chromosome arm 6RS into wheat. *Front. Plant Sci.* **2018**, *9*, 1040. [CrossRef] [PubMed]
17. Dundas, I.S.; Frappell, D.E.; Crack, D.M.; Fisher, J.M. Deletion mapping of a nematode resistance gene on rye chromosome 6R in wheat. *Crop Sci.* **2001**, 1771–1778. [CrossRef]
18. Feuillet, C.; Langridge, P.; Waugh, R. Cereal breeding takes a walk on the wild side. *Trends Genet.* **2008**, *24*, 24–32. [CrossRef]
19. Tester, M.; Langridge, P. Breeding technologies to increase crop production in a changing world. *Science* **2010**, *327*, 818–822. [CrossRef]
20. Qiu, L.; Tang, Z.X.; Li, M.; Fu, S.L. Development of new PCR-based markers specific for chromosome arms of rye (*Secale cereal* L.). *Genome* **2016**, *59*, 159–165. [CrossRef]
21. Tang, Z.X.; Yang, Z.J.; Fu, S.L. Oligonucleotides replacing the roles of repetitive sequences pAs1, pSc119.2, pTa-535, pTa71, CCS1, and pAWRC.1 for FISH analysis. *J. Appl. Genet.* **2014**, *55*, 313–318. [CrossRef]
22. Fu, S.L.; Chen, L.; Wang, Y.Y.; Li, M.; Yang, Z.J.; Qiu, L.; Yan, B.J.; Ren, Z.L.; Tang, Z.X. Oligonucleotide probes for ND-FISH analysis to identify rye and wheat chromosomes. *Sci. Rep.* **2015**, *5*, 10552. [CrossRef] [PubMed]
23. Han, F.P.; Lamb, J.C.; Birchler, A. High frequency of centromere inactivation resulting in stable dicentric chromosomes of maize. *Proc. Natl. Acad. Sci. USA* **2006**, *103*, 3238–3243. [CrossRef] [PubMed]

24. Duan, Q.; Wang, Y.Y.; Qiu, L.; Ren, T.H.; Li, Z.; Fu, S.L.; Tang, Z.X. Physical location of new PCR-based markers and powdery mildew resistance gene(s) on rye (*Secale cereale* L.) chromosome 4 using 4R dissection lines. *Front. Plant Sci.* **2017**, *8*, 1716. [CrossRef] [PubMed]
25. Li, R.; Yu, C.; Li, Y.; Lam, T.W.; Yiu, S.M.; Kristiansen, K.; Wang, J. SOAP2: an improved ultrafast tool for short read alignment. *Bioinformatics* **2009**, *25*, 1966–1967. [CrossRef] [PubMed]

# 15

# Overexpression of Arabidopsis *OPR3* in Hexaploid Wheat (*Triticum aestivum* L.) Alters Plant Development and Freezing Tolerance

**Alexey V. Pigolev [1,†], Dmitry N. Miroshnichenko [1,2,†], Alexander S. Pushin [1,2], Vasily V. Terentyev [1], Alexander M. Boutanayev [1], Sergey V. Dolgov [2] and Tatyana V. Savchenko [1,*]**

1 Institute of Basic Biological Problems RAS, Pushchino 142290, Russia; alexey-pigolev@rambler.ru (A.V.P.); miroshnichenko@bibch.ru (D.N.M.); aspushin@gmail.com (A.S.P.); v.v.terentyev@gmail.com (V.V.T.); boutanaev@mail.ru (A.M.B.)

2 Branch of Shemyakin and Ovchinnikov Institute of Bioorganic Chemistry RAS, Pushchino 142290, Russia; dolgov@bibch.ru

* Correspondence: savchenko_t@rambler.ru.

† These authors contributed equally to this work.

**Abstract:** Jasmonates are plant hormones that are involved in the regulation of different aspects of plant life, wherein their functions and molecular mechanisms of action in wheat are still poorly studied. With the aim of gaining more insights into the role of jasmonic acid (JA) in wheat growth, development, and responses to environmental stresses, we have generated transgenic bread wheat plants overexpressing *Arabidopsis 12-OXOPHYTODIENOATE REDUCTASE 3* (*AtOPR3*), one of the key genes of the JA biosynthesis pathway. Analysis of transgenic plants showed that *AtOPR3* overexpression affects wheat development, including germination, growth, flowering time, senescence, and alters tolerance to environmental stresses. Transgenic wheat plants with high *AtOPR3* expression levels have increased basal levels of JA, and up-regulated expression of *ALLENE OXIDE SYNTHASE*, a jasmonate biosynthesis pathway gene that is known to be regulated by a positive feedback loop that maintains and boosts JA levels. Transgenic wheat plants with high *AtOPR3* expression levels are characterized by delayed germination, slower growth, late flowering and senescence, and improved tolerance to short-term freezing. The work demonstrates that genetic modification of the jasmonate pathway is a suitable tool for the modulation of developmental traits and stress responses in wheat.

**Keywords:** transgenic wheat; 12-oxophytodienoate reductase; jasmonates; freezing tolerance

## 1. Introduction

Plant growth, development and responses to environmental signals are coordinated by complex multicomponent signaling networks. Jasmonates, derived from fatty acids, are important components of this regulatory system. Jasmonates biosynthesis and signal transduction pathways have been thoroughly studied in model plant Arabidopsis [1–4]. The biosynthesis is initiated in chloroplasts, where lipoxygenases oxidize the 13th carbon atom of the α-linolenic acid, forming 13(S)-hydroperoxy-9,11,15-octadecatrienoic acid (13-HPOT). The 13-HPOT is further converted into (9S,13S)-12-oxo-phytodienoic acid (12-OPDA) by the consequent action of allene oxide synthase and allene oxide cyclase, and then 12-OPDA is transported to peroxisomes. Inside the peroxisomes, the double bond of the cyclopentenone ring in 12-OPDA is reduced by 12-OPDA reductase [5], and three cycles of β-oxidation of the carboxylic acid side chain occur, resulting in the formation of JA. JA itself can be further modified in the cytoplasm through the formation of a conjugate with the amino acid isoleucine, the signaling ligand, which is responsible for the regulation of majority of JA-dependent processes. Several metabolites of the pathway, 12-OPDA, JA, jasmonate–isoleucine conjugate, and methyl jasmonate display biological activity with only a partial overlap in functions. To date, most of the information on jasmonates biosynthesis and signaling comes from studies of dicotyledonous species, mainly Arabidopsis [1–5]. In monocots, this pathway is less well studied, and the information is mainly limited to maize and rice [6]. Biosynthesis, functions, and molecular mechanisms of action of jasmonates in wheat, one of the dominant cereal crops worldwide, remain mostly uncharacterized.

Unfavorable environmental conditions affect wheat crop yield globally [7]; therefore, the improvement of wheat tolerance to environmental stresses is of particular importance. The modification of individual steps of oxylipin biosynthesis pathway results in the alteration of plant tolerance to environmental challenges, as shown for Arabidopsis [8–12] and other species, including a few monocots [6,13,14]. For wheat, a protective effect of exogenously applied jasmonates under conditions of biotic and abiotic stresses was demonstrated in numerous studies [15–17], suggesting that the modulation of the activity of jasmonate-based signaling systems is a promising approach for the improvement of wheat tolerance to environmental stresses. The effect of the genetic modification of the jasmonate pathway on wheat development and stress response has not been sufficiently investigated, probably due to the fact that hexaploid wheat remains recalcitrant in tissue culture [18]. With the purpose of altering the level of jasmonic acid and studying the effect of such alterations on plant growth, development, and stress responses, we have generated transgenic hexaploid (*Triticum aestivum* L.) wheat plants constitutively overexpressing the *12-OXOPHYTODIENOATE REDUCTASE 3* gene from *Arabidopsis thaliana* (*AtOPR3*). 12-Oxophytodienoate reductases are classified into OPRI and OPRII subgroups, and only OPRII subgroup genes take part in the biosynthesis of jasmonic acid. Out of three oxophytodienoate reductases in Arabidopsis, only *OPR3* belongs to the OPRII subgroup, and it is involved in the jasmonate biosynthesis pathway [19]. *AtOPR3* is a well-characterized gene, coding for the enzyme controlling the levels of major metabolites of the jasmonate pathway, the upstream substrate 12-OPDA, and downstream products of the biosynthetic pathway, JA, and jasmonic acid derivatives [5,12,20,21]. The role of *AtOPR3* in plant growth and stress responses has been demonstrated [5,22,23], and the crystal structure of the OPR3 protein has been determined [24].

## 2. Results

### *2.1. Generation of Wheat Plants Overexpressing* AtOPR3

The hexaploid spring bread wheat variety 'Saratovskaya-60' (Sar-60) was previously thoroughly characterized and used as a parent line for the generation of a number of commercial wheat varieties grown in Russia [25]. In the present study, Sar-60 was genetically transformed with the pBAR-GFP.UbiOPR3 vector, enabling the expression of the *Arabidopsis thaliana OPR3* gene under the control of a strong constitutive promoter (Figure 1A). A number of transgenic plants of various generations (T0–T3) were produced to study the effect of *AtOPR3* overexpression in wheat. A total of 23 putative transgenic plantlets produced from 20 independent explants, designated as Tr-1 to Tr-20, were established in the greenhouse. In a few cases, two plants were regenerated from one explant (1a and 1b, 3a and 3b, 6a and 6b). Sixteen primary independent lines were identified as being transgenic by PCR, with a pair of primers targeting a 600 bp fragment of *GFP* (Figure 1B, upper panel), and a 295 bp fragment of *AtOPR3* (Figure 1B, lower panel). *AtOPR3* was detected in all transgenic lines, except line 10. End-point RT-PCR was performed on leaf samples collected from plants at the heading stage, to confirm the expression of the introduced genes in *GFP*-positive primary transgenic lines. The *TaWIN1* gene [26] was used as a control for the expression analysis (Figure 1C, upper panel). The *GFP* transcript was detected in all studied plants except for non-transformed Sar-60 control (Figure 1C, middle panel). The amplification bands corresponding to *AtOPR3* were specifically detected in the nine primary transgenic lines, but not in non-transformed Sar-60 plants and several green fluorescent protein (GFP)-positive transgenic lines (Figure 1C, lower panel).

T0 plants from eight *AtOPR3*-positive events produced T1 seeds, and segregation analysis was conducted on their progeny via GFP fluorescence, as described in [27]. The embryos produced by self-pollination of individual T1 plants were used for the detection of homozygous populations. At first, we analyzed the pollen produced by T1 plants, collected immediately after the anthers appeared (Figure S1A). If the pollen displayed a homozygous pattern, T2 zygotic embryos were additionally analyzed for GFP fluorescence (Figure S1B). Using this method, we were able to identify homozygous offspring, derived from six self-pollinated independent primary (T0) lines. The successful expression of transgenes in homozygous plants of the selected T2 population was confirmed by end-point RT-PCR (Figure S1C). Homozygous lines with confirmed expression of *AtOPR3* in the T2 generation, which produced sufficient amount of seeds, including Tr-3 (the progeny from 3b line), Tr-15, Tr-18, Tr-19, and Tr-20, were used for further analysis.

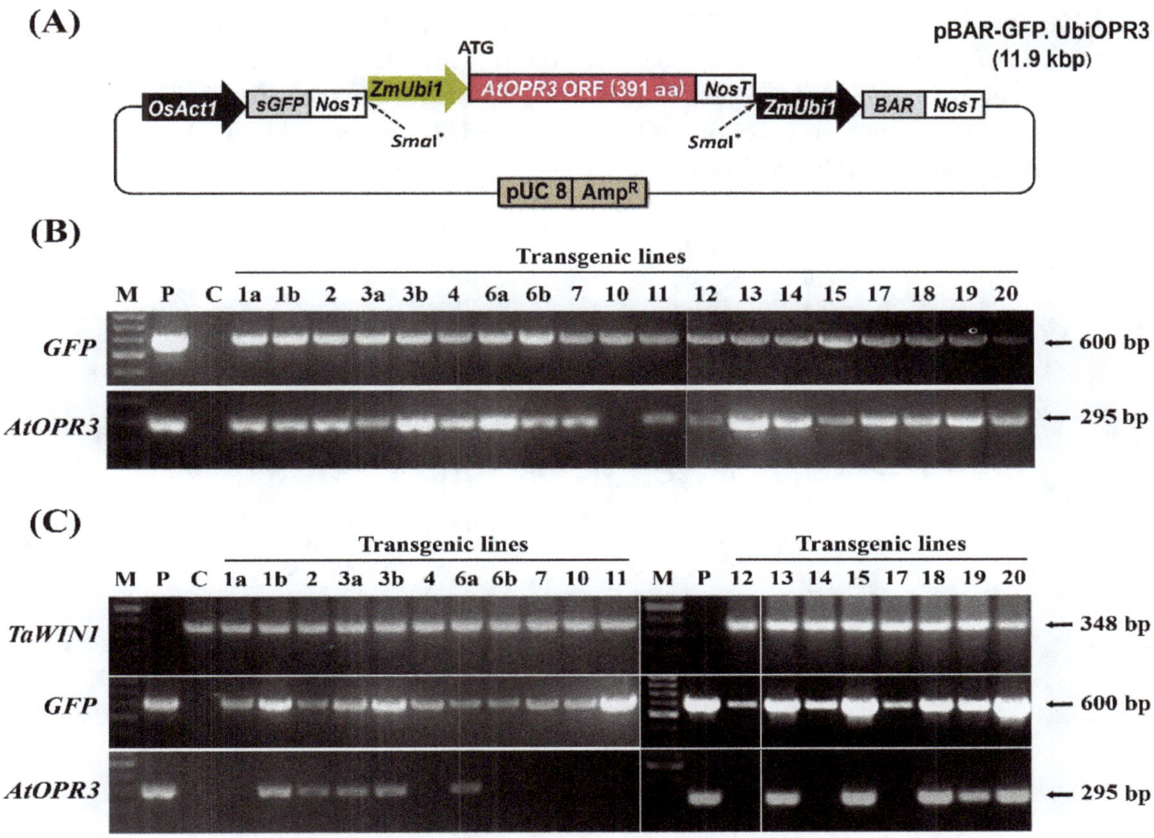


**Figure 1.** Generation and molecular analysis of transgenic plants. (**A**) Schematic representation of the pBAR-GFP.UbiOPR3 expression cassette used for Sar-60 transformation. *OsAct1*, rice Actin 1 promoter; *ZmUbi1*, maize ubiquitin 1 promoter; *NosT*, nopaline synthase terminator; *GFP*, Green Fluorescent Protein gene; *BAR*, BASTA resistance gene (phosphinothricin acetyl transferase); *AmpR*, ampicillin resistance gene. (**B**) PCR analysis performed on the genomic DNA of T0 plants for the insertion of *GFP* gene (upper panel) and *AtOPR3* (bottom panel). (**C**) End-point RT-PCR analysis on the total RNA of T0 plants (at heading stage) for the expression of the reference gene *TaWIN1* (top panel), *GFP* gene (middle panel) and *AtOPR3* (bottom panel). Lane M, DNA ladder as a molecular weight marker; Lane P, plasmid DNA pBAR-GFP.UbiOPR3; Lane C, non-transgenic wheat plant Sar-60; Lanes Tr-1–Tr-20 represent putative transgenic wheat plants.

### 2.2. *Transgenic Wheat Plants Overexpressing the Arabidopsis* OPR3 *Gene Display an Altered Growth and Development Phenotype*

To find out if there is an effect of *AtOPR3* gene overexpression on wheat plants growth and development, we carefully examined the germination, growth, flowering time, and senescence of generated transgenic lines. The observation allowed us to identify two distinct phenotypes among stable transgenic lines. The lines Tr-3 and Tr-18 (and to a lesser extent Tr-19) were characterized by delayed seed germination, slower growth, delayed flowering, and senescence, while the plants of Tr-15 and Tr-20, on the contrary, are characterized by earlier germination, faster growth, earlier flowering, and senescence (Figures 3–5, and Figures S2–S4). Quantitative analysis of *AtOPR3* gene expression in T3 homozygous plants revealed a significant difference in the expression levels between transgenic lines (Figure 2A). *AtOPR3* expression level was highest in the Tr-3 line; in Tr-18, this level was lower by almost three times in comparison to Tr-3, and in Tr-20, it was lower by 40 times. The JA level was also highest in Tr-3 and lowest in Tr-20 and Tr-15 (Figure 2B). Interestingly, JA levels in Tr-20 and Tr-15 were decreased, even in comparison to Sar-60. The JA level in Sar-60 leaves was about 3 ng/g fresh weight; in Tr-3, this value was increased by 70%, in line Tr-18 the increase was about 21%, in lines Tr-20 and Tr-15 JA level was decreased by 23% (Table S1). We also analyzed the expression levels of three genes involved in jasmonates biosynthesis and signaling, *ALLENE OXIDE SYNTHASE* (*TaAOS*) (GenBank: KJ001800.1), *12-OPDA REDUCTASE 2* (*TaOPR2*) (TraesCS7B02G311600), and *CORONATINE INSENSITIVE 1-LIKE* (*TaCOI-1*) (TraesCS1A02G279100) (Figure 2C). The analysis revealed a significant increase in the level of *TaAOS* gene expression in Tr-3 (16-fold increase) and Tr-18 (4-fold increase) in comparison to Sar-60. Expression level of *TaOPR2* was slightly increased in Tr-3, while *TaCOI-1* levels remained unchanged in all analyzed lines.

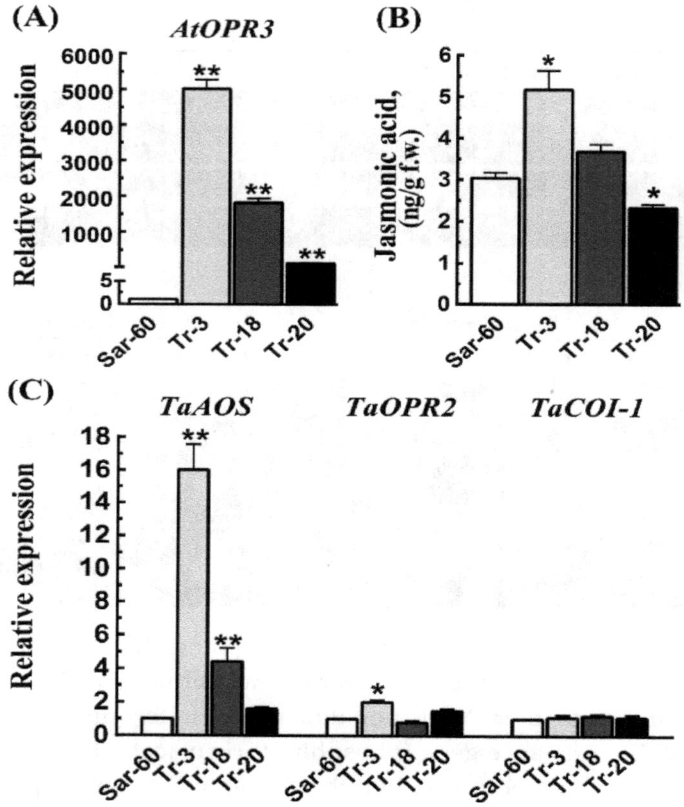


**Figure 2.** Gene expression analysis and JA levels in transgenic plants. The relative expression of *AtOPR3* (**A**), and endogenous wheat jasmonate biosynthesis and signaling pathways genes *TaAOS*, *TaOPR2*, and *TaCOI-1* (**C**). Total RNA was extracted from leaves (the second leaf of T3 plants at four leaves stage) and subjected to real-time quantitative RT-PCR analysis. The transcript levels of each gene were normalized to *TaWIN1* and *TaUbi*, as measured in the same samples. Data are means of four biological replicates represented by 6–8 leaves ± SE. (**B**) JA levels in second leaves of T3 plants collected at the four-leaf stage. Stars above the graphs indicate statistically significant differences (* $p \leq 0.05$, ** $p \leq 0.001$).

We examined germination and early growth of immature embryos of Tr-3 and Tr-20, the lines with pronounced phenotype on the synthetic medium. The analysis showed that Tr-3 embryos germinated one day later than non-transformed Sar-60, while Tr-20 germinated one day earlier. The representative Petri plates with seedlings (four days after culture initiation) are shown in Figure 3A. The significant difference between the seedlings was further confirmed by measurements of the length of coleoptiles, radicles, and lateral seminal roots (Figure 3B). To confirm that observed differences between the transgenic lines were indeed associated with altered JA level, we studied the effect of exogenous JA treatment on the germination of non-transformed Sar-60 seeds. The analysis showed that JA treatment indeed led to delays in seed germination, and the slower growth of coleoptiles and roots (Figure 3C).

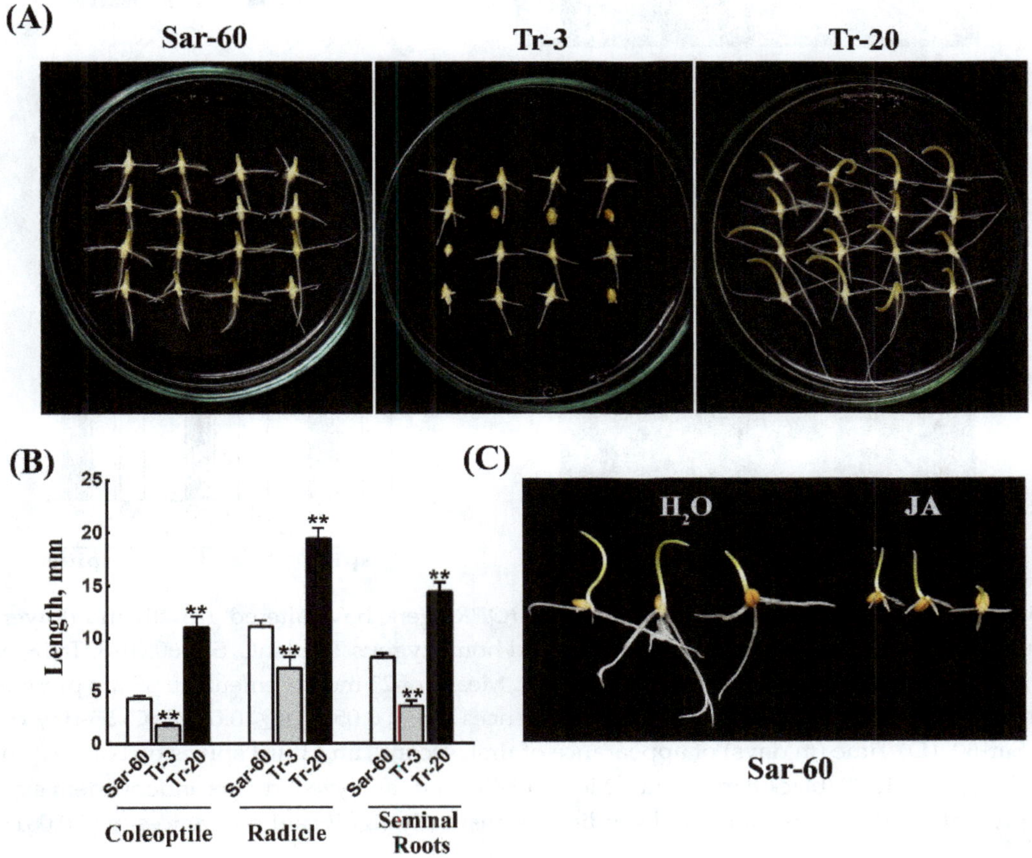


**Figure 3.** Jasmonate suppresses germination and early growth of wheat. (**A**) The representative Petri plates with seedlings grown from immature embryos of Sar-60, Tr-3, and Tr-20 T3 seeds, four days after culture initiation. (**B**) Length of coleoptiles, radicle and lateral seminal roots of four-day-old seedlings of Sar-60 (white bars), Tr-3 (grey bars), and Tr-20 (black bars). The means of 16 measurements ±SE are presented; stars above the graphs indicate statistically significant difference between genotypes ($p \leq 0.001$). (**C**) Effects of exogenously applied MeJA on the germination and growth of Sar-60 seeds. Seeds were placed on filter paper soaked in distilled water (left side) or a 0.1 mM solution of MeJA (right side); the picture was taken after four days of seed imbibition.

The difference in the growth rate between the transgenic plants and the non-transgenic Sar-60 was obvious in soil-grown plants (Figure 4). As seen in Figure 4A,B, the growth of the above-ground parts of Tr-3 and Tr-18 plants was retarded, while Tr-20 plants grew faster. The difference in roots growth was not observed at this growth stage under the given experimental conditions. The difference in plants height was also noticeable during later developmental stages (Figure S2). Further, we found that the flowering time was also affected in the transgenic lines (Figure 4C,D). Anthesis occurred later in Tr-3 plants in comparison to Sar-60, as observed in first, second, and third spikes. As expected, Tr-20 plants flowered earlier.

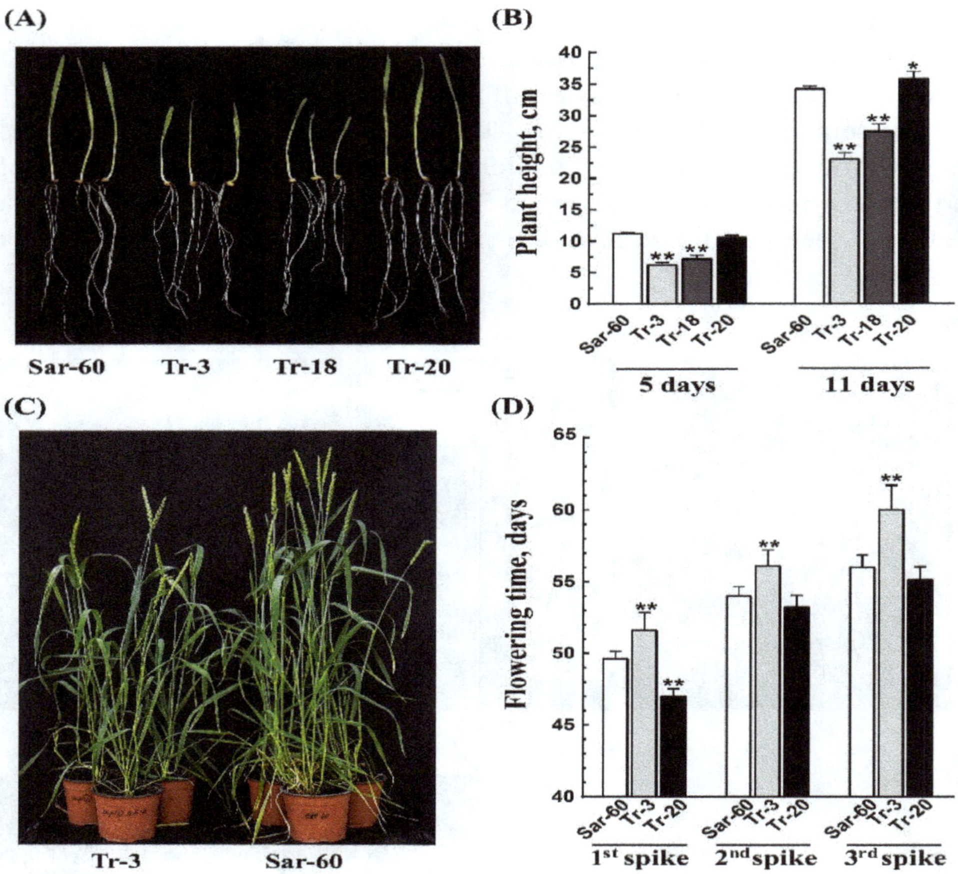


**Figure 4.** Transgenic wheat plants overexpressing *AtOPR3* gene have altered growth and flowering time phenotype. (**A**) The picture of soil-grown five-day-old homozygous T2 plants, Sar-60, Tr-3, Tr-18, and Tr-20. (**B**) The height of plants in 5 and 11 days after sowing. Means of 25 measurements $\pm$ SE are presented; stars above graphs indicate statistically significant differences (* $p \leq 0.05$, ** $p \leq 0.001$). (**C**) 55-day-old plants, Tr-3 and Sar-60. (**D**) Time (in days) of appearance of first, second, and third spikes on Sar-60 (white bars), Tr-3 (grey bars), and Tr-20 (black bars) plants. Means of 45 plants analyzed in three independent experiments $\pm$SE are presented. Stars above the graphs indicate statistically significant differences ($p \leq 0.001$).

Leaf senescence symptoms (leaf yellowing) also appear earlier in Tr-20, and later in Tr-3 plants, in comparison to Sar-60. For the quantitative estimation of senescence in the studied plant lines, we have performed the measurements of variable chlorophyll fluorescence parameters, reflecting functional activity of Photosystem II (PS II). The analysis of the functional activity of PS II is a simple and reliable method for the evaluation of leaf senescence when performed with the necessary precautions. It is important to remember that photosynthetic parameters show heterogeneous spatial patterns within the leaf blade, and that individual leaf senescence starts at the tip of the leaf, and then the senescent area enlarges progressively toward the leaf sheath [28]. For the measurements of light-induced chlorophyll fluorescence in leaves of studied plants the area located 10 cm away from the tip of leaf blade was chosen. In preliminary experiments, it has been established that the measured parameter does not differ in young leaves of different genotypes. Results of the analysis confirmed that the senescence time was altered in the studied transgenic lines (Figure 5). Analysis of PS II operating efficiency in the fourth and flag leaves of plants (at maturation stage) in combination with visual assessment (yellowness of leaves) confirmed delayed senescence in Tr-3 and accelerated senescence in Tr-20. Interestingly, detached same-aged leaves of Tr-3 plants also displayed later senescence in comparison to Sar-60 (Figure 5B). Similar trends were observed in independent transgenic lines with altered JA levels: growth and development lines with increased JA levels were slowed down, while in lines with decreased JA level, they were accelerated (Figures S3 and S4).

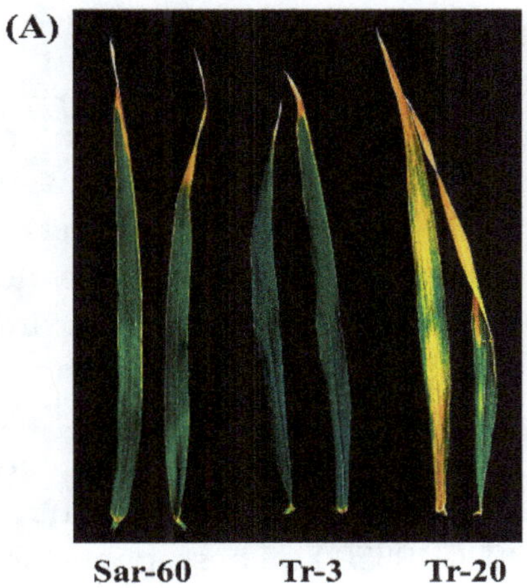


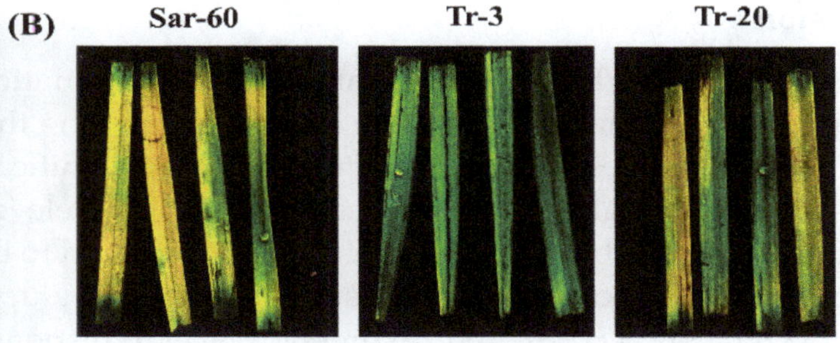


(C)

| | Sar-60 | Tr-3 | Tr-20 |
|---|---|---|---|
| Y(II) flag leaf | 0.54 ± 0.06 | 0.69 ± 0.08* | 0.37 ± 0.07* |
| Y(II) 4th leaf | 0.30 ± 0.07 | 0.68 ± 0.07** | 0.12 ± 0.07* |
| Y(II) in detached 2nd leaf | 0.25 ± 0.12 | 0.71 ± 0.02* | 0.36 ± 0.11 |

**Figure 5.** The timing of senescence is altered in wheat plants overexpressing *AtOPR3*. (**A**) Representative flag leaves of two-month-old homozygous T2 plants, Sar-60, Tr-3, and Tr-20. (**B**) The senescence of detached leaves. Same-aged detached leaves of Sar-60, Tr-3, and Tr-20 were kept floating on distilled water in Petri dishes under photoperiodic illumination 16/8 h (light/dark). The picture was taken on the fifth day of the experiment. (**C**) Analysis of PS II operating efficiency in flag (upper line) and fourth (middle line) leaves of two-month-old Sar-60, Tr-3, and Tr-20 plants growing in the soil, and detached leaves kept for five days under photoperiodic illumination (bottom line). Means of 13–16 measurements ± SE are presented. Stars indicate statistically significant differences from Sar-60 (* $p \leq 0.05$, ** $p \leq 0.001$).

### 2.3. AtOPR3 *Overexpression Alters Wheat Tolerance to Freezing Stress*

To estimate the stress tolerance of transgenic plants with altered JA level, we have performed short-term freezing treatment of plants of different transgenic lines as described in the Materials and Methods section. The freezing-induced damages were estimated by two methods—analysis of PS II activity by measuring light-induced chlorophyll fluorescence parameters and estimation of electrolytes leakage (EL), reflecting the degree of cell membranes damages in leaves. Both analyses showed that stress treatment causes the least amount damage to the leaves of Tr-3 and Tr-18 lines, while leaves of Tr-20 and Tr-15 plants displayed higher sensitivities to given stresses, even in comparison to Sar-60 (Figure 6A,B, and Figure S5). At the same time, no significant differences between the studied lines were observed when plants have been grown under low positive temperature conditions for prolonged periods of time (up to 30 days) (Figure S6).

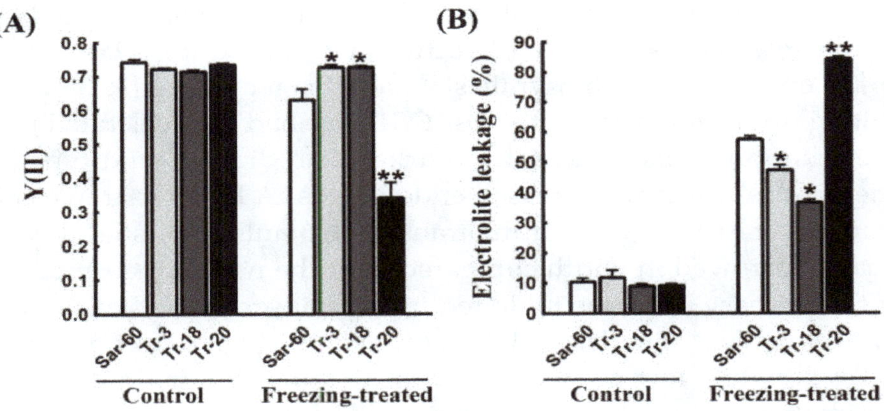


**Figure 6.** Freezing tolerance test of transgenic wheat plants overexpressing *AtOPR3*. Freezing treatments were performed on homozygous T2 plants, as described in the Materials and Methods section, and then the degree of damage was assessed by measurements of PS II operating efficiency (**A**) and electrolytes leakage (**B**) in control and treated leaves of Sar-60, Tr-3, Tr-18, and Tr-20. Means of 20 measurements ± SE are presented. Stars indicate statistically significant differences from Sar-60 (* $p \leq 0.05$, ** $p \leq 0.001$).

## 3. Discussion

In this work, we have implemented the transformation of hexaploid bread wheat with the Arabidopsis *OPR3* gene coding for the enzyme catalyzing the reduction of the cyclopentenone ring of 12-OPDA. The expression level of *AtOPR3* varied significantly between different transgenic lines, wherein plants with high *AtOPR3* expression levels have elevated levels of JA. Besides, the expression of *TaAOS*, a jasmonate biosynthesis pathway gene known to be regulated by a positive feedback loop that maintains and boosts JA levels, is also significantly up-regulated in these plants. Increase in *ALLENE OXIDE SYNTHASE* gene expression level in response to the up-regulation of the jasmonate pathway was demonstrated previously [29]. There are three homeologs of *12-OPDA REDUCTASE*, coding for OPRII subgroup proteins, which are potentially involved in the biosynthesis of endogenous jasmonic acid in the wheat genome (one homeolog in each of the A, B, and D subgenomes), TraesCS7A02G412400, TraesCS7B02G311600, and TraesCS7D02G405500. These genes share high similarities at the amino acid level with each other and *AtOPR3* (Figure S7). The expression level of wheat *12-OPDA REDUCTASE* (*TaOPR2*) encoded in the B subgenome, previously shown to be induced by methyl jasmonate [30], was slightly increased in Tr-3, and remained unaltered in other transgenic lines (Figure 2C). We could not detect the expression of *TaOPR2* homeolog genes from the A and D genomes, either because of the low expression level of these genes in leaf tissue, or because of the sequence difference between Saratovskaya-60 and the sequenced wheat varieties. The expression level of the jasmonate receptor COI-1 (*TaCOI-1*) [31] was not altered in transgenic lines (Figure 2C). The gene expression analysis demonstrates that *AtOPR3* gene overexpression selectively affects some genes of the endogenous jasmonate system, while the expression of other genes remains unaltered. The growth and development phenotypes of transgenic wheat plants with high *AtOPR3* expression levels were altered noticeably. These plants are characterized by delayed germination, slower growth, and delayed flowering time and senescence. The obtained results are in accordance with several previously published studies demonstrating the role of jasmonates in the regulation of plant growth and development, including seed germination, root growth, blossoming, fruit ripening, and senescence [13,32,33]. It is known, that JA-mediated suppression of plant growth is implemented through the regulation of cell cycle progression [34]. In Arabidopsis, jasmonate caused a touch-induced rosette diameter reduction and delay, in flowering [35]. The decreased size was also observed in rice plants with increased JA levels caused by the inactivation of the parallel competing hydroperoxide lyase branch of the oxylipin biosynthesis pathway [36]. Transgenic rice plants accumulating high levels of methyl jasmonate (MeJA) as a result of overexpression of the Arabidopsis jasmonic acid carboxyl methyltransferase gene were slightly smaller than the non-transformed controls. Flowering time was not affected in these plants, but grain yield was reduced [37]. Rice mutants *cpm2* (*coleoptile photomorphogenesis2*)/*hebiba* (defective in allene oxide cyclase) and *eg1* (*extra glume1*) (defective in a lipase involved in JA biosynthesis) have longer leaves [38,39]. A rice mutant with low JA levels resulting from a mutation in the *OsAOS1* gene had an accelerated juvenile–adult phase change, and flowered approximately five days earlier than did the wild type [40]. Leaf size was also affected in the rice *aos1* mutant. Increased endogenous JA levels in transgenic soybeans led to alterations in organogenesis: the leaves of the transgenic plants were slightly elongated in length, but they dramatically narrowed in width compared with the non-transformed plants. In addition, the elongation of the primary root was inhibited in the transgenic soybean plantlets, whereas the development of lateral root was stimulated [41]. In *Pharbitis nil*, high concentrations of MeJA inhibited root and shoot growth, while low concentrations of this metabolite, in contrast, enhanced plant growth [42]. MeJA treatment also inhibited growth and flowering in *Chenopodium rubrum* plants [43]. Transgenic potato plants overexpressing the gene coding for jasmonic acid carboxyl methyltransferase, in contrast, showed a significant increase in size and tuber yield [44]. A role for MeJA in the modulation of vernalization and flowering time in einkorn wheat was demonstrated [45]. Bread wheat plants overexpressing the allene oxide cyclase *TaAOC1* developed shorter roots and exhibited enhanced tolerance to salinity [13].

We estimated the senescence in growing plants and detached leaves by measuring the activity of PS II. A decline in PS II activity is one of the earliest events during leaf senescence, and visible senescence observed on individual leaves correlates with a decline in PSII activity [46]. Results demonstrate that increased JA levels in wheat plants leads to delay in senescence (brightly manifested in Tr-3), while the decreased level of the hormone leads to earlier senescence (Figure 5). Jasmonate roles in leaf senescence have been studied extensively [47]. Similarly to our observation, JA involvement in the regulation of barley leaf senescence was previously described [48].

As expected, the alteration of JA level also affected wheat tolerance to abiotic stress. It was established in our experiments that Tr-3 and Tr-18 have improved tolerances to short-term freezing treatment (Figure 6), although these plants' growth at low positive temperatures is not altered (Figure S6). Freezing temperatures are one of the most important abiotic factors controlling the geographical distribution of plants and restricting the use of agricultural lands. Short spring freezing episodes may cause injuries to young wheat plants, affecting the yield of this economically important crop. The role of jasmonates in the regulation of plant responses to abiotic stresses has been established [18,49]. An increase in JA level in response to stress factors was demonstrated for many plant species, including those phylogenetically closely related to wheat barley [50]. Exogenous application of jasmonates also improves plant tolerance to abiotic stresses [51,52]. Numerous studies have shown that jasmonates protect plants from damage caused by low temperatures when it accumulates in tissue or it is applied exogenously [53–56]. Molecular mechanisms underlying the protective effect of jasmonates have also been characterized. It is known that the transcription pathway *ICE-CBF/DREB1* (Inducer of CBF Expression—C-Repeat Binding Factor/DRE Binding Factor 1), playing a key role in plant adaptation to low temperatures, is regulated by jasmonates [57,58].

Interestingly, in two independent transgenic lines, Tr-15 and Tr-20, constitutive expression of *AtOPR3* leads to a decrease in basal JA level in comparison to Sar-60. Gene expression analysis showed that the *AtOPR3* expression level is significantly lower in these plants (about a 40-fold difference between Tr-20 and Tr-3) (Figure 2A). In all assays, these plants displayed phenotypes that were opposite to Tr-3 and Tr-18. Tr-20 showed an earlier germination phenotype, faster growth and development, and higher susceptibility to short-term freezing (Figures 3–6). Similar traits, such as faster growth, early senescence, and sensitivity to short-term freezing were also observed in Tr-15 (Figures S3–S5). Currently, we are unable to explain the mechanisms leading to a decrease in JA level upon overexpression of the *AtOPR3* gene, but similar results have been previously obtained in several studies describing transgenic plants overexpressing jasmonates biosynthesis pathway genes [12,59]. Transgenic tobacco plants overexpressing few JA pathway genes from Arabidopsis, including *AtOPR3*, had a decreased level of jasmonate–isoleucine conjugate, the most active metabolite of the pathway [59]. Interestingly, the expression level of jasmonate-regulated genes involved in nicotine biosynthesis was increased in these plants, despite the lack of an increase in the JA–isoleucine conjugate level known to regulate the expression of the mentioned genes. Our previous attempts to manipulate the jasmonate branch via overexpression of *AtAOS* in Arabidopsis led to the silencing of the pathway and the generation of sterile plants [12], a clear physiological response in plants depleted of jasmonates [5,60]. In another research, an attempt to overexpress *AtAOS* in tobacco resulted in the reduction of basal JA level by nearly 10-fold [61]. For transgenic wheat plants overexpressing allene oxide cyclase *TaAOC1*, another gene of the jasmonate biosynthesis pathway, authors reported the increase in wounding-induced levels of JA, while the basal levels of the hormone remained unaltered [13]. The dimerization of OPR proteins leading to the inhibition of enzymatic activity could be another possible explanation for the decrease in JA level upon *AtOPR3* overexpression in wheat, as shown previously for OPR3 from tomato [62]. The jasmonate level in plant tissue is controlled by the intricate regulatory system with positive and negative feedback loops [63–65]. Reduction of cyclopentenone catalyzed by AtOPR3 is also a highly regulated process [19]. In wheat, the regulation of the pathway activity may be even more complicated, due to polyploidy [66]. Apparently, jasmonate pathway-related genes are represented by large gene families in hexaploid wheat. According to WheatExp, a homologue-specific database of gene expression profiles [67], there are 12 expressed allene oxide synthase genes; based on the available published data, there are 18 COI genes [68] and 14 JASMONATE-ZIM DOMAIN transcription repressor genes in the wheat genome [69]. Further expression profiling study of the generated transgenic lines may shed light onto the jasmonate-dependent regulation of homolog gene expression in wheat.

Altogether, the results obtained in this study prove that the modulation of jasmonate pathway activity can be used as a tool for the modification of developmental characteristics and tolerance to freezing in wheat. Constitutive *AtOPR3* expression causes a pleiotropic phenotype in transgenic wheat, including the manifestation of beneficial traits and characteristics, which could be disadvantageous to wheat growers. On the one hand, transgenic plants with high *AtOPR3* expression levels are characterized by freezing tolerance and reduced height, which is usually associated with increased resistance against lodging, and enables plants to support high grain yields [70]. On the one hand, these plants display late germination and slower developmental phenotypes, which under field conditions could lead to plant exposure to adverse temperatures and water deficiency during late development [71]. The use of development stage-specific and inducible promoters [72,73] will allow the desired combination of characteristics to be achieved through the fine-tuning of the activity of the jasmonate-based system.

## 4. Materials and Methods

### 4.1. Plants and Growth Conditions

Non-transgenic and transgenic plants of spring bread wheat 'Saratovskaya-60' (Sar-60) (*Triticum aestivum* L. $2n = 6x = 42$, ABD genome) were grown at greenhouse conditions (25 ± 2 °C during the day and 20 ± 2 °C at night). Plants were grown in potted soil under a 16-h photoperiod with additional lighting during the winter period for providing light intensity of up to 150 μmol/m$^2$s. For the germination test and segregation analysis, mature seeds or embryos of immature seeds were plated under sterile conditions on Murashige and Skoog medium, or on wet filter paper in a Petri dish.

### 4.2. Generation of Transgenic Wheat Plants

The full-length coding sequence for the *AtOPR3* gene (AT2G06050, UniProt ID Q9FUP0) (including the sequence for peroxisome targeting signal) was obtained by reverse transcription polymerase chain reaction (RT-PCR) using total RNA isolated from *Arabidopsis thaliana* leaves as a template. *Sma*I and *Sac*I restriction sites were added to gene-specific primer sequences for cloning (Table S2). The *AtOPR3* gene sequence was introduced into pUC19-based vectors between the *Ubi1* promoter from maize and the *Tnos* terminator from *Agrobacterium tumefaciens*. the fragment with the sequence of the target gene with regulatory sequences was then cut out using *Pvu*II, and the PUbi-*AtOPR3*-Tnos fragment was transferred into the *Sma*I site of the psGFP-BAR vector [74], containing two selective marker genes, *GFP* and *BAR*, coding respectively for green fluorescent protein, and phosphinothricin acetyl transferase, conferring resistance to herbicide phosphinotricin. It was established in our previous work that wheat transformation with the psGFP-BAR vector does not result in any phenotypical alterations [27]. For all constructs, sequences were checked after cloning to ensure fidelity and integrity. The generated construct pBAR-GFP.UbiOPR3 was used for biolistic delivery by a particle inflow gun (PIG) into cells of wheat morphogenic callus induced from immature embryos. The vector DNA was bombarded into 377 immature embryo-derived calli of Sar-60 to generate primary transgenic wheat plants (T0). Primary transformants (T0) were selected on medium supplemented with herbicide glufosinate ammonium, and by monitoring the fluorescence of GFP, as described [75]. The herbicide-resistant plantlets with a height of at least 10 cm were transferred to the soil, as described [76], and grown in an environmentally controlled greenhouse, as described in Section 4.1.

Leaf samples collected from plants at heading stage were analyzed for the presence of the insert by PCR. The expression of the transferred genes was confirmed by end-point RT-PCR using RNA isolated from the studied plants as a template. RNA isolation and reverse transcription was carried out as described below (Section 4.4). The primers used for the target sequences amplification are shown in Table S2.

T1 and T2 progenies of *AtOPR3* overexpressing plants produced by self-pollination of primary T0 transgenic plants were assessed for transgene inheritance. To speed up the segregation analysis and the selection of homozygous transgenic progeny, we analyzed T0→T1→T2 plants, as was described earlier [27,75]. Zygotic T1 embryos were isolated from immature T0 spikes, placed on germination medium for 7–9 days, and then GFP fluorescence was monitored in the seedlings using a ZEISS SteREO Discovery.V12 microscope equipped with a PentaFluar S 120 vertical illuminator. Two commercially available filter sets 38 GFP BP (EX BP 470/40, BS FT 495, EM BP 525/50) and 57 GFP BP (EX BP 470/40, BS FT 495, EM LP 550) (Carl Zeiss MicroImaging GmbH, Jena, Germany) were used to examine GFP expression. Seedlings with a high level of GFP fluorescence were transferred to soil and placed in greenhouse conditions where mature plants were grown. To identify homozygous plants, the pollen from individual mature T1 plants and immature embryos were analyzed for GFP fluorescence (Figure S1A,B), as described previously [27]. Individual homozygous T1 plants were grown in a greenhouse to produce T2 homozygous wheat seeds. T2 homozygous transgenic plants and the T3 progeny obtained from T2 transgenic plants were used for further analysis.

### 4.3. *DNA Extraction and PCR Analysis*

Total genomic DNA was extracted from leaf tissues at heading stage, in accordance with the method described [77]. PCR analysis was carried in a thermal cycler, MJ Mini (BIO-RAD, Hercules, CA, USA). Each PCR reaction mixture of 25 μL consisted of 2.5 μL of 10 × buffer for Encyclo DNA polymerase (Evrogen, Moscow, Russia), 0.5 μL of 10 mM dNTPs, 1.25 μL of reverse and forward primers each at 10 μM, 0.5 μL of Encyclo DNA polymerase (Evrogen, Russia), 1 μL DMSO (≥99.5%), 90 ng (1.5 μL) of DNA template, and 16.5 μL of deionized water. Primers are listed in Table S2.

### 4.4. *RNA Extraction and Gene Expression Analysis*

Real-time quantitative reverse transcription PCR was used for the analysis of the expression of selected genes: *ALLENE OXIDE SYNTHASE* (*TaAOS*) (GenBank: KJ001800.1), *12-OPDA REDUCTASE 2* (*TaOPR2*) (TraesCS7B02G311600), and *CORONATINE INSENSITIVE 1-LIKE* (*TaCOI*-1) (TraesCS1A02G279100). Gene sequences were obtained from the GenBank and Ensemble Plants databases [78]. Total RNA from the second leaf of T3 plants at the four-leaf stage was isolated by Extract RNA reagent (Evrogen, Moscow, Russia), and further purified using the Clean RNA Standard kit (Evrogen, Moscow, Russia) with on-column DNase I treatment (Thermo Fisher Scientific, Welhem, MA, USA) to eliminate DNA contamination. RNA was reverse-transcribed using the MMLV Reverse Transcriptase kit (Eurogen, Russia). Real-time quantitative PCR was conducted in 20 μL reactions containing complementary DNA (cDNA) synthesized from 10 ng of total RNA, SYBR Green I qPCR Master Mix, (Syntol, Moscow, Russia), and 200 nM for each primer. Amplification was performed using Roche LightCycler®96 System according to the manufacturer's instructions. *TaWIN1* and *TaUbi* genes were used as reference genes for the internal controls, as previously described for transcript normalization [26]. Primers are listed in Table S2.

### 4.5. *Jasmonic Acid Extraction and Quantification*

The second leaves of homozygous T3 plants at the four-leaf stage were used for the analysis of JA level. Jasmonic acid extraction was carried out similarly to the procedure previously described [8]. Dihydrojasmonic acid (Merck KGaA, Darmstadt, Germany) was used as an internal standard. The formed methyl ester derivatives of jasmonic acid (JA-ME) and dehydrojasmonic acid were analyzed on a gas chromatograph Bruker 436-GC coupled with a mass spectrometer detector Bruker Scion SQ operating in electron ionization mode. One microliter volume samples were injected by autosampler Bruker CP-8400 in splitless mode at 250 °C, and separated on a column DB-5MS UI (length 30 m, inner diameter 0.25 mm, film thickness 0.25 μm) with helium being used as a carrier gas (constant flow

0.7 mL/min). Oven temperature programming was as follows: hold at 40 °C for 1 min after injection, ramp at 15 °C/min to 150 °C, then increase at 10 °C/min to 250 °C, hold for 10 min. Mass spectral analysis was done in selective ion monitoring mode (SIM). The fragment ions monitored were as follows: JA-ME 224, 151, 83; dihydro-JA-ME 96, 83. The Bruker Workstation 8.2.1. program was used for the data analysis.

### *4.6. Electrolyte Leakage*

The second leaves of homozygous T2 plants at the four-leaf stage were used for the analysis. A total of 2 g of leaves cut into 4 cm pieces were rinsed with deionized water, placed into 50 mL of deionized water, and left on an orbital shaker at room temperature. The conductivity of the solution was measured after 20 h of gentle shaking using the conductivity meter OK-102/1 (Radelkis, Budapest, Hungary). Total conductivity was determined on the same samples after they were autoclaved for 30 min at 121 °C. The samples were cooled down to room temperature prior to measurements. Results were expressed as a percentage of total conductivity.

### *4.7. Chlorophyll Fluorescence Measurements*

Light-induced chlorophyll fluorescence measurements were carried out on leaves of growing plants, or on detached leaves, using a FluorPen FP 100 fluorimeter (Photon Systems Instruments, Drasov, Czech Republic). Photosystem II operating efficiency was measured in light-adapted samples according to the FluorPen FP 100 fluorimeter manual, and calculated according to equations: Y(II) = (F′m − Fs)/F′m, where Fs—stationary level and F′m—light-induced maximum level of chlorophyll fluorescence in light-adapted leaves [79]. Taking into account the heterogeneity in photosynthetic parameters within the leaf blade, all measurements were carried out in the area 10 cm away from the tip of the leaf blade. Light-induced chlorophyll fluorescence measurements were performed on the flag and fourth leaves of homozygous T2 plants at the maturation stage to estimate senescence. For the detached leaf senescence analysis, the second leaves of homozygous T2 plants at the four-leaf stage were used.

### *4.8. Exogenous Methyl Jasmonate Treatment*

For exogenous methyl jasmonate treatment, 100 μM solution of MeJA was applied to filter paper lining Petri dishes with germinating seeds of Sar-60.

### *4.9. Freezing-Tolerance Test*

Three-week-old homozygous T2 plants at the four-leaf stage growing in pots were placed into a growth chamber at 4 °C for 24 h for cold acclimation. Then, leaves of the low temperature-adapted plants were cut, wrapped into plastic bags, and placed into a chilled vacuum-insulated vessel. Leaves in the vessel were subjected to decreasing temperatures for 30, 60, or 90 min (the temperature was controlled by a wireless thermometer with a remote sensor; the decrease rate was 1 °C/10 min), then leaves in the vacuum-insulated vessel were placed back at 4 °C for another 24 h prior to assessment of leaf damage caused by the freezing treatment by the means of light-induced chlorophyll fluorescence and electrolyte leakage measurements.

### *4.10. Statistical Analysis*

To determine statistically significant differences between genotypes, a one-way analysis of variance (ANOVA) followed by post hoc comparisons by Tukey's test and the Student's t-test for independent means were performed.

**Author Contributions:** T.V.S., S.V.D., A.V.P., and D.N.M. conceived the research, A.V.P., D.N.M., A.S.P., V.V.T., A.M.B., and T.V.S. carried out the experiments, T.V.S. wrote the manuscript with support from A.V.P. and D.N.M.

**Acknowledgments:** We thank Alexey Nezvetckiy (ANO "Test-Pushchino") for the analysis of jasmonates.

## Abbreviations

| | |
|---|---|
| 13-HPOT | 13(S)-hydroperoxy-9,11,15-octadecatrienoic acid |
| 12-OPDA | (9S,13S)-12-oxo-phytodienoic acid |
| JA | Jasmonic acid |
| MeJA | Methyl jasmonate |

## References

1. Wasternack, C.; Song, S. Jasmonates: Biosynthesis, metabolism, and signaling by proteins activating and repressing transcription. *J. Exp. Bot.* **2017**, *68*, 1303–1321. [CrossRef]
2. Wasternack, C.; Strnad, M. Jasmonates: News on occurrence, biosynthesis, metabolism and action of an ancient group of signaling compounds. *Int. J. Mol. Sci.* **2018**, *19*, 2539. [CrossRef]
3. Schaller, A.; Stintzi, A. Enzymes in jasmonate biosynthesis—Structure, function, regulation. *Phytochemistry* **2009**, *70*, 1532–1538. [CrossRef]
4. Feussner, I.; Wasternack, C. The lipoxygenase pathway. *Annu. Rev. Plant Biol.* **2002**, *53*, 275–297. [CrossRef] [PubMed]
5. Stintzi, A.; Browse, J. The Arabidopsis male-sterile mutant, opr3, lacks the 12-oxophytodienoic acid reductase required for jasmonate synthesis. *Proc. Natl. Acad. Sci. USA* **2000**, *97*, 10625–10630. [CrossRef] [PubMed]
6. Lyons, R.; Manners, J.M.; Kazan, K. Jasmonate biosynthesis and signaling in monocots: A comparative overview. *Plant Cell Rep.* **2013**, *32*, 815–827. [CrossRef]
7. Afzal, F.; Chaudhari, S.K.; Gul, A.; Farooq, A.; Ali, H.; Nisar, S.; Sarfraz, B.; Shehzadi, K.J.; Mujeeb-Kazi, A. Bread wheat (*Triticum aestivum* L.) under biotic and abiotic stresses: An overview. In *Crop Production and Global Environmental Issues*; Hakeem, K., Ed.; Springer International Publishing: Cham, Switzerland, 2015; pp. 293–317.
8. Chehab, E.W.; Kaspi, R.; Savchenko, T.; Rowe, H.; Negre-Zakharov, F.; Kliebenstein, D.; Dehesh, K. Distinct roles of jasmonates and aldehydes in plant-defense responses. *PLoS ONE* **2008**, *3*, e1904. [CrossRef]
9. Savchenko, T.; Dehesh, K. Insect herbivores selectively mute GLV production in plants. *Plant Signal. Behav.* **2013**, *8*, e24136. [CrossRef]
10. Savchenko, T.; Yanykin, D.; Khorobrykh, A.; Terentyev, V.; Klimov, V.; Dehesh, K. The hydroperoxide lyase branch of the oxylipin pathway protects against photoinhibition of photosynthesis. *Planta* **2017**, *245*, 1179–1192. [CrossRef] [PubMed]
11. Xiao, Y.; Savchenko, T.; Baidoo, E.E.; Chehab, W.E.; Hayden, D.M.; Tolstikov, V.; Corwin, J.A.; Kliebenstein, D.J.; Keasling, J.D.; Dehesh, K. Retrograde signaling by the plastidial metabolite MEcPP regulates expression of nuclear stress-response genes. *Cell* **2012**, *149*, 1525–1535. [CrossRef] [PubMed]
12. Savchenko, T.; Kolla, V.A.; Wang, C.Q.; Nasafi, Z.; Hicks, D.R.; Phadungchob, B.; Chehab, W.E.; Brandizzi, F.; Froehlich, J.; Dehesh, K. Functional convergence of oxylipin and abscisic acid pathways controls stomatal closure in response to drought. *Plant Physiol.* **2014**, *164*, 1151–1160. [CrossRef]
13. Zhao, Y.; Dong, W.; Zhang, N.; Ai, X.; Wang, M.; Huang, Z.; Xiao, L.; Xia, G. A wheat allene oxide cyclase gene enhances salinity tolerance via jasmonate signaling. *Plant Physiol.* **2014**, *164*, 1068–1076. [CrossRef] [PubMed]
14. Sharma, V.K.; Monostori, T.; Gobel, C.; Hansch, R.; Bittner, F.; Wasternack, C.; Feussner, I.; Mendel, R.R.; Hause, B.; Schulze, J. Transgenic barley plants overexpressing a 13-lipoxygenase to modify oxylipin signature. *Phytochemistry* **2006**, *67*, 264–276. [CrossRef]
15. Qiu, Z.; Guo, J.; Zhu, A.; Zhang, L.; Zhang, M. Exogenous jasmonic acid can enhance tolerance of wheat seedlings to salt stress. *Ecotoxicol. Environ. Saf.* **2014**, *104*, 202–208. [CrossRef] [PubMed]
16. Liu, H.; Carvalhais, L.C.; Schenk, P.M.; Dennis, P.G. Effects of jasmonic acid signalling on the wheat microbiome differ between body sites. *Sci. Rep.* **2017**, *7*, 41766. [CrossRef]
17. Ahmad, P.; Rasool, S.; Gul, A.; Sheikh, S.A.; Akram, N.A.; Ashraf, M.; Kazi, A.M.; Gucel, S. Jasmonates: Multifunctional roles in stress tolerance. *Front. Plant Sci.* **2016** *7*, 813. [CrossRef]

18. Alikina, O.; Chernobrovkina, M.; Dolgov, S.; Miroshnichenko, D. Tissue culture efficiency of wheat species with different genomic formulas. *Crop Breed. Appl. Biotechnol.* **2016**, *16*, 307–314. [CrossRef]
19. Strassner, J.; Schaller, F.; Frick, U.B.; Howe, G.A.; Weiler, E.W.; Amrhein, N.; Macheroux, P.; Schaller, A. Characterization and cDNA-microarray expression analysis of 12-oxophytodienoate reductases reveals differential roles for octadecanoid biosynthesis in the local versus the systemic wound response. *Plant J.* **2002**, *32*, 585–601. [CrossRef]
20. Chehab, E.W.; Kim, S.; Savchenko, T.; Kliebenstein, D.; Dehesh, K.; Braam, J. Intronic T-DNA insertion renders Arabidopsis opr3 a conditional jasmonic acid-producing mutant. *Plant Physiol.* **2011**, *156*, 770–778. [CrossRef]
21. Wasternack, C.; Hause, B. A bypass in jasmonate biosynthesis—The OPR3-independent formation. *Trends Plant Sci.* **2018**, *23*, 276–279. [CrossRef]
22. Stintzi, A.; Weber, H.; Reymond, P.; Browse, J.; Farmer, E.E. Plant defense in the absence of jasmonic acid: The role of cyclopentenones. *Proc. Natl. Acad. Sci. USA* **2001**, *98*, 12837–12842. [CrossRef] [PubMed]
23. Zheng, H.; Pan, X.; Deng, Y.; Wu, H.; Liu, P.; Li, X. AtOPR3 specifically inhibits primary root growth in Arabidopsis under phosphate deficiency. *Sci. Rep.* **2016**, *6*, 24778. [CrossRef] [PubMed]
24. Han, B.W.; Malone, T.E.; Kim, D.J.; Bingman, C.A.; Kim, H.-J.; Fox, B.G.; Phillips, G.N. Crystal structure of Arabidopsis thaliana 12-oxophytodienoate reductase isoform 3 in complex with 8-iso prostaglandin A(1). *Proteins* **2011**, *79*, 3236–3241. [CrossRef] [PubMed]
25. Gagarinsky, E.L.; Stepanov, S.A.; Signaevsky, V.D. Microevolution of elements productivity of shoot spring soft wheat of the Saratov selection. *Izv. Saratov Univ. New Ser. Ser. Earth Sci.* **2015**, *13*, 171–181. (In Russian)
26. Tenea, G.N.; Peres Bota, A.; Cordeiro Raposo, F.; Maquet, A. Reference genes for gene expression studies in wheat flag leaves grown under different farming conditions. *BMC Res. Notes* **2011**, *4*, 373. [CrossRef] [PubMed]
27. Miroshnichenko, D.N.; Poroshin, G.N.; Dolgov, S.V. Genetic transformation of wheat using mature seed tissues. *Appl. Biochem. Microbiol.* **2011**, *47*, 767–775. [CrossRef]
28. Feller, U.; Keist, M. Senescence and nitrogen metabolism in annual plants. In *Fundamental, Ecological and Agricultural Aspects of Nitrogen Metabolism in Higher Plants. Developments in Plant and Soil Sciences*; Lambers, H., Neeteson, J.J., Stulen, I., Eds.; Springer: Dordrecht, The Netherlands, 1986; Volume 19, pp. 219–234.
29. Liu, H.; Carvalhais, L.C.; Kazan, K.; Schenk, P.M. Development of marker genes for jasmonic acid signaling in shoots and roots of wheat. *Plant Signal. Behav.* **2016**, *11*, e1176654. [CrossRef]
30. Wang, Y.; Yuan, G.; Yuan, S.; Duan, W.; Wang, P.; Bai, J.; Zhang, F.; Gao, S.; Zhang, L.; Zhao, C. TaOPR2 encodes a 12-oxo-phytodienoic acid reductase involved in the biosynthesis of jasmonic acid in wheat (*Triticum aestivum* L.). *Biochem. Biophys. Res. Commun.* **2016**, *470*, 233–238. [CrossRef]
31. Liu, X.; Wang, J.; Fan, B.; Shang, W.; Sun, Y.; Dang, C.; Xie, C.; Wang, Z.; Peng, Y. A *COI1* gene in wheat contributes to the early defence response against wheat powdery mildew. *J. Phytopathol.* **2018**, *166*, 116–122. [CrossRef]
32. Wasternack, C. Jasmonates: An update on biosynthesis, signal transduction and action in plant stress response, growth and development. *Ann. Bot.* **2007**, *100*, 681–697. [CrossRef]
33. Creelman, R.A.; Mullet, J.E. Biosynthesis and action of jasmonates in plants. *Annu. Rev. Plant Physiol. Plant Mol. Biol.* **1997**, *48*, 355–381. [CrossRef] [PubMed]
34. Noir, S.; Bomer, M.; Takahashi, N.; Ishida, T.; Tsui, T.L.; Balbi, V.; Shanahan, H.; Sugimoto, K.; Devoto, A. Jasmonate controls leaf growth by repressing cell proliferation and the onset of endoreduplication while maintaining a potential stand-by mode. *Plant Physiol.* **2013**, *161*, 1930–1951. [CrossRef] [PubMed]
35. Chehab, E.W.; Yao, C.; Henderson, Z.; Kim, S.; Braam, J. Arabidopsis touch-induced morphogenesis is jasmonate mediated and protects against pests. *Curr. Biol.* **2012**, *22*, 701–706. [CrossRef] [PubMed]
36. Liu, X.; Li, F.; Tang, J.; Wang, W.; Zhang, F.; Wang, G.; Chu, J.; Yan, C.; Wang, T.; Chu, C.; et al. Activation of the jasmonic acid pathway by depletion of the hydroperoxide lyase OsHPL3 reveals crosstalk between the HPL and AOS branches of the oxylipin pathway in rice. *PLoS ONE* **2012**, *7*, e50089. [CrossRef] [PubMed]
37. Kim, E.H.; Kim, Y.S.; Park, S.H.; Koo, Y.J.; Choi, Y.D.; Chung, Y.Y.; Lee, I.J.; Kim, J.K. Methyl jasmonate reduces grain yield by mediating stress signals to alter spikelet development in rice. *Plant Physiol.* **2009**, *149*, 1751–1760. [CrossRef] [PubMed]
38. Riemann, M.; Muller, A.; Korte, A.; Furuya, M.; Weiler, E.W.; Nick, P. Impaired induction of the jasmonate pathway in the rice mutant hebiba. *Plant Physiol.* **2003** *133*, 1820–1830. [CrossRef] [PubMed]

39. Cai, Q.; Yuan, Z.; Chen, M.; Yin, C.; Luo, Z.; Zhao, X.; Liang, W.; Hu, J.; Zhang, D. Jasmonic acid regulates spikelet development in rice. *Nat. Commun.* **2014**, *5*, 3476. [CrossRef]
40. Hibara, K.; Isono, M.; Mimura, M.; Sentoku, N.; Kojima, M.; Sakakibara, H.; Kitomi, Y.; Yoshikawa, T.; Itoh, J.; Nagato, Y. Jasmonate regulates juvenile-to-adult phase transition in rice. *Development* **2016**, *143*, 3407–3416. [CrossRef] [PubMed]
41. Xue, R.; Zhang, B. Increased endogenous methyl jasmonate altered leaf and root development in transgenic soybean plants. *J. Genet. Genom.* **2007**, *34*, 339–346. [CrossRef]
42. Maciejewska, B.; Kopcewicz, J. Inhibitory effect of methyl jasmonate on flowering and elongation growth in Pharbitis nil. *J. Plant. Growth. Regul.* **2002**, *21*, 216–223. [CrossRef]
43. Albrechtova, J.T.P.; Ullmann, J. Methyl jasmonate inhibits growth and flowering in *Chenopodium-Rubrum*. *Biol. Plant.* **1994**, *36*, 317–319. [CrossRef]
44. Sohn, H.B.; Lee, H.Y.; Seo, J.S.; Jung, C.; Jeon, J.H.; Kim, J.H.; Lee, Y.W.; Lee, J.S.; Cheong, J.J.; Do Choi, Y. Overexpression of jasmonic acid carboxyl methyltransferase increases tuber yield and size in transgenic potato. *Plant Biotechnol. Rep.* **2011**, *5*, 27–34. [CrossRef]
45. Diallo, A.O.; Agharbaoui, Z.; Badawi, M.A.; Ali-Benali, M.A.; Moheb, A.; Houde, M.; Sarhan, F. Transcriptome analysis of an mvp mutant reveals important changes in global gene expression and a role for methyl jasmonate in vernalization and flowering in wheat. *J. Exp. Bot.* **2014**, *65*, 2271–2286. [CrossRef] [PubMed]
46. Wingler, A.; Mares, M.; Pourtau, N. Spatial patterns and metabolic regulation of photosynthetic parameters during leaf senescence. *New Phytol.* **2004**, *161*, 781–789. [CrossRef]
47. Hu, Y.; Jiang, Y.; Han, X.; Wang, H.; Pan, J.; Yu, D. Jasmonate regulates leaf senescence and tolerance to cold stress: Crosstalk with other phytohormones. *J. Exp. Bot.* **2017**, *68*, 1361–1369. [CrossRef] [PubMed]
48. Weidhase, R.A.; Kramell, H.M.; Lehmann, J.; Liebisch, H.W.; Lerbs, W.; Parthier, B. Methyl jasmonate-induced changes in the polypeptide pattern of senescing barley leaf segments. *Plant Sci.* **1987**, *51*, 177–186. [CrossRef]
49. Kazan, K. Diverse roles of jasmonates and ethylene in abiotic stress tolerance. *Trends Plant Sci.* **2015**, *20*, 219–229. [CrossRef]
50. Walia, H.; Wilson, C.; Condamine, P.; Liu, X.; Ismail, A.M.; Close, T.J. Large-scale expression profiling and physiological characterization of jasmonic acid-mediated adaptation of barley to salinity stress. *Plant Cell Environ.* **2007**, *30*, 410–421. [CrossRef]
51. Sharma, M.; Laxmi, A. Jasmonates: Emerging players in controlling temperature stress tolerance. *Front. Plant Sci.* **2016**, *6*, 1129. [CrossRef]
52. Wang, S.Y. Methyl jasmonate reduces water stress in Strawberry. *J. Plant Growth Regul.* **1999**, *18*, 127–134. [CrossRef]
53. Yoshikawa, H. Effect of low-temperature stress on abscisic acid, jasmonates, and polyamines in apples. *Plant Growth Regul.* **2007**, *52*, 199–206. [CrossRef]
54. Ding, C.K.; Wang, C.Y.; Gross, K.C.; Smith, D.L. Jasmonate and salicylate induce the expression of pathogenesis-related-protein genes and increase resistance to chilling injury in tomato fruit. *Planta* **2002**, *214*, 895–901. [CrossRef] [PubMed]
55. Kondo, S.; Jitratham, A.; Kittikorn, M.; Kanlayanarat, S. Relationships between jasmonates and chilling injury in mangosteens are affected by spermine. *HortScience* **2004**, *39*, 1346–1348.
56. Pedranzani, H.; Sierra-de-Grado, R.; Vigliocco, A.; Miersch, O.; Abdala, G. Cold and water stresses produce changes in endogenous jasmonates in two populations of *Pinus pinaster* Ait. *Plant Growth Regul.* **2007**, *52*, 111–116. [CrossRef]
57. Hannah, M.A.; Wiese, D.; Freund, S.; Fiehn, O.; Heyer, A.G.; Hincha, D.K. Natural genetic variation of freezing tolerance in Arabidopsis. *Plant Physiol.* **2006**, *142*, 98–112. [CrossRef] [PubMed]
58. Hu, Y.; Jiang, L.; Wang, F.; Yu, D. jasmonate regulates the inducer of cbf expression-C-repeat binding factor/DRE binding factor1 cascade and freezing tolerance in Arabidopsis. *Plant Cell* **2013**, *25*, 2907–2924. [CrossRef]
59. Chen, H.; Wang, B.; Geng, S.; Arellano, C.; Chen, S.; Qu, R. Effects of overexpression of jasmonic acid biosynthesis genes on nicotine accumulation in tobacco. *Plant Direct* **2018**, *2*, e00036. [CrossRef]
60. Park, J.H.; Halitschke, R.; Kim, H.B.; Baldwin, I.T.; Feldmann, K.A.; Feyereisen, R. A knock-out mutation in allene oxide synthase results in male sterility and defective wound signal transduction in Arabidopsis due to a block in jasmonic acid biosynthesis. *Plant J.* **2002**, *31*, 1–12. [CrossRef]

61. Laudert, D.; Schaller, F.; Weiler, E.W. Transgenic Nicotiana tabacum and Arabidopsis thaliana plants overexpressing allene oxide synthase. *Planta* **2000**, *211*, 163–165. [CrossRef]
62. Breithaupt, C.; Kurzbauer, R.; Lilie, H.; Schaller, A.; Strassner, J.; Huber, R.; Macheroux, P.; Clausen, T. Crystal structure of 12-oxophytodienoate reductase 3 from tomato: Self-inhibition by dimerization. *Proc. Natl. Acad. Sci. USA* **2006**, *103*, 14337–14342. [CrossRef]
63. Hickman, R.; Van Verk, M.C.; Van Dijken, A.J.H.; Mendes, M.P.; Vroegop-Vos, I.A.; Caarls, L.; Steenbergen, M.; Van der Nagel, I.; Wesselink, G.J.; Jironkin, A.; et al. Architecture and dynamics of the jasmonic acid gene regulatory network. *Plant Cell* **2017**, *29*, 2086–2105. [CrossRef] [PubMed]
64. Woldemariam, M.G.; Onkokesung, N.; Baldwin, I.T.; Galis, I. Jasmonoyl-L-isoleucine hydrolase 1 (JIH1) regulates jasmonoyl-L-isoleucine levels and attenuates plant defenses against herbivores. *Plant J.* **2012**, *72*, 758–767. [CrossRef] [PubMed]
65. Abdelrahman, M.; Suzumura, N.; Mitoma, M.; Matsuo, S.; Ikeuchi, T.; Mori, M.; Murakami, K.; Ozaki, Y.; Matsumoto, M.; Uragami, A.; et al. Comparative de novo transcriptome profiles in *Asparagus officinalis* and *A. kiusianus* during the early stage of *Phomopsis asparagi* infection. *Sci. Rep.* **2017**, *7*, 2608. [CrossRef] [PubMed]
66. Powell, J.J.; Fitzgerald, T.L.; Stiller, J.; Berkman, P.J.; Gardiner, D.M.; Manners, J.M.; Henry, R.J.; Kazan, K. The defence-associated transcriptome of hexaploid wheat displays homoeolog expression and induction bias. *Plant Biotechnol. J.* **2017**, *15*, 533–543. [CrossRef] [PubMed]
67. Pearce, S.; Vazquez-Gross, H.; Herin, S.Y.; Hane, D.; Wang, Y.; Gu, Y.Q.; Dubcovsky, J. WheatExp: An RNA-seq expression database for polyploid wheat. *BMC Plant Biol.* **2015**, *15*, 299. [CrossRef]
68. Bai, J.F.; Wang, Y.K.; Wang, P.; Yuan, S.H.; Gao, J.G.; Duan, W.J.; Wang, N.; Zhang, F.T.; Zhang, W.J.; Qin, M.Y.; et al. Genome-wide identification and analysis of the COI gene family in wheat (*Triticum aestivum* L.). *BMC Genom.* **2018**, *19*, 754. [CrossRef] [PubMed]
69. Wang, Y.; Qiao, L.; Bai, J.; Wang, P.; Duan, W.; Yuan, S.; Yuan, G.; Zhang, F.; Zhang, L.; Zhao, C. Genome-wide characterization of JASMONATE-ZIM DOMAIN transcription repressors in wheat (*Triticum aestivum* L.). *BMC Genom.* **2017**, *18*, 152. [CrossRef] [PubMed]
70. Piñera-Chavez, F.J.; Berry, P.M.; Foulkes, M.J.; Molero, G.; Reynolds, M.P. Avoiding lodging in irrigated spring wheat. II. Genetic variation of stem and root structural properties. *Field Crop. Res.* **2016**, *196*, 64–74. [CrossRef]
71. Harris, D.; Joshi, A.; Khan, P.A.; Gothkar, P.; Sodhi, P.S. On-farm seed priming in semi-arid agriculture: Development and evaluation in maize, rice and chickpea in India using participatory methods. *Exp. Agric.* **1999**, *35*, 15–29. [CrossRef]
72. Tang, W.; Luo, X.; Samuels, V. Regulated gene expression with promoters responding to inducers. *Plant Sci.* **2004**, *166*, 827–834. [CrossRef]
73. Muthusamy, S.K.; Dalal, M.; Chinnusamy, V.; Bansal, K.C. Differential regulation of genes coding for organelle and cytosolic ClpATPases under biotic and abiotic stresses in wheat. *Front. Plant Sci.* **2016**, *7*, 9. [CrossRef] [PubMed]
74. Richards, H.A.; Rudas, V.A.; Sun, H.; McDaniel, J.K.; Tomaszewski, Z.B.; Conger, V. Construction of a GFP-BAR plasmid and its use for switchgrass transformation. *Plant Cell Rep.* **2001**, *20*, 48–54. [CrossRef]
75. Miroshnichenko, D.; Ashin, D.; Pushin, A.; Dolgov, S. Genetic transformation of einkorn (*Triticum monococcum* L. ssp. *monococcum* L.), a diploid cultivated wheat species. *BMC Biotechnol.* **2018**, *18*, 68. [CrossRef] [PubMed]
76. Miroshnichenko, D.; Chernobrovkina, M.; Dolgov, S. Somatic embryogenesis and plant regeneration from immature embryos of *Triticum timopheevii* Zhuk. and *Triticum kiharae* Dorof. et Migusch, wheat species with G. genome. *Plant Cell Tiss. Org. Cult.* **2016**, *125*, 495–508. [CrossRef]
77. Rogers, S.O.; Bendich, A.J. Extraction of DNA from milligram amounts of fresh herbarium and mummified plant tissues. *Plant Mol. Biol.* **1985**, *5*, 69–76. [CrossRef] [PubMed]
78. Bolser, D.M.; Staines, D.M.; Perry, E.; Kersey, P.J. Ensembl Plants: Integrating tools for visualizing, mining, and analyzing plant genomic data. *Methods Mol. Biol.* **2017**, *1533*, 1–31. [CrossRef]
79. Genty, B.; Harbinson, J.; Cailly, A.; Rizza, F. *Fate of Excitation at PS II in Leaves: The Non-Photochemical Side. Proceedings of the Third BBSRC Robert Hill Symposium on Photosynthesis, 31 March–3 April 1996*; University of Sheffield, Department of Molecular Biology and Biotechnology, Western Bank: Sheffield, UK, 1996; Abstract no. P28.

# Permissions

All chapters in this book were first published in MDPI; hereby published with permission under the Creative Commons Attribution License or equivalent. Every chapter published in this book has been scrutinized by our experts. Their significance has been extensively debated. The topics covered herein carry significant findings which will fuel the growth of the discipline. They may even be implemented as practical applications or may be referred to as a beginning point for another development.

The contributors of this book come from diverse backgrounds, making this book a truly international effort. This book will bring forth new frontiers with its revolutionizing research information and detailed analysis of the nascent developments around the world.

We would like to thank all the contributing authors for lending their expertise to make the book truly unique. They have played a crucial role in the development of this book. Without their invaluable contributions this book wouldn't have been possible. They have made vital efforts to compile up to date information on the varied aspects of this subject to make this book a valuable addition to the collection of many professionals and students.

This book was conceptualized with the vision of imparting up-to-date information and advanced data in this field. To ensure the same, a matchless editorial board was set up. Every individual on the board went through rigorous rounds of assessment to prove their worth. After which they invested a large part of their time researching and compiling the most relevant data for our readers.

The editorial board has been involved in producing this book since its inception. They have spent rigorous hours researching and exploring the diverse topics which have resulted in the successful publishing of this book. They have passed on their knowledge of decades through this book. To expedite this challenging task, the publisher supported the team at every step. A small team of assistant editors was also appointed to further simplify the editing procedure and attain best results for the readers.

Apart from the editorial board, the designing team has also invested a significant amount of their time in understanding the subject and creating the most relevant covers. They scrutinized every image to scout for the most suitable representation of the subject and create an appropriate cover for the book.

The publishing team has been an ardent support to the editorial, designing and production team. Their endless efforts to recruit the best for this project, has resulted in the accomplishment of this book. They are a veteran in the field of academics and their pool of knowledge is as vast as their experience in printing. Their expertise and guidance has proved useful at every step. Their uncompromising quality standards have made this book an exceptional effort. Their encouragement from time to time has been an inspiration for everyone.

The publisher and the editorial board hope that this book will prove to be a valuable piece of knowledge for researchers, students, practitioners and scholars across the globe.

# List of Contributors

**Xiaojiang Guo**
Triticeae Research Institute, Sichuan Agricultural University, Chengdu 611130, China

**Yongrui Wu**
National Key Laboratory of Plant Molecular Genetics, CAS Center for Excellence in Molecular Plant Sciences, Institute of Plant Physiology and Ecology, Shanghai Institutes for Biological Sciences, Chinese Academy of Sciences, Shanghai 200032, China

**Yuxin Fu**
Triticeae Research Institute, Sichuan Agricultural University, Chengdu 611130, China
National Key Laboratory of Plant Molecular Genetics, CAS Center for Excellence in Molecular Plant Sciences, Institute of Plant Physiology and Ecology, Shanghai Institutes for Biological Sciences, Chinese Academy of Sciences, Shanghai 200032, China

**Mengping Cheng and Maolian Li**
Triticeae Research Institute, Sichuan Agricultural University, Chengdu 611130, China
Key Laboratory for Crop Genetic Resources and Improvement in Southwest China, Sichuan Agricultural University, Ministry of Education, Chengdu 611130, China

**Jirui Wang**
Triticeae Research Institute, Sichuan Agricultural University, Chengdu 611130, China
Key Laboratory for Crop Genetic Resources and Improvement in Southwest China, Sichuan Agricultural University, Ministry of Education, Chengdu 611130, China
State Key Laboratory of Crop Gene Exploration and Use in Southwest China, Sichuan Agricultural University, Chengdu 611130, China

**Piotr T. Bednarek and Renata Orłowska**
Department of Plant Physiology and Biochemistry, Plant Breeding and Acclimatization Institute – National Research Institute, Radzików, 05-870 Błonie, Poland

**Jacek Zebrowski**
Institute of Biology and Biotechnology, University of Rzeszow, Al. Rejtana 16c A, 35-959 Rzeszow, Poland

**Muhammad Amjad Ali, Mahpara Shahzadi and Adil Zahoor**
Department of Plant Pathology, Faculty of Agriculture, University of Agriculture, Faisalabad 38040, Pakistan

**Abdelfattah A. Dababat**
International Maize and Wheat Improvement Center (CIMMYT), Ankara 06511, Turkey

**Halil Toktay**
Department of Plant Production and Technologies, Faculty of Agricultural Sciences and Technologies, Nigde Omer Halisdemir University, Nigde 51240, Turkey

**Allah Bakhsh**
Department of Agricultural Genetic Engineering, Faculty of Agricultural Sciences and Technologies, Nigde Omer Halisdemir University, Nigde 51240, Turkey

**Muhammad Azher Nawaz**
College of Agriculture, University of Sargodha, Sargodha 40100, Pakistan

**Hongjie Li**
National Engineering Laboratory for Crop Molecular Breeding, Institute of Crop Sciences, Chinese Academy of Agricultural Sciences, Beijing 100081, China

**Harsimardeep S. Gill, Jagdeep S. Sidhu and Sunish K. Sehgal**
Department of Agronomy, Horticulture & Plant Science, South Dakota State University, Brookings, SD 57006, USA

**Chunxin Li**
National Engineering Laboratory of Wheat, Wheat Research Institute, Henan Academy of Agricultural Sciences, Zhengzhou 450002, China

**Wenxuan Liu**
The State Key Laboratory of Wheat and Maize Crop Science, College of Life Sciences, Henan Agricultural University, Zhengzhou 450002, China

**Duane Wilson and Bikram S. Gill**
Wheat Genetics Resource Center, Department of Plant Pathology, Kansas State University, Manhattan, KS 66506, USA

**Guihua Bai**
USDA-ARS, Hard Winter Wheat Genetic Research Unit, Manhattan, KS 66506, USA

**Guoshu Gong, Jizhi Yang, Yuting Hu, Liya Luo, Tao Li, Lu He and Min Zhang**
College of Agronomy & Key Laboratory for Major Crop Diseases, Sichuan Agricultural University, Chengdu 611130, China

**Madison Cahill and Ye Xia**
Department of Plant Pathology, The Ohio State University, Columbus, OH 43202, USA

**Yinping Liang**
College of Agronomy & Key Laboratory for Major Crop Diseases, Sichuan Agricultural University, Chengdu 611130, China
Department of Plant Pathology, The Ohio State University, Columbus, OH 43202, USA

**Xiaoli Chang**
College of Agronomy & Key Laboratory for Major Crop Diseases, Sichuan Agricultural University, Chengdu 611130, China
Institute of Ecological Agriculture, Sichuan Agricultural University, Chengdu 611130, China

**Veronika Kapustová, Zuzana Tulpová, Helena Toegelová, Miroslava Karafiátová, Eva Hřibová, Jaroslav Doležel and Hana Šimková**
Institute of Experimental Botany, Centre of the Region Haná for Biotechnological and Agricultural Research, Šlechtitelů 31, CZ-78371 Olomouc, Czech Republic

**Petr Novák and Jiří Macas**
Biology Centre, Czech Academy of Sciences, Institute of Plant Molecular Biology, Branišovská 31, CZ-37005 České Budějovice, Czech Republic

**Claus Krogh Madsen and Henrik Brinch-Pedersen**
Department of Molecular Biology and Genetics, Research Center Flakkebjerg, Aarhus University, 4200-Slagelse, Denmark

**Ryo Nishijima, Kentaro Yoshida, Kohei Sakaguchi, Shin-ichi Yoshimura and Shigeo Takumi**
Graduate School of Agricultural Science, Kobe University, Rokkodai 1-1, Nada, Kobe 657-8501, Japan

**Kazuhiro Sato**
Institute of Plant Science and Resources, Okayama University, Kurashiki, Okayama 710-0046, Japan

**Cyrine Robbana**
Faculté des Sciences de Bizerte, Jarzouna, 7021 Bizerte, Tunisia
Banque Nationale de Gènes, Boulevard du Leader Yasser Arafat Z. I Charguia 1, 1080 Tunis, Tunisia

**M'barek Ben Naceur**
Banque Nationale de Gènes, Boulevard du Leader Yasser Arafat Z. I Charguia 1, 1080 Tunis, Tunisia

**Zakaria Kehel, Filippo Bassi and Ahmed Amri**
International Center for Agricultural Research in the Dry Areas (ICARDA), 10112 Rabat, Morocco

**Carolina Sansaloni**
Centro Internacional de Mejoramiento de Maíz y Trigo (CIMMYT), El Batán, 56237 Texcoco, Mexico

**Iris Koeppel, Christian Hertig, Robert Hoffie and Jochen Kumlehn**
Plant Reproductive Biology, Leibniz Institute of Plant Genetics and Crop Plant Research (IPK) Gatersleben, 06466 Seeland, Germany

**Kateřina Perničková, Veronika Koláčková, Jan Vrána and David Kopecký**
Institute of Experimental Botany, Centre of the Region Haná for Biotechnological and Agricultural Research, Šlechtitelů 31, 78371 Olomouc, Czech Republic

**Adam J. Lukaszewski and Chaolan Fan**
Department of Botany and Plant Sciences, University of California, Riverside, CA 92521, USA

**Martin Duchoslav and Michaela Sedlářová**
Department of Botany, Faculty of Science, Palacký University Olomouc, Šlechtitelů 27, 783 71 Olomouc, Czech Republic

**Glyn Jenkins and Dylan Phillips**
Institute of Biological, Environmental and Rural Sciences, Aberystwyth University, Aberystwyth, Ceredigion, Wales SY23 3DA, UK

**Olga Šamajová and Jozef Šamaj**
Department of Cell Biology, Centre of the Region Haná for Biotechnological and Agricultural Research, Faculty of Science, Palacký University Olomouc, Šlechtitelů 27, 783 71 Olomouc, Czech Republic

**Jianxin Bian, Pingchuan Deng, Haoshuang Zhan, Xiaotong Wu, Mutthanthirige D. L. C. Nishantha, Xianghong Du and Xiaojun Nie**
State Key Laboratory of Crop Stress Biology in Arid Areas, College of Agronomy and Yangling Branch of China Wheat Improvement Center, Northwest A&F University, Yangling, Shaanxi 712100, China

**Zhaogui Yan**
Huazhong Agricultural University, Wuhan 430070, China

**Weining Song**
State Key Laboratory of Crop Stress Biology in Arid Areas, College of Agronomy and Yangling Branch of China Wheat Improvement Center, Northwest A&F University, Yangling, Shaanxi 712100, China
Joint Research Center for Agriculture Research in Arid Areas, Yangling, Shaanxi 712100, China

**Malika Ayadi, Faiçal Brini and Khaled Masmoudi**
Biotechnology and Plant Improvement Laboratory, Centre of Biotechnology of Sfax (CBS), University of Sfax, B.P 1177, 3018 Sfax, Tunisia

**Haimei Du and Shulan Fu**
College of Agronomy, Sichuan Agricultural University, Wenjiang, Chengdu 611130, China

**Zongxiang Tang, Qiong Duan and Shuyao Tang**
College of Agronomy, Sichuan Agricultural University, Wenjiang, Chengdu 611130, China
Institute of Ecological Agriculture, Sichuan Agricultural University, Wenjiang, Chengdu 611130, China

**Alexey V. Pigolev, Vasily V. Terentyev, Alexander M. Boutanayev and Tatyana V. Savchenko**
Institute of Basic Biological Problems RAS, Pushchino 142290, Russia

**Sergey V. Dolgov**
Branch of Shemyakin and Ovchinnikov Institute of Bioorganic Chemistry RAS, Pushchino 142290, Russia

**Dmitry N. Miroshnichenko and Alexander S. Pushin**
Institute of Basic Biological Problems RAS, Pushchino 142290, Russia
Branch of Shemyakin and Ovchinnikov Institute of Bioorganic Chemistry RAS, Pushchino 142290, Russia

# Index

Printed in the USA
CPSIA information can be obtained
at www.ICGtesting.com
LVHW080042211023
761596LV00010BA/1868

9 781641 168236